Computational Biology and Bioinformatics
Gene Regulation

Gene | RNA | Protein | Epigenetics

Computational Biology and Bioinformatics

Gene Regulation

Gene | RNA | Protein | Epigenetics

Editor

Ka-Chun Wong

Department of Computer Science
City University of Hong Kong
Kowloon Tong, Hong Kong

CRC Press
Taylor & Francis Group
Boca Raton London New York

CRC Press is an imprint of the
Taylor & Francis Group, an **informa** business

A SCIENCE PUBLISHERS BOOK

CRC Press
Taylor & Francis Group
6000 Broken Sound Parkway NW, Suite 300
Boca Raton, FL 33487-2742

First issued in paperback 2021

Version Date: 20160115

ISBN-13: 978-0-367-78297-9 (pbk)
ISBN-13: 978-1-4987-2497-5 (hbk)

Library of Congress Cataloging-in-Publication Data

Names: Wong, Ka-Chun.

Title: Computational biology and bioinformatics / Ka-Chun Wong.

Description: Boca Raton : Taylor & Francis, 2016. | "A CRC title." | Includes bibliographical references and index

Identifiers: LCCN 2015048051 | ISBN 9781498724975 (hardcover : alk. paper)

Subjects: LCSH: Computational biology. | Bioinformatics. | Gene expression.

Classification: LCC QH324.2 .W66 2016 | DDC 572.80285--dc23

LC record available at http://lccn.loc.gov/2015048051

Visit the Taylor & Francis Web site at
http://www.taylorandfrancis.com

and the CRC Press Web site at
http://www.crcpress.com

Preface

The field of computational biology and bioinformatics has been rapidly evolving in recent years. It holds tremendous potential to have a good impact on understanding genetics, developing personalized medicine solutions. Hundreds of books have been written on it. In particular, there are only few bioinformatics books with a focus on gene regulation. Therefore, this book has been compiled to fill the void. The book is an open-chapter-called collection of peer-reviewed chapters contributed by international experts who have passed through rigorous blind reviews. Following the central dogma of molecular biology, the book content is divided into six sections: Genes (Chapters 1 to 3), RNAs (Chapters 4 and 5), Proteins (Chapters 6 and 7), Epigenetics (Chapters 8 and 9), Case Study (Chapter 10), and Advanced Topics (Chapters 11 to 17).

(Section 1: Genes) In Chapter 1, Cao and Yip compile a comprehensive survey on enhancers which are central to gene transcription. In Chapter 2, Wei et al. introduce a software package for detecting differential gene expression. In Chapter 3, Fujita et al. describe how Granger causality can be applied to infer gene regulatory networks from gene expression data.

(Section 2: RNAs) In Chapter 4, Keservani et al. present a reader-friendly tutorial on RNA sequencing and its relation to gene regulation. In Chapter 5, Samantarrai et al. focus on non-coding RNAs and its network applications from the view of system biology.

(Section 3: Proteins) In Chapter 6, Mishra et al. compare different algorithms for the annotation of hypothetical proteins. In Chapter 7, Hahn and Muley review the computational protein-protein functional linkage prediction methods. They also propose a gene co-regulation based method for functional linkage prediction.

(Section 4: Epigenetics) In Chapter 8, Wang et al. give us a review on bioinformatics methodologies and tools for performing epigenomic analysis of chromatin organization and DNA methylation. In Chapter 9, Lou focuses on the statistical modeling of DNA methylation with sharp insights into its role in coding regions and gene regulation.

(Section 5: Case Study) In Chapter 10, Wang et al. describe a case study of EGFR-related gene mutations which are linked to drug resistance in non-small-cell lung carcinoma treatments.

(Section 6: Advanced Topics) In Chapter 11, Giannoulatou et al. review contemporary approaches for quality assurance in genome-scale bioinformatics analyses. In Chapter 12, Issa reports the recent trends in biological sequence alignment methods. In Chapter 13, Mansouri et al. introduce a mathematical book chapter on state estimation and process monitoring of nonlinear biological phenomena. In Chapter 14, Neelapu and Challa provide us a review on metagenomics with an emphasis on gene regulation. In Chapter 15, Thukral and Hasija describe how gene regulation can be altered in the context of metabolic engineering. In Chapter 16, Gupta et al. review general bioinformatics tools for phylogenetic analysis and ortholog identification which can shape gene regulation mechanisms of different species throughout the evolutionary history. In Chapter 17, Shikhin et al. have compiled a book chapter about protein model ranking through topological assessment.

In summary, state-of-the-art computational biology and bioinformatics studies are described and summarized in this book. This book provides updated reviews on the computational biology and bioinformatics studies for gene regulation. I hope that this book can help accelerate the advancement of the scientific community in computational biology and bioinformatics, benefiting the human beings and other species on the earth.

Ka-Chun Wong
City University of Hong Kong
July 2015

List of Reviewers

Alan Moses
André Fujita
Antonio Sze-To
Ashish Runthala
Bibekanand Mallick
Bjorn Wallner
Bo Hu
Carlos Lijeron
Daniel Alex
Debby D. Wang
Eleni Giannoulatou
Hafeez Ur Rehman
Harishchander Anandaram
Hong Yan
Huiluo Cao
Joshua Ho
Konstantinos Krampis
Salim Bougouffa
Shailendra K. Gupta
Shaoke Lou
Sun Jin
Vijaykumar Yogesh Muley
Wenye Li
Yanjie Wei
Yasha Hasija
Yingying Wei
Yu Chen
Yue Li

Contents

Section 3: Proteins

Section 4: Epigenetics

Section 1
Genes

1

A Survey of the Computational Methods for Enhancers and Enhancer-target Predictions

Qin Cao[1] and Kevin Y. Yip[1]*

Abstract

Enhancers are important cis-regulatory elements that play critical roles in a wide range of cellular processes by enhancing expression of target genes through promoter-enhancer loops. There are many interesting biological questions about enhancers, including their evolution and the relationships between their dysregulation and genetic diseases. The recent developments of experimental methods such as high-throughput reporter assays and ChIA-PET have enabled large-scale identification of enhancers and their targets. However, the current lists of identified enhancers and enhancer targets remain incomplete and unreliable due to the high noise level or low resolution of these methods. As a result, computational methods have emerged as an alternative for predicting the genomic locations of enhancers and their target genes. These methods have used a variety of features for predicting enhancers, including sequence motifs and epigenomic modifications. Potential enhancer targets have been predicted using activity correlations, distance constraints, and other features. Both prediction tasks are non-trivial due to cell-type specificity of enhancer activities, lack of definite orientation and distance of an enhancer from its target genes, insufficient known examples for training computational models, and other complexities. In this survey, we discuss the current computational methods for these two prediction tasks and analyze their pros and cons. We also point out obstacles of computational prediction of enhancers and enhancer targets in general, and suggest future research directions.

1. Introduction

Enhancers are important transcriptional regulatory DNA elements that can enhance transcription of target genes by recruiting transcription factors (TFs), which bring an enhancer close to the promoter of its target gene and trigger interactions with RNA polymerase II.

[1.] Department of Computer Science and Engineering, The Chinese University of Hong Kong, Shatin, New Territories, Hong Kong Tel: (852) 39438418; Fax: (852) 26035024
[*] Corresponding author : kevinyip@cse.cuhk.edu.hk

Strong sequence conservation at a non-coding region is a strong indicator of a potential enhancer (Pennacchio et al. 2006), especially when conservation is measured in ways related to the function, such as clustering or protein binding sites (Berman et al. 2004). Active enhancers are usually enriched in the histone mark H3K27ac, while both active and poised enhancers are enriched in H3K4me1, and latent enhancers lack these marks in general (Shlyueva et al. 2014). A typical enhancer is several hundred base pairs long as defined by transcription factor binding signals, while much longer enhancers called super-enhancers, which are bound by the Mediator complex and master transcription factors, have been found to be important in the control of cell identity (Whyte et al. 2013; Hnisz et al. 2013).

Previous studies have uncovered that enhancer dysregulation could cause abnormal gene expressions and lead to genetic diseases (Carroll 2008; Visel et al. 2009; Dawson & Kouzarides 2012; Shlyueva et al. 2014), making enhancers an important study topic for both conceptual and practical values. Understanding the sophisticated operational mechanisms of enhancers has become a crucial part towards a complete understanding of the landscape of gene regulation.

In this chapter, we shall describe computational methods for identifying enhancers and their targets. We start with a brief introduction of the current experimental approaches to these two tasks, based on which we shall discuss their main limitations and introduce computational methods as a key alternative. We shall then discuss the current state-of-the-art method of computational enhancer prediction, from the features used in both unsupervised and supervised methods. We shall next discuss computational methods for predicting enhancer-target promoter associations. Finally, we shall conclude the chapter and discuss future research directions on these two problems.

1.1   Introduction to current experimental approaches to testing enhancer activities and enhancer-target associations

Enhancers can be tested experimentally by different kinds of reporter assays (Shlyueva et al. 2014; ENCODE Project Consortium et al. 2012; Kwasnieski et al. 2014), including *in vivo* systems such as embryos of transgenic mice (Visel et al. 2007). To scale up reporter assays for testing many enhancers at the same time, high-throughput multiplexed reporter assays have been developed (Kwasnieski et al. 2012; Melnikov et al. 2012; Patwardhan et al. 2012; Sharon et al. 2012). These methods have been applied to test previously predicted enhancers. For example, a recent study (Kwasnieski et al. 2014) has tested human enhancers predicted by the ENCODE consortium (ENCODE Project Consortium et al. 2012), and found that around 26% of these enhancer predictions have regulatory activities in the K562 cell line.

Another high-throughput method that can test the enhancer activities of millions of candidates simultaneously is STARR-seq (Arnold et al. 2013). The main novelty of this method is placing each enhancer to be tested downstream of the reporter gene, such that the enhancer sequence itself becomes part of the resulting RNA transcript. Standard RNA-sequencing (RNA-seq) can then be applied to measure quantitatively the activity of each enhancer by counting the number of reads containing the enhancer sequence.

A common limitation of these methods is that they do not preserve the whole native context of the predicted enhancers. For example, if an enhancer is predicted to be active in a context (cell/tissue type, development stage, disease state, etc.) but is tested in another context or even in another species, the chromatin state around the enhancer could be different, the TFs that bind the enhancer may not be expressed, and the genome structure required for enhancer-promoter looping could be altered. This means an enhancer that could be active in certain contexts may not show activities in a reporter assay, and even if it shows activities in a reporter assay, in which natural contexts it would be active is still unknown.

It is also important to note that these high-throughput experimental methods have been mainly used for testing enhancer candidates already defined by some other means, but not for discovering enhancers *ab initio*. In theory it should be possible to tile a major portion of a genome for testing the enhancer activities of the involved genomic regions using these high-throughput experimental methods. Such large-scale datasets are remained to be seen.

Many experimental approaches to enhancer-promoter association predictions rely on techniques that can capture chromosome conformations based on chromosome conformation capture (3C) (Dekker et al. 2002). There are many extended versions of 3C, such as circularized chromosome conformation capture (4C) (Zhao et al. 2006), chromosome conformation capture carbon copy (5C) (Dostie et al. 2006), genome-wide chromosome conformation capture (Hi-C) (Lieberman-Aiden et al. 2009) and chromatin interaction analysis with paired-end tag sequencing (ChIA-PET) (Fullwood et al. 2009). Hi-C and ChIA-PET have facilitated whole-genome identification of DNA regions that are in close proximity in the three-dimensional genome structure but are not necessarily adjacent to each other in the primary DNA sequence, without requiring an input set of candidates. Among these two techniques, ChIA-PET further requires that a chosen factor, such as RNA polymerase II, is involved in the DNA contacts. If a promoter and a predicted enhancer are found to interact based on these chromosome conformation data, the promoter would be predicted as a target of the enhancer.

In order to study enhancer-promoter contacts, the chromosome conformation data need to have a very high (<10kb) resolution. Correspondingly, a large amount of sequencing data needs to be produced to ensure statistical stability at such a high data resolution, since the contact map matrix could be very sparse and unstable without sufficient data. Several recent studies have used Hi-C and ChIA-PET to study DNA contacts in human cell lines at sub-10kb resolutions (Jin et al. 2013; Heidari et al. 2014; Rao et al. 2014). These studies represent the current state-of-the-art techniques in studying DNA long-range interactions.

While high-throughput chromosome conformation data have provided various insights about enhancer-promoter associations, they are still unable to comprehensively and accurately determine the targets of all enhancers for a number of reasons. First, having a physical interaction does not necessarily imply a functional relationship. In particular, many DNA contacts observed in Hi-C data may not be relevant to promoter-enhancer interactions (Shlyueva et al. 2014). Second, these high-throughput data could be noisy and are subject to different types of bias (DeMare et al. 2013; Duan et al. 2010; Li et al. 2010). Third, enhancer-promoter associations are also context-specific, and thus experimental data from a given context may not be relevant to other contexts.

Due to these limitations of current experimental approaches, the numbers of experimentally proven enhancers and enhancer-target associations are still limited, both in general and in particular contexts. As a result, computational methods have been widely used as an alternative in identifying enhancers and their targets. The advantage of using computational methods is that they can utilize different types of available data to make predictions in an inexpensive way as compared to their experimental counterparts. In the past 15 years, many computational methods have been proposed, using ideas and data ever more advanced. The last few years have seen a rapid adaptation of high-throughput data originally generated not specifically for studying enhancers in these methods. As of today, both computational enhancer prediction and enhancer target prediction are still very active areas of research with new discoveries being constantly published.

1.2   Difficulties in computational predictions of enhancer and enhancer-promoter associations

Before going into the details of these computational methods, we first discuss the difficulties of the corresponding problems that explain the continuous need for better methods. These difficulties lie in several aspects, mainly related to the intrinsic properties of enhancers and the lack of high-confidence examples of experimentally validated enhancers and enhancer targets.

First, there is no simple rule governing the relative location of an enhancer from a gene that it targets. It can be positioned either upstream or downstream of the transcription start site (TSS) of its target gene. It can reside in an intergenic region, an intron, or even an exon of another gene. It can be as close as ten kilobases or as far as hundreds of kilobases or more from the target promoter. A recent study has suggested that the median distance between enhancers and their target promoters is 124kb (Jin et al. 2013). All these flexibility in enhancer location makes them much harder to identify than some other types of sequence elements, such as promoters, which are right upstream of the target genes.

Second, up to now, no single feature or combination of features have been found that can perfectly locate enhancers or determine enhancer-promoter associations (Shlyueva et al. 2014). The different features used by existing computational methods all have their pros and cons, which we will discuss in detail in the next section.

Third, enhancer activities and enhancer-promoter associations are both context specific. A recent study that analyzed data from twelve human cell lines has suggested that among the two, enhancer-promoter associations have relatively stronger cell type specificity (He et al. 2014). Context-specificity implies that computational methods using static features that do not change with the context, such as DNA sequence patterns, can only predict whether a genomic region could be an enhancer but not the contexts in which it is active, and only whether an enhancer could target a gene, but not the contexts in which the enhancer actually regulates the gene. This property implies that computational methods need to incorporate information from the context of interest in their predictions (Yip et al. 2013).

Fourth, enhancers and promoters could associate with each other in a multiple-to-multiple manner. In other words, one enhancer can target multiple promoters and one promoter can be targeted by multiple enhancers (He et al. 2014). As a result, some standard computational methods that deal with one object at a time may not be suitable for predicting enhancers and enhancer targets.

Lastly, the lack of comprehensive lists of experimentally tested enhancers and enhancer targets means that there are limited examples for computational methods to reference. Some computational methods, especially those based on machine learning, require adequate positive and negative examples for modeling the general features of enhancers and enhancer targets. As a result, different studies have used a variety of ways to define "gold-standard" enhancers and enhancer targets for training their methods. A lot of these "gold-standard" examples are either not experimentally tested, or are taken from another context that may not be relevant to the context of interest. The devoid of experimentally tested examples also means that computational predictions cannot be easily validated without performing additional experiments.

Owing to all these difficulties, computational methods should be considered a supplement to experimental methods rather than a replacement. Computational predictions of existing methods all need to be experimentally tested to confirm their correctness.

2. Computational methods for enhancer prediction

The problem of computational prediction of enhancers is defined as follows. Given a set of genomic regions, each of which is described by a set of features, the goal is to identify the regions that correspond to enhancers based on the features.

This definition requires an input list of genomic regions the status of which (enhancer or non-enhancer) is to be predicted. In many cases, one only wants to predict an approximate location of each enhancer, in which case it is common to divide the whole genome into bins of a fixed size, and predict whether each bin overlaps an enhancer or not. On the other hand, if the predicted enhancers are to be tested experimentally, it is necessary to make sure that an enhancer candidate includes the core part of the enhancer, such as the TF binding sites (TFBSs). In this scenario, the raw predictions need to be further refined.

Many computational methods have been proposed for this prediction task. They differ from each other by the features they use and the way the features are used to make the predictions. In the following section, we first describe the features considered by different enhancer prediction methods, and then move on to discuss these methods themselves.

2.1 Features used in enhancer prediction

Many types of features have been considered in predicting enhancers (Table 1 and Fig. 1). Before the boom of high-throughput sequencing data that probe different types of features related to enhancers in a context-specific manner, researchers predicted cis-regulatory modules (CRMs), enhancers included, largely based on evolutionary conservation and sequence motifs (Su et al. 2010). Evolutionary conservation signifies regions with functional importance. Non-coding regions, including intergenic regions and introns, with unexpectedly strong evolutionary conservation could be CRMs. On the other hand, some functionally conserved enhancers do not have high sequence conservation (Su et al. 2010; Meireles-Filho & Stark 2009). This could indicate that conservation is not sufficient for identifying enhancers, or that the way to measure conservation needs to be improved (Berman et al. 2004).

TABLE 1 A summary of features used in computational enhancer prediction

Feature	Advantages	Potential drawbacks
TF binding motifs	Widely available	Presence of a motif does not guarantee binding of a TF in a given context; A TF could bind regions without a canonical sequence motif; Many TFBSs are not within enhancers
Evolutionary conservation	Widely available	Some functional enhancers do not have high sequence-level conservation; Cannot distinguish between different types of conserved DNA elements; Does not provide context-specific information
TFBSs based on ChIP-seq or ChIP-exo	Directly measured from the context of interest	Many TFBSs are not within enhancers; Requires many ChIP-seq experiments to obtain a comprehensive list of binding sites for many TFs
HMs	Provides information about both poised and active enhancers; There are both positive and negative HMs for enhancers; Only a small number of ChIP-seq experiments is needed for each context	No single HMs or their combinations have been found to correlate perfectly with enhancer activities
Chromatin accessibility	Only a single type of features is required for each context	Regions with high chromatin accessibility do not necessarily correspond to enhancers
eRNA	One of the most accurate single features for enhancer prediction; Transcriptome data are widely available	The detailed mechanisms of eRNA remain to be explored; Active enhancers may not produce eRNAs; Regions producing eRNA-like RNAs may not be enhancers; Many produced RNA-seq data are poly-A enriched, which may not contain eRNA signals
DNA methylation	Provides complementary information to the other features	Quantitative relationship between enhancer methylation and target gene expression is still unclear; Different types of DNA methylation may play different roles in enhancer regulation

Regions with a good match to a sequence motif could be binding sites of the TF. Excluding binding sites at annotated regions such as promoters, the remaining could be CRMs, especially for regions with a high density of motif matches (Su et al. 2010). Since TF binding also depends on factors other than the sequence, sequence motifs can be considered a weak feature for enhancer prediction.

As discussed above, using these static features to predict enhancers could at best identify regions with a potential to be an enhancer, without telling the contexts in which the enhancers are actually active. It is also hard to use conservation and

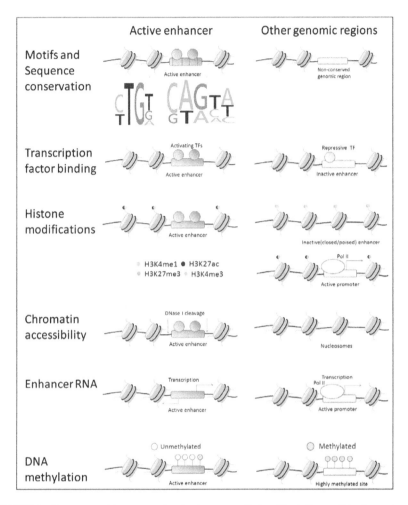

FIGURE 1 Features used in computational enhancer prediction. The left part of the figure shows features of active enhancers while the right part shows the corresponding features of inactive enhancers, other regulatory elements (such as promoters), or other genomic regions.

sequence motifs alone to distinguish enhancers from other types of regulatory elements such as silencers and insulators without a thorough understanding of the factors that bind these different types of elements.

Later on, the development of ChIP-seq (Park 2009) made it easy to measure DNA-binding affinity of transcription factors genome-wide (Bailey & MacHanick 2012). Compared to sequence motifs, the TFBSs identified by ChIP-seq are directly measured in the context of interest. They were thus used to predict enhancers in a context specific manner (Yip et al. 2012). Again, some of these binding sites may correspond to other types of functional enhancers (Shlyueva et al. 2014; Li et al. 2008). Moreover, there is a limited number of TFs with ChIP-seq data available, making it impossible to rely on ChIP-seq data alone to identify all TFBSs for enhancer prediction. Standard ChIP-seq data also have limited data resolution. This problem has been tackled by a new method called ChIP-exo (Rhee & Pugh 2011; Rhee & Pugh 2012), which provides close to single nucleotide precision of TFBSs by enzymatically digesting unbound portions of the pulled-down DNA.

ChIP-seq experiments were also used extensively in studying various types of histone modifications (HMs) at whole-genome scales. Some HMs were found to be highly related to enhancers, including H3K4me1 that marks both poised and active enhancers, and H3K27ac that marks active enhancers (Rada-Iglesias et al. 2011). These HMs provide a way to distinguish enhancers from other types of regulatory elements, such as promoters, which are marked by H3K4me3. On the other hand, while H3K4me1 and H3K27ac have been well-recognized as important enhancer marks, there has not been a consensus as to whether they are sufficient or necessary for identifying active enhancers. For instance, a recent study has found that H3K4me3 (as a negative feature for enhancers), H3K4me1 and H3K4me2 are the top three HMs for enhancer prediction while H3K27ac was not selected as one of the most important predictors (Rajagopal et al. 2013), although H3K27ac is widely used in many other studies as an indicator of active enhancers. Some previous studies have also shown that no single types of HM or a combination of several HMs could predict enhancers perfectly (Arnold et al. 2013), and some active enhancers do not have typical active marks (Bonn et al. 2012). Despite these complications, HMs still represent a cost-effective set of features in identifying context-specific enhancers, in that only a small set of ChIP-seq experiments are sufficient for identifying a fairly accurate set of active enhancers in a context.

Pushing this idea further is to use one single context-specific feature in identifying enhancers. One popular choice is chromatin accessibility as measured by DNase I hypersensitivity using DNase-Seq (Boyle et al. 2008) or FAIRE-Seq (Giresi et al. 2007). These data indicate genomic regions with high accessibility of chromatin where DNA sequences are depleted of nucleosomes, which signify functional activities of these regions. Active enhancers were found to overlap with DNase hypersensitive sites (DHSs), but obviously not all highly accessible genomic regions are enhancers. Chromatin accessibility data can thus be used to limit the search space of active enhancers to only the DHSs, and let the precise locations be identified with the help of other features such as TF sequence motifs.

Recently, it has been discovered that active enhancers produce short (<2kb) noncoding RNAs called eRNAs in a bi-directional manner (Kim et al. 2010). Based on this idea, a recent study has identified enhancers as regions with some bi-directional transcription patterns (Andersson et al. 2014), according to the abundant CAGE-based TSS data from FANTOM5 (FANTOM Consortium and the RIKEN PMI and CLST (DGT) et al. 2014). Currently, knowledge about eRNAs, including their functional mechanisms, is still limited. It is not yet clear whether active enhancers must produce eRNAs, and whether genomic regions producing eRNA-like RNAs must be enhancers. Nevertheless, the idea of using eRNAs to identify enhancers has become popular due to the wide availability of transcriptome data.

Less popularly, some studies have attempted to use DNA methylation level to predict enhancers (Aran et al. 2013). The role of DNA methylation in marking repressed promoters has long been recognized. Many inactive genomic regions are also marked by DNA methylation. Due to the diverse types of regions marked by DNA methylation, data about DNA methylation in a single context can hardly be used to identify enhancers. However, if two contexts are being compared (e.g. tumor vs. normal tissue), sites with differential DNA methylation could have differential activities in the two contexts, and some of them could correspond to functional elements such as enhancers. Currently, the degree of enhancer activities reflected by their DNA methylation levels is still unknown. The roles of different types of DNA methylation, such as 5-mC and 5-hmC (Xu et al. 2011), in regulating enhancer activities are also unclear.

Some studies have used correlation information between enhancer candidates and promoters to predict enhancers (Thurman et al. 2012). The main idea is that some activity indicators of enhancers (such as H3K27ac) are believed to correlate strongly with the transcription of their target genes across multiple contexts. If a non-promoter genomic region is found to exhibit such a correlation with a gene, the region could be an enhancer that regulates the gene. This idea is also commonly used in identifying enhancer targets. It has some limitations as we will discuss later.

Table 1 summarizes the features used in current computational methods for enhancer predictions discussed above. A detailed discussion on the pros and cons of some of these features in identifying enhancers from the perspective of biological experiments can be found in a recently published review (Shlyueva et al. 2014).

It should be noted that there are some additional enhancer features that have been more commonly used to define "gold-standard" enhancers instead of being used in the prediction process. For example, previous studies have shown that a large fraction of the binding sites of transcriptional co-activator proteins P300 and CBP are enhancers (Blow et al. 2010; May et al. 2011; Ramos et al. 2010). As a result, they have been used in some studies to define gold-standard enhancers (He et al. 2014; Rajagopal et al. 2013). One likely reason that binding sites of these proteins have not been as popularly used as enhancer predictors is that they are found in only a subset of active enhancers. This means although their presence stronger indicates an enhancer, using them as the only features could lead to a lot of false negatives.

2.2   Unsupervised methods for enhancer prediction

Many computational methods have been proposed for predicting enhancers using the features described above. Traditional methods that mainly use non-context-specific features have been discussed in detail in another review (Su et al. 2010). Here we focus on more recent methods that incorporate the different types of context-specific features. These methods can be broadly grouped into two categories, namely unsupervised methods and supervised methods. Unsupervised methods do not require any known enhancers and non-enhancers as examples. Some of these methods define simple filtering rules to identify the most likely enhancers based on the observed features. Some other methods cluster genomic regions according to these features, and identify clusters that are likely enhancers. In contrast, supervised methods require known enhancers and non-enhancers as inputs and derive models for enhancers using machine learning techniques. A more detailed discussion of the use of unsupervised and supervised methods (and also semi-supervised methods) in identifying genomic elements can be found in a recent review (Yip et al. 2013). In this section we first discuss the unsupervised methods for enhancer prediction.

Thurman et al. (Thurman et al. 2012) defined distal DHSs separated from a TSS by at least one other DHS as enhancer candidates. The DNase I hypersensitivity signals of each enhancer candidate in different cell types were correlated with those of each promoter within 500 kb from it. Any candidate with a resulting Pearson correlation of 0.7 or above was predicted as an enhancer.

Andersson et al. (Andersson et al. 2014) identified enhancers based on a directionality score of eRNA. They superimposed CAGE tags on H3K27ac-marked enhancers defined by ENCODE (the methods of which will be discussed below) and found that CAGE tags showed a bimodal distribution flanking the central P300 peak with divergent transcription from the enhancer. In contrast, the transcripts at promoters were strongly biased towards the sense direction. With this distinct difference, a directionality score was calculated for every 200bp window genome-wide, and loci with low, non-promoter-like directionality scores were selected as enhancer candidates, among which the ones located far away from TSSs and exons of protein-coding and noncoding genes were predicted as enhancers. To validate these predicted enhancers, they further selected strong, moderate and low-activity enhancers defined by CAGE tag frequency in HeLa cells and conducted enhancer reporter assays. They found that 73.9%, 70.7% and 67.4% of the strong, moderate and low-activity CAGE-defined enhancers showed significant signals in the reporter assays, respectively, demonstrating that eRNA could be an intrinsic characteristic of active enhancers.

The above two methods are simple unsupervised methods based on thresholding on a single feature. ChromHMM (Ernst & Kellis 2010; Ernst & Kellis 2012; Ernst et al. 2011) and Segway (Hoffman et al. 2012) utilize more complex machine-learning models and dozens of features each in predicting enhancers.

ChromHMM characterizes chromatin states including enhancers by learning a multivariate hidden Markov model (HMM) with the largest data set available at the time it was proposed (Barski et al. 2007; Wang et al. 2008) containing various HMs, histone variants and protein binding ChIP-seq signals (e.g. H2AZ, RNA polymerase II and CTCF) (Ernst & Kellis 2010). This method involves five key steps. First, the whole genome was divided into 200bp intervals. The signals of different HMs in an interval were then binarized, and thus each interval was described by a binary vector of the presence/absence of HM signals. Third, the number of states and the model parameters were determined by an exhaustive comparison of the cluster number from 2 to 80, with three different types of random initialization of parameters. The best model was selected by a Bayesian Information Criterion (BIC) score. Intuitively, the procedure attempted to find the minimum number of states that could still distinguish genomic regions exhibiting distinct HM patterns into different states. Finally, a 51-state model was selected. The fourth step was to associate each genomic interval with the state that maximizes the posterior probability using the forward-backward algorithm. The last step was to interpret the states biologically. This step involved both analyses of additional data (including expression, sequence motif, gene ontology, SNP and GWAS) and manual annotations. Based on the annotation results, several states were found to be related to enhancers (States 20, 29, 30, 31, 32 and 33). For instance, genomic regions in States 29 and 30 were interpreted as strong distal enhancers with characteristic high DNase I hypersensitivity and TF binding signals.

The authors subsequently applied ChromHMM to nine human cell types and identified 15 states that showed distinct enrichments of different types of biological signals (Ernst et al. 2011). Eight predicted strong enhancers (State 4) and seven predicted weak/poised enhancers (State 7) from the Hep-G2 cell line and seven predicted weak/poised enhancers specific to the GM12878 cell line were tested in

Hep-G2 using luciferase reporter assays. Only strong enhancers from Hep-G2 were observed to show strong luciferase signals.

Segway, based on Dynamic Bayesian Network (DBN), is similar to ChromHMM in the underlying mechanisms. In fact, a standard HMM can be represented by a DBN (Koller & Friedman 2009). The main differences between the original applications of Segway and ChromHMM lie in the following aspects: First, Segway used HMs and TF binding as features while ChromHMM mainly used HMs; second, Segway worked at single base pair resolution while ChromHMM worked on 200bp bins; third, Segway accepted continuous features while ChromHMM dealt with binary features; fourth, Segway had an explicit indicator variable for missing values while ChromHMM considers them as 0s. The first two differences were mainly choices made in the corresponding studies, but the ChromHMM method itself could incorporate TF binding signals and work at a higher resolution. When applying to the dataset from ENCODE, Segway identified 25 labels (analogous to the "states" in ChromHMM) and marked enhancers by the E-label. In a later study, the authors of ChromHMM and Segway collaborated and integrated these two methods to identify sequence elements from ENCODE data (Hoffman et al. 2013).

Yip et al. (Yip et al. 2012) defined two pipelines for predicting enhancers. Both pipelines start from all genome regions, and apply a series of filters to retain only regions likely to be enhancers. The first pipeline involves ChIP-seq signal shapes, gene annotations and HM signals. The second pipeline involves sequence features, TF binding active regions (BARs), gene annotations, conservation scores, sequence motifs and TF expression levels. BARs were determined using ENCODE TF binding data. Although ChIP-seq data of more than 100 TRFs were collected, this number of TRFs is still only a small portion of the estimated 1,700 to 1,900 human TFs (Vaquerizas et al. 2009). Therefore, instead of defining BARs by the binding sites directly observed in the limited data, a statistical model of BARs was constructed using these directly observed binding regions as positive examples and various types of ENCODE data as the features, including DNase I hypersensitivity and HMs.

Predictions of the two pipelines were combined, and the integrated predictions underwent two rounds of experimental validations. In the first round, among six predictions randomly selected from the top 50 predictions, five were found to have enhancer activities in various tissues in mouse embryo with good reproducibility. In the second round, the goal was to predict all enhancers in the human genome. Therefore a large number of predictions were made, among which about 50 were experimentally tested in mouse and Medaka fish. Overall, 42 unique regions could be successfully tested, among which 28 showed enhancer activities in at least one assay.

Overall, the five methods described above represent some of the latest unsupervised methods for computational enhancer prediction. It should be noted that the first two methods were specially designed for enhancer prediction while the other three were designed to discover various types of chromatin states in general, but with enhancers as some of the states in particular.

2.3 Supervised methods of enhancer prediction

As explained above, supervised methods for enhancer prediction require known enhancers and non-enhancers as input examples. Since the numbers of experimentally tested positive and negative examples are limited, different methods have used a variety of strategies to define these input examples. The different methods also differ from each other by the features being used and the statistical models constructed.

Heintzman et al. (2007) used an correlation-based methodology to predict enhancers based on their similarity to the enhancer examples. Enhancer examples were defined as regions with P300 binding sites. The genome was divided into 10kb windows, where an HM profile was constructed for each window based on the average ChIP-seq signals of different HMs. Enhancers were then predicted as those windows having an HM profile highly correlated with a P300-defined enhancer. In total, around 700 enhancers were predicted in this way. They were found to be significantly enriched in predicted transcriptional regulatory modules and DHSs. A large fraction of these predictions were also found to contain highly conserved sequences.

Won et al. (2008) presented an HMM-based methodology integrating HMs to predict enhancers. The positive examples were again defined by P300 binding sites. A simulated annealing procedure was used to search for the most informative combination of HMs and the optimal window size. The procedure identified a set of 6 HMs as the most informative, and a window size of 2kb to be optimal. A 3-state HMM model was then trained on a subset of the enhancer examples, and tested on another subset. The prediction results were found to be more accurate than the predictions by the Heintzman et al. method (Heintzman et al. 2007) in terms of positive predictive value and sensitivity.

Firpi et al. (2010) developed a method called CSI-ANN based on a time-delayed neural network (TDNN) framework to predict enhancers in HeLa and Human CD4+ T cells. In the case of T cells, the whole genome was divided into 2.5kb windows with consecutive windows overlapped by 1.25kb. Windows that contain gene-distal and narrow P300 binding peaks in human T cells and overlap computationally predicted enhancers in the PReMod database (Ferretti et al. 2007) were defined as enhancer examples, leading to a positive set of 213 enhancers. The negative set was composed of random windows 10 times the number of positive examples. For each window, the average signals of 39 HMs in T cells, or an energy function of them (D'Alessandro et al. 2003) were computed as its features. Fisher discriminant analysis (FDA) was then performed to reduce these $39 \times 2 = 78$ features to a one-dimensional feature. Finally this feature was fed into a TDNN classifier. 36,769 predictions were made and 13.1% of them were found to overlap P300 sites and DHSs in T cells. 22.1% of the predictions were found to be conserved across 17 vertebrate genomes and 24.6% were enriched for TF binding motifs.

Rajagopal et al. (2013) developed a vector-random-forest-based supervised model called RFECS for enhancer prediction. Gene-distal P300 binding sites overlapping DHS were defined as positive enhancer examples, while TSSs overlapping DHS and random 100bp bins distal from P300 binding sites or TSS were defined as negative enhancer examples. For each 100bp genomic region, the average signal of each of 24 HMs was computed. However, instead of taking only these average signals as the features of a genomic region like what was commonly done, each region also took the signal values from the adjacent regions within the 1kb upstream and downstream window as its own features. Therefore for each genomic region, each HM produced a 20-dimensional feature vector of numeric values. The reason for doing that was to capture the local signal pattern, which could be useful for identifying enhancers. To handle these vector features, RFECS constructed a linear classifier using the Fisher Discriminant approach inside each decision tree node.

This method was applied to the H1 embryonic stem cells and the IMR90 lung fibroblasts. To validate the predictions, some "gold standard" enhancer regions were defined by combining DHS, P300 binding sites and a few sequence specific transcription factors known to function in each of these two cell types. The validation rate of the predicted enhancers was 80% in H1, which was highly significant when compared to the 18.43% validation rate of randomly predicted enhancers. 5% of the predicted enhancers overlapped with TSSs, which were considered misclassified. The validation and misclassification rates in IMR90 were 85% and 4%, respectively. It should be noted that since the criteria used for defining the enhancer examples in the training set and the criteria used to define the validation set were not mutually exclusive, the accuracy of the model needs to be further confirmed by independent data sets.

Another contribution of this work was its proposed set of HMs optimal for enhancer predictions. The top three HMs were found to be H3K4me3, H3K4me1 and H3K4me2 in H1, while H3K27ac, commonly believed to mark active enhancers, seemed not very predicative.

In summary, due to the increasing number of experimentally validated enhancers and the availability of high-throughput features, supervised methods have become increasingly popular. It is expected that more supervised enhancer prediction methods will be proposed in the coming years.

3. Computational methods for enhancer target prediction

3.1 Features used in enhancer target prediction

Compared to enhancer prediction, less feature types have been considered in predicting enhancer targets (Table 2). The first and simplest feature considered is whether a promoter is the nearest one from an enhancer. A slight variation of this idea is to consider the distance between an enhancer and a promoter, assuming a higher possibility

TABLE 2 A summary of features used in computational enhancer target prediction

Feature	Advantages	Potential drawbacks
Closest promoter	Easy to identify	An enhancer does not always regulate the closest promoter (Andersson et al. 2014; He et al. 2014)
Distance between enhancer and promoter	Easy to compute	There may not be a single threshold suitable for all cases; An enhancer does not always regulate the closest promoters
Co-conservation	Easy to compute; Utilizes information from multiple species	Both enhancers and enhancer-promoter associations are not necessarily highly conserved
Correlation of molecular signals	Utilizes context-specific information	No signal correlates perfectly between enhancers and promoters; Correlation coefficients could be strongly affected by outliers; Requires a large number of context to reach statistical significance

that the enhancer regulates a promoter if they are closer to each other. Some previous studies have considered enhancers between 125kb (Ernst et al. 2011) and 1Mb (Fu et al. 2014) from potential target promoters. As discussed, chromosome conformation data have suggested that the median distance between an enhancer and a target promoter is 124kb (Jin et al. 2013). One drawback of using distance to predict enhancer targets is that very distal associations could be missed if the distance threshold is set too low. Conversely, if the distance threshold is set too high, many false positives could be produced. One way to avoid setting an arbitrary distance threshold is to consider only enhancer-promoter pairs within same topologically associating domains (TAD) (Dixon et al. 2012; Nora et al. 2012), which are genomic blocks separated from other blocks by the genome structure.

Sequence co-conservation is another feature that has been used in enhancer-promoter association prediction (He et al. 2014). The rationale is that if an enhancer regulates a promoter, there would be selective pressure against independent evolution of them, and thus they may exhibit co-conservation patterns. Some previous studies (Ahituv et al. 2005; Kikuta et al. 2007) also suggested that a real enhancer-promoter association is more likely to be maintained in a conserved synteny block (Larkin et al. 2009), which could be used as a soft distance constraint.

As high-throughput sequencing data became widely available, the correlations between certain molecular signals at an enhancer and its candidate target promoters across multiple contexts were considered. As discussed above, the main idea is that if the activity of an enhancer correlates with the activity of a promoter, the enhancer could be regulating the promoter. The molecular signals considered and the potential issues of using correlation features have been discussed above when discussing the features used in enhancer prediction. An additional issue is that if correlations are computed between all enhancer-promoter pairs without any pre-filtering, there would be a very large number of pairs being considered. As a result, a very large number of contexts are needed to reach statistical significance after considering the issue of multiple hypothesis testing. We also note that to what extent enhancer activities can quantitatively correlate with promoter activities is still not clear. In fact, some studies (Andersson et al. 2014) have observed enhancer-promoter associations with low activity correlations.

Among these features, only signal correlations consider context-specific information. A tricky point is that depending on how this feature is used, it may still be unable to identify context-specific enhancer targets. For instance, if a single correlation value is computed based on all the contexts, this correlation value only tells whether the enhancer appears to regulate the promoter in general, but not exactly the contexts in which the regulation happens.

3.2 Unsupervised methods for enhancer target prediction

Similar to enhancer prediction, most methods for enhancer target predictions are unsupervised, due to the limited number of experimentally validated enhancers and enhancer targets.

As discussed, the most straightforward method is to predict the closest promoter as the only target of each enhancer. This is a simple but imperfect method. Several studies (Andersson et al. 2014; He et al. 2014) have shown that only a fraction (e.g., 40% (Andersson et al. 2014)) of enhancers recognize the nearest promoter as their targets, and one enhancer could regulate multiple promoters. A variation of this method is to predict the nearest promoter within a certain distance range (e.g. between 5kb and 50kb from the enhancer (Ernst et al. 2011)) from an enhancer as its target.

Most current unsupervised methods extract all promoters within a certain distance range from an enhancer as candidate targets, and then use activity correlations to identify the most likely targets. A practical problem is finding a proper correlation threshold. Some studies (Andersson et al. 2014) use a rather low threshold of 0.2 while some other studies (Thurman et al. 2012) use a much higher value of 0.7. If a value-based correlation function such as Pearson correlation is used, the correlation values can be easily affected by a few outlier points. On the other hand, if a rank-based correlation function such as Spearman correlation is used, the correlation value can become quite arbitrary if the activity values in many contexts are similar and their ranks are sensitive to small differences. Multiple hypothesis testing is also a critical issue, because without a proper distance cutoff, many enhancer-promoter pairs would be considered and it is easy to get some strong correlation values merely by chance. Table 3 compares some of these unsupervised methods.

TABLE 3 A summary of correlation-based unsupervised methods for enhancer target prediction

Reference	Distance	Features denoting activity/ inactivity (A/I: enhancer-promoter)	Correlation function(s)	Threshold
(Thurman et al. 2012)	Within 500kb	A: DNase I hypersensitivity -DNase I hypersensitivity	Pearson	0.7
(Andersson et al. 2014)	Within 500kb	A: CAGE(eRNA)-CAGE(mRNA)	Pearson	0.2
(Fu et al. 2014)	Within 1Mb	A: H3K4me1/H3K27ac-mRNA I: DNA methylation-mRNA	Pearson and Spearman	User-defined

Some studies used a further step to validate their predictions. Thurman et al. (Thurman et al. 2012) profiled chromatin interactions using 5C for the phenylalanine hydroxylase (PAH) gene in hepatic cell and found the chromatin interactions measured by 5C closely paralleled the correlations of the corresponding predicted associations. They also overlapped their predictions with 5C and ChIA-PET data in K562, and discovered that their predictions were markedly enriched in the DNA long-range interactions. Andersson et al. (Andersson et al. 2014) found that 15.3% of their predictions could be validated by ChIA-PET data from multiple cell types. Moreover, their predictions were enriched in conserved sequence motifs and ChIP-seq peaks.

Ernst et al. (2011) selected all TSSs between 5kb and 125kb from an enhancer as its potential targets. To identify the more likely ones, these enhancer-TSS pairs were first assumed to be the positive examples, and a set of negative examples was formed by randomly assigning expression values of the same pairs. For each (positive or negative) enhancer-TSS pair, the correlation between the HM signals at the enhancer and the expression levels of the TSS across multiple contexts was computed. A logistic regression classifier was then constructed to distinguish the positive and negative examples based on the activity correlations. The classifier was then used to computer a link score for each enhancer-TSS pair, defined as the ratio of the positive association probability to the negative association probability. The pairs with a link score larger than 2.5 were predicted as real associations. This is another example that even a supervised model (logistic regression) was used, but since the positive examples were not really known examples but just a set of examples more likely to be positive due to the proximity of the corresponding enhancers and TSSs, the overall method should be considered an unsupervised one for predicting enhancer targets.

Corradin et al. (2014) developed a method called PreSTIGE for cell-type specific enhancer-promoter association prediction. Enhancers were defined as H3K4me1 sites across 12 cell types. First, a specificity score was assigned to each enhancer and to each transcript separately in the 12 cell types based on Shannon's entropy (Schug et al. 2005). Thresholds were set to define cell-type specific enhancers and transcripts based on the specificity scores. For example, enhancers with high specificity to a certain cell type were considered to be active in this cell type but not in the others. The next step was to link cell-type specific enhancers to their target cell-type specific genes. Several linear domain models for setting the distance thresholds were compared, based on which a model called 100kb/CTCF was selected to link enhancers and genes. In this model, all TSSs closer to an enhancer than the closest CTCF binding site, or 100kb at most, were predicted as the targets of the enhancer. This model identified over 226,000 and 113,000 enhancer-target predictions across the 12 cell types with low and high thresholds, respectively. The predictions were further overlapped with existing 3C, ChIA-PET, eQTL, 5C and colon cancer specific enhancer alteration data and showed significant intersections.

3.3   Supervised methods for enhancer target prediction

There have not been a lot of supervised methods proposed for enhancer target prediction, due to the limited number of validated examples. In this section, we introduce one supervised method that uses chromosome conformation data to define the examples.

A sophisticated Random Forest based supervised method called IM-PET was developed by He et al. (2014). The positive examples were selected from enhancer-promoter pairs with ChIA-PET connections in K562 and MCF-7 cells, with the additional requirements that there were at least 5 PET counts, at least one of the two interacting sites contained P300 binding, and the other contained a promoter of RPKM larger than 0. A naïve way to define the negative examples would be to draw random enhancer-promoter pairs. However, if the promoters in these pairs were very far away from their enhancers, which would likely be the case if enhancers and promoters were drawn uniformly from the whole genome or the same chromosome, the positive and negative examples could be easily separated by a simple model that considers only the distance between the enhancer and promoter. Therefore, IM-PET instead used random enhancer-promoter pairs with a distance that follows a background distribution of non-interacting genomic loci in a chromatin fiber (Dekker et al. 2002). The negative examples were also required not to have 3 or more PET counts in the ChIA-PET data. Four features were then used to train a supervised Random Forest model for enhancer-target associations. The first feature was the activity correlation between an enhancer and a promoter, with enhancer activities defined by H3K4me1, H3K4me3 and H3K27ac signals, and promoter activities defined by its

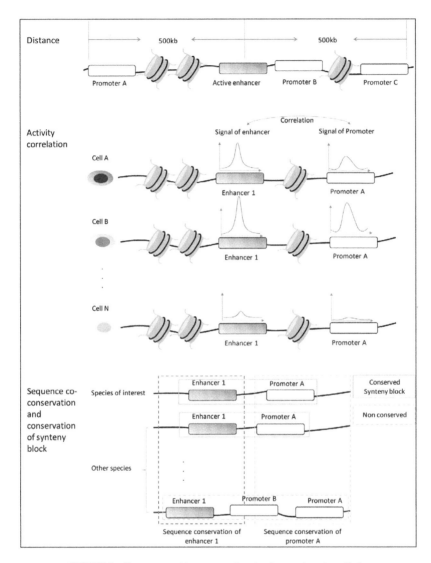

FIGURE 2 Features used in computational enhancer target prediction.

expression value. The second feature was similar to the first one, but the enhancer activity score was replaced by the expression levels of TFs that bind the enhancer. The third feature was the co-conservation of the enhancer and promoter sequences and the conservation of the synteny block across multiple species. The last feature was the genomic distance between the enhancer and promoter.

The trained model was applied to 12 human cell types by first identifying active enhancers in each cell type followed by extracting all promoters within a 2Mb window centered on the enhancer as their candidate targets. At a false discovery rate of 0.01, the resulting model predicted more than 440,000 unique enhancer-promoter associations in the 12 cell types in total. To validate the predictions, chromosome conformation capture coupled with quantitative PCR (3C-qPCR) was performed for 16 predictions and 13 of them could be validated. The predictions were also compared with interactions obtained from Hi-C and ChIA-PET, and reported eQTL-gene pairs. The results showed that IM-PET performed the better as compared to four other methods, namely nearest promoter, Ernst et al. (2010), Thurman et al. (2012) and PreSTIGE (Corradin et al. 2014).

The four features used in this work appear reasonable and biologically meaningful. The careful selection criteria for the training sets probably contributed to the good prediction results. Nevertheless, it should be noted that all the four features were not context-specific, including the activity correlation feature since only a single correlation was produced from each pair, as discussed above. Therefore, the method was unable to identify enhancer-target associations that are specific to particular contexts.

4. Databases useful for enhancer and enhancer-promoter association prediction

After discussing the features and latest methods used in computational prediction of enhancers and enhancer targets, here we list in Table 4 some of the popular databases that contain computationally predicted or experimentally validated enhancers and enhancer targets.

TABLE 4 Some databases that contain predicted or experimentally tested enhancers and enhancer targets

Database	Species	Description
dbSUPER (Khan and Zhang 2015)	Human and mouse	The first database of super-enhancers, containing a catalog of 66033 super-enhancers in 96 human and 5 mouse tissue/cell types. Provides a browser for functional analyses.
EI (Pennacchio et al. 2007)	Human and mouse	A database containing computational predicted tissue-specific enhancers based on TFBSs.
FANTOM5 Transcribed Enhancer Atlas (Andersson et al. 2014)	Human	An atlas of predicted enhancers based on eRNA; Contains 43011 computational predicted enhancers in total; Contains cell/organ/tissue-specific computational predicted enhancers; Contains computational predicted enhancer-promoter associations.
PEDB (Kumaki et al. 2008)	Human and mouse	A database of computational predicted enhancers based on conserved non-coding regions, TSSs and TFBSs.
PReMod (Ferretti et al. 2007)	Human and mouse	A computationally predicted CRM database based on TFBSs.

REDfly (Gallo et al. 2011)	Drosophila	A curated collection of known Drosophila CRMs and TFBSs; Contains enhancers *in vivo*; Contains enhancer-promoter associations *in vivo*.
VISTA (Visel et al. 2007)	Human and mouse	Tested predicted human enhancers in mouse; Contains information on 2192 *in vivo* tested elements; 1154 elements with enhancer activity. (As of 4/15/2015)
ZEnBase (Navratilova et al. 2009)	Zebrafish	A database containing computational predicted enhancers based on conservation information.

5. Conclusions and discussions

5.1 Data processing

A fundamentally important but usually neglected topic in both enhancer prediction and enhancer target prediction is data processing. As with many problems in bioinformatics, different data processing strategies could result in huge differences in the results. For instance, Andersson et al. (Andersson et al. 2014) computed enhancer eRNA directionality scores based on the normalized CAGE data across 808 samples, which were normalized by converting tag counts to tags per million mapped reads (TPM) followed by normalization by relative log expression (RLE) between samples. Our own analysis of this dataset shows that if a different normalization strategy is used, the resulting set of enhancers could become very different. In enhancer target prediction, whether taking log on gene expression levels could have big effects, especially when engaging a Pearson-correlation based measurement. Unfortunately, there is not a gold-standard normalization method that works best in all cases. Simple statistical analyses and plots of the data would help in the selection of the proper normalization method.

5.2 Feature usage

Good features play crucial roles in the prediction performance of machine learning methods, which we have discussed comprehensively above. Here, we discuss three important aspects of feature usage in enhancer and enhancer target predictions. First, as context specificity is an intrinsic characteristic of both enhancer activities and enhancer-target associations, we stress the importance of including context specific features. In the history of enhancer prediction, motifs and conservation were first used. These are "static" features, which means we could only use these features to judge whether a genomic region is an enhancer in some contexts, but not when (e.g. which developmental stage) and where (e.g. which cell type, cellular process) it would become active. Later, thanks to the boom of ChIP-seq data in a wide range of cell types, context-specific features such as HM and TF binding signals made it possible to perform cell-type specific enhancer predictions. In contrast, most current methods for enhancer target prediction use only static features. If the active enhancer-target associations in a given context are to be identified, one common strategy is to consider only the pairs involving an enhancer predicted/proved to be active in the context. Due to changes of chromosome conformation or other reasons, it is possible that an enhancer active in two different contexts regulates different genes in the two contexts. Novel methods that can utilize more context-specific information in directly predicting enhancer targets in a given context are called for.

A second interesting aspect is the relative importance of different features. When investigating a context with insufficient experimental data, and one is to perform additional experiments to get data for predicting enhancers or enhancer targets, it would be desirable to know what experiments are most cost-effective. Rajagopal et al. (2013) found a set of HMs that resulted in the best prediction accuracy, which partially answered this question. More generally, the relative importance of different types of features such as HM, TF binding, eRNA and DNA methylation is yet to be studied.

Another aspect is that a feature could be used for defining positive/negative examples, constructing the prediction model, or evaluating the performance of a model. For instance, P300 binding has been used in a number of studies for defining positive examples; Some studies use the enrichment of P300 binding signals as a way to partially validate the predictions; P300 could as well be used as a feature for building a model for predicting enhancer. One major current challenge is that given the limited number of features, one needs to determine which of them should be used in each of these three tasks, so that prediction accuracy can be maximized while there is no "leakage" of information in the prediction process, i.e., having some information used both in training and validating a model. This problem is expected to be mitigated as more experimentally validated enhancers and enhancer targets become available.

5.3   Prediction validation

Prediction validation is a crucial part of every prediction task in bioinformatics. However, among the studies discussed, only a very small portion of the predictions made were tested experimentally. Obviously it is difficult to validate all predictions using highly accurate, low-throughput experimental assays due to the prohibitive cost. Another type of validations commonly performed is cross-checking the predictions with previously published experimental results such as ChIA-PET, Hi-C, 5C and eQTL-gene pairs for enhancer-target associations. One potential problem is that the predictions could be made in a context different from the one from which these public data were produced. Noise in these experimental data could be another issue. Also, some of these data only provide supporting evidence, but cannot completely prove the correctness of a prediction. For instance, a predicted enhancer-promoter association with *in vitro* ChIA-PET data support does not necessarily mean the enhancer-promoter interaction must have a regulatory role; It does not even guarantee the enhancer and the promoter are in contact *in vivo*. Having these shortcomings notwithstanding, including independent experimental supports would definitely help in evaluating and improving existing computational prediction methods. High-throughput assays such as STARR-seq, which has higher data variability but lower relative cost than low-throughput assays, could be a good choice for large-scale validations of computational predictions.

5.4 Training set design in supervised methods

The careful selection of training examples is key to the success of machine learning methods. In many bioinformatics problems, the design of a suitable negative training set is far from trivial. For instance, in enhancer prediction, the negative examples cannot be simply defined as randomly-selected regions not known to be enhancers, for these examples are too different from the positive examples in many aspects, and any model that distinguishes active regions in the genome from the inactive ones would probably separate the positive and negative examples well. In other words, the resulting model may not be useful for predicting enhancers, but just general active genomic regions including gene bodies and other types of regulatory elements. The rule of thumb is that the negative examples should not be "too negative", i.e., they should share as many features as the positive examples as possible, except for the ones very unique to the positive examples. Alternatively, including a mixture of different types of negative examples could make it more robust.

5.5 Multiple-to-multiple relationships

After reviewing the current methods for association prediction, we notice that there are no existing methods that explicitly handle multiple-to-multiple relationships between enhancers and promoters. Every enhancer-promoter pair was considered independently in all the surveyed methods. Though the mechanisms of enhancer targeting are not completely clear yet, previous studies have shown that multiple enhancers (called shadow enhancers) controlling the same promoter could ensure the robust expression of the corresponding genes (Meireles-Filho & Stark 2009; Perry et al. 2011). New computational methods are needed to study the significance of modeling the effects of multiple enhancers and/or targets simultaneously.

5.6 Future outlook

Overall, we predict that context specificity and multiple-to-multiple relationships would be two important aspects that should be incorporated in future enhancer and enhancer target predictions.

Among all the features considered for the two tasks, eRNA is a promising feature for both tasks for two reasons: First, CAGE experiment is mature and economical and thus can be applied to many samples; second, both eRNA and promoter activity are quantified in the same way based on CAGE tags, making the corresponding data easily comparable.

Since there are experimentally validated enhancers and enhancer targets, but the numbers are small, semi-supervised prediction methods that make use of both labeled examples and properties of unlabeled points could be more suitable than purely unsupervised or purely supervised methods.

Active learning is another direction worth pursuing. The active learning setting aims at acquiring new examples that can benefit the overall learning process most. In enhancer and enhancer-target predictions, ambiguous cases (such as enhancers with intermediate levels of H3K27ac) could be most informative in refining prediction models.

Finally, we hope to see more collaboration between computer scientists and biologists in studying enhancers and enhancer targets, since the validation process is of utmost importance for evaluating the computational methods and providing insights for improving the methods.

References

Ahituv, N. et al., 2005. Mapping cis-regulatory domains in the human genome using multi-species conservation of synteny. *Human Molecular Genetics*, 14(20):3057–63.

Andersson, R. et al., 2014. An atlas of active enhancers across human cell types and tissues. *Nature*. 507(7493):455–61.

Aran, D., Sabato, S. and A. Hellman, 2013. DNA methylation of distal regulatory sites characterizes dysregulation of cancer genes. *Genome Biology*. 14(3):R21.

Arnold, C.D. et al., 2013. Genome-wide quantitative enhancer activity maps identified by STARR-seq. *Science*, 339(6123):1074–77.

Bailey, T.L. and P. MacHanick, 2012. Inferring direct DNA binding from ChIP-seq. *Nucleic Acids Research*. 40(17):e128.

Barski, A. et al., 2007. High-Resolution Profiling of Histone Methylations in the Human Genome. *Cell*. 129(4):823–37.

Berman, B.P. et al., 2004. Computational identification of developmental enhancers: conservation and function of transcription factor binding-site clusters in *Drosophila melanogaster* and *Drosophila pseudoobscura*. *Genome Biology*. 5(9):R61.

Blow, M.J. et al., 2010. ChIP-Seq identification of weakly conserved heart enhancers. *Nature Genetics*. 42(9):806–10.

Bonn, S. et al., 2012. Tissue-specific analysis of chromatin state identifies temporal signatures of enhancer activity during embryonic development. *Nature Genetics*. 44(2):148–56.

Boyle, A.P. et al., 2008. High-Resolution Mapping and Characterization of Open Chromatin across the Genome. *Cell*. 132(2):311–22.

Carroll, S.B., 2008. Evo-Devo and an Expanding Evolutionary Synthesis: A Genetic Theory of Morphological Evolution. *Cell*. 134(1):25–36.

Corradin, O. et al., 2014. Combinatorial effects of multiple enhancer variants in linkage disequilibrium dictate levels of gene expression to confer susceptibility to common traits. *Genome Research*. 24(1):1–13.

D'Alessandro, M. et al., 2003. Epileptic seizure prediction using hybrid feature selection over multiple intracranial EEG electrode contacts: a report of four patients. *IEEE Transactions in Biomedical Engineering*. 50(5):603-15.

Dawson, M.A. and T. Kouzarides, 2012. Cancer epigenetics: From mechanism to therapy. *Cell*. 150(1):12–27.

Dekker, J. et al., 2002. Capturing chromosome conformation. *Science*. 295(5558):1306–11.

DeMare, L.E. et al., 2013. The genomic landscape of cohesin-associated chromatin interactions. *Genome Research*, 23(8):1224–34.

Dixon, J.R. et al., 2012. Topological domains in mammalian genomes identified by analysis of chromatin interactions. *Nature*, 485(7398):376–80.

Dostie, J. et al., 2006. Chromosome Conformation Capture Carbon Copy (5C): A massively parallel solution for mapping interactions between genomic elements. *Genome Research*, 16(10):1299–09.

Duan, Z. et al., 2010. A three-dimensional model of the yeast genome. *Nature*, 465(7296):363–67.

ENCODE Project Consortium et al., 2012. An integrated encyclopedia of DNA elements in the human genome. *Nature*, 489(7414):57–74.

Ernst, J. et al., 2011. Mapping and analysis of chromatin state dynamics in nine human cell types. *Nature*, 473(7345):43–49.

Ernst, J. and M., Kellis, 2012. ChromHMM: automating chromatin-state discovery and characterization. *Nature Methods*, 9(3):215–16.

Ernst, J. & Kellis, M., 2010. Discovery and characterization of chromatin states for systematic annotation of the human genome. *Nature Biotechnology*, 28(8):817–25.

FANTOM Consortium and the RIKEN PMI and CLST (DGT) et al., 2014. A promoter-level mammalian expression atlas. *Nature*, 507(7493):462–70.

Ferretti, V. et al., 2007. PReMod: A database of genome-wide mammalian cis-regulatory module predictions. *Nucleic Acids Research*, 35(SUPPL. 1):D122-D126.

Firpi, H.A., D. Ucar, and K., Tan, 2010. Discover regulatory DNA elements using chromatin signatures and artificial neural network. *Bioinformatics*, 26(13):1579–86.

Fu, Y. et al., 2014. FunSeq2 : a framework for prioritizing noncoding regulatory variants in cancer. *Genome Biology*, 15(10):480.

Fullwood, M.J. et al., 2009. An oestrogen-receptor-alpha-bound human chromatin interactome. *Nature*, 462(7269):58–64.

Gallo, S.M. et al., 2011. REDfly v3.0: Toward a comprehensive database of transcriptional regulatory elements in Drosophila. *Nucleic Acids Research*, 39(SUPPL. 1):D118-D123.

Giresi, P.G. et al., 2007. FAIRE (Formaldehyde-Assisted Isolation of Regulatory Elements) isolates active regulatory elements from human chromatin. *Genome Research*, 17(6):877–85.

He, B. et al., 2014. Global view of enhancer-promoter interactome in human cells. *Proceedings of the National Academy of Sciences of the United States of America*, 111(21):E2191–E2199.

Heidari, N. et al., 2014. Genome-wide map of regulatory interactions in the human genome. *Genome Research*, 24(12):1905–17.

Heintzman, N.D. et al., 2007. Distinct and predictive chromatin signatures of transcriptional promoters and enhancers in the human genome. *Nature Genetics*, 39(3):311–18.

Hnisz, D. et al., 2013. Super-enhancers in the control of cell identity and disease. *Cell*, 155(4):934–47.

Hoffman, M.M. et al., 2013. Integrative annotation of chromatin elements from ENCODE data. *Nucleic Acids Research*, 41(2):827–41.

Hoffman, M.M. et al., 2012. Unsupervised pattern discovery in human chromatin structure through genomic segmentation. *Nature Methods*, 9(5):473–76.

Jin, F. et al., 2013. A high-resolution map of the three-dimensional chromatin interactome in human cells. *Nature*, 503(7475):290–94.

Khan, A. & X., Zhang, 2015. dbSUPER: an integrated database of super-enhancers in mouse and human genome. bioRxiv doi: http://dx.doi.org/10.1101/014803.

Kikuta, H. et al., 2007. Genomic regulatory blocks encompass multiple neighboring genes and maintain conserved synteny in vertebrates. *Genome Research*, 17(5):545–555.

Kim, T.-K. et al., 2010. Widespread transcription at neuronal activity-regulated enhancers. *Nature*, 465(7295):182–87.

Koller, D. and N., Friedman, 2009. *Probabilistic Graphical Models: Principles and Techniques*, MIT Press.

Kumaki, Y. et al., 2008. Analysis and synthesis of high-amplitude Cis-elements in the mammalian circadian clock. *Proceedings of the National Academy of Sciences of the United States of America*, 105(39):14946–51.

Kwasnieski, J.C. et al., 2012. Complex effects of nucleotide variants in a mammalian cis-regulatory element. *Proceedings of the National Academy of Sciences of the United States of America*, 109(47):19498–03.

Kwasnieski, J.C. et al., 2014. High-throughput functional testing of ENCODE segmentation predictions. *Genome Research*, 24(10):1595-02.

Larkin, D.M. et al., 2009. Breakpoint regions and homologous synteny blocks in chromosomes have different evolutionary histories. *Genome Research*, 19(5):770–77.

Li, G. et al., 2010. ChIA-PET tool for comprehensive chromatin interaction analysis with paired-end tag sequencing. *Genome Biology*, 11(2):R22.

Li, X.Y. et al., 2008. Transcription factors bind thousands of active and inactive regions in the Drosophila blastoderm. *PLoS Biology*, 6(2):0365–88.

Lieberman-Aiden, E. et al., 2009. Comprehensive mapping of long-range interactions reveals folding principles of the human genome. *Science*, 326(5950):289–93.

May, D. et al., 2011. Large-scale discovery of enhancers from human heart tissue. *Nature Genetics*, 44(1):89–93.

Meireles-Filho, A.C. and A., Stark, 2009. Comparative genomics of gene regulation-conservation and divergence of cis-regulatory information. *Current Opinion in Genetics and Development*, 19(6):565–70.

Melnikov, A. et al., 2012. Systematic dissection and optimization of inducible enhancers in human cells using a massively parallel reporter assay. *Nature Biotechnology*, 30(3):271–77.

Navratilova, P. et al., 2009. Systematic human/zebrafish comparative identification of cis-regulatory activity around vertebrate developmental transcription factor genes. *Developmental Biology*, 327(2):526–40.

Nora, E.P. et al., 2012. Spatial partitioning of the regulatory landscape of the X-inactivation centre. *Nature*, 485(7398):381–85.

Park, P.J., 2009. ChIP-seq: advantages and challenges of a maturing technology. *Nature Reviews. Genetics*, 10(10):669–80.

Patwardhan, R.P. et al., 2012. Massively parallel functional dissection of mammalian enhancers *in vivo*. *Nature Biotechnology*, 30(3):265–70.

Pennacchio, L.A. et al., 2006. *In vivo* enhancer analysis of human conserved non-coding sequences. *Nature*, 444(7118):499–02.

Pennacchio, L.A. et al., 2007. Predicting tissue-specific enhancers in the human genome. *Genome Research*, 17(2):201–11.

Perry, M.W., A.N. Boettiger, and M., Levine, 2011. Multiple enhancers ensure precision of gap gene-expression patterns in the Drosophila embryo. *Proceedings of the National Academy of Sciences of the United States of America*, 108(33):13570–75.

Rada-Iglesias, A. et al., 2011. A unique chromatin signature uncovers early developmental enhancers in humans. *Nature*, 470(7333):279–83.

Rajagopal, N. et al., 2013. RFECS: A Random-Forest Based Algorithm for Enhancer Identification from Chromatin State. *PLoS Computational Biology*, 9(3):e1002968.

Ramos, Y.F.M. et al., 2010. Genome-wide assessment of differential roles for p300 and CBP in transcription regulation. *Nucleic Acids Research*, 38(16):5396–408.

Rao, S.S.P. et al., 2014. A 3D Map of the Human Genome at Kilobase Resolution Reveals Principles of Chromatin Looping. *Cell*, 159(7):1665–80.

Rhee, H.S. and B.F., Pugh, 2012. ChiP-exo method for identifying genomic location of DNA-binding proteins with near-single-nucleotide accuracy. *Current Protocols in Molecular Biology*, 21:21.24.

Rhee, H.S. and B.F., Pugh, 2011. Comprehensive genome-wide protein-DNA interactions detected at single-nucleotide resolution. *Cell*, 147(6):1408–19.

Schug, J. et al., 2005. Promoter features related to tissue specificity as measured by Shannon entropy. *Genome Biology*, 6(4):R33.

Sharon, E. et al., 2012. Inferring gene regulatory logic from high-throughput measure-ments of thousands of systematically designed promoters. *Nature Biotechnology*, 30(6):521–530.

Shlyueva, D., G. Stampfel, and A., Stark, 2014. Transcriptional enhancers: from properties to genome-wide predictions. *Nature Reviews. Genetics*, 15(4):272–86.

Su, J., S.A. Teichmann, and T.A., Down, 2010. Assessing computational methods of cis-regulatory module prediction. *PLoS Computational Biology*, 6(12):e1001020.

Thurman, R.E. et al., 2012. The accessible chromatin landscape of the human genome. *Nature*, 489(7414):75–82.

Vaquerizas, J.M. et al., 2009. A census of human transcription factors: function, expression and evolution. *Nature Reviews. Genetics*, 10(4):252–63.

Visel, A. et al., 2007. VISTA Enhancer Browser - A database of tissue-specific human enhancers. *Nucleic Acids Research*, 35(SUPPL. 1):D88-D92.

Visel, A., E.M. Rubin, and L.A., Pennacchio, 2009. Genomic views of distant-acting enhancers. *Nature*, 461(7261):199–05.

Wang, Z. et al., 2008. Combinatorial patterns of histone acetylations and methylations in the human genome. *Nature Genetics*, 40(7):897–903.

Whyte, W.A. et al., 2013. Master transcription factors and mediator establish super-enhancers at key cell identity genes. *Cell*, 153(2):307–19.

Won, K.-J. et al., 2008. Prediction of regulatory elements in mammalian genomes using chromatin signatures. *BMC Bioinformatics*, 9:547.

Xu, Y. et al., 2011. Genome-wide Regulation of 5hmC, 5mC, and Gene Expression by Tet1 Hydroxylase in Mouse Embryonic Stem Cells. *Molecular Cell*, 42(4):451–64.

Yip, K.Y. et al., 2012. Classification of human genomic regions based on experimentally determined binding sites of more than 100 transcription-related factors. *Genome Biology*, 13(9):R48.

Yip, K.Y., C. Cheng, and M., Gerstein, 2013. Machine learning and genome annotation: a match meant to be? *Genome Biology*, 14(5):205.

Zhao, Z. et al., 2006. Circular chromosome conformation capture (4C) uncovers exten-sive networks of epigenetically regulated intra- and interchromosomal interactions. *Nature Genetics*, 38(11)1341–47.

2

Cormotif: An R Package for Jointly Detecting Differential Gene Expression in Multiple Studies

Yingying Wei[1]*, Toyoaki Tenzen[2] and Hongkai Ji[3]

Abstract

With the rapid decrease of costs, the high-throughput gene expression data are accumulating at exponential rate in larger public repositories. Nevertheless, usually only very few replicates are available in each experiments, which make differential gene expression detection suffer from low sample size. On the other hand, multiple similar studies conducted by different groups are accessible now. The standard algorithms for detecting differential genes from microarray data are mostly designed for analyzing a single dataset. Separately analyzing each study may fail to detect some key genes showing low fold changes consistently in all studies. Rather, jointly modeling all data allows one to borrow information across studies to improve statistical inference. However, the simple concordance model, which assumes that differential expression occurs in either all studies or none of the studies, fails to capture study-specific differentially expressed genes. In contrast, a model that naively enumerates and analyzes all possible differential patterns across studies can deal with study-specificity and allow information pooling, but the complexity of its parameter space grows exponentially as the number of studies increases.

In this chapter, we describe our correlation motif approach to address this dilemma. Our approach searches for a small number of latent probability vectors called correlation motifs to capture the major correlation patterns among multiple studies. The

[1] Department of Statistics, The Chinese University of Hong Kong, Shatin NT, Hong Kong Tel: (852) 3943-7922.

[2] Center for Regenerative Medicine, Cardiovascular Research Center, Massachusetts General Hospital, Boston, MA 02114, USA.

[3] Department of Biostatistics, The Johns Hopkins University Bloomberg School of Public Health, Baltimore, MD, USA. Tel: (001) 410-955-3517.

* Corresponding author : ywei@sta.cuhk.edu.hk

Sections of this chapter have been taken from Wei, Y, Toyoaki T, and Ji, H. (2015) Joint analysis of differential gene expression in multiple studies using correlation motifs. *Biostatistics* 16.1: 31-46.

approach has the flexibility to handle all possible study-specific differential patterns, improves detection of differential expressions and overcomes the barrier of exponential model complexity. This chapter provides description of the method as well as instructions on using the corresponding Bioconductor R package *Cormotif.*

1. Introduction

Detecting differentially expressed genes is a basic task in the analysis of gene expression data. The state-of-the-art solutions to this problem, such as *limma* (Smyth, 2004), *SAM* (Tusher et al. 2001), edgeR (Robinson and Smyth, 2007, 2008), and DESeq (Anders and Huber, 2010), are mostly designed for analyzing data from a single experiment or study. With 1,000, 000+ samples stored in public databases such as Gene Expression Omnibus (GEO), it is now very common for scientists to have data from multiple related experiments or studies. An emerging problem is how one can integrate data from multiple studies to more effectively analyze differential expression.

One example that motivated this article is a study of the vertebrate sonic hedgehog (SHH) signaling pathway. SHH is a signaling protein that can bind to patched 1 (PTCH1), a receptor protein in cell membrane [Fig 1(a)]. PTCH1 can interact with another membrane protein smoothened (SMO) to repress its activity. In the absence of SHH, PTCH1 keeps SMO inactive. The presence of SHH will repress PTCH1 and activate SMO. The active SMO triggers a signaling cascade to modulate activities of three transcription factors, GLI1, GLI2, and GLI3, which in turn induce or repress the expression of hundreds of downstream target genes. SHH pathway is a core signaling pathway in vertebrate (Ingham and McMahon, 2001). To elucidate the underlying mechanisms linking this pathway to development and diseases, multiple studies have been conducted in different contexts to identify genes whose transcriptional activities are modulated by SHH signaling. Some studies perturb the SHH signal in different tissues by knocking out or over-expressing the pathway's key signal transduction components such as SHH, PTCH1, and SMO, while others compare

TABLE 1 SHH microarray data description: 8somites and 13somites indicate two different developmental stages of embryos; smo indicates mice with mutant Smo; ptc stands for mice with mutant Ptch1; wt means wild type; shh represents Shh mutant. Medulloblastoma and BCC are two types of tumors.

Study ID	Condition 1 (case)	Sample No.	Condition 2 (control)	Sample No.	Reference
1	8somites smo	3	8somites wt	3	Tenzen et al. (2006)
2	8somites ptc	3	8somites wt	3	Tenzen et al. (2006)
3	13somites ptc	3	13somites wt	3	Tenzen et al. (2006)
4	head shh	3	head wt	3	Tenzen et al. (2006)
5	limb shh	3	limb wt	3	Tenzen et al. (2006)
6	Medulloblastoma tumor	3	Medulloblastoma control	2	Mao et al. (2006)
7	BCC tumor	3	BCC control	3	Mao et al. (2006)
8	13somites smo	3	13somites wt	3	Tenzen et al. (2006)

disease samples with corresponding controls. Table 1 contains eight such datasets in mouse originally collected by Tenzen et al. (2006) and Mao et al. (2006). Each dataset involves a comparison of genome-wide expression profiles between two different sample types. These data were all generated usin Affymetrix Mouse Expression Set 430 arrays. The questions of biological interest include (i) which genes are controlled by the SHH signal in each dataset, (ii) which genes are the core targets that respond to the SHH signal irrespective of tissue type and developmental stage, and (iii) which genes are context-specific targets and are modulated by the SHH signal only in certain conditions.

For simplicity, below each dataset is a *study*. One simple approach to analyze these data is to analyze each study separately using existing state-of-the-art methods such as *limma* (Smyth, 2004) or *SAM* (Tusher et al. 2001). This approach is not ideal as it may fail to detect genes with low-fold changes but consistently differential in many or all studies.

Modeling all data jointly may allow one to borrow information across studies to improve the analysis. A simple model to combine data is to assume that each gene is either differential in all studies or non-differential in all studies (Conlon et al. 2006). This concordance model may help with identifying genes with small but consistent expression changes in all studies. However, it ignores the reality that activities of many important genes are tissue- or time-specific. This method will only produce a single gene list that reports and ranks genes in the same way for all studies. It cannot prioritize genes differently for different studies to account for context-specificity.

A more flexible approach is to consider all possible differential expression patterns. Suppose there are D studies and each gene can either be differential or non-differential in each study, there will be 2^D possible differential expression patterns. One can model the data as a mixture of 2^D different gene classes. This allows one to deal with context-specificity. However, an obvious drawback is that as the number of studies increases, the number of possible patterns increases exponentially. Thus, the model does not scale well with the increasing D.

In this chapter, we describe our method, *CorMotif*, for jointly analyzing multiple studies to improve differential expression detection. This method is both flexible for handling context-specificity and scalable to increasing study number. The key idea is to use a small number of latent probability vectors called "correlation motifs" to model the major correlation patterns among the studies. The motifs essentially group genes into clusters based on their differential expression patterns, and the differential gene detection is coupled with the clustering. Unlike *CorMotif*, many methods developed previously for analyzing differential expression in multiple studies or conditions have exponential model complexity and therefore limited scalability.

Previously, Kendziorski et al. (2003) proposed an Empirical Bayes approach (called "eb1" in this chapter) for analyzing differential expression involving multiple biological conditions. This approach requires users to specify all possible differential patterns, and the data are then modeled accordingly. If a user applies this method to detect differential expression between two conditions in multiple studies and wants to accommodate all possible differential patterns, the user has to enumerate all 2^D possible patterns, leading to the exponential complexity problem. Similar to Kendziorski et al. (2003), Jensen et al. (2009) developed a hierarchical Bayesian model and a Markov Chain Monte Carlo (MCMC) algorithm to analyze multiple conditions, again with exponential complexity due to requirement of enumerating

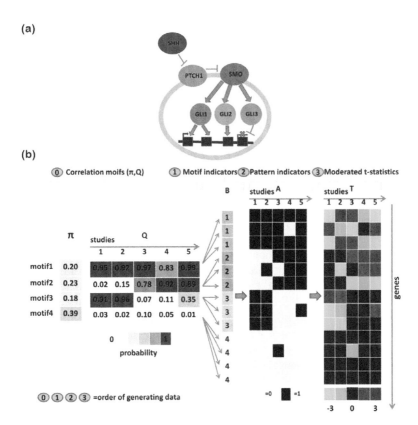

FIGURE 1 (a) A cartoon illustration of SHH pathway. (b) A numerical example of the data generating model. There exist four motifs in the dataset, with the abundance $\pi = (0.2, 0.23, 0.18, 0.39)$. Each row of the Q matrix represents a motif and each column corresponds to a study. Thus, q_{kd} indicates the probability for genes belonging to motif k to be differentially expressed in study d. For example, the probability for genes belonging to motif 1 to be differentially expressed in study 4 is 0.83. The gray scale of the cells in π and Q illustrates the probability value. The probability increases from 0 to 1 as the color changes from light to dark. Given π and Q, each gene is assigned a motif indicator bg. For instance, the fifth gene belongs to motif 2 (indicated by a cell with a number "2"). Next, the configuration of the fifth gene, $[a_{51}, a_{52}, a_{53}, a_{54}, a_{55}]$, is generated according to $q_2 = (0.02, 0.15, 0.78, 0.92, 0.89)$. As a result, the fifth gene is differentially expressed in study 2, 4, and 5. Finally, the moderated t-statistic t_5d within each study d is produced according to the configuration a_5d.

all possible patterns. Ruan and Yuan (2011) generalize Kendziorski et al. (2003) to a model that can integrate information from multiple studies where each study may involve comparisons of multiple conditions. Within each study, this method enumerates all possible combinatorial patterns among multiple conditions, again resulting in exponential complexity. Moreover, differential expression patterns are assumed to be concordant across studies, that is, each gene is assumed to have the same differential pattern in all studies. The concordance assumption does not allow study-specific differential expression.

Scharpf et al. (2009) proposed a fully Bayesian framework, XDE, for cross-study differential expression analysis. It offers two implementations. The "Single-Indicator" implementation uses a concordance model by assuming that each gene's differential state is the same across all studies. The "Multiple-Indicator" implementation allows study-specific differential expression. However, it assumes that all genes have the same prior probability to be differential within the same study, and the differential states of each gene in different studies are a priori independent. Conceptually, these assumptions are similar to a *CorMotif* model with a single cluster, which often is insufficient to capture the heterogeneity among genes since the cross-study correlation pattern may vary from one gene to another. XDE does not have the exponential complexity problem, but it uses MCMC for posterior inference and is very slow computationally.

To capture the heterogeneity among genes, Yuan and Kendziorski (2006) developed a method for simultaneous clustering and differential expression analysis. Similar to *CorMotif*, this method also assumes that genes belong to multiple clusters, and different clusters have different propensities to show differential expression. However, Yuan and Kendziorski (2006) only considered detecting differential expression between two conditions in one study. Although one may conceptually extend this approach to handle multiple studies by combining it with the model developed by Kendziorski et al. (2003), such a simple extension would lead to the model "eb10best" in which genes are assumed to fall into multiple clusters and each cluster is a mixture of 2^D differential patterns. As a result, the complexity of the parameter space would become $O(K*2^D)$ where K is the number of clusters. Once again, the model complexity explodes as the dataset number increases.

Compared with these methods, *CorMotif* offers a unique data integration solution in that it addresses study-specificity, heterogeneity among genes, and exponential complexity simultaneously. Below we focus on discussing *CorMotif* for microarray data since it was motivated by the microarray analysis in the SHH study. However, the idea behind *CorMotif* is general, and it should be straightforward to develop a similar framework for RNA-seq data.

2.   Methods

2.1   Data structure and preprocessing

Suppose there are G genes and D microarray studies. Each study d compares two biological conditions (e.g. cancer *vs.* normal), and each condition l has n_{dl} replicate samples. Different studies may be related, but they can compare different biological conditions. Let x_{gdlj} be the normalized and appropriately transformed expression value of gene g in study d, condition l, and replicate j . In this book chapter, all data were normalized and log-transformed using RMA (Irizarry et al. 2003). The ensemble of observed data is $X = \{x_{gdlj}: g=1, \ldots, G; d=1, \ldots, D; l=1, 2; j=1, \ldots, n_{dl} \}$.

Each gene can be differentially expressed in some, all, or none of the studies. Let $a_{gd}=1$ or 0 indicate whether gene g is differentially expressed in study d or not. $A=(a_{gd})_{G\times D}$ is a $G \times D$ matrix that contains all a_{gd} s. Given the observed data X, one is interested in inferring A.

CorMotif first applies limma (Smyth, 2004) to each study separately. In other words, the moderated t-statistic (Smyth, 2004) t_{gd} was calculated for each gene g within each study d according to their normalized and log-transformed data $X_d = \{x_{gdlj}; g=1,\ldots,G; l=1,2; j=1,\ldots,n_{dl}\}$. the moderated t-statistic summarizes gene g's differential expression information in study d. Under the limma model, when gene g is not differentially expressed in study d (i.e. $a_{gd}=0$), t_{gd} follows a t-distribution f_{d0}; when $a_{gd}=1$, t_{gd} follows a scaled t-distribution f_{d1} (Smyth, 2004). Readers may refer to Smyth 2004 for details. Next, we arrange all t_{gd} s into a matrix $T=(t_{gd})_{G\times D}$. *CorMotif* will then use T instead of the raw expression values X to infer A.

2.2   Correlation motif model

Organize the differential expression states of gene g into a vector $a_g=[a_{g1}, a_{g2}, \ldots, a_{gD}]$. For D studies, a_g has 2^D possible configurations. A simple way to describe the correlation among studies is to document the empirical frequency of observing each of the 2^D configurations of a_g among all genes. This is because $f(a_g)$, the joint distribution of $=[a_{g1}, a_{g2}, \ldots, a_{gD}]$, is known once the probability of observing each configuration is given. This joint distribution will determine how a_{gd} s from different studies are correlated. While simple, this approach is not scalable since it requires $O(2^D)$ parameters and the parameter space expands exponentially with increasing D.

To avoid this limitation, *CorMotif* adopts a hierarchical mixture model (Fig. 1(b)). The model assumes that genes fall into K different classes ($K<<2^D$ for big D), and the moderated t-statistics $T=(t_{gd})_{G\times D}$ are viewed as generated as follows:

(1) Each gene g is randomly and independently assigned a class label b_g according to probability $\pi=(\pi_1, \ldots, \pi_K)$. Here, $\pi_k \equiv \Pr(b_g = k)$ is the prior probability that a gene belongs to class k.

(2) Given genes' class labels (i.e. b_gs), genes' differential expression states a_{gd} s are generated independently according to probabilities $q_{kd} \equiv \Pr(a_{gd} = 1|b_g=k)$. For genes in the same class k, a_gs are generated using the same probabilities $q_k=(q_{k1}, \ldots, q_{kD})$.

(3) Given the differential expression states a_{gd} s, genes' moderated t-statistics t_{gd} s are generated independently according to $f_{d1}(t_{gd})$ or $f_{d0}(t_{gd})$.

Let $B=(b_1, \ldots, b_G)$ be the class membership for all genes. Organize q_k into a matrix $Q = (q_1^T, \ldots, q_K^T)^T = (q_{kd})_{K\times D}$. Let $\delta(\cdot)$ be an indicator function: $\delta(\cdot)=1$ if its argument is true, and $\delta(\cdot)=0$ otherwise. Based on the above model, the joint probability distribution of A, B, and T conditional on π and Q is

$$\Pr(\pi,Q,A,B\mid T) \propto \prod_{g=1}^{G}\prod_{k=1}^{K} \left\{ \pi_k \prod_{d=1}^{D}\left[q_{kd}f_{d1}(t_{gd})\right]^{a_{gd}} \right.$$

$$\left. \left[(1-q_{kd})f_{d0}(t_{gd})\right]^{1-a_{gd}} \right\}^{\delta(b_g=k)}$$

In this model, each gene class k is associated with a vector q_k whose elements are the prior probabilities of a gene in this class to be differential in studies $1, \ldots, D$. Each

q_k represents a probabilistic differential expression pattern and therefore is called a "motif". Since q_{kd} s are probabilities, genes in the same class can have different a_g configurations. On the other hand, genes from the same class share the same q_k, and hence their differential expression configuration a_gs tend to be similar. Genes in different classes have different q_ks, and their a_gs also tend to be different. Essentially, our model groups genes into K clusters based on a_g. However, unlike a usual clustering algorithm, here a_gs are unknown.

Despite the assumption that a_{gd}s are *a priori* independent conditional on the class label b_g, a_{gd}s are no longer independent once the class label b_g is integrated out. To see this, consider the prior probability that a gene is differentially expressed in all studies. Based on our model, $\Pr(a_g = [1,...,1]) = \Sigma_k (\pi_k \Pi_d q_{kd})$ which is different from the product of the marginal $\Pi_d \Pr(a_{gd} = 1) = \Pi_d \Sigma_k \pi_k q_{kd}$. This explains why the hierarchical mixture model above can be used to describe the correlation among multiple studies. Since the mixture of q_ks provides the key to model the cross-study correlation, each vector q_k is also called a "correlation motif".

A model with K correlation motifs requires $O(KD)$ parameters in total. Usually, a small K ($\ll 2^D$ when D is big) is sufficient to capture the major correlation structure in the real data. Therefore, our method can be easily scaled up to deal with large D scenarios. When $0 < q_{kd} < 1$, each q_k will be able to generate all 2^D configurations with non-zero probabilities. Thus, our model also retains the flexibility to allow all 2^D configurations of a_g to occur at individual gene level.

2.3 Statistical inference

In reality, only T is observed. π and Q are unknown parameters. A and B are unobserved missing data. To infer the unknowns from T, we first assume that K is given and introduce a Dirichlet prior Dir(2, . . . , 2) for π and a Beta prior $B(2, 2)$ for q_{kd}. As a result,

$$\Pr(\pi, Q, A, B \mid T) \propto \prod_{g=1}^{G} \prod_{K=1}^{K} \left\{ \pi_k \prod_{d=1}^{D} \left[q_{kd} f_{d1}(t_{gd}) \right]^{a_{gd}} \right.$$

$$\left. [(1 - q_{kd}) f_{d0}(t_{gd})]^{1-a_{gd}} \right\}^{\delta(b_g = k)} \prod_{k=1}^{K} \pi_k \prod_{k=1}^{K} \prod_{d=1}^{D} q_{kd} (1 - q_{kd})$$

Based on the above posterior distribution, an expectation–maximization (EM) algorithm (Gelman et al. 2004) can be derived to search for the posterior mode of π and Q.

Using the estimated $\hat{\pi}$ and $\hat{Q}$, one can then compute $\Pr(a_{gd}=1|T, \hat{\pi}, \hat{Q})$, the posterior probability that gene g is differentially expressed in study d after integrating out the motif membership b_g. Next, we rank-order genes in each study separately using $\Pr(a_{gd}=1|T, \hat{\pi}, \hat{Q})$. The ranked lists can be used to choose follow-up targets. Users can also provide a posterior probability cutoff to dichotomize genes into *differential* or *non-differential* genes in each study. The default cutoff is 0.5. Users have the option to set the cutoff to other values. In order to choose the motif number K, we use Bayesian Information Criterion (BIC).

CorMotif improves the differential expression detection by integrating informa-tion both across studies and across genes. $\Pr(a_{gd}=1|T, \hat{\pi}, \hat{Q})$ can be decomposed as $\Sigma_{k=1}^{K} \Pr(a_{gd}=1|T, \hat{\pi}, \hat{Q}, b_g = k) * \Pr(b_g = k|T, \hat{\pi}, \hat{Q})$. Here, $\Pr(b_g = k|T, \hat{\pi}, \hat{Q})$ is determined by jointly evaluating gene g's data in all studies, and $\Pr(a_{gd}=1|T, \hat{\pi}, \hat{Q}, b_g = k)$ contains information specific to study d. According to Bayes' theorem, $\Pr(a_{gd}=1|T, \hat{\pi}, \hat{Q}, bg = k) \, \alpha \, \Pr(t_{gd}|a_{gd}=1, \hat{\pi}, \hat{Q}, b_g = k) \times \Pr(a_{gd}=1|\hat{\pi}, \hat{Q}, b_g = k)$. t_{gd} in the first term contains expression information for a given gene g in study d. To compute its denominator, the limma approach also utilized information across genes to help with estimating the variance. Meanwhile, the second term $\Pr(a_{gd}=1|\hat{\pi}, \hat{Q}, b_g = k)$ involves prior probabilities given by the correlation motifs which are estimated using data from all genes. Owing to this two-way information pooling (i.e. across both studies and genes), *CorMotif* uses information more effectively than methods based on only a single gene or a single study. This is especially useful for analyzing studies with relatively weak signal-to-noise ratio.

3. Simulations

3.1 Compared methods

We compared *CorMotif* with six other methods: *separate limma, all concord, full motif, SAM, ebl,* and *eb10best*. We did not compare the method in Jensen et al. (2009) as no software was available for this method. The *separate limma* approach ana-lyzes each study separately using limma. The moderated *t*-statistics in each study are assumed to be a mixture of f_{d0} and f_{d1}. To better evaluate the gain from data integration, we matched this analysis to *CorMotif* as much as possible by running an EM algorithm similar to *CorMotif* to compute the posterior probability for dif-ferential expression using 0.5 as default cutoff. Conceptually, this makes *separate limma* equivalent to *CorMotif* with a single cluster ($K=1$), and the analysis produces the same gene ranking as limma in each study. *All concord* assumes that a gene is either differential in all studies or non-differential in all studies (i.e. $a_g =[1, 1, \ldots, 1]$ or $[0, 0, \ldots, 0]$). Conditional on a_g, the model for t_{gd} remains the same as *CorMotif* and limma. *Full motif* assumes that genes fall into 2^D classes, corresponding to the 2^D possible a_g configurations. It can be viewed as a saturated version of *CorMotif*. All the other methods are applied to x_{gdlj} s directly. *SAM* (Tusher et al. 2001) processes each study separately, whereas *ebl* and *eb10best* analyze all studies jointly. The *ebl* method corresponds to the R package EBarrays with lognormal–normal (LNN) and one cluster assumption (Kendziorski et al. 2003). The *eb10best* method is EBarrays with LNN and multiple cluster assumption, and the cluster number is chosen by EBarrays as the one with the lowest AIC (Yuan and Kendziorski, 2006). We also tried XDE (Scharpf et al. 2009). However, it is based on Markov Chain Monte Carlo (MCMC) and took extremely long computing time, usually 24 h on a machine with 2.7 GHz CPU and 4 Gb RAM for 1000 iterations, for an analysis involving four studies which was the smallest data we analyzed here. Moreover, 1000 iterations usually were not enough for XDE to converge. Therefore, XDE will not be compared hereinafter. *eb10best* failed to work when it was used to jointly analyze ≥7 studies. *Full motif* and *eb1* failed when there were 20 studies.

We first tested *CorMotif* using simulations. For a complete set of comprehensive simulation studies with varying study numbers and differential expression patterns, please refer to our journal paper (Wei et al. 2015). Here, we just present simulation 1 for illustrative purpose regarding the work flow. In simulation 1, we generated 10,000 genes and four studies according to the four differential patterns in Figure 2(a): 100 genes were differentially expressed in all four studies (a_g =[1, 1, 1, 1]); 400 genes were differential only in studies 1 and 2 ([1, 1, 0, 0]); 400 genes were differential only in studies 2 and 3 ([0, 1, 1, 0]); 9100 genes were non-differential ([0, 0, 0, 0]). Each study had six samples: three cases and three controls. The variances $\sigma_{gd}^2 s$ were simulated from a scaled inverse χ^2 distribution $n_{0d} s_{0d}^2 / \chi^2(n_{0d})$, where $n_{0d} = 4$ and $s_{0d}^2 = 0.02$. Given σ_{gd}^2, the expression values were generated using $x_{gdlj} \sim N(0, \sigma_{gd}^2)$. Whenever $a_{gd} = 1$, we drew μ_{gd} from $N(0, w_{0d*} \sigma_{gd}^2)$, where $w_{0d} = 4$, and μ_{gd} was then added to the expression values of the three cases (i.e. x_{gdlj} s).

CorMotif was fit with varying motif number K. As Figure 2(c) shows, the minimal BIC was achieved at K =4. As a result, four motifs were reported (Figure 2(b)). The reported motifs were very similar to the true underlying differential patterns in Figure 2(a).

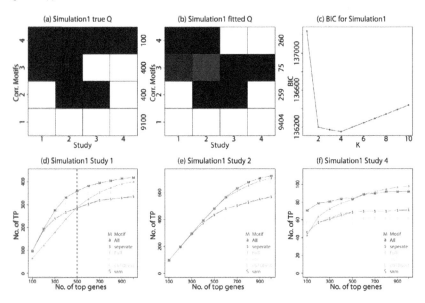

FIGURE 2   Results for Simulations 1. (a) True motif patterns for simulation 1. The Q of the true motifs is shown. Each row indicates a motif pattern and each column represents a study. The actual number of genes belonging to each motif (i.e. $\pi * G$) is displayed at the right end of each row. The gray scale of the cell (k, d) demonstrates the probability of differential expression in study d for pattern k. Black means 1 and white means 0. (b) The estimated $\hat{Q}$ from the learned motifs with $\pi * G$ annotated at the end of each row. (c) BIC plots. It can be seen that motif patterns reported by *CorMotif* under the minimal BIC are similar to the true underlying motif patterns. (d)–(f) Gene ranking performance of different methods in simulations 1. TPd (r), the number of genes that are truly differentially expressed in study d among the top r ranked genes by a given method, is plotted

TABLE 2 Confusion matrix for simulation 1. The column labels indicate the true underlying patterns and the row labels represent the reported configurations at gene level. For CorMotif, separate limma, all concord, full motif, eb1, and eb10best, differential expression in each study is determined using their default posterior probability cutoff 0.5. For SAM, q-value cutoff 0.1 was used to call differential expression. This yields similar number of correct classifications for pattern [0, 0, 0, 0] compared with CorMotif.

Method	Differential config.	c(0, 0, 0, 0)	c(0, 1, 1, 0)	c(1, 1, 0, 0)	c(1, 1, 1, 1)
CorMotif	c(0, 0, 0, 0)	9072	161	165	16
	c(0, 1, 1, 0)	3	168	3	7
	c(1, 1, 0, 0)	3	2	151	6
	c(1, 1, 1, 1)	0	1	0	33
	other	22	68	81	38
separate limma	c(0, 0, 0, 0)	9035	144	144	16
	c(0, 1, 1, 0)	0	68	0	5
	c(1, 1, 0, 0)	0	0	57	6
	c(1, 1, 1, 1)	0	0	0	4
	other	65	188	199	69
all concord	c(0, 0, 0, 0)	9095	236	236	20
	c(0, 1, 1, 0)	0	0	0	0
	c(1, 1, 0, 0)	0	0	0	0
	c(1, 1, 1, 1)	5	164	164	80
	other	0	0	0	0
full motif	c(0, 0, 0, 0)	9072	161	164	16
	c(0, 1, 1, 0)	4	172	4	7
	c(1, 1, 0, 0)	3	2	155	6
	c(1, 1, 1, 1)	0	1	0	35
	other	21	64	77	36
eb1	c(0, 0, 0, 0)	62	0	2	0
	c(0, 1, 1, 0)	2178	30	22	3
	c(1, 1, 0, 0)	569	7	12	0
	c(1, 1, 1, 1)	753	34	32	64
	other	5538	329	332	33
eb10best	c(0, 0, 0, 0)	0	0	0	1
	c(0, 1, 1, 0)	316	220	16	10
	c(1, 1, 0, 0)	180	23	226	10
	c(1, 1, 1, 1)	5789	77	52	63
	other	2815	80	106	16
SAM	c(0, 0, 0, 0)	9099	256	279	48
	c(0, 1, 1, 0)	0	20	0	3
	c(1, 1, 0, 0)	0	0	9	2
	c(1, 1, 1, 1)	0	0	0	1
	other	1	124	112	46

against the rank cutoff r. Results for a few representative studies are shown. Each plot is for one study.

Different methods were then compared in terms of how good they rank differential genes in each individual study (Figure 2(d)–(f)) as well as how accurate they can infer each gene's differential configuration a_g in all studies (Table 2). For each study d, *CorMotif* ranks genes using the posterior probability $Pr(a_{gd}=1|\, T, \hat{\pi}, \hat{Q})$ which is obtained after integrating out the motif membership b_g. A gene was called differential in study d if $\Pr(a_{gd}=1|\,T,\,\hat{\pi}\,,\hat{Q}) > 0.5$. Both the gene rankings and differential expression calls were different for different studies since $\Pr(a_{gd}=1|\,T,\,\hat{\pi},\,\hat{Q})$ depends on d and can change across studies. This is a desirable property as in reality the sets of true differential genes may be different in different studies due to study-specific differential expression, and ultimately one wants to know which genes are differential in each study. Using a similar approach, we obtained gene rankings and differential calls for *full motif*, *eb1* and *eb10best* which were also study-specific. *Separate limma* and *SAM* analyze each study separately and naturally produce study-specific gene ranking and differential calls. For all the methods above, we did not combine differential calls of a gene in D studies into a single call to indicate whether the gene is differential in any study, nor did we use such a combined call to rank genes, since the combined call would fail to capture study-specificity. Unlike the other methods, *all concord* assumes common differential states in all studies, therefore its gene ranking and differential calls remain the same across studies.

To examine if *CorMotif* can improve gene ranking, in each study and for each method we counted the number of true differential genes (true positives), TPd (r), among the top r ranked genes, and we plotted TPd (r) versus r in Figure 2(d)–(f). *CorMotif* consistently performed among the best in all studies. For instance, Figure 2(d) shows the results for study 1. *CorMotif* identified 361 true differential genes among its top 500 gene list. This performance was almost the same as the saturated model *full motif* which identified 362 true positives among the top 500 genes. Among the other methods, *eb10best* identified 341, *all concord* identified 292, and the others identified fewer than 292 true positives among the top 500 genes. Thus, *CorMotif* detected at least 23.6% more true positives compared with any other method except *full motif* and *eb10best*. Similarly, among the top 1000 genes, *CorMotif* and *full motif* both identified 419 true positives, *all concord* identified 401, *eb10best* identified 360, and the other methods identified fewer than 337. *CorMotif* and *full motif* detected 4.5% more true positives compared with *all concord* and improved the ranking by at least 16.4% compared with *eb10best* and other methods. Both *full motif* and *eb10best* have the problem of exponentially growing parameter space. As we will show later, they both will break down when the study number D is large.

To test whether *CorMotif* can more accurately determine a gene's differential configuration, we constructed the confusion matrix in Table 2. For each gene, its binary differential calls a_{gd}s based on $\Pr(a_{gd}=1|\,T,\,\hat{\pi}\,,\hat{Q})$ in different studies were arranged into a vector to represent its estimated differential configuration $\boldsymbol{a}_g$. For *CorMotif*, *separate limma*, *all concord*, *full motif*, *eb1* and *eb10best*, differential expression was called using their default posterior probability cutoff 0.5. For *SAM*, q-value cutoff 0.1 was used to call differential expression. At this cutoff, *SAM* correctly identified similar number of genes with $\boldsymbol{a}_g=[0, 0, 0, 0]$ (i.e. non-differential in all studies) compared with *CorMotif*. This allowed us to meaningfully compare *SAM* and *CorMotif* in terms of their ability to find differential genes. Table 2 shows that *CorMotif* was better at characterizing genes' true differential configurations compared with most other methods. For instance, among the 400 [0, 1, 1, 0], 400 [1, 1, 0, 0], and 100 [1, 1, 1, 1] genes, *CorMotif* correctly reported differential label *agd* in all four studies for 168, 151, and 33 genes, respectively. In contrast, *separate limma* only unmistakenly

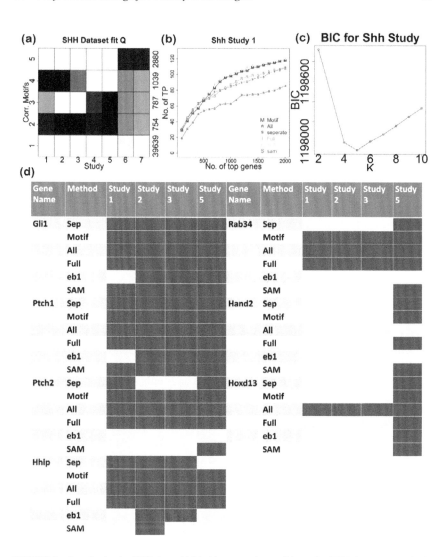

FIGURE 3 Results for the SHH data. (a) Motif patterns learned from the SHH data composed of 7 studies. (b) BIC plots for the SHH data. (c) Gene ranking performance for SHH study 1. The genes differentially expressed in dataset 8 (13somites smo versus 13somites wt) were obtained using *separate limma*. They were used as the gold standard. TPd (r), the number of genes in dataset 1 that are truly differentially expressed among the top r ranked genes by each method, is plotted against the rank cutoff r. (d) Differential status claimed by each method for known SHH pathway genes. Dark color indicates differential expression and light color represents non-differential expression.

labeled 68, 57, and 4 genes, respectively. Here, the increased power by *CorMotif* was purely due to the use of correlation motifs to integrate multiple studies, since all other model assumptions made by *CorMotif* and *separate limma* are the same. *All concord* requires genes to have the same differential status in all studies. As such, it is powerful at identifying concordant signals across studies but lacks the flexibility to handle study-specific differential expression: it correctly identified 80 out of 100 [1,

1, 1, 1] genes, but none of the [0, 1, 1, 0] and [1, 1, 0, 0] genes were correctly labeled as study-specific. With the default cutoff, *eb1* and *eb10best* only labeled 62 and 0 out of 9100 [0, 0, 0, 0] genes as completely non-differential, compared with 9072 labeled by *CorMotif.* In other words, *eb1* and *eb10best* reported more false-positive differential events. Both were anti-conservative. At the same time, fewer [0, 1, 1, 0] and [1, 1, 0, 0] genes were correctly identified by *eb1* (30 and 12 versus 168 and 151 by *CorMotif*). *SAM* was also poor at identifying the differential patterns [1, 1, 1, 1], [1, 1, 0, 0], and [0, 1, 1, 0] but behaved more conservatively by labeling many of them as [0, 0, 0, 0]. Among all the methods, only *full motif* performed slightly better than *CorMotif.* Even so, *CorMotif* was able to perform close to this saturated model. Adding up the diagonal elements in the confusion matrix, *CorMotif* unmistakenly assigned ag labels to 9424 genes, whereas this number was 9164 for *separate limma*, 9175 for *all concord*, 9434 for *full motif*, 168 for *eb1*, 509 for *eb10best*, and 9129 for *SAM*.

3.2   Application to the SHH signaling data sets

We used *CorMotif* to analyze the SHH data in Table 1. Datasets 1 and 2 compare SMO mutant mice with wild type mice (wt) and PTCH1 mutant with wild type, respectively, in the 8 somite stage of developing embryos. Dataset 3 compares the PTCH1 mutant with wild type in 13 somite stage. Datasets 4 and 5 compare the SHH mutant with wild type in developing head and limb, respectively. Datasets 6 and 7 study gene expression changes in two SHH-related tumors, medulloblastoma and basal cell carcinoma (BCC), compared with normal samples (control). Dataset 8 compares SMO mutant with wild type in the 13 somite stage of developing embryos. *CorMotif* was applied to datasets 1–7. Dataset 8 was reserved for testing.

Five motifs were discovered [Fig. 3 (a) and (b)]. Motif 1 mainly represents background. Motif 2 contains genes that have high probability to be differential in all studies. Genes in motif 3 tend to be differential in most studies except for the two involving PTCH1 mutant (i.e. studies 2 and 3). Most genes in motif 4 are not differential in the two studies involving the SHH mutant (i.e. studies 4 and 5) but tend to be differential in all other studies. Motif 5 mainly represents genes differential in tumors (i.e. studies 6 and 7) but not in embryonic development (i.e. studies 1–5). In general, looking at the columns in Figure 3(a), the two studies involving tumors (6, 7) are more similar to each other compared with other studies. The two PTCH1 mutant studies (2, 3) are also relatively similar, and the same trend holds true for the two SHH mutant studies (4,5).

In this real data analysis, no comprehensive truth is available for evaluating differential expression calls. Without comprehensive knowledge about the true differential expression states of all genes in all cell types, we can only perform a partial evaluation based on existing knowledge. In this regard, we used dataset 8 as a test. Similar to dataset 1, this dataset compares SMO mutant with wild type. One expects that differential genes in these two datasets should be largely similar. Therefore, we used the top 217 differentially expressed genes detected by *separate limma* (at the posterior probability cutoff 0.5) in dataset 8 as gold standard to evaluate the gene ranking performance of different methods in dataset 1. Figure 3(c) shows that

CorMotif again performed similar to *full motif* and outperformed all other methods. *ebl0best* failed to run here. We note that since dataset 8 and datasets 2–7 represent more different biological contexts, one cannot use it as gold standard for evaluating these other datasets.

Finally, we examined well-studied SHH responsive target genes. Gli1, Ptch1, Ptch2, Hhip, and Rab34 are known to be regulated by SHH in somites and developing limb (Vokes et al. 2007, 2008). Therefore, we expect them to be differential in studies 1, 2, 3, and 5. Figure 3(d) shows that *CorMotif, allconcord* and *full motif* were able to correctly identify differential expression of these genes in all these studies, whereas *separate limma, SAM,* and *ebl* failed to do so (they missed some cases). Hand2 is known to be a SHH target in developing limb but not in somites (Vokes et al. 2008). While *separate limma, CorMotif, full motif,* and *SAM* can correctly identify this, *all concord* and *ebl* failed to do so. For *all concord*, since Hand2 was not differential in studies 1–4, 6, and 7, the method thinks that this gene is not differential in any study. Similarly, Hoxd13 is a limb specific target of SHH signaling (Vokes et al. 2008). While the other methods correctly identified this, *all concord* failed again by claiming it to be differential in all studies. In all the genes examined, only *CorMotif* and *full motif* were able to correctly identify all known differential states.

4. Discussion

Together, our analyses show that *CorMotif* offers unique advantage over the other methods in the integrative analysis of multiple gene expression studies. Besides its ability to increase statistical power by combining information across studies, *CorMotif* is also flexible and scalable. Using a few probability vectors instead of 2^D dichotomous vectors to characterize the differential expression patterns provides the key to avoid the exponential growth of parameter space as the study number increases. At the same time, the probabilistic nature of the motifs allows all 2^D differential patterns to occur in the data at individual gene level.

The motif matrix Q can be viewed in two different ways. Each row of Q represents a cluster of genes with similar differential expression patterns across studies. Having many different motifs in Q is an indication that a concordance model, such as *all concord*, may not be enough to describe the correlation structure in the data. On the other hand, each column of Q represents differential expression propensities of different gene classes in a given study. If two columns are similar, the corresponding studies share similar differential expression profiles (e.g. studies 6 and 7 in the SHH data are more similar to each other compared with the other studies).

As we use probability vectors to serve as motifs, it is possible that multiple weak patterns can be merged into a single motif. For instance, two complementary patterns [1,1,0,0] and [0,0,1,1] each with n genes can be absorbed into a single motif with $q_k = (0.5, 0.5, 0.5, 0.5)$ having $2n$ genes. According to extensive simulation studies, we observed that in general weaker patterns were more likely to be merged than patterns with abundant data support. In all cases, however, *CorMotif* still provided the best gene ranking results compared with other methods. Moreover, the higher the proportions of study-specific motifs (e.g. [1,1,0,0] and [0,0,1,1]), the better *CorMotif* will perform compared with the concordance analysis (i.e. *all concord*) in terms of ranking genes in each study. All in all, the correlation motifs only represent a parsimonious representation of the correlation structure supported by the available data. One should not expect *CorMotif* to always recover all the true underlying clusters exactly. In spite of this, our simulations show that *CorMotif* can still effectively utilize the correlation among studies to improve differential gene detection even when the chosen K is not the same as the underlying true pattern number.

Currently, *CorMotif* first computes moderated t-statistics T and then applies the correlation motif model to T. We used this two-stage approach for considerations of effective presentation, computational efficiency, and clean method comparison. Instead of using this two-stage approach, a potential future extension is to introduce a single coherent Bayesian model that fully integrates the correlation motifs with a model directly describing the raw expression values x_{gdlj}. In the present version of *CorMotif*, we chose to use the two-stage approach for several reasons. First, it allows us to better present the core idea of the method, that is, how to use correlation motifs to integrate multiple studies. By taking advantage of the well-documented *limma* approach, the two-stage approach allows us to simplify the presentation of some of the model details and focus on discussing the core idea of correlation motifs. Moreover, the two-stage approach as presented now also represents a very general framework. Conceptually, one can modify f_{d0} and f_{d1} to accommodate other data types. Because of the two-stage design, this will not change the correlation motif model and the corresponding EM algorithm. Second, using the two-stage framework, one can develop a simple EM algorithm to fit the model. This approach is computationally more efficient than running a Markov Chain Monte Carlo (MCMC) algorithm on a fully Bayesian model with many levels of unknown parameters (e.g., mean and variances of x_{gdlj}s and parameters in their prior distributions, missing indicators A and B, and motif parameters π and Q). Third, the present design also allowed us to perform a well-controlled comparison with the state-of-the-art approach *limma*. In our two-stage design, the first stage of *CorMotif* uses the same model as *limma* to compute the moderated t-statistics. The only difference between *CorMotif* and *limma* is in the second stage, that is, the correlation motif part. For this reason, the comparison between *CorMotif* and *limma* can unambiguously demonstrate the gain of using correlation motifs to integrate multiple studies. This gain is not confounded with other factors such as differences in the data distributions f_{d0} and f_{d1}. By contrast, differences in performance between *CorMotif* and other methods such as *SAM* and *eb1* , etc., can be caused by a number of different factors such as differences in models for data x_{gdlj} . The two-stage design therefore has helped us to perform a clean comparison to show the effectiveness of correlation motifs. As a result, we were able to contribute a general tool with proven effectiveness (i.e., the correlation motif framework for data integration) to the toolbox other people can use to build future data analysis methods.

In the future, *CorMotif* may be extended in multiple ways. For example, instead of using moderated t-statistics and the two-stage design, one may develop a single coherent model that couples correlation motifs with a more sophisticated model for the raw data X. Also, it remains to be investigated whether the problem of choosing motif number can be better dealt with by a fully Bayesian approach such as by imposing a Dirichlet Process prior for K or using a variant of Dirichlet Process prior instead of using BIC. A fully Bayesian model, however, may require MCMC in the implementation, and this may pose additional challenges for developing computationally efficient algorithms capable of handling large datasets.

5. Software

CorMotif is freely available as an R package in Bioconductor: http://www.biocon-ductor.org/packages/release/bioc/html/Cormotif.html. Here we provide a short tutorial to demonstrate how to use the package for analysis of real data.

5.1 Data preparation

In order to fit the *correlation motif* model, one needs to call the function *cormotiffit*. The first requirement exprs is the matrix containing the gene expression data that needs to be analyzed. Each row of the matrix corresponds to a gene and each column of the matrix corresponds to a sample. The data should be normalized, for example by RMA, on log2 scale.

The second argument, groupid, identifies the group label of each sample. Here we use data *simudata2* as an illustration. *simudata2* are combined from four studies sharing the same 3,000 genes, each having two experimental conditions and three samples for each condition.

```
> library(Cormotif)
> data(simudata2)
> colnames(simudata2)
```

```
[1] "gene" "R1" "R2" "R3" "S1" "S2" "S3" "T1" "T2" "T3"
[11] "U1" "U2" "U3" "V1" "V2" "V3" "W1" "W2" "W3" "X1"
[21] "X2" "X3" "Y1" "Y2" "Y3"
> exprs.simu2<-as.matrix(simudata2[,2:25])
> data(simu2_groupid)
> simu2_groupid
R1 R2 R3 S1 S2 S3 T1 T2 T3 U1 U2 U3 V1 V2 V3 W1 W2 W3 X1 X2 X3 Y1 Y2 Y3
 1  1  1  1  2  2  2  3  3  3  4  4  4  5  5  5  6  6  6  7  7  7  8  8  8
```

The third argument, compid, represents the study design and hence the comparison pattern. In *simudata2*, R1, R2, R3 are samples from condition 1 in study1 and S1, S2, S3 are from condition 2 in study 1. Similarly, T1,T2,T3 represent condition 1 in study 2 and U1,U2,U3 represent condition 2 in study 2, and so on so forth. We aim at detecting the differential expression pattern of a gene under two different experimental conditions in each study, so we make up the comparison matrix simu2_compgroup as following:

```
> data(simu2_compgroup)
> simu2_compgroup
   Cond1 Cond2
1    1     2
2    3     4
3    5     6
4    7     8
```

5.2   Model fitting

Once we have specified the group labels and the study design, we are able to fit the *CorMotif* model. We can fit the data with varying motif numbers and use information criterion, such as AIC or BIC, to select the best model. Here for *simudata2*, we fit 5 models with total motif patterns number varying from 1 to 5, and as we can see later from the BIC plot, using BIC criterion, the best model is the one with 3 motifs.

```
>motif.fitted<-cormotiffit(exprssimu2,simu2_groupid,simu2_compgroup,
 + K=1:5,max.iter=1000,BIC=TRUE)
 [1] "We have run the first 50 iterations for K=2"
 [1] "We have run the first 50 iterations for K=3"
 [1] "We have run the first 100 iterations for K=3"
 [1] "We have run the first 50 iterations for K=4"
 [1] "We have run the first 100 iterations for K=4"
 [1] "We have run the first 150 iterations for K=4"
 [1] "We have run the first 200 iterations for K=4"
 [1] "We have run the first 50 iterations for K=5"
 [1] "We have run the first 100 iterations for K=5"
 [1] "We have run the first 150 iterations for K=5"
 [1] "We have run the first 200 iterations for K=5"
```

After fitting the *CorMotif* model, we can check the BIC values obtained by all cluster numbers:

```
> motif.fitted$bic
K bic
[1,] 1 44688.73
[2,] 2 44235.62
[3,] 3 44210.74
[4,] 4 44236.05
[5,] 5 44247.30

> plotIC(motif.fitted)
```

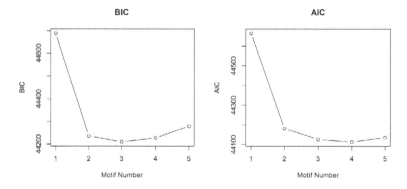

FIGURE 4   BIC plots for *simudata2*.

To picture the motif patterns learned by the algorithm, we can use function *plot-Motif*. Each row in both graphs corresponds to the same one motif pattern. We call the left graph *pattern graph* and the right bar chart *frequency graph*. In the pattern graph, each row indicates a motif pattern and each column represents a study. The grey scale of the cell (*k, d*) demonstrates the probability of differential expression in study *d* for pattern *k*, and the values are stored in motif.fitted$bestmotif$motif. prior. Each row of the frequency graph corresponds to the motif pattern in the same row of the left pattern graph. The length of the bar in the frequency graph shows the number of genes of the given pattern in the dataset, which is equal to motif. fitted$bestmotif$motif.prior multiplying the number of total genes.

```
> plotMotif(motif.fitted)
```

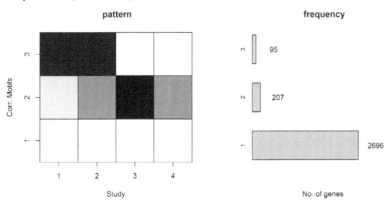

FIGURE 5   Motif pattern plots for *simudata2*.

The posterior probability of differential expression for each gene in each study is saved in motif.fitted$bestmotif$p.post:

```
> head(motif.fitted$bestmotif$p.post)
           [,1]        [,2]       [,3]        [,4]
[1,] 0.98054515 0.76640793 0.2484231 0.77337505
[2,] 0.99945249 0.35361152 0.9948324 0.99931734
[3,] 0.98047218 0.15768466 0.6450539 0.99841369
[4,] 0.01887596 0.02799143 0.3243530 0.29013282
[5,] 0.99959700 0.97149171 0.9986405 0.37861772
[6,] 0.10755781 0.94486423 0.9975589 0.04995237
```

At 0.5 cutoff for the posterior distribution, the differential expression pattern can be obtained as following:

```
> dif.pattern.simu2<-(motif.fitted$bestmotif$p.post>0.5)
> head(dif.pattern.simu2)
       [,1]   [,2]    [,3]   [,4]
[1,] TRUE TRUE FALSE TRUE
```

We can also order the genes in each study according to their posterior probability of differential expression:

```
> topgenelist<-generank(motif.fitted$bestmotif$p.post)
> head(topgenelist)
[,1] [,2] [,3] [,4]
[1,] 117 394  59  221
[2,]  31  23 330 238
[3,]  73  97  38 288
[4,] 196  63  96 249
[5,] 454 355 355 286
[6,] 177 333 319 230
```

In summary, we hope our package can provide an easy, accessible way for readers to apply our *CorMotif* method to jointly analyze multiple microarray experiments to improve differential gene expression detection as well as explore differential expression pattern changes across studies.

References

Anders, S. and W. Huber, 2010. Differential expression analysis for sequence count data. *Genome Biology.* 11: R106.

Conlon, E. M., J. J. Song, and J. S. Liu, 2006. Bayesian models for pooling microarray studies with multiple sources of replications. *BMC Bioinformatics.* 7: 1979–85.

Gelman, A., Carlin, J. B., Stern, H. S. and D. B. Rubin, 2004. *Bayesian Data Analysis*, 2nd edition. New York, NY: Chapman Hall/CRC.

Ingham, P. W. A. P. McMahon, 2001. Hedgehog signaling in animal development: paradigms and principles. *Genes and Development.* 15: 3059–3087.

Irizarry, R. A., Hobbs, B., Collin, F., Beazer-Barclay, Y. D., Antonellis, K. J., Scherf, U. and T. P. Speed, 2003. Exploration, normalization, and summaries of high density oligonucleotide array probe level data. *Biostatistics.* 4(2): 249–64.

Jensen, S. T., Erkan, I., Arnardottir, E. S. and D. S. Small, 2009. Bayesian testing of many hypothesis*many genes: a study of sleep apnea. *Annals of Applied Statistics* 3(3): 1080–101.

KendziorskiI, C. M., Newton, M. A., Lan, H. and Gould, M. N. (2003). On parametric empirical bayes methods for comparing multiple groups using replicated gene expression profiles. *Statistics in Medicine* 22: 3899–14.

Mao, J., Ligon, K. L., Rakhlin, E. Y., Thayer, S. P., Bronson, R. T., Rowitch, D. A. P. McMahon, 2006. A novel somatic mouse model to survey tumorigenic potential applied to the hedgehog pathway. *Cancer Research:* 66(20), 10171–78.

Robinson, M. D. and G. K. Smyth, 2007. Moderated statistical tests for assessing differences in tag abundance. *Bioinformatics.* 23: 2881–87.

Robinson, M. D. and G. K. Smyth, 2008. Small-sample estimation of negative binomial dispersion, with applications to sage data. *Biostatistics* 9: 321–32.

Ruan, L. and M. Yuan, (2011). An empirical bayes approach to joint analysis of multiple microarray gene expression studies. *Biometrics* 67: 1617–26.

Scharpf, R. B., Tjelmeland, H., Parmigiani, G. and A. B. Nobel, (2009). A Bayesian model for cross-study differential gene expression. *Journal of the American Statistical Association* 104(488): 1295–10.

Smyth, G. K. (2004). Linear models and empirical Bayes methods for assessing differential expression in microarray experiments. *Statistical Applications in Genetics and Molecular Biology.* 3: 3.

Tenzen, T, Allen, B. L., Cole, F., Kang, J. S., Krauss, R. S. and A. P. McMahon. 2006. The cell surface membrane proteins cdo and boc are components and targets of the hedgehog signaling pathway and feedback network in mice. *Developmental Cell.* 10(5): 647–56.

Tusher, V. G., Tibshirani, R. and G. Chu, 2001. Significance analysis of microarrays applied to the ionizing radiation response. *Proceedings of the National Academy of Sciences.* 98(9): 5116–21.

Vokes, S. A., Ji, H., McCuine, S., Tenzen, T., Giles, S., Zhong, S., Longabauch, W. J. R., Davidson, E. H. and A. P. McMahon. 2007. Genomic characterization of gli-activator targets in sonic hedgehog-mediated neural patterning. *Development.* 134: 1977–89.

Vokes, S. A., Ji, H., Wong, W. H. and A. McMahon, P. 2008. Whole genome identification and characterization of gli cis-regulatory circuitry in hedgehog-mediated mammalian limb development. *Genes Development.* 22: 2651–63.

Wei, Y, Toyoaki T, and Ji, H. 2015 Joint analysis of differential gene expression in multiple studies using correlation motifs. *Biostatistics* 16.(1): 31-46.

Yuan, M. C. M. Kendziorsk. 2006. A unified approach for simultaneous gene clustering and differential expression identification. *Biometrics.* 62: 1089–98.

3

Granger Causality for Time Series Gene Expression Data

André Fujita[1]*, Patricia Severino[2], Paulo Moises Raduan Alexandrino[1], Fernando Cipriano de Andrade Oliveira[1] and Satoru Miyano[3]

Abstract

Molecular processes underlying cellular behavior may be comprehended through the analysis of gene expression profiles. This approach is complex and encompasses the identification of which genes are expressed at any given time and how their products interact in so called gene regulatory networks (GRN). High-throughput technologies, such as DNA micro arrays and next generation sequencing are the technologies of choice to quantify gene expression that will be used to model GRN. Mathematical models aim to infer the structure of GRN, possibly identifying which genes relate to which other genes. Among such models, Granger causality allows for the identification of directionality at the edges of GRN through the analysis of time series gene expression data. The intuitive concept underlying Granger causality is the idea that an effect never occurs before its cause. This concept was introduced by Norbert Wiener in 1956 but it was Clive Granger who proposed a statistical method to identify Granger causality between two time series in 1969. In 1982 John Geweke generalized Granger's idea to a multivariate form, a more interesting methodology for dealing with biological data sets generated by high-throughput technologies. In this chapter we review Granger causality concepts and we describe recently obtained results using a generalization of the multivariate Granger causality to identify Granger causality between gene clusters. Detailed descriptions of the concept, algorithms to identify and statistically test Granger causality between sets of time series are described.

[1] Rua do Matão, 1010 – Cidade Universistária, São Paulo – SP, 05508-090, Brazil.
  E-mail: pauloalexandrino@usp.br; fandrade@ime.usp.br
[2] Av. Albert Einstein, 627, Morumbi, São Paulo – SP, 05652-000, Brazil.
  E-mail: psever@einstein.br
[3] 4-6-1 Shirokanedai, Minato-ku, Tokyo, 108-8639, Japan. E-mail: miyano@ims.u-tokyo.ac.jp
* Corresponding authors: fujita@ime.usp.br

1. Introduction

Time series gene expression data analysis is a powerful tool to assess the relationship between genes within a cellular system and may be used to comprehend biological processes or to tackle the effect of therapeutics within a time frame. High-throughput technologies, including gene expression micro arrays and next generation sequencing, provide the appropriate scenario for such applications due to the broadness of their results. One of the biggest challenges, however, lies in finding appropriate mathematical models to infer relationship between time series gene expression data and, consequently, to be able to derive biological meaning from such complex datasets. The identification of relationships between genes or gene sets may be addressed through the study of gene regulatory networks (GRNs) structures. However, time series gene expression data are auto correlated. Since this intrinsic dependency in data is not appropriately addressed by classical statistical methods, models that identify Granger causality have been proposed and seem to overcome this short come when the aim is to identify temporal association between time series gene expression data (Fujita et al. 2009, Guo et al. 2008, Mukhopadhyay and Chatterjee 2007).

Granger causality (Granger 1969) is a widely used approach for the detection of putative interactions between variables in a data-driven framework based on temporal precedence. Due to both simplicity and flexibility, Granger causality analysis application can be found in areas as diverse as economics (Wayne 1986, Hiemstra and Jones 1994), geophysics (Kaufmann et al. 2004, Elsner 2006), bioinformatics (Fujita et al. 2007, 2008), and neuroscience (Freiwald et al. 1999, Sameshima and Baccalá 1999, Baccalá and Sameshima 2001, Valdés-Sosa et al. 2005, Schelter et al. 2006, 2009, Sato et al. 2006, 2007, 2009, Hemmelmann et al. 2009). The intuitive idea underlying Granger causality is that an effect never occurs before its cause. This idea was originally proposed by Norbert Wiener in 1956 (Wiener 1956), when he stated that the prediction of one time series could be improved by incorporating the information of past values of a second one. If this proved to be true, then the latter time series was said to have a "causal" influence on the former.

The original idea was philosophical and not applicable directly to empirical data. But in 1969, Clive Granger formalized Wiener's idea and proposed a mathematically tractable concept of causality defined in terms of forecasting power. The intuitive concept is that temporal precedence does not imply, but may help to identify causal relationships (since a cause never occurs after its effects). In other words, if the variance of the auto regressive prediction error of one time series at the present time is statistically reduced by inclusion of past measurements from another time series, then the latter is said to have a Granger causal influence on the former. From this definition it is easy to see that the time flow is important to infer causality in time series data.

Later, John Geweke (Geweke 1982) generalized the bivariate Granger causality to a multivariate model in order to identify conditional Granger causality. To illustrate the differences between bivariate and multivariate Granger causalities, suppose that there are three processes namely *A*, *B*, and *C*, where *A* drives *B* and *C* with one and two time delays, respectively. A pair wise analysis (bivariate Granger causality) would indicate a causal influence from process *B* (that receives an early input) to process *C* (that receives a late input) while conditional (multivariate) Granger causality may be useful to identify that, in fact, A drives both B and C. Multivariate Granger causality is thus able to discriminate whether the interaction between two time series is direct or is mediated by another time series present in the model. In contrast to Structural Equation Models (SEM) and Dynamical Causal Models (DCM), the methods focused on the identification of Granger causality gather information that may expose the presence of temporal relationships between observed signals. In this sense, the analysis is hypothesis-free since it does not require any a priori knowledge about the edges constituting a network or the directions of these edges.

Vector Auto Regressive (VAR) models are usually the choice for the identification of multivariate Granger causality. This is due to the fact that both the statistical theory behind VAR models and the computational algorithms are well understood and that they are particularly suitable to describe processes composed of locally interacting components.

Nagarajan and Upreti (2008) and Nagarajan (2009) investigated the use of bivariate VAR models for acyclic approximations in the case of networks composed of two genes. They explored parameters defined as transcriptional noise variance, auto regulatory feedback, and transcriptional coupling strength, which may influence some measures of Granger Causality. These authors have shown that under some specific conditions, VAR parameters may influence the statistical tests used for identifying significant Granger causality, therefore leading to bias that need to be considered during data interpretation.

Several extensions of the standard VAR model, namely Dynamic VAR (DVAR – to model time-varying structural changes in GRN) (Fujita et al. 2007a), Sparse VAR (SVAR – to model GRNs in high dimensional cases such as when the number of parameters (genes) is greater than the number of observations (time points)) (Fujita et al. 2007b, Lozano et al. 2009, Opgen-Rhein et al 2008, Shimamura et al. 2009) and Nonlinear VAR (NVAR, to identify nonlinear Granger causality) (Chen et al 2004, Fujita et al. 2008, Guo et al. 2008, Marinazzo et al 2008, have been proposed to model GRNs.

However, the theoretical generalization of Granger causality for sets of time series has not been sufficiently explored. Recently, Fujita et al. (2010) proposed a method to identify Granger causality between sets of time series (i.e. if a set containing n genes Granger causes another set with m genes). Their main goal was to create networks representing pathway-level connections that could help the understanding of molecular mechanisms underlying biological processes. In this chapter we describe this approach. We define Granger causality between sets of time series while describing a method for its identification and we also describe two statistical tests, one based on non parametric bootstrap and another based on likelihood ratio test, that verify Granger causality results between sets of time series. Finally, we demonstrate a particular application of Granger causality between sets of time series (Granger causality between gene clusters, CGC): its use for gene clustering.

---

2.   Granger Causality for Sets of Time Series

Consider that $\mathbf{Y}_t = \left\{ y_t^1, y_t^2, y_t^3 ... y_t^q \right\}$ represents a set of q time series with $t = 1, ..., T$ and F_t a set containing all relevant information available up to and including time point t. Let $\mathbf{Y}_t^i$, and $\mathbf{Y}_t^j$ be two disjoint (each time series belongs to either $\mathbf{Y}_t^i$ or $\mathbf{Y}_t^j$ but not to both of them) subsets of $\mathbf{Y}_t$ ($\mathbf{Y}_t^i \cap \mathbf{Y}_t^j = \{\ \}$), containing m and n time series each, respectively, and $q \geq m + n$. In this scenario we would like to verify whether the subset $\mathbf{Y}_t^j$ Granger causes the subset $\mathbf{Y}_t^i$, where the subset $\mathbf{Y}_t^j$ is the set of time series from which past values will be used to predict the future values of a function of the subset $\mathbf{Y}_t^i$. According to Fujita and collaborators (2010) and Sato and collaborators (2010) Granger causality for *sets* of time series can be defined as: let $\mathbf{Y}_t$ ($h \mid F_t$) be the optimal h-step predictor of a function f (i.e., the one which produces the minimum mean squared error (MSE) prediction) of the set of m time series $\mathbf{Y}_t^i$ from the time point t, based on the information in F_t. In most cases, the predictor of a function f is assumed to be linear, and only one-step-ahead prediction is considered. The forecast MSE of a function of $\mathbf{Y}_t^i$ will be denoted by $\Omega_Y(h \mid F_t)$. The set of n time series $\mathbf{Y}_t^j$ is said to Granger-cause the set of m time series $\mathbf{Y}_t^i$ if

$$\Omega_Y(h \mid F_t) < \Omega_Y(h \mid F_t \setminus \{\mathbf{Y}_s^j \mid s \leq t\}) \text{ for at least one } h = 1, 2, \ ...,$$

where $F_t / \{\mathbf{Y}_s^j \mid s \leq t\}$ is the set containing all relevant information except for the information in the past and present of $\mathbf{Y}_t^j$.

In other words, if a function of $\mathbf{Y}_t^i$ can be predicted more accurately when the information in $\mathbf{Y}_t^j$ is taken into account, then the set $\mathbf{Y}_t^j$ is said to Granger cause $\mathbf{Y}_t^i$.

The application of CGC to GRNs can be interpreted as follows: a set of gene expression time series $\mathbf{Y}_t^j$ Granger-causes another set of gene expression time series $\mathbf{Y}_t^i$, if linear combinations of $\mathbf{Y}_t^j$ provide statistically more significant information about future values of linear combinations of $\mathbf{Y}_t^i$ than considering only the past values of $\mathbf{Y}_t^i$. Thus, past gene expression values of $\mathbf{Y}_t^j$ allow the prediction of more accurate gene expression values of $\mathbf{Y}_t^i$. Another interpretation is that there is an information flow between these two sets of genes (Baccalá and Sameshima 2001). It is important to highlight that Granger causality is based on quantitative criteria and its results are, therefore, only an indication of causal relationships.

Notice that the definition of CGC generalizes both the original bivariate Granger causality (where $\mathbf{Y}_t^i$ and $\mathbf{Y}_t^j$ are 1-dimensional) and also, the multivariate Granger causality (where $\mathbf{Y}_t^j$ is m-dimensional and $\mathbf{Y}_t^i$ is 1-dimensional). In other words, bivariate and multivariate cases are particular cases of the CGC.

3. Canonical Correlation Analysis and Granger Causality

The main challenge in the identification of CGC consists in dealing with two sets of time series. Harold Hotelling (Hotelling 1935; Hotelling 1936) was the first to propose an approach to identify, measure, and maximize linear relationships between two *sets* of random variables, namely canonical correlation analysis (CCA). CCA can be interpreted as the bivariate correlation of two synthetic variables that are the linear combinations of the two sets of original (observed) variables. The original variables of each set are linearly combined to produce pairs of synthetic variables that have maximal correlation. Mathematically, let $\mathbf{X} = (\mathbf{X}_1, \mathbf{X}_2, ..., \mathbf{X}_m)$ and

$\mathbf{Y} = (\mathbf{Y}_1, \mathbf{Y}_2,..., \mathbf{Y}_n)$ be two sets with m and n variables each one, respectively. Then, canonical correlation analysis calculates a linear combination of the $\mathbf{x}$'s and the $\mathbf{y}$'s (i.e., $\mathbf{a'X}$ and $\mathbf{b'Y}$), which present maximum correlation with each other.

In the linear case, the Granger causality for *sets* of variables may be identified based on this idea initially proposed by Hotelling. Notice that the choice of linear combinations may provide new variables for each set of time series potentially increasing the power to identify CGC. However, the direct application of CCA to the gene expression time series is not suitable for quantifying temporal relationships for two reasons. First because the direct application is not taking into account the temporal precedence information that is the key to identify Granger causality. Consequently, it would return instantaneous correlations between $\mathbf{Y}_t^j$ and $\mathbf{Y}_t^i$. Second, it is necessary to remove the effects from auto correlation relationships within the target time series set itself and from other time series not involved in the examination of prediction. Note that the past values of the target set may contain information to predict its own future. Thus, if the target time series set is instantaneously correlated with the predictor and there is auto correlation within the target set, this may lead to a spurious Granger causality from the predictor to the target set. In other words, if we are interested in evaluating the Granger causality from set $\mathbf{Y}_t^j$ to $\mathbf{Y}_t^i$, we must remove the influences of past values of $\mathbf{Y}_t^i$ (and possibly other sets of time series) from the correlation measure.

Thus, it is necessary to modify the use of CCA to take into account the time flow information and also identify *partial* canonical correlations, since Granger causality needs to be distinguished from auto correlation relationships within the target time series. In its original proposal, CCA does not consider the partialization by a third set of time series. The partialization of CCA was described by Rao (1969) and is explained in more details in section 3.1. The bootstrap procedure and the likelihood ratio test for hypothesis testing are described in sections 3.2.1 and 3.2.2, respectively.

3.1   Identification

The problem presented here consists in verifying whether $\mathbf{Y}_t^j$ Granger-causes $\mathbf{Y}_t^i$. For the linear case, let $\mathbf{Y}_t^i$ (m-dimensional) and $\mathbf{Y}_t^j$ (n-dimensional) be two separate subsets of $\mathbf{Y}_t$, where $\mathbf{Y}_t$ is a k-dimensional set of stationary time series ($q \geq m + n$), $\mathbf{Y}_{t-1}^j$ be the single time point lagged values of the time series in the set $\mathbf{Y}_t^j$, and $\mathbf{Y}_{t-1} \setminus \mathbf{Y}_{t-1}^j$ be the set $\mathbf{Y}_{t-1}$ subtracted by the set $\mathbf{Y}_{t-1}^j$. For simplicity, only a single time-point lag (i.e., only a correlation between the observations at time t and $t-1$) is considered, which, in practice, should be enough for analyzing most gene expression data since they are usually constituted of short and time intervals repeated over time. However, other time lags may be evaluated with a very straight forward extension of this method. Considering the correlation as an indication of predictive power, $\mathbf{Y}_t^j$ is Granger non-causal for $\mathbf{Y}_t^i$ partialized by the set $\mathbf{Y}_{t-1} \setminus \mathbf{Y}_{t-1}^j$ if the following condition holds:

$$\text{CCA}(\mathbf{Y}_t^i, \mathbf{Y}_{t-1}^j \mid \{\mathbf{Y}_{t-1} \setminus \mathbf{Y}_{t-1}^j\}) = \hat{\rho} = 0 \tag{1}$$

Notice that CCA is applied to time lags in a partial manner and not to the instantaneous time point. It is important to highlight that information regarding past values of $\mathbf{Y}_t^i$ belong to the set $\mathbf{Y}_{t-1} \setminus \mathbf{Y}_{t-1}^j$. Correlation may be interpreted as the square root of R^2 of a linear regression model, where R^2 represents the variance of the response variable explained by the regressor. This means that when partial CCA is applied between the past values of one subset and present values of another subset, it identifies Granger causality between the two groups of time series.

In order to calculate ρ, set:

$$\mathbf{u} = \mathbf{a}' \mathbf{Y}_t^i , \tag{2}$$

and

$$\mathbf{v} = \mathbf{b}' \mathbf{Y}_{t-1}^j , \tag{3}$$

for some pair of coefficient vectors $\mathbf{a}$ and $\mathbf{b}$. Then, we obtain

$$\mathrm{Var}(\mathbf{u}) = \mathbf{a}' \, \mathrm{Cov}(\mathbf{Y}_t^i)\mathbf{a} = \mathbf{a}' \sum\nolimits_{\mathbf{Y}_t^i \mathbf{Y}_t^i} \mathbf{a} , \tag{4}$$

$$\mathrm{Var}(\mathbf{v}) = \mathbf{b}' \, \mathrm{Cov}(\mathbf{Y}_{t-1}^j)\mathbf{b} = \mathbf{b}' \sum\nolimits_{\mathbf{Y}_{t-1}^j \mathbf{Y}_{t-1}^j} \mathbf{b} , \tag{5}$$

$$\mathrm{Var}(\mathbf{u},\, \mathbf{v}) = \mathbf{a}' \, \mathrm{Cov}(\mathbf{Y}_t^i,\, \mathbf{Y}_{t-1}^j)\mathbf{b} = \mathbf{a}' \sum\nolimits_{\mathbf{Y}_{t-1}^i \mathbf{Y}_{t-1}^j} \mathbf{b} . \tag{6}$$

We shall then seek coefficient vectors $\mathbf{a}$ and $\mathbf{b}$ that maximizes

$$\mathrm{Corr}\,(\mathbf{u}, \mathbf{v}) = \frac{\mathbf{a}' \sum_{\mathbf{Y}_t^i \mathbf{Y}_{t-1}^j} \mathbf{b}}{\sqrt{\mathbf{a}' \sum_{\mathbf{Y}_t^i \mathbf{Y}_t^i} \mathbf{a}} \sqrt{\mathbf{b}' \sum_{\mathbf{Y}_{t-1}^j \mathbf{Y}_{t-1}^j} \mathbf{b}}} . \tag{7}$$

Equation (7) represents the Pearson correlation applied to the synthetic variables $\mathbf{u}$ and $\mathbf{v}$. Notice that the Pearson correlation coefficient is exactly the squared root of the R^2 measure in univariate linear regression models, and it can be interpreted as the percentage of variance in the variable linearly predicted by the regressor (Seibold and McPhee 1979). Thus, in time series analysis, we apply linear regression to study how the past values of a time series can improve the prediction of the present values of another variable, which is the main idea of using vector auto regressive models to identify Granger causality (Lütkepohl, 2005). Then, since correlation is associated with predictive power, we suggest building sets of variables and their correspondent time lags, so that canonical correlation analysis can be applied to identify Granger causality between the sets of time series. The calculations described up to this point are not able to identify Granger causality involving more than two sets of time series (conditional Granger causality), but only pair wise correlations between lagged sets of time series $\left(\mathrm{CCA}\left(\mathbf{Y}_t^i, \mathbf{Y}_{t-1}^j\right) \right)$. In order to include a third set of time series and consequently identify partialized Granger causality, it is necessary to develop a partial canonical correlation analysis.

For partial canonical correlation analysis (Rao 1969) it is imperative to derive linear coefficients for combining original variables into canonical variables from variance-covariance matrices. The residual variance-covariance matrices of $\mathbf{Y}_t^i$ and $\mathbf{Y}_{t-1}^j$ obtained after partializing out the effect of vector $\mathbf{X} = \mathbf{Y}_{t-1} \setminus \mathbf{Y}_{t-1}^j$ from both $\mathbf{Y}_t^i$ and $\mathbf{Y}_{t-1}^j$, provide the solution. More specifically, suppose that, when three vectors of variables are combined, $\mathbf{Y}_t^i$, $\mathbf{Y}_{t-1}^j$ and $\mathbf{X}$, the following partitioned variance-covariance matrix is built:

$$\hat{\Sigma}_{\mathbf{Y}_t^i \mathbf{Y}_{t-1}^j \mathbf{X}} = \begin{pmatrix} \hat{\Sigma}_{\mathbf{Y}_t^i \mathbf{Y}_t^i} & \hat{\Sigma}_{\mathbf{Y}_t^i \mathbf{Y}_{t-1}^j} & \hat{\Sigma}_{\mathbf{Y}_t^i \mathbf{X}} \\ \hat{\Sigma}_{\mathbf{Y}_{t-1}^j \mathbf{Y}_t^i} & \hat{\Sigma}_{\mathbf{Y}_{t-1}^j \mathbf{Y}_{t-1}^j} & \hat{\Sigma}_{\mathbf{Y}_{t-1}^j \mathbf{X}} \\ \hat{\Sigma}_{\mathbf{X} \mathbf{Y}_t^i} & \hat{\Sigma}_{\mathbf{X} \mathbf{Y}_{t-1}^j} & \hat{\Sigma}_{\mathbf{X} \mathbf{X}} \end{pmatrix}, \tag{8}$$

where the (r, s)th entry of $\hat{\Sigma}_{\mathbf{UV}}$ is given by $(T-1)^{-1} \Sigma_{K=1}^{T} (\mathbf{U}_{kr} - \bar{\mathbf{U}}_r)(\mathbf{V}_{ks} - \bar{\mathbf{V}}_s)$ (T: time series' length).

The matrices may be defined as:

$$\hat{\Sigma}_{\mathbf{Y}_t^i \mathbf{Y}_t^i | \mathbf{X}} = \hat{\Sigma}_{\mathbf{Y}_t^i \mathbf{Y}_t^i} - \hat{\Sigma}_{\mathbf{Y}_t^i \mathbf{X}} \hat{\Sigma}_{\mathbf{XX}}^{-1} \hat{\Sigma}_{\mathbf{X} \mathbf{Y}_t^i}, \tag{9}$$

$$\hat{\Sigma}_{\mathbf{Y}_t^i \mathbf{Y}_{t-1}^j | \mathbf{X}} = \hat{\Sigma}_{\mathbf{Y}_t^i \mathbf{Y}_{t-1}^j} - \hat{\Sigma}_{\mathbf{Y}_t^i \mathbf{X}} \hat{\Sigma}_{\mathbf{XX}}^{-1} \hat{\Sigma}_{\mathbf{X} \mathbf{Y}_{t-1}^j}, \tag{10}$$

$$\hat{\Sigma}_{\mathbf{Y}_{t-1}^j \mathbf{Y}_t^i | \mathbf{X}} = \hat{\Sigma}_{\mathbf{Y}_{t-1}^j \mathbf{Y}_t^i} - \hat{\Sigma}_{\mathbf{Y}_{t-1}^j \mathbf{X}} \hat{\Sigma}_{\mathbf{XX}}^{-1} \hat{\Sigma}_{\mathbf{X} \mathbf{Y}_t^i}, \tag{11}$$

$$\hat{\Sigma}_{\mathbf{Y}_{t-1}^j \mathbf{Y}_{t-1}^j | \mathbf{X}} = \hat{\Sigma}_{\mathbf{Y}_{t-1}^j \mathbf{Y}_{t-1}^j} - \hat{\Sigma}_{\mathbf{Y}_{t-1}^j \mathbf{X}} \hat{\Sigma}_{\mathbf{XX}}^{-1} \hat{\Sigma}_{\mathbf{X} \mathbf{Y}_{t-1}^j}. \tag{12}$$

Then, the conditional variance-covariance matrices of $\mathbf{Y}_t^i$ and $\mathbf{Y}_{t-1}^j$, partializing out the effect of $\mathbf{X}$ is given as described by Anderson (1984), Johnson and Wichern (2002) and Timm (1975):

$$\hat{\Sigma}_{\mathbf{Y}_t^i \mathbf{Y}_{t-1}^j | \mathbf{X}} = \begin{pmatrix} \hat{\Sigma}_{\mathbf{Y}_t^i \mathbf{Y}_t^i | \mathbf{X}} & \hat{\Sigma}_{\mathbf{Y}_t^i \mathbf{Y}_{t-1}^j | \mathbf{X}} \\ \hat{\Sigma}_{\mathbf{Y}_{t-1}^j \mathbf{Y}_t^i | \mathbf{X}} & \hat{\Sigma}_{\mathbf{Y}_{t-1}^j \mathbf{Y}_{t-1}^j | \mathbf{X}} \end{pmatrix}. \tag{13}$$

The matrices $\mathbf{A}$ and $\mathbf{B}$ are defined as follows:

$$\mathbf{A} = \hat{\Sigma}_{\mathbf{Y}_t^i \mathbf{Y}_t^i | \mathbf{X}}^{-1/2} \hat{\Sigma}_{\mathbf{Y}_t^i \mathbf{Y}_{t-1}^j | \mathbf{X}} \hat{\Sigma}_{\mathbf{Y}_{t-1}^j \mathbf{Y}_{t-1}^j | \mathbf{X}}^{-1} \hat{\Sigma}_{\mathbf{Y}_{t-1}^j \mathbf{Y}_t^i | \mathbf{X}} \hat{\Sigma}_{\mathbf{Y}_t^i \mathbf{Y}_t^i | \mathbf{X}}^{-1/2}, \tag{14}$$

$$\mathbf{B} = \hat{\Sigma}_{\mathbf{Y}_{t-1}^j \mathbf{Y}_{t-1}^j | \mathbf{X}}^{-1/2} \hat{\Sigma}_{\mathbf{Y}_{t-1}^j \mathbf{Y}_t^i | \mathbf{X}} \hat{\Sigma}_{\mathbf{Y}_t^i \mathbf{Y}_t^i | \mathbf{X}}^{-1} \hat{\Sigma}_{\mathbf{Y}_t^i \mathbf{Y}_{t-1}^j | \mathbf{X}} \hat{\Sigma}_{\mathbf{Y}_{t-1}^j \mathbf{Y}_{t-1}^j | \mathbf{X}}^{-1/2}, \tag{15}$$

and let $\lambda_1 \geq \lambda_2 \geq \cdots \lambda_{\min(m,n)}$ be the ordered eigenvalues of matrices $\mathbf{A}$ and $\mathbf{B}$. Similar to regular canonical correlation analysis described by Johnson and Wichern (2002), the eigenvalues λ_d ($d = 1, ..., \min(m, n)$) from matrices $\mathbf{A}$ and $\mathbf{B}$ will be the squared partial canonical correlation coefficients for the dth canonical functions and the eigenvectors $\mathbf{a}_d$ and $\mathbf{b}_d$ associated with the eigenvalue λ_d will be the linear coefficient vectors which combine the original variables into synthetic canonical variables. Therefore,

$$\mathrm{CCA}\,(\mathbf{Y}_t^i, \mathbf{Y}_{t-1}^j \mid \mathbf{Y}_t \setminus \{\mathbf{Y}_{t-1}^j\}) = \hat{\rho} = \sqrt{\lambda_1}.$$

As in regular canonical correlation analysis, the two matrices $\mathbf{A}$ and $\mathbf{B}$ have the same eigenvalues but with different eigenvectors. This means that each eigenvector $\mathbf{b}_d$ is proportional to $\hat{\Sigma}_{\mathbf{Y}_{t-1}^j \mathbf{Y}_{t-1}^j | \mathbf{X}}^{-1/2} \hat{\Sigma}_{\mathbf{Y}_{t-1}^j \mathbf{Y}_t^i | \mathbf{X}} \hat{\Sigma}_{\mathbf{Y}_t^i \mathbf{Y}_t^i | \mathbf{X}}^{-1/2} \mathbf{a}_d$.

In summary, the algorithm to compute $\mathrm{CCA}\,(\mathbf{Y}_t^i, \mathbf{Y}_{t-1}^j \mid \{\mathbf{Y}_{t-1} \setminus \mathbf{Y}_{t-1}^j\}) = \hat{\rho}$ can be described as follows:

Input: three sets of time series namely, $\mathbf{Y}_t^j, \mathbf{Y}_t^i$, and $\mathbf{Y}_{t-1} \setminus \mathbf{Y}_{t-1}^j$.

1. Estimate the covariance matrices $\widehat{\Sigma}_{\mathbf{Y}_t^j \mathbf{Y}_t^i \mid \mathbf{x}}, \widehat{\Sigma}_{\mathbf{Y}_t^i \mathbf{Y}_{t-1}^j \mid \mathbf{x}}, \widehat{\Sigma}_{\mathbf{Y}_{t-1}^j \mathbf{Y}_t^i \mid \mathbf{x}}$, and $\widehat{\Sigma}_{\mathbf{Y}_{t-1}^j \mathbf{Y}_{t-1}^j \mid \mathbf{x}}$ by using the equations (9-12), respectively;

2. Compute matrices $\mathbf{A}$ and $\mathbf{B}$ by using equations (14) and (15), respectively;

3. Compute the eigenvalues $\lambda_1 \geq \lambda_2 \geq \cdots \lambda_{\min(m,n)}$ of matrix $\mathbf{A}$ (or $\mathbf{B}$);

Output: $\mathrm{CCA}(\mathbf{Y}_t^i, \mathbf{Y}_{t-1}^j \mid \mathbf{Y}_t \setminus \{\mathbf{Y}_{t-1}^j\}) = \hat{\rho} = \sqrt{\lambda_1}$.

The value of $\hat{\rho}$ varies from zero to one, where estimates close to zero indicate Granger non-causality. Although instantaneous correlations between gene expression time series cannot be directly used for the computation of Granger causality, cross-correlation between the predictor and target time series, where the predictor series is the one that may Granger cause the target time series, is the basis behind the calculations of "causality" (Granger, 1969). As described in the previous sections, CCA is applied to the time lags in a partialized manner and not to the instantaneous time step.

It is important to emphasize that CGA cannot be implemented in an equivalent Vector Auto regressive (VAR) model because the former is based on maximizing the correlation between two sets of variables, i.e., between predictor and target. Thus, the dependent variable (target) in VAR would be a linear combination of variables, whose coefficients differ for each predictor set. This is a fundamental difference between CGA and standard VAR models.

In the next section, a statistical test to verify the existence of Granger causality is described.

3.2. Statistical test

CCA may be used as a tool to identify linear Granger causality since the existence of the latter is implied by the presence of a non-null canonical correlation. The main issue is, therefore, to define a measure $\hat{\rho}$, that, when above a threshold to be defined by an adequate statistical test, implies that the non-causality can be rejected. In other words, if $\hat{\rho}$ is statistically different from zero, one cannot neglect Granger causality.

The hypothesis test to verify the existence of Granger causality between *sets* of time series is defined as follows:

$$\mathrm{H}_0 : \mathrm{CCA}(\mathbf{Y}_t^i, \mathbf{Y}_{t-1}^j \mid \mathbf{X}) = \rho = 0 \text{ (Granger non causality)}$$

$$\mathrm{H}_1 : \mathrm{CCA}(\mathbf{Y}_t^i, \mathbf{Y}_{t-1}^j \mid \mathbf{X}) = \rho \neq 0 \text{, (Granger causality)}$$

A statistical test is necessary since the partial CCA focuses on maximizing the correlation between two sets of time series and this correlation be artificially increased due to different kinds of noise in the data. An appropriate statistical test is crucial and it will control type I errors. In the next two subsections we present two approaches to test Granger non causality between sets of time series, namely bootstrap and likelihood ratio test (LRT).

3.2.1. Bootstrap procedure

The bootstrap procedure presented here is based on the block bootstrap approach (Lahiri 2003). It consists of splitting the data into blocks of observations presenting some overlapping and then sampling the blocks randomly with replacement. The bootstrap procedure can be described as follows (Fig. 1):

Input: Set of time series $\mathbf{Y}_t$, and subsets $\mathbf{Y}_t^i$ and $\mathbf{Y}_t^j$.

1. With overlapping blocks of length l block 1 is observations $\mathbf{Y}_h : h = 1,$ ..., l, block 2 is observations $\mathbf{Y}_{h+1} : h = 1, \cdots, l$, block 3 is observations $\mathbf{Y}_{h+2} : h = 1, \cdots, l$, and so forth. The bootstrap sample $\mathbf{Y}_t^{i*}$ is obtained by sampling blocks randomly with replacement from $\mathbf{Y}_t^i$ and laying them end-to-end in the order sampled. The bootstrap sample $\mathbf{Y}_t^{j*}$ is obtained in an analogous way. This block resampling is carried out in order to capture the dependence structure of neighborhood observations, i.e., auto correlation. Moreover, the resampling of all time series of $\mathbf{Y}_t^i$ is carried out together, in order to capture contemporaneous correlations between time series. The same is performed for the time series of $\mathbf{Y}_t^j$. However, $\mathbf{Y}_t^i$ and $\mathbf{Y}_t^j$ are resampled independently, in order to break the relationship between the response and predictor variables.

2. After constructing the bootstrap samples $\mathbf{Y}_t^{i*}$ and $\mathbf{Y}_t^{j*}$, calculate $\mathrm{CCA}(\mathbf{Y}_t^{i*}, \mathbf{Y}_{t-1}^{j*} \mid \mathbf{X}^*) = \widehat{\rho}^*$ where $\mathbf{X}^* = \mathbf{Y}^*_{t-1} \setminus \mathbf{Y}_{t-1}^{j*}$.

3. Repeat these steps (1) and (2) until the desired number of bootstrap samples is obtained.

4. Use the empirical distribution of ρ^* to test whether $\widehat{\rho} = 0$ (gather the information from the empirical distribution of ρ^* to obtain a p-value for $\widehat{\rho} = 0$, by analyzing the probability of obtaining values equal or greater than $\widehat{\rho}$).

Output: a p-value for $\widehat{\rho}$ under the null hypothesis
$$(\mathrm{H}_0 : \mathrm{CCA}\ (\mathbf{Y}_t^i, \mathbf{Y}_{t-1}^j \mid \mathbf{X}) = \rho = 0).$$

Regardless of the block bootstrap that is used, the block length l must increase with increasing time series length T to render bootstrap estimators of moments and distribution functions consistent (Carlstein 1986, Hall 1985, Künsch 1989). Similarly, the block length must increase with increasing sample size to enable the block bootstrap to achieve asymptotically correct coverage probabilities for confidence intervals and rejection probabilities for hypothesis tests. For the special case of an auto regressive process of order one (Carlstein 1986), it was shown that the block length l that minimizes the asymptotic mean-square error of the variance estimator increases at the rate of $l \propto \mathrm{T}^{\frac{1}{3}}$. Since time series gene expression data are generally short, it is unfeasible to fit an AR model of higher orders. However, if a longer time series data becomes available, one may use the algorithm proposed by Bühlmann and Künsch (1999) in order to select the block length for the bootstrap procedure.

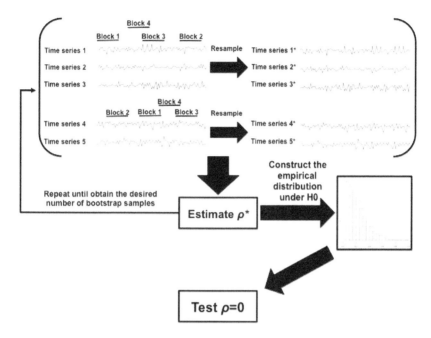

FIGURE 1 Diagram describing the bootstrap procedure. Time series 1, 2, and 3 belong to group Y_t^i while time series 4 and 5 belong to group Y_t^j. Notice that blocks of group Y_t^i and Y_t^j are different and may be overlapped among them. The estimative ρ^* is calculated by using the resampled time series.

3.2.2. Likelihood Ratio Test (LRT)

Despite the usefulness of the bootstrap procedure to test Granger non-causality, it is computationally intensive, becoming particularly slow when accurate *p*-values are required. This is due to the fact that it becomes necessary to increase the number of bootstrap samples. To overcome this drawback, an analytical approach based on likelihood ratio test (LRT) may be useful.

Testing Granger non causality as presented in section 3.2 is equivalent to test $\mathbf{a}'\Sigma_{Y_t^i Y_{t-1}^j|\mathbf{X}} = 0$, where $\Sigma_{Y_t^i Y_{t-1}^j|\mathbf{X}}$ is the covariance between Y_t^j and Y_{t-1}^j partialized by $\mathbf{X}$, and $\mathbf{a}$ and $\mathbf{b}$ are the coefficient vectors that maximize the correlation between $\mathbf{a}'Y_t^j$ and $\mathbf{b}'Y_{t-1}^j$ given $\mathbf{X}$. Therefore, the Cluster Granger non-causality test may be set as $H_0 : \Sigma_{Y_t^i Y_{t-1}^j|\mathbf{X}} = 0$ (Granger non-causality) versus $H_1 : \Sigma_{Y_t^i Y_{t-1}^j|\mathbf{X}} \neq 0$ (Granger causality) for the following statistics:

$$-2\ln\Lambda = (T - 1 - (k - n))\ln\left(\left|\hat{\Sigma}_{Y_t^i Y_t^i}\right|\left|\hat{\Sigma}_{Y_{t-1}^j Y_{t-1}^j|\mathbf{X}}\right| / \left|\hat{\Sigma}_{Y_t^i Y_{t-1}^j|\mathbf{X}}\right|\right), \qquad (16)$$

where |.| is the determinant of the matrix. For large *T*, the statistic test is approximately distributed as a χ^2 random variable. For relatively small *T* (e.g., less than 100), which is very common in gene expression data, we suggest the use of Bartlett correction (Bartlett 1939) to improve the asymptotic approximation of the statistic.

Timm and Carlson (1976) showed that the Bartlett correction under our problem arises by replacing the multiplicative factor $T - 1 - (k - n)$ in the likelihood ratio statistic (equation 16) by the factor $r = T - 2 - (k - n) - \frac{1}{2}(m + n + 1)$ to improve the χ^2 approximation to the sampling distribution of equation (16). Thus, we reject H_0 at significance level α if $r \ln \left(\left| \widehat{\Sigma}_{\mathbf{Y}_t^i \mathbf{Y}_t^i} \right| \left| \widehat{\Sigma}_{\mathbf{Y}_{t-1}^j \mathbf{Y}_{t-1}^j | \mathbf{X}} \right| / \left| \widehat{\Sigma}_{\mathbf{Y}_t^i \mathbf{Y}_{t-1}^j | \mathbf{X}} \right| \right) > \chi_{m \times n}^2(\alpha)$ where $\chi_{m \times n}^2(\alpha)$ is the upper $(100\,\alpha)$th percentile of a χ^2 distribution with $m \times n$ degrees of freedom.

Fujita et al. (2010b) carried out Monte Carlo simulations to compare the performance of LRT with Bartlett correction-based test and the bootstrap-based test. The results show that: (i) the LRT is equivalent to Wald's test in the multivariate model in terms of statistical power; (ii) the LRT is much faster and has a higher statistical power than the bootstrap method (when analyzing sets of time series data); and (iii) both the LRT and bootstrap can control the rate of false positives even under non-Normal noises.

4. Functional Clustering in Terms of Granger Causality

In the previous sections, we studied how to identify Granger causality between sets of genes. The first step in this process is the definition of which sets of genes, or clusters, will be used to construct the GRN and, therefore, to identify which genes belong to which cluster. One possible manner to perform this clustering is to use a priori biological information, such as Gene Ontology or Pathway analysis for functional clustering. Here we describe a data-driven approach based on the gene expression levels that are ultimately responsible for the structural proximity of genes in the network. There are numerous definitions for network clusters in the literature (Edachery et al. 1999). A functional cluster in terms of Granger causality can be defined as a subset of genes that strongly interact with each other but interact weakly with the rest of the network. For our data-driven approach for gene clustering, we use an extension of the concept of functional clustering initially proposed by Tononi et al. (1998) in neuroscience. The authors applied mutual information to group the most co-activated brain regions. Similarly, we used the concept of information flow (Baccala and Sameshima 2001) between sets of time series (Fujita et al. 2010a) for gene clustering. The gene expression time series are grouped by the spectral clustering algorithm (Ng et al., 2002) in a way that genes which are structurally close in terms of Granger causality are clustered. We use the concept of Granger causality for sets of times series (Fujita et al. 2010a, 2010b) described in section 2 in order to define distance, degree and flow in gene expression time series. This method is, therefore, able to propose gene clusters based on the Granger causality between genes, and genes within a given cluster may be functionally related in time. The definition of the optimal number of clusters for a given data set is important since it may facilitate biological data interpretation. In the next sections we detail an approach that may be used for the identification of the optimum number of clusters for a given data set.

4.1. Functional clustering

Let $\mathbf{y}_t^1, \mathbf{y}_t^2, \cdots, \mathbf{y}_t^q$ be a set of q time series and $w_{ij} \geq 0$ $(i \neq j)$ a definition of similarity between all pairs of time series $\mathbf{y}_t^i$ and $\mathbf{y}_t^j$. The purpose of clustering is to obtain highly connected time series within groups while seeing little connectivity, in terms of Granger causality, between distinct groups. This connectivity between time series may be represented in the form of a graph $G = (V, E)$, where V is the set of vertices (genes or time series) and E is the set of edges (Granger causality) connecting two elements of V. Thus, each vertex $v_i \in V$ represents a gene expressions time series $\mathbf{y}_t^i$. Two vertices v_i and v_j are connected if the similarity w_{ij} between the corresponding time series $\mathbf{y}_t^i$ and $\mathbf{y}_t^j$ is not zero (the edge of the graph is weighted by w_{ij}). In other words, a $w_{ij} > 0$ represents existence of Granger causality between time series $\mathbf{y}_t^i$ and $\mathbf{y}_t^j$. On the hand, $w_{ij} = 0$ represents absence of Granger causality. When the problem is described in this way, the task of clustering time series in terms of Granger causality can now be reformulated using the graph, i.e., we want to find a partition of the graph such that there is less Granger causality between different groups and more Granger causality within the group.

Consider $G = (V, E)$ as an undirected graph with set of vertices $V = \{v_1, ..., v_q\}$ (where each vertex represents one time series) and a set of weighted edges E. In the following we assume that the graph G is weighted, i.e., each edge between two vertices v_i and v_j carries a non-negative weight $w_{ij} \geq 0$. Thus, the weighted adjacency matrix of the graph G is the matrix $W = w_{ij}$; $i, j = 1, ..., q$. If $w_{ij} = 0$, this means that the vertices v_i and v_j are not connected by an edge. Since G is undirected, the matrix $\mathbf{W}$ is symmetric, i.e., $w_{ij} = w_{ji}$. Therefore, in terms of Granger causality, w_{ij} can be defined as the distance between two times series $\mathbf{y}_t^i$ and $\mathbf{y}_t^j$. The *distance between two (sets of) times series* $\mathbf{y}_t^i$ *and* $\mathbf{y}_t^j$ can be defined as follows:

$$\text{dist}(\mathbf{y}_t^i, \mathbf{y}_t^j) = 1 - \frac{|\text{CCA}(\mathbf{y}_t^i, \mathbf{y}_{t-1}^j) + \text{CCA}(\mathbf{y}_t^j, \mathbf{y}_{t-1}^i)|}{2}. \tag{17}$$

In this representation CCA $(\mathbf{y}_t^i, \mathbf{y}_{t-1}^j)$ is the Granger causality from time series $\mathbf{y}_t^j$ to $\mathbf{y}_t^i$. In the case sets of time series are considered, one can replace $\mathbf{y}_t^i$ and $\mathbf{y}_t^j$ by the considered set of time series $\mathbf{y}_t^i$ and $\mathbf{y}_t^j$ (Fujita et al. 2010a, b). The absolute value of CCA ranges from zero to one and higher CCA values correspond to an increased quantity of information flow and, thus, a shorter distance. It is necessary to point out that since the distance must be symmetric, the average between CCA $(\mathbf{y}_t^i, \mathbf{y}_{t-1}^j)$ and CCA $(\mathbf{y}_t^j, \mathbf{y}_{t-1}^i)$ is calculated. It becomes obvious that the higher is the CCA's coefficient, the lower the distance between the time series (or set of time series).

Additionally, the CCA corresponds to the Pearson correlation after the dimension reduction. As a consequence, dist $(\mathbf{y}_t^i, \mathbf{y}_t^j)$ satisfies three out of four criteria for distances: (i) non-negativity; (ii) dist $(\mathbf{y}_t^i, \mathbf{y}_t^j) = 0$ if and only if $\mathbf{y}_t^i = \mathbf{y}_t^j$; and (iii) symmetry; but does not satisfy the (iv) triangular inequality. Therefore, Pearson correlation (and CCA) is not a real metric. Despite this fact, Pearson correlation is frequently used as a distance measure in several gene expression data analysis (Bhattacharya and De 2008, Ihmels et al. 2005). The main advantage of considering the proposed definition of distance is the fact that it is possible to interpret the clustering process by a Granger causality concept. Without loss of generality, it is possible to extend the concept of distance of a vertex v_i (time series $\mathbf{y}_t^i$) to a set of vertices / time series (sub-network) $\mathbf{Y}_t^u$, where $u = 1, \ldots, k$ and k is the number of sub-networks.

A necessary concept that needs to be introduced is the idea of *degree* of a time series $\mathbf{y}_t^i$ (vertex v_j). It can be defined as:

$$\text{degree } (\mathbf{y}_t^i) = \frac{\text{in-degree } (\mathbf{y}_t^i) + \text{out-degree } (\mathbf{y}_t^i)}{2}, \tag{18}$$

where *in-degree* and *out-degree* are respectively

$$\text{in-degree } (\mathbf{y}_t^i) = |\, \text{CCA } (\mathbf{y}_t^i, \mathbf{y}_{t-1} \,|\, \mathbf{Y}_t \setminus \{\mathbf{Y}_{t-1}\}) \,|, \tag{19}$$

$$\text{out-degree } (\mathbf{y}_t^i) = |\, \text{CCA } (\mathbf{Y}_t, \mathbf{y}_{t-1}^i \,|\, \mathbf{Y}_t \setminus \{\mathbf{Y}_{t-1}\}) \,|. \tag{20}$$

The concept of in-degree and out-degree represent the total information flow that "enters" and "leaves" the vertex v_i, respectively. Hence, the degree of vertex v_i represents the total information flow passing through vertex v_i.

Without risking oversimplification, one can extend the concept of degree of a vertex v_i (time series $\mathbf{y}_t^i$) to a set of time series (sub-network) $\mathbf{Y}_t^u$, where $u = 1, \ldots, k$ and k is the number of sub-networks. In this scenario, the *degree of sub-network* $\mathbf{Y}_t^u$ *is defined by:*

$$\text{degree } (\mathbf{Y}_t^u) = \frac{\text{in-degree } (\mathbf{Y}_t^i) + \text{out-degree } (\mathbf{Y}_t^i)}{2}, \tag{21}$$

where *in-degree* and *out-degree* are respectively

$$\text{in-degree}(\mathbf{y}_t^i) = |\, \text{CCA } (\mathbf{y}_t^u, \mathbf{Y}_{t-1} \,|\, \mathbf{Y}_t \setminus \{\mathbf{Y}_{t-1}\}) \,|, \tag{22}$$

$$\text{in-degree } (\mathbf{y}_t^i) = |\, \text{CCA } (\mathbf{Y}_t, \mathbf{Y}_{t-1}^u \,|\, \mathbf{Y}_t \setminus \{\mathbf{Y}_{t-1}\}) \,|. \tag{23}$$

Straightaway, while keeping the definitions of distance and degrees for time series and sets of time series in terms of Granger causality, it is possible to develop a spectral clustering-based algorithm to identify sub-networks consisting of sets of time series that are highly connected, in terms of Granger causality, within a group and poorly connected between groups, in the regulatory networks. The algorithm based on spectral clustering is described as follows (Ng et al. 2002):

Input: The q time series $(\mathbf{y}_t^i; i = 1, \cdots, q)$ and the number k of sub-networks to be constructed.

1. Let $\mathbf{W}$ be the $(q \times q)$ symmetric weighted adjacency matrix where $w_{ij} = w_{ji} = 1 - \text{dist}(\mathbf{y}_t^i; \mathbf{y}_t^j)$, $i, j = 1, \ldots, q$.

2. Compute the non-normalized $(q \times q)$ Laplacian matrix $\mathbf{L}$ as (Mohar 1991):

   $\mathbf{L} = \mathbf{D} - \mathbf{W}$ (10)

   where $\mathbf{D}$ is the $(q \times q)$ diagonal matrix with the degrees $d_1, \cdots, d_q$ (degree $(\mathbf{y}_t^i)$ = d_i; $i = 1, \cdots, q$) on the diagonal.

3. Compute the k eigenvectors $\{\mathbf{e}_1, \ldots, \mathbf{e}_k\}$ (corresponding to the k smallest eigenvalues) of $\mathbf{L}$.

4. Let $\mathbf{U} \in \mathbb{R}^{q \times k}$ be the matrix containing the vectors $\{\mathbf{e}_1, ..., \mathbf{e}_k\}$ as columns.

5. For $i = 1, ..., q,$ let $\mathbf{y}_i \in \mathbb{R}^{q \times k}$ be the vector corresponding to the ith row of $\mathbf{U}$.

6. Cluster the points $(\mathbf{y}_i)_{i=1,\cdots,q} \in \mathbb{R}^k$ with the k-means algorithm into clusters $\{\mathbf{Y}_t^1, \cdots, \mathbf{Y}_t^k\}$. For k-means, one may select a large number of initial values to achieve (or to be closer) the global optimum configuration.

Output: Sub-networks $\{\mathbf{Y}_t^1, \cdots, \mathbf{Y}_t^k\}$.

It must be clear, however, that this clustering approach does not require the construction of the entire network.

4.2. Estimation of the number of clusters

As presented in Section 4.1, structural distances in terms of Granger causality can be used as a framework for clustering genes (time series) or time series. Nevertheless, similarly to the majority of the clustering algorithms, including the spectral clustering also requires the desired or "optimum" number of clusters k as input. Consequently, the objective determination of the number of sub-networks k constitutes a major challenge. The most adequate number of sub-networks k depends on a diversity of parameters such as the biological system being studied and the number of genes to be evaluated in downstream wet lab experiments. For the purpose of this chapter, we describe an approach to identify clusters based on the density of the connectivity between time series included in the cluster and between clusters, where the first should be high and the second low.

The most suitable number of clusters in this specific context may be determined using the slope statistic (Fujita et al. 2014). This method is based on the silhouette statistic (Rousseeuw 1987) but may be considered more robust than the silhouette when clusters differ in sizes and variances.

In order to exemplify, a cluster index s_i in the case of dissimilarities should be defined. Consider a time series $\mathbf{y}_t^i$ in the data set, and denote by $\mathbf{A}$ the sub-network to which it has been assigned. If the sub-network $\mathbf{A}$ contains several time series other than $\mathbf{y}_t^i$, we can compute: $a_i = \mathrm{dist}\,(\mathbf{y}_t^i, \mathbf{A})$, which is the average dissimilarity of $\mathbf{y}_t^i$ to $\mathbf{A}$. Considering another sub-network $\mathbf{C}$, different from $\mathbf{A}$, we can compute: $\mathrm{dist}\,(\mathbf{y}_t^i, \mathbf{C})$ which is the dissimilarity of $\mathbf{y}_t^i$ to $\mathbf{C}$. After computing $\mathrm{dist}\,(\mathbf{y}_t^i, \mathbf{C})$ for every sub-networks $\mathbf{C} \neq \mathbf{A}$, we select the smallest of the results and express it by $b_i = \min_{\mathbf{C} \neq \mathbf{A}} \mathrm{dist}\,(\mathbf{y}_t^i, \mathbf{C})$. The sub-network called $\mathbf{B}$ for which this minimum value was attained (that is, $b_i = \mathrm{dist}\,(\mathbf{y}_t^i, \mathbf{B})$ will be called the neighbor sub-network, or cluster of $\mathbf{y}_t^i$ (Fig. 2). This means that this neighbor cluster would be the second-best cluster for time series $\mathbf{y}_t^i$ meaning that if $\mathbf{y}_t^i$ could not belong to sub-network $\mathbf{A}$, the best sub-network for it to belong to would be $\mathbf{B}$. Therefore, once we obtain b_i we can know the best alternative cluster for the time series in the network. But the calculation of b_i depends on the existence of other sub-networks apart from $\mathbf{A}$, within the data set. It is necessary, thus, to assume that there is more than one sub-network k within a given network (Rousseeuw 1987).

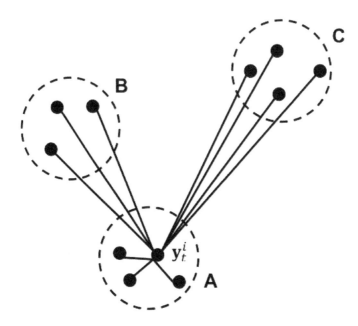

FIGURE 2 An illustration of the time series involved in the computation of s_i. Notice that, in this example, time series $\mathbf{y}_t^i$ belongs to cluster **A**. Cluster **B** is the neighbor cluster of $\mathbf{y}_t^i$ (the second-best cluster for time series $\mathbf{y}_t^i$).

After the computation of a_i and b_i, the cluster index s_i can be obtained by combining them as follows:

$$S_i = \frac{b_i - a_i}{\max(a_i, b_i)}. \tag{24}$$

It is clear from equation (24) that $-1 \leq s_i \leq 1$ for each time series $\mathbf{y}_t^i$. Thus, there are at least three cases to be analyzed:

1. $s_i \approx 1$: For cluster index s_i to be close to one we require $a_i \ll b_i$. As a_i is a measure of how dissimilar $\mathbf{y}_t^i$ is to its own sub-network, a small value means it is well matched. Furthermore, a large b_i implies that $\mathbf{y}_t^i$ is badly matched to its neighboring sub-network. Thus, a cluster index s_i close to one means that the gene is appropriately clustered.

2. $s_i \approx 0$: If s_i is close to negative one, then by the same logic we see that $\mathbf{y}_t^i$ would be more appropriate if it was clustered in its neighboring sub-network.

3. $s_i \approx -1$: A cluster index s_i near zero means that the gene is on the border of two sub-networks.

In other words, the cluster index s_i can be interpreted as the fitness of the time series $\mathbf{y}_t^i$ to the assigned sub-network.

For each number of clusters k = 2, 3, ..., q, the average cluster index $s(k)$ can be computed as $s(k) = \dfrac{1}{q} \sum_{i=1}^{q} s_i$. The average cluster index $s(k)$ of the entire data set is a measurement of how tightly grouped all the genes in the network are, or how appropriately the genes have been clustered in a structural point of view and in terms of Granger causality. Fujita et al. (2014) suggest the use of the estimator $\hat{k} = \arg\max_{k \in (2,\cdots,q)} - [s(k+1) - s(k)]\, s(k)^p$ where p is a positive integer value that can be tuned to interpolate between a criterion where $s(k+1) - s(k)$ is more important (small p) and a criterion where the silhouette value has more weight (large p). The slope method identifies the maximum number of clusters that breaks down the structure of the data set. The difference between slope and silhouette statistic is the fact that, by maximizing the silhouette statistic, the number of clusters is estimated correctly only when the within-cluster variances are equal. In the general case in which within-cluster variances are unequal, maximizing the slope statistic yields an optimal number of clusters for separating within cluster and between cluster distances. The intuition behind the slope statistic is as following: suppose a scenario that there is a dominant cluster (one sub-network presenting high variance).When the number of identified sub-networks is equal or lower than the adequate number of sub-networks, the cluster index values are very similar. However, when the number of identified sub-networks becomes higher than the adequate number of sub-networks, the cluster index value s decreases abruptly. This is due to the fact that one of the highly connected sub-networks is split into two new sub-networks. For more details, refer to (Fujita et al. 2014).

5.　Network Construction from Large Datasets

When analyzing data generated from high-throughput technologies, gene networks would have to be constructed from large number of data sets (time series) obtained from a significantly smaller number of experimental points (observations). One possible solution to this problem is to reduce the dimensionality of the data prior to the statistical analysis. The dimension reduction procedure consists in clustering similar (highly correlated) time series and to remove the redundancy since time series belonging to the same cluster may share the same biological information.

Clustering may be performed using the Classification Expectation Maximization (CEM) algorithm– see Appendix (Celeux and Govaert 1992) and redundancy may be removed by the Principal Component Analysis (PCA). PCA allows us to keep only the most significant components leading to variability in the data set, thus reducing the number of variables for subsequent processing. We suggest retaining only components accounting for more than 5% of the temporal variance in each cluster (Celeux and Govaert 1992). By applying the PCA, it is possible to extract the eigen-time series from each cluster. The eigen-time series are then clustered in terms of Granger causality as described in the section 4 and the network (CGC) can be inferred from clusters composed of the eigen-time series by applying the method described in Section 3. Figure 3 illustrates the entire process.

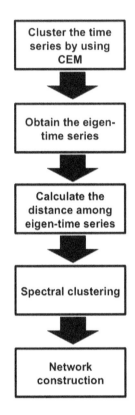

FIGURE 3 Pipeline to construct GRNs from large datasets. Similar time series are clustered by using CEM. PCA analysis is carried out to remove redundancy and obtain the eigen-time series. Then, the spectral clustering algorithm is applied to the eigen-time series to obtain the clusters (sub-networks). Notice that the clusters obtained here are based on the Granger causality among eigen-time series. The CGC is inferred among the clusters composed of the eigen-time series.

6. Software

Granger is an R package containing all the methods described in this chapter for Granger causality identification, including the bootstrap procedure and the LRT with Bartlett correction. It can be downloaded from: http://dnagarden.hgc.jp/afujita/en/doku.php?id=ggranger.

Appendix

Classification Expectation Maximization Algorithm (CEM)

Consider the case we have q times series $\mathbf{Y}_t = \{\mathbf{y}_t^1, \mathbf{y}_t^2, \mathbf{y}_t^3, \cdots, \mathbf{y}_t^q\}$, and the unknown associated label variables (k groups) $\mathbf{z} = (\mathbf{z}_1, \mathbf{z}_2, \ldots, \mathbf{z}_k)$, where $\mathbf{z}_i = (z_{i1}, z_{i2}, \ldots, z_{iq})$, and $z_{ig} = 1$ if time series $\mathbf{y}_t^i$ belongs to group g or 0 otherwise.

The log-likelihood is given by:

$$l(\theta \mid \mathbf{Y}_t, \mathbf{z}) = \sum_{i=1}^{k} \sum_{g=1}^{q} z_{ig} \log(p_g f(\mathbf{z}_i \mid \lambda_g)),$$

where θ is a vector containing the mixtures parameters p_g, λ_g, and $f(.)$ is the probability density function of $\mathbf{Y}_t$.

The first step is to set initial values to the mixture parameters $\theta^{(0)}$ which contain proportions, means, and variances of the Gaussian mixture. Then, the current conditional probabilities of each observation belong to the group g are estimated considering the current estimates of the vector of mixture parameters $\theta^{(m-1)}$, where $m = 1, 2, ..., M$ denotes the iteration number. The conditional probabilities (expectation step) are estimated using the formula $w_{ig}^{(m)} = \dfrac{p_g^{(m-1)} f(\mathbf{y}_t^i \mid \lambda_g^{(m-1)})}{\sum_{i=1}^{k} p_l^{(m-1)} f(\mathbf{y}_t^i \lambda_l^{(m-1)})}$, for g in $1, 2, ..., q$.

The following step (classification) is to allocate each observation to the group with the highest probability of containing it, i.e., label the data considering

$$\hat{z}_{ig}^{(m)} = \begin{cases} 1, \text{ if } g = \arg \max_{l=1,...,k} w_{ig}^{(m)} \\ 0, \text{ otherwise.} \end{cases},$$

The maximization step is then achieved by obtaining maximum likelihood estimates for the mixture parameters in $\theta^{(m)}$. The algorithm then returns to the expectation step until convergence of all parameters and labels has been achieved within allowed limits.

Note that as CEM is based on likelihood maximization, it requires additional information about the probability distribution of the data, which in most cases is assumed to be multivariate Gaussian. For automatic selection of the number of clusters k, one may use the slope statistic as aforementioned or the Bayesian Information Criterion (BIC) introduced by Schwarz (1978). The asymptotic properties and consistency of BIC are well known and have been described extensively in the literature (Hannan and Quinn 1979, Haughton 1988). The optimum number of clusters minimizes the quantity $BIC_k = -2l(\mathbf{Y}_t \mid k, \hat{\theta}) + v_k \log(q)$, where v_k is the number of parameters in a model with k clusters.

Acknowledgments

This work was supported by FAPESP (2014/09576-5; 2013/03447-6), CNPq (304020/2013-3; 473063/2013-1), NAP eScience – PRP – USP, and CAPES.

References

Anderson, T.W., 1984. An introduction to multivariate statistical analysis. Wiley, New York.

Baccalá, L.A. and K. Sameshima, 2001. Partial directed coherence: a new concept in neural structure determination. Biological Cybernetics. 84: 463–474.

Bartlett, M.S., 1939. A note on tests of significance in multivariate regression. Proc. Camb. Phil. Soc. 35: 180–185.

Bhattacharya, A. and R.K. De, 2008. Divisive correlation clustering algorithm (DCCA) for grouping of genes: detecting varying patterns in expression profiles. Bioinformatics 24: 1359–1366.

Bühlmann, P. and R. Künsch, 1999. Block length selection in the bootstrap for time series. Computational Statistics & Data Analysis 31: 295–310.

Carlstein, E., 1986. The use of subseries methods for estimating the variance of a general statistic from a stationary time series. Annals of Statistics 14: 1171–1179.

Celeux, G. and G. Govaert, 1992. A classification EM algorithm for clustering and two stochastic versions. Comput. Stat. Data Anal. 14: 315–332.

Chen, Y., G. Rangarajan, J. Feng, M. Ding, 2004. Analyzing multiple nonlinear time series with extended Granger causality. Physics Letters A 324: 26–35.

Edachery, J., A. Sen, F. Brandenburg, 1999. Graph clustering using distance-k cliques. Lecture Notes in Computer Science. 1731: 98–106.

Elsner, J.B., 2006. Evidence in support of the climate change–Atlantic hurricane hypothesis. Geophys. Res. Lett. 33: L16705.

Freiwald, W.A., P. Valdes, J. Bosch, R. Biscay, J.C. Jimenez, L.M. Rodriguez, V. Rodriguez, A.K. Kreiter and W. Singer, 1999. Testing non-linearity and directedness of interactions between neural groups in the macaque inferotemporal cortex. J Neurosci Methods. 94: 105–119.

Fujita, A., J.R. Sato, C.E. Ferreira and M.C. Sogayar, 2007. GEDI: a user-friendly toolbox for analysis of large-scale gene expression data. BMC Bioinformatics 8: 457.

Fujita, A., J.R. Sato, H. Garay-Malpartida, M.C. Sogayar, C.E. Ferreira and S. Miyano, 2008. Modeling nonlinear gene regulatory network from time series gene expression data. Journal of Bioinformatics and Computational Biology 6: 961–979.

Fujita, A., A.G.Patriota, J.R. Sato and S. Miyano, 2009. The impact of measurement errors in the identification of regulatory networks. BMC Bioinformatics 10: 412.

Fujita, A., J.R. Sato, K. Kojima, L.R. Gomes, M. Nagasaki, M.C. Sogayar and S. Miyano, 2010. Identification of Granger causality between gene sets. Journal of Bioinformatics and Computational Biology 8: 679.

Fujita, A., K. Kojima, A.G. Patriota, J.R. Sato, P. Severino and S. Miyano, 2010. A fast and robust statistical test based on likelihood ratio with Bartlett correction to identify Granger causality between gene sets. Bioinformatics 26: 2349–2351.

Fujita, A., D.Y. Takahashi and A.G. Patriota 2014. A non-parametric method to estimate the number of clusters. Computational Statistics & Data analysis 73: 27–39.

Geweke, J., 1984. Measures of conditional linear dependence and feedback between time. Series. J Am Stat Assoc. 79: 907–915.

Granger, C.W.J., 1969. Investigating causal relationships by econometric models and cross-spectral methods. Econometrica 37: 424–438.

Guo, S., J. Wu, M. Ding and J. Feng, 2008. Uncovering interactions in the frequency domain. PLoS Computational Biology 4: e1000087.

Hall, P. 1985. Resampling a coverage process. Stochastic Process Applications 19: 259–269.

Hannan, E.J. and B.G. Quinn, 1979. The determination of the order of an autoregression. J. Roy. Statis Soc Ser B 41: 190–195.

Haughton, D., 1988. On the choice of model to fit data from an exponential family. Ann Statist 16: 342–355.

Hemmelmann, D., M. Ungureanu, W. Hesse, T. Wüstenberg, J.R. Reichenbach, O.W. Witte, H. Witte, L. Leistritz, Modelling and analysis of time-variant directed interrelations between brain regions based on BOLD-signals. Neuroimage 45: 722–737.

Hiemstra, C. and J.D. Jones, 1994. Testing for linear and nonlinear Granger causality in the stock price–volume relation. J. Finance 49: 1639–1664.

Hotelling, H., 1935. The most predictable criterion. J. Educ. Psychol. 26: 139–142.

Hotelling, H., 1936. Relations between two sets of variates. Biometrika 28: 321–377.

Ihmels, J., S. Bergmann, J. Berman and N. Barkai, 2005. Comparative gene expression analysis by differential clustering approach: applications to the Candida albicans transcription program. PLoS Genet 1: e39.

Johnson, R.A. and D.W. Wichern, 2002. Applied Multivariate Statistical Analysis. Prentice Hall, New Jersey.

Kaufmann, R.K., R.D. D'Arrigo, C. Laskowski, R.B. Myneni, L. Zhou and N.K. Davi, 2004. The effect of growing season and summer greenness on northern forests. Geophys. Res. Lett. 31: L09205.

Künsch, H.R., 1989. The jacknife and the bootstrap for general stationary observations. Annals of Statistics 17: 1217–1241.

Lahiri, S.N., 2003. Resampling Methods for Dependent Data. Springer-Verlag, New York, 2003.

Lozano, A.C., N. Abe, Y. Liu and S. Rosset, 2009. Grouped graphical Granger modeling for gene expression regulatory networks discovery. Bionformatics 25: i110 – i118.

Lütkepohl, H., 2005. New introduction to multiple time series analysis. Springer, Berlin.

Marinazzo, D., M. Pellicoro and S. Stramaglia, 2008. Kernel Granger causality and the analysis of dynamical networks. Physical Review E 77: 056215.

Mohar, B. 1991. The Laplacian spectrum of graphs. pp. 871–898. In Y. Alavi, G. Chartrand, O.R. Oellermann, A.J. Schwenk (eds.). Graph Theory, Combinatorics, and Applications. Wiley.

Mukhopadhyay, N.D. and S. Chatterjee, 2007. Causality and pathway search in microarray time series experiment. Bioinformatics 23: 442–449.

Nagarajan, R. and M. Upreti, 2008. Comment on causality and pathway search in microarray time series experiment. Bioinformatics 24: 1029–1032.

Nagarajan, R., 2009. A note on inferring acyclic network structures using Granger causality tests. The International Journal of Biostatistics 5: 10.

Ng, A., M.I. Jordan, Y. Weiss, 2002. On spectral clustering: analysis and an algorithm. Advances in Neural Information Processing Systems. New York: MIT Press.

Opgen-Rhein, R., K. Strimmer, 2008. Learning causal networks from systems biology time course data: an effective model selection procedure for the vector autoregressive process. BMC Bioinformatics 8: S3.

Rao, B.R. 1969. Partial canonical correlations. Trabajos de Estadistica y de Investigacion operative. 20: 211–219.

Rousseeuw, P.J. 1987. Silhouettes: a graphical aid to the interpretation and validation of cluster analysis. J. Comput. Appl. Math. 20: 53–65.

Sato, J.R., D.Y. Takahashi, E. Cardoso, M.G.M. Martin, E. Amaro Jr, and P.A. Morettin, 2006. Intervention models in functional connectivity identification applied to fMRI. Int. J. Biomed. Imaging 1: 1–7.

Sato, J.R., P.A. Morettin, P. Arantes and E. Amaro Jr, 2007. Wavelet based time-varying vector autoregressive modeling. Computational Statistics & Data Analysis 51: 5847–5866.

Sato, J.R., D.Y. Takahashi, S.M. Arcuri, K. Sameshima, P.A. Morettin, and L.A. Baccalá, 2009. Frequency domain connectivity identification: an application of partial directed coherence in fMRI. Human Brain Mapping 30: 452–461.

Sato, J.R., A. Fujita, E.F. Cardoso, C.E. Thomaz, M.J. Brammer and E. Amaro Jr, 2010. Analyzing the connectivity between regions of interest: an approach based on cluster Granger causality for fMRI data analysis. NeuroImage 52: 1444–1455.

Schelter, B., M. Winterhalder, M. Eichler, M. Peifer, B. Hellwig, B. Guschlbauer, C.H. Lucking, R. Dahlhaus and J. Timmer, 2006. Testing for directed influences among neural signals using partial directed coherence. J. Neurosci. Meth. 152: 210–219.

Schelter, B., J. Timmer and M. Eichler, 2009. Assessing the strength of directed influences among neural signals using renormalized partial directed coherence. J. Neurosci. Meth. 179: 121–130.

Schwarz, G. 1978. Estimating the dimension of a model. Ann Statist 6: 461–464.

Seibold, D. and R.D. McPhee, 1979. Commonality analysis: a method for decomposing explained variance in multiple regression analyses. Human Comm. Res. 5: 355–365.

Shimamura, T., S. Imoto, R. Yamaguchi, A. Fujita, M. Nagasaki and S. Miyano, 2009. Recursive regularization for inferring gene networks from time-course gene expression profiles. BMC Systems Biology. 3: 41.

Timm, N.H. and J.E. Carlson, 1976. Part and bipartial canonical correlation analysis. Psychometrika 41: 159–176.

Tononi, G., A.R. McIntosh, D.P. Russel and G.M. Edelman, 1998. Functional clustering: identifying strongly interactive brain regions in neuroimaging data. NeuroImage 7: 133–149.

Valdés-Sosa, P.A., J.M. Sánchez-Bornot, A. Lage-Castellanos, M. Vega-Hernández, J. Bosch-Bayard, L. Melie-García and E. Canales-Rodríguez, 2005. Estimating brain functional connectivity with sparse multivariate auto regression. Philos. Trans. R. Soc. Lond. B Biol. Sci. 360: 969–981.

Wayne, J. 1986. Economic growth and defense spending: Granger causality. J. Dev. Econ. 21: 35–40.

Wiener, N., 1956. The Theory of Prediction. Modern Mathematics for Engineers, McGraw-H, New York.

Section 2
RNAs

4

RNA Sequencing and Gene Expression Regulation

Mirza Sarwar Baig[1], Raj K. Keservani[2]*, Anil K. Sharma[3]

Abstract

RNA is synthesized using one of the DNA strands as a template and has the same chemical structure except that thymine is replaced by uracil (U). Some RNA molecules can be the end product in themselves and some can in turn be used as a template for the creation of other molecules, proteins, by a process called 'translation'. Gene expression is the process by which information from a gene is used in the synthesis of a functional gene product. Bacterial genomes usually contain several thousand different genes. Some of the gene products are required by the cell under all growth conditions and are called 'housekeeping genes'. Gene regulation can occur at three possible places in the production of an active gene product. First, the transcriptional regulation second, the translational regulation and third the post-transcriptional or Post-translational regulation mechanisms. The regulatory mechanisms controlling gene expression are typically discovered by mutational analysis. *Cis*-acting molecules act upon and modulate the expression of physically adjacent, operably linked polypeptide-encoding sequences. *Trans*-acting factors affect the expression of genes that may be physically located very far away, even on different chromosomes. The expression of a particular gene may be regulated by the concerted action of both *cis* and *trans*-acting elements. Since changes in gene expression levels are thought to underlie many of the phenotypic differences between species, identifying and characterizing the regulatory mechanisms responsible for these changes is an important goal of molecular biology. The main aim of this chapter is to provide scattered information of RNA sequencing and gene expression in gathered form to the scientist and researchers.

[1] Department of Biosciences, Jamia Millia Islamia (A Central University), New Delhi-110025, India.
[2] School of Pharmaceutical Sciences, Rajiv Gandhi Proudyogiki Vishwavidyalaya, Bhopal, India-462033; Mob.: +917897803904; E-mail: rajksops@gmail.com.
[3] Department of Pharmaceutics, Delhi Institute of Pharmaceutical Sciences and Research, New Delhi, 110017, India.
* Corresponding author : rajksops@gmail.com

## 1.	The Fundamentals of DNA, RNA and Gene

The deoxyribonucleic acid (DNA) and ribonucleic acid (RNA) are linear chains of polynucleotide molecules linked by nucleotide bonds. A single nucleotide consists of one cyclic penta-carbo sugar (pentose sugar), one nitrogenous base and one phosphate group. The basic chemical difference in DNA and RNA is due to sugar and a nitrogenous base. The DNA contains deoxyribose sugar while ribose sugar is present in RNA. The ribose sugar in RNA has a hydroxyl group (-OH) on the 2' carbon and deoxyribose sugar of DNA does not. The lacking of hydroxyl group on the 2' carbon makes DNA more stable than RNA. The nitrogenous bases are of two main type's purine and pyramidine derivatives, in which adenine (A) and guanine (G) are purine derivative while cytosine (C), thymine (T) and uracil (U) are pyramidine derivatives. Adenine, guanine and cytosine are found in both DNA and RNA, but thymine is only found in DNA and uracil is uniquely present only in RNA. Uracil is simply an un-methylated form of thymine. The carbons found in penta-carbo sugar are numbered 1' to 5' (pronounced as 1-prime to 5-prime) only to make difference between the numbering of atoms of nitrogenous bases (1 to 6 or 1 to 9). The atoms of the purine ring are numbered from 1 to 9, and those of the pyrimidine ring are numbered from 1 to 6. The phosphate group is linked to the 5' carbon of the sugar in both RNA and DNA. Nucleotide is a combination of nitrogenous base-sugar-phosphate; where 1' and 5' carbons of a sugar molecule are attached to a nitrogenous base and a phosphate group respectively. Two nucleotides of a single DNA or RNA strand are linked together by a versatile phosphodiester bond (phosphate with two ester linkages). In this bond a phosphate group is shared between 3' carbon of one sugar to the next 5' carbon of adjoining sugar of another nucleotides. This alternating sugar and phosphate molecules forms the structural framework of DNA and RNA termed as the "sugar-phosphate backbone". It defines the directionality (5' to 3' or 3' to 5' depending on the convention) of nucleic acids, as well as the directionality of molecular processes like, replication, transcriptional and translational initiation.

The two strands of nucleotides of DNA molecules runs anti-parallel and twisted together to form a ladder called "double helix" (Watson and Crick 1953a). The outer side of this double helix is a negatively charged and hydrophilic sugar-phosphate backbone, while in the inner core nitrogenous bases are found which hold the two strands together by hydrogen bonds. Contrary, mostly RNA consists of single strand of nucleotides linked together by phosphodiester bonds only and no as such hydrogen bonds exists, due to which RNA is structurally and chemically less stable than DNA. Therefore, we can say that there are two versatile properties of DNA, one is complementation and another is anti-parallelism. The RNA molecules don't have these two exclusive properties of complementation and anti-parallelism, except transfer RNA (tRNA), which forms intra-strand double helix but to a limited extent (Gerald 1989).

Due to the properties of complementary base pairing and anti-parallelism (run in opposite directions), two strands of DNA are complementary to one another, where A always pairs with T by two hydrogen bonds; G always pairs with C by three hydrogen bonds (Chargaff et al. 1951, Chargaff et al. 1952, Zamenhoff et al. 1952). For example: If one strand is 5'-AAGGCTTC-3' the reverse complementary strand will be 3'-TTCCGAAG-5'. Therefore, complementation and anti-parallelism is important for storage and the transmittance of genetic information. In a double stranded DNA molecule, by convention one strand is denoted as coding or forward or sense or plus or non-template strand and the other strand is non-coding or anti-sense or minus or template strand, depending on the ease of author. The directionality of DNA or RNA strand is assigned on the basis of free phosphate (at 5') or free hydroxyl (at 3') group attached to the 5' or 3' carbon of the sugar found at the end of the strands. A gene is a sequence of nucleotides or a combination of meaningful genetic codes that transcribes and translates a functional product. A gene can be present on a DNA strand in any one of two orientations, which may be on a sense strand or on an anti-sense strand.

But in this chapter, we will focus mostly on chemical, sequence, structural, functional, temporal and spatial features of RNA, which makes it a very attractive candidate for sequencing and expression. Also, RNA (especially messenger RNA) is a connecting link between DNA and proteins, so to analyze the protein we must have a library of full set of the transcriptome. Therefore, to get the full set of the transcriptome, sequencing of RNA is an inevitable step.

1.1 Staring from Biggest to Smallest RNA

RNA is a short form of **R**ibo**N**ucleic **A**cid. Like DNA, RNA molecules are manufactured in the nucleus of the cell. However, unlike DNA, RNA is not restricted to the nucleus. It can migrate into other parts of the cell. On the basis of sequence (like coding and non-coding RNA, big and small RNA), structure (messenger and transfer RNA), function (regulatory and non-regulatory RNA), spatial (nuclear and extra-nuclear RNA, cell and tissue specific RNA) and temporal (RNA expressed at different developmental stages) attributes RNA can be classified into many categories. like, messenger RNA (mRNA), transfer RNA (tRNA), ribosomal RNA (rRNA), small nuclear RNA (snRNA), small nucleolar RNA (snoRNA), microRNAs (miRNA), small interfering RNA (siRNA), small non-coding RNA (tsnRNA), small modeling RNA (smRNA), PIWI RNA (piRNA) etc.

Some RNA, called messenger RNA (mRNA) communicates the genetic message found in the DNA out to the rest of the cell for the purpose of promoting the synthesis of proteins (Gilbert 1987).

Some RNA, Small nuclear RNA (snRNA), also commonly referred to as U-RNA (uridylate RNA), is a class of small RNA molecules that are found within the cell nucleus in eukaryotic cells (Hodnett and Harris 1968). The length of snRNA ranges from 80 to 350 nucleotides. Their primary function is in the processing of pre-messenger RNA (hnRNA) in the nucleus. They have also been shown to aid in the regulation of transcription factors (7SK RNA) or RNA polymerase II (B2 RNA), and maintaining the telomeres.

The non-coding RNA (ncRNA) does not encode a protein, but this does not mean that such RNAs do not contain information nor have function. These ncRNAs include microRNAs (miRNA) and small nucleolar RNA (snoRNA), as well as likely other classes of yet-to-be-discovered small regulatory RNAs, and tens of thousands of longer transcripts, most of whose functions are unknown (Mattick and Makunin 2006).

A large group of ncRNAs are known as small nucleolar RNA (snoRNA). These are small RNA (60-300 nucleotides long) molecules that play an essential role in RNA biogenesis and guide chemical modifications of ribosomal RNAs (rRNAs) and other RNA genes (tRNA and snRNAs) but never translated into protein (Mattaj et al. 1993). They are located in the nucleolus and the Cajal bodies of eukaryotic cells where they are called scaRNA (small Cajal body-specific RNAs). More than thirty kinds are known.

MicroRNAs (miRNAs) are 19-23 nucleotide long non-coding RNAs that regulate gene expression at the post-transcriptional level, either by endonucleolytic cleavage or by translational inhibition (Carrington and Ambros 2003, Rhoades and Bartel 2004). They play major roles in key aspects of plant development and their response to environmental stresses. miRNAs and siRNAs are ~21–26-nucleotide (nt) RNA molecules. Although both types of molecules can be functionally equivalent, they are distinguished by their mode of biogenesis (Carmell and Hannon 2004, Kim 2005). miRNAs are generated from ss-precursor transcripts that fold into imperfectly base-paired hairpin structures. Usually only a single mature, stable miRNA is liberated from each stem- loop precursor. Strikingly, bioinformatics analyses suggest that up to 30% of human genes may be regulated by miRNAs (Lewis et al. 2005). Bioinformatics analyses indicate that target genes in vertebrate species may number in the thousands.

Short interfering RNAs (siRNAs) are the canonical participants in RNA interference (RNAi) and are generated from perfectly base-paired dsRNA precursors. The regulatory pathways mediated by these small RNAs are usually collectively referred to as RNAi or RNA silencing. miRNAs and siRNAs can silence cytoplasmic mRNAs either by triggering an endonuclease cleavage, by promoting translation repression, or possibly by accelerating mRNA decapping (Doench et al. 2003, Allen et al. 2005, Simard and Hutvagner 2005) .

Repeat-associated siRNA (rasiRNAs) are major classes of siRNAs which are encoded by heterochromatic regions including centromeres and telomeres that contain many repetitive elements (transposons or retro-elements) within the genome (Volpe et al. 2002). The fascinating world of small RNAs includes another RNA known as PIWI RNA (piRNA) that belongs to the complex of related pathways termed RNA silencing. The piRNAs are short single stranded RNAs arising from a Dicer-independent pathway, which are found in germ cells and associate with the PIWI subfamily of Argonaute proteins. In many organisms piRNAs are derived from repetitive sequences (Aravin et al. 2007, Houwing et al. 2007).

Promoter associated small RNAs (PASRs) are another class of small RNAs of unknown size that were identified by high-throughput screening. These PASRs have an unknown length but cover a region of about -500 to +500 base pairs (bp) relative to transcriptional start sites. They apparently differ from small RNAs implicated in RNA silencing because conventional miRNA cloning procedures failed to identify them. PASRs are typically found within 0.5 kb of transcription start sites and about 40% of them map to 5' expressed sequence tags. Their function is unknown so far but they possibly regulate transcription as exogenous PASRs can reduce expression of genes with homologous promoter sequences (Han et al. 2007, Ryan et al. 2009).

1.2   Sequencing of Gene to Genomes and Transcript to Transcriptomes and Beyond

This chapter is written in a broad view to familiarize the reader, why and how to sequence a single DNA and RNA transcript to the whole genome and transcriptome expressed in an organism at particular instance of time? What are the goals of sequencing the transcriptome? How sequencing technologies are helpful for molecular biologists and clinicians in molecular cloning, genetic engineering, identifying pathogen resistant genes, comparative and evolution studies? First, a brief history of DNA Sequencing will be introduced, and then every aspect of RNA sequencing will be discussed. Previously, every types of structural and functional RNA, whether they are protein coding or non-coding have been discussed. Finally, we will increase our horizon of RNA sequencing methods and technologies involved.

An ideal sequencing technologies should be accurate, economical and high-through put in nature. In the past few years, it has evolved with a tremendous pace and acts as the driving force of genomic, meta-genomic and transcriptomic research and development. Sequencing is a gateway for molecular, phylogenetic and biostatistical characterization of vast amount of genomic and transcriptomic data. The major aims of sequencing of the transcriptome are annotation and quantification. Annotation involves identification of novel genes or transcripts, exons, splicing junctions, ncRNAs etc. While, quantification is the process of identification of abundance of transcripts in a genomic or transcriptomic sample. Therefore, genome and transcriptome sequencing projects subsequently opens door for new research areas and several modern applications.

Tremendous development of biological technologies over the last decade has resulted in the whole genome sequencing of many important model organisms of different categories belonging from lower plants to higher plants and similarly from lower animals to higher animals. Luckily, *Haemophilus influenzae* is the first prokaryote to be sequenced having genome size 1,830,137 base pairs (Fleischmann et al. 1995). A Fungi, *Saccharomyces cerevisiae* is the first eukaryotic genome sequence with 12,068 kilo bases (Goffeau et al. 1996). A Nematode, *Caenorhabditis elegans* is the first multicellular organism to be complete sequenced (The *C. elegans* Sequencing Consortium 1998). An insect, *Drosophila melanogaster* (120 mega bases) (Adams et al. 2000), an important plant model Arabidopsis thaliana (Meinke et al. 1998), and many more microbial genomes are being sequenced. By 2001, the first draft version of the sequence of base pairs in human DNA had been released that means, human chromosome number 22 in 1999 (Dunham et al. 1999) and chromosome number 21 in 2000 (Hattori et al. 2000). We are now moving from the pregenomic era characterized by the effort to sequence the human genome, to a post-genomic era that concentrates on harvesting the fruits hidden in the genomic text.

An overarching challenge in this post-genomic era is the management and analysis of enormous quantities of DNA and RNA sequence data. These data are now a challenge for molecular biologists, bioinformaticians, biotechnologists, biostaticians, clinicians and computer programmers to analyze and deduce their purposely research. Understanding the biological systems with hundreds of thousands of DNA and RNA genes will require a good cataloging, better planning and best organizing skills at diverse levels such as: primary DNA sequence in coding and regulatory regions; temporal and spatial RNA expression during development; polymorphistic variation within a species or subgroup; sub-cellular localization and intermolecular interaction of protein products and finally physiological response and disease.

1.3 Methods and Technologies of RNA/Transcriptome Sequencing

There are two most common methods of DNA sequencing, one is Maxam and Gilbert and another is Sanger and Coulson. The Maxam and Gilbert method is a chemical degradation of a double stranded DNA to identify the order of A, C, G, and T (Maxam and Gilbert 1977). The Sanger and Coulson developed an enzymetic synthesis of a second strand of DNA, complementary to an existing template popularly known as chain termination sequencing (Sanger and Coulson 1978). Since, Sanger sequencing is highly efficient, involves no hazardous chemicals and low radioactivity, therefore it was adopted in several laboratories and whole genome sequencing projects. In recent times, fluorescent labels are usually attached to the dNTPs, leading to new horizons for automated sequencing. However, these DNA sequencing method cannot be implemented on RNA because of different chemical composition and more prone to nuclease attack experimentally. Also, eukaryotic RNA molecules are not necessarily co-linear with their DNA template, as introns have to be excised by alternative splicing.

The Direct Chemical Method for RNA Sequencing.

A chemical method in which different base-specific chemical reactions directly sequences 3' end-labelled RNA with ^{32}P. Here purified RNA molecules are labelled at their 3' termini with T4RNA ligase [5'-^{32}P] pCp. The RNA reaction mixture contains purified RNA, [5'-^{32}P] pCp, ATP, and RNA ligase. After a partial, specific modification of each kind of RNA base (Adenine, Cytosine, Guanine and Uracil), an amine-catalyzed strand scission generates labelled fragments whose lengths determine the position of each nucleotide in the sequence. Five different chemical reactions are performed to modify the RNA bases; Guanosine (G) reaction, Adenosine (A) > Guanosine (G) Reaction, Uridine (U) Reaction, Cytidine (C) > Uridine (U) Reaction and Aniline Reaction.

In all reactions, aniline induces a subsequent uniform strand scission of the phosphate backbone along the length of the RNA molecule. This strand scission at the site of a chemical attack generates a relatively even distribution of radioactive labelled fragments (^{32}P). These fragments can be easily resolved according to length by a polyacrylamide gel electrophoresis (PAGE). The guanosine and uridine reactions are each monospecific where dimethyl sulfate modifies guanosine and hydrazine modifies uridine. While, the dispecific adenosine > guanosine reaction is primarily an adenosine reaction where adenosine is modified by diethyl pyrocarbonate. Similarly, the dispecific cytidine > uridine reaction is mainly a cytidine reaction where cytidine is also modified by hydrazine. The electrophoretic fractionation of the labeled fragments on a polyacrylamide gel, followed by autoradiography, determines the RNA sequence with the known sequence of RNA (Miyazaki 1974, Gilbert et al., 1977). Each band on autoradiograph of PAGE gel displays a series of nucleotide fragment of discrete length generated by cleavage at a specific base, and the four lanes correspond to cleavages at the bases as follows: guanosine (G), adenosine (A > G), cytidine (C > U), and uridine (U). Therefore, radioactive nucleotide fragments of RNA labelled at the 3' end yields clean cleavage patterns for each purine and pyrimidine and allows a determination of the entire RNA sequence out to 100-200 bases from the labelled terminus.

This chemical sequencing method does not depend on secondary structures of RNA molecules; otherwise it can hinder the RNA sequencing. We can directly correlate the primary structure of long non-coding RNA (lncRNAs) from the sequencing data. This method is also useful for sequencing the purines of 5' end-labeled RNA molecules. This RNA sequencing allows detection of even very low abundance transcripts in a very precise and accurate manner using only picomole amounts of RNA sample. There are certain clearcut advantages of this technique which complements other RNA sequencing methods; (i) the strand scission reactions does not depends on the anomalies due to sequence and secondary structure; (ii) it does not require primers for adjunct cDNA synthesis, as cDNA synthesis dictates the transcript classes to be captured and represents the material-limiting and most length bias-prone step in the experimental pipeline; and (iii) the RNA substrates for sequence-specific recognition are not limited to those synthesized *in vitro* (Chang et al. 1977, Baralle 1977, Cartwright et al. 1977, Kramer and Mills 1978, Lockard et al. 1978, McGeoch and Turnball 1978, Ross and Brimacombe 1978, Zimmern and Kaesberg 1978). These RNA sequencing techniques are very helpful in identifying the secondary structure of RNA molecules as well. These techniques in conjunction with an enzymatic base-specific attack, we can location and determine the sequence of tight hairpin loops. Because tight secondary structures are generally enzymatic non-hydrolyzed at temperature 90°C or pH >13 (Keller et al. 1977, Simoncsits et al. 1977, Stanley and Vassilenko 1978). Thus, these techniques are useful for determining the structures of RNA molecules in solution as well as RNA-protein complexes (or RNPs).

The first RNA to be sequenced was the alanine transfer RNA (tRNA), which was purified by Holley et al., in 1965. They applied similar sequencing methods with some modifications that were used for protein sequencing: partial hydrolysis with enzymes, and fractionation of the products on ion exchange columns. Since every nucleotide contains a phosphorous atom, so RNA is most suitable candidate for ^{32}P- labelling *in vivo*, therefore Sanger and his colleagues applied this pre-labelling technique for the sequencing of ^{32}P-labeled RNA. It has great advantages both in the simpler fractionation of oligoribonucleotide fragments and sensitive detection by autoradiography (Sanger et al. 1965). Later, several post-labelling techniques for sequencing RNA by high resolution PAGE were developed. These techniques include enzymatic digestion (Keller et al. 1977), chemical degradation (Peattie 1979, Tanaka et al. 1980) and the wandering-spot method (Lockard et al. 1978).

The use of radioisotopes for labelling is one of the most risky, skill intensive and expensive parts of these methods, but still they are highly sensitive. Since fluorescence-labelling for sequencing nucleic acid is least hazardous, therefore it is preferred over radioisotopes-labelling techniques for DNA sequencing (Smith et al. 1986, Prober et al. 1987). However for sequencing RNA, very sensitive and specific fluorescent-labelling technique has not been commercialized yet.

Small RNA/Non-coding RNA Sequencing.

Relative to protein coding RNA, structural RNAs are more difficult to identify because of their reduced alphabet size, lack of open reading frames, and short length. When sequencing small RNA other than the cellular RNA (mRNA), it is first required to select the desired type and size of small RNA/non-coding RNA. For instance, select the target RNA species e.g. 17-23nt for microRNAs, 25-32nt for piRNAs to be sequenced as 17-32 nt transcript/read can encompass both populations. This can be performed with a size exclusion gel, through size selection magnetic beads, or with a commercially developed kit. Once isolated, fluorescent or radioactive linkers are added to the 3' and 5' end of purified RNA. Direct ligation of linkers to RNA molecules and then sequence into the adapters which reveals strand specificity.

Radiolabeling and fluorescent labelling are widely used in DNA and RNA sequencing techniques. Fluorescent dyes conjugated with nucleic acids offer clear cut advantages over a radioactive label, first it can be detected in real time manner with high resolution and second, fluorescent dyes are non-hazardous therefore no need of intensive safety measures for waste disposal. Fluorescent dyes have ability to label different parts of RNA molecules to study the dynamics, structural motifs and motions of RNA molecules. Fluorescent labeling of nucleic acids through enzymatic reactions are generally preferred in which organic fluorophores are chemically introduced into primers or dNTPs. They are then incorporated into nucleic acids either through PCR amplification or using DNA/RNA polymerases or terminal polynucleotide transferase. Apart from this, there are several means of direct incorporation of fluorescent dye into nucleic acids by chemical methods like, fluorescent labeling of RNA with tetramethylrhodamine (TMR) hydrazine and through ethylenediamine attachment. Liu *et al.* proposed RNA sequencing by fluorescence-labelling and sequenced oligoribonucleotide fragments by partial digestion (Liu et al. 1980). They used sodium periodate to oxidize the 3' terminus of RNA into dialdehyde, and then the fluorescent dye (fluorescein-5-thiosemicarbazide) was added to label the 3' terminus of RNA through the condensation reaction between carbazide and aldehyde. The sequence of the terminally labeled RNA was partial digested with very base specific ribonucleases enzymes viz., RNase T1 (Gp N), RNase U2 (Ap N), RNase *B.cerus* (Up N and Cp N) or RNase *Phy*I (Gp N, Ap N and Up N). Then these labeled RNA fragments were separated on 15% polyacrylamide gel (containing 15% N-dimethylformamide for short RNA fragments) electrophoresis (PAGE). Finally PAGE gel was observed under ultraviolet (UV) light to detect the clear fluorescent bands on the gel to sequence the 3' labelled RNA fragments. Therefore, in spite of the traditional radioisotope labeling method, these fluorescence-labeling and fluorescent photograph techniques avoids the hazards of radioactivity; also it is convenient and safe to sequencing big as well as small RNAs.

RNA/Transcriptome Sequencing Technologies.

The availability of DNA sequencing methods and technologies has rapidly produce vast amounts of sequence information this triggered a paradigm shift in genomics and transcriptomics. The diversity of applications of these technologies has opened new doors to identify the sequence of smallest to the biggest DNA and RNA of any type, any form and any nature. There are shotgun methods for detection of alternative splicing, full-length RNA sequencing for the determination of complete transcript structures, and targeted methods for studying the process of transcription and translation.

For mRNA sequencing utilize full read length for alignment to compare with the known genome map. Reads map to individual transcript components to identify exons, UTRs and non-coding RNAs. Also ascertain alternative splice variation as well as gene expression to refine existing annotation of genomic components. These include genome re-sequencing and polymorphism discovery, mutation mapping, DNA methylation, histone modifications, transcriptome sequencing, gene discovery, alternative splicing identification, small RNA profiling, DNA-protein and possibly even protein-protein interactions. Technological advances in the sequencing field support in-depth characterization of the transcriptome. There are now many genome-wide RNA sequencing methods used to investigate specific aspects of gene expression and its regulation, from transcription to RNA processing and translation. Tag-based sequencing approaches, Like Serial Analysis of Gene Expression (SAGE), Cap Analysis of Gene Expression (CAGE) and Massively parallel signature sequence (MPSS) are applied for studying transcription, alternative initiation and polyadenylation events. These SAGE, CAGE and MPSS all use Sanger method of sequencing.

Isolate the total RNA samples that are treated by DNase with standardized protocol. Avoid protein contamination during RNA isolation. For plant samples, total RNA≥20 µg; for animal samples (human and mouse), total RNA≥5 µg; for other animal samples, total RNA≥10 µg. For all samples, concentration should be ≥ 200 ng/µl. Complicated library of all different types of RNA are constructed having short-insert (up to 200 nucleotides) as well as long-insert (more than 1000 nucleotides) library. Following sequencing strategies are opted for sequencing of RNA and RNA transcriptomes having different types of RNAs. For mRNA sequencing, construct simpler library not limited to 17nt reads.

Currently, there are three widely deployed deep sequencing platforms in hundreds of research labs and in some core facilities worldwide, the Solexa Sequencing, 91 or 101 Paired-End Illumina sequencing, and Applied Biosystems SOLiD sequencing. Each instrument essentially massively parallelizes individual reactions, sequencing hundreds of thousands to hundreds of millions of distinct, relatively short (50 to 400 bases) RNA sequences in a single run. In 1987, the first automatic sequencing machine (namely AB370) was introduced by Applied Biosystems (AB) which caused the paradigm shift in sequencing. Subsequently in year 1998, the automated sequencers based on Sanger sequencing technology and associated software became the driven force of human genome project (HGP) (Collins et al. 2003). In 2005, since vast amount of sequencing data was generated by HGP, for assembling and solving this puzzle 454 pyrosequencing was launched by 454 Pyrosequencing. Immediately in 2006, Solexa released Genome Analyzer, followed by Sequencing through Oligo Ligation Detection (SOLiD) introduced by Agencourt. These founder companies were then purchased by other companies: Agencourt was purchased by Applied Biosystems, and 454 was purchased by Roche, while Solexa was purchased by Illumina. These are three most massively parallel sequencing systems in the next generation sequencing (NGS), but all are having some advantages and some limitations. For example, Illumina GA/HiSeq System is low throughput and fast turnaround in nature, while Roche 454 system is even lower throughput in nature. AB SOLiD system has a low error rate. Therefore, one NGS system is very high performance in terms of read length, accuracy, throughput and infrastructures, another NGS system has some other features.

When the whole transcriptome is sequenced by these three technologies: Roche 454 System, AB SOLiD system and Illumina GA/HiSeq system, standard bioinformatics procedures are applied to annotate small RNA sequencing libraries and to identify novel RNA transcripts. For re-annotating the sequenced transcriptome, first identify expressed transcripts from trace read alignments to the target genome, then screen for well known structural RNAs (e.g. ribosomal RNA, tRNAs, snoRNAs, etc), finally determine which types of small RNA components are present. Now align transcripts to current version of miRbase to identify expressed microRNAs, similarly align transcripts to piRNA database built from recently published candidate piRNA sequences and set remaining unknown transcript population aside, examine for potentially novel RNAs. The transcribed regions fall into several categories viz., correlate well with annotated (coding) gene loci, non-coding RNAs and novel transcripts. To further characterize the novel transcript, RNA secondary structure prediction is performed starting from thousands of candidate sequences. First, select a suitable RNA secondary structure prediction algorithm like, RNAfold (Vienna package), Mfold (Zucker lab). Then look for favorable energy conformations (ΔG), Mean free energy (MFE), mean free energy index etc. Now, visualize the putative secondary structures drawn by RNAfold algorithm in RNAplot (Vienna) and StructureLab (Shapiro lab). Finally, indentify homology across multiple species to manually inspect the highly conserved regions of the putative secondary structures of RNA.

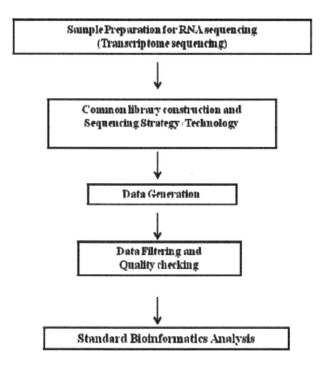

FIGURE 1   Workflow of RNA/ transcriptome sequencing

1.4   Applications of RNA and Transcriptome Sequencing in Modern Research

Since, genome and transcriptome sequencing projects have opened doors or even corridors of research and vast array of advance applications starting from the annotation, quantification and characterization of the novel genome and transcriptome. Although there are several advances made so far in characterizing several eukaryote transcriptomes but still many challenges are associated with its application. RNA sequencing will undoubtedly be valuable for understanding transcriptomic dynamics during development and normal physiological changes, and in the analysis of biomedical samples, where it will allow robust comparison between diseased and normal tissues, as well as the sub-classification of disease states. The RNA sequencing provides dynamic range to quantify gene expression level from a few-hundred fold to more than 8,000-fold. Transcriptomic data insights a researcher an ability to distinguish different isoforms and to distinguish allelic expression. Another clear cut advantage of RNA-Seq compared with other transcriptomics methods is reduction in the cost of mapping transcriptomes of large genomes. Other broad applications are, Annotation of Protein-Coding Gene, Gene Expression Profiling, Transcriptome shotgun sequencing, Noncoding RNA Discovery and Detection, Transcript Rearrangement Discovery, Single-Nucleotide Variation Profiling. We will discuss them below in very concise manner.

Annotation of Protein Coding Gene.

The transcriptome of any organism exclusively eukaryotes, includes protein-coding RNAs and non-protein-coding RNAs, both of which can have many alternative splice variants, transcription start sites, and termination signals. Most protein-coding genes of higher eukaryotes contain one or more intervening complex sequences (introns), which are spliced out during pre-mRNA processing to generate a combination of exons, which are reassembled to form mature mRNA, and subsequently translated into protein. This alternative splicing in higher eukaryotes is responsible for the production of different protein isoforms from the same gene (Croft et al. 2000). The genomic and transcriptomic data of human and other organisms which are generated by so many genomic projects are still not completely analyzed, puzzled out or not fully well understood (Brent 2008). The transcriptome sequences can be aligned to the genome of either the same species (cis-alignment) or a related species (trans alignment) if a reference genome sequence is not available. To date, next-generation sequencing technologies have been used to generate EST libraries for many model organism species and human tissues. Expressed sequence tags (EST) are very useful for annotating protein-coding RNAs for organisms who's genomic and transcriptomic sequences are not available. In such cases, annotations are done by comparative analysis based on homology searching of the derived EST sequences with reference genomes of related species (trans alignment). Since, annotation of protein-coding gene would require availability of information of all transcription start and stop codons, polyadenylation sites, exon-intron junctions, GT-AG donor and acceptor sites, splice variants, and regulatory sequences. If these valuable information of a fully characterized genome are available, then one can easily annotate the protein coding genes of a newly sequenced genome. Sanger-based transcriptome sequencing in the form of ESTs or full length cDNAs (FLcDNAs) is an accurate and effective means for annotating protein-coding genes (Adams et al. 1991, Hillier et al. 1996, Seki et al. 2002).

Gene Expression Profiling.

Large-scale gene expression profiling is the need of hour in present scientific era. It is extensively studied for mapping 3' ends by serial analysis of gene expression (SAGE), mapping 5' ends by cap analysis of gene expression (CAGE) and microarray technologies, which are based on transcriptome sequencing. SAGE and CAGE are more advantageous than microarrays, such as the ability to detect novel transcripts, the ability to obtain direct measures of transcript abundance, and the discovery of novel alternative splice isoforms (Wang 2007). Gene Expression profiling is helpful to snapshot the active and passive components of total transcriptome of a cell, tissue or organism at a given time to distinguish between different cell types, disease states, developmental stages etc.

Transcriptome Shotgun Sequencing.

Transcriptome sequences, produced by next generation technologies, achieve sufficient sequencing depth to provide an adequate representation of the cellular transcriptome. With the elimination of the cloning step and common use of random priming, next-generation EST sequencing data became indistinguishable from those generated by transcriptome shotgun sequencing. In this approach, mRNA is reverse transcribed into cDNA, which is then fragmented and sequenced using a next-generation technology to generate reads covering the full length of a transcript.

Non-coding RNA Discovery and Detection.

By computational algorithms, we can identify the small noncoding RNAs (ncRNAs) from the transcriptome data. These ncRNAs have recently identified as crucial regulators of vast array of cellular functions starting from development and cell fate determination. These 18-30 nucleotide long RNA molecules such as siRNA and miRNA are transcribed from genomic DNA, especially intronic in origin but never non-protein coding. These RNAs serve as posttranscriptional regulators of gene expression by translational inhibition, translational cleavage or by both mechanism in several organisms (Filipowicz et al. 2008, Watanabe et al. 2008). Mature miRNAs and siRNAs bind to complementary sequences of exons especially un-translated regions (UTR) of genes and causes inhibition or cleavage of target mRNAs, thereby regulating their translation rates (Filipowicz et al. 2008).

Transcript Rearrangement Discovery.

Genome rearrangements are common features in case of human cancers, which are caused by an aberrant transcriptional events or due to aberrant use of alternative promoters (Hanahan and Weinberg 2000; Pal et al., 2011). Such rearrangements may include translocations, inversions, small insertions/deletions (indels), and copy number variants (CNVs) and may occur in all or a fraction of cancer cells within a tumor. While cytogenetics and microarray-based methods have been developed to identify genome rearrangements, most of them are suitable for the detection of only particular types of rearrangements and have limited resolution. Next-generation sequencing technologies offer important advantages over conventional methods such as microarrays or array comparative genomic hybridization (array CGH) for high-throughput detection of genome aberrations (Morozova and Marra 2008). In particular, high throughput transcriptome sequencing can efficiently identify all types of genome rearrangements such as fusion transcripts derived from chromosomal structure variations (Eichler 2001; Zelent et al. 2004), trans-splicing (Mayer and Floeter-Winter 2005; Horiuchi and Aigaki 2006), transcription-induced chimerism (Akiva et al. 2006; Parra et al. 2006), and pseudogenes (Balakirev and Ayala 2003; Zheng et al.

2005). Since, these aberrant modifications affect coding sequences therefore it can be detect potentially a single nucleotide resolution. Also, transcriptome sequencing is enough sensitive to detect variants that are present in a subpopulation of cells (Thomas et al. 2007). Therefore, analyzing aberrant genome rearrangements events are particularly important while studying tumor genome pathology. These genome rearrangements can be studied by transcript End Sequence Profiling (tESP) techniques (Al-Hajj 2007; Volik et al. 2006).

Single Nucleotide Polymorphism Profiling.

Human genome sequencing has revealed an abundance of single nucleotide variation/polymorphism (>1.42 million) distributed throughout an individual human genome, due to which all the races and all the faces exists (Wang et al. 2008, Wheeler et al. 2008). These single nucleotide polymorphism (SNP) may occur in the germline or somatic cells, such as those that comprise human tumors genome (Greenman et al. 2007, Wood et al. 2007). Although all types of genetic polymorphisms can be identified by resequencing of whole genomes, or via transcriptome sequencing studies focusing only to analyze coding parts of the genome. As the cost of sequencing continues to fall, RNA-Seq is expected to replace microarrays for many applications that involve determining the structure and dynamics of the transcriptome. The sequencing redundancy thus contributes to increased costs associated with finding rare sequence variants and the need for subsequent validation work, often using Sanger-based resequencing.

1.5 Concluding Remarks

In summary, we have discussed so many different types of RNAs in a very brief manner to familiarize our reader in one go. Since, the basic composition of every RNA molecule is same then why there is so much diversity in structure and function? Reasonably, we can say that different structural and functional capabilities are engraved in the building blocks of RNA or they are encoded into the corresponding genomic DNA fragments. Therefore, understanding the combinations of these building blocks and the resulting secondary structures are basically dependent on the exact sequencing of RNA. As a rule of thumb, we can say that the function of any macromolecule is dependent on the structure, and its structure is dependent on the sequence. So, if we determine the sequence correctly, then we can correlate or simulate their function as well. Although, there has been significant progress in technologies and sophisticated computational algorithms for elucidating the sequence, structure, and functions of RNAs.

1.6 Future dimensions

We are still in infancy for understanding RNAs structure relative to DNA and protein structures. There are still many non coding RNAs for which the cellular, metabolic, developmental, or physiological roles are not clear. We have to elucidate whether some or many RNAs act by more than one mechanism and whether any RNAs, apart from messenger RNA have intrinsic functions to encode proteins. Since, microarray technology, transcriptome sequencing (RNA-Seq), and high-throughput next-generation (second generation) sequencing technologies are all having their own limitations. There should be multilevel filtering and rigorous experimental and computational parameters to identify and screen different types of RNA from the whole set of the transcriptome. Although there is a serious need in technological enhancement (third or fourth generation sequencing) before interpreting whether an RNA is a ncRNA, mRNA, tRNA, rRNA or solely a byproduct of transcription.

2. The Fundamentals of Gene Expression and Regulation

In this section we will discuss about a gene, then how a gene expresses itself into a protein, and what are the regulatory check points and regulation mechanism to control the over and under expression of a gene? Similarly, how a gene expresses itself with the aid of some specific types of proteins or protein complexes to produce another type of proteins? Can a DNA gene be regulated by an RNA gene or non coding RNAs? How do different cell types in a multicellular eukaryotic organism get signals to which types of proteins they must prepare? All the answers to such fascinating questions lie in the study of gene expression and its regulation.

Since there is no such jacketed definition of a gene, but we can simply say that a gene is an array of exons and introns located in a nucleus or cytoplasm or mitochondria matrix or chloroplast of an organism, and responsible for coding a protein. There are limited ways of gene expression mechanism but unlimited ways of gene control or regulation mechanism depending on the organism from viruses to angiospermic plants to mammals. According to the "central dogma of molecular biology" (Crick 1958, 1970) gene expression means "DNA makes RNA and RNA makes protein", which includes four to five step process like - replication, transcription or reverse transcription, RNA processing, translation and post-translational modifications.

While gene regulation varies from organism to organism at every regulatory process and points. For example, gene regulation can act on replication level, transcription or reverse transcription level, RNA processing level, translation and post-translational modification levels. Studies on these regulations from replication to translation level of gene expression play a pivotal role in many areas of biology. Therefore, gene regulation in insects is totally different from gene regulation in bacteria, gene regulation in mammals is totally different from gene regulation in insects, even the gene regulation in mammals one cell type is totally different from gene regulation in another cell type. In multicellular organisms, different cell types express characteristic sets of transcriptional regulators, as a result of which, specific combinations of regulators are turned on and off. Such developmental and gene regulatory patterns, causes variety of cell types in the mature multicellular organisms. For example, gene in plasma cell's continuously regulates the gene expression responsible for the antibody it synthesizes, while the gene in thyroid cells regulates the gene expression responsible for hormone at the demand of the situation.

2.1. Analysis of Gene expression and regulation in Prokaryotes and Eukaryotes

This section begins to illuminate the readers about, how a well-decorated array of DNA (gene) is transcribed to make mRNA, and how the mRNA is translated into protein. Next, we will discuss about the regulation of genes expression in prokaryotes and eukaryotes. Before discussing the detailed processes of gene expression and regulation in prokaryotes and eukaryotes, first understand the deoxyribonucleic acid (DNA) as a genetic messenger and decoder of RNA and protein. DNA is the only molecule where the fate of all living processes of a cell is sealed, unless and until there is no genetic rather genomic alteration (addition or deletion of nucleotides).

Oswald Theodore Avery in 1944 demonstrated that DNA is the genetic material (Avery et al. 1944). Then, James D. Watson and Francis Crick in 1953 deciphered the double helical stranded structure of DNA, leading to the central dogma of molecular biology (Watson and Crick 1953, Crick 1954, Crick and Watson 1954, Crick 1956). In most of the organisms, genomic DNA (in terms of size, composition and percentage similarity of nucleotides) defines the species and individuals. Therefore, expression and regulation of DNA or gene sequence is the fundamental point of the research to know the structures and functions of cells for decoding the life mysteries. Gene expression is the process of transcription of DNA into mRNA and finally translation of mRNA into protein. Genes encode proteins and proteins dictate cell function. Simply, a gene is a well-decorated array of DNA in terms of arrangements of components like, a 5' upstream promoter, 5' untranslated region, exon with a start codon (AUG), intron with GT or GU and AG dinucleotide, an exon with stop codon, a 3' untranslated region and finally a 3' terminator signal. The upstream promoter site is a region on the 5' side of the DNA strand which indicates that other vital components of a gene is forthcoming. The start codon (mostly AUG), which codes for methionine, serves as the signal for the anticodon loop of tRNA to start translation of a mRNA gene into protein. In most of the eukaryotes, the intron with GT or GU and AG dinucleotides signals for the alternative splicing of introns. The terminal exon with stop codon (UAA, UAG, UGA any one of them) signal for the termination of protein synthesis, as they do not encode any aminoacid.

Therefore, thousands of genes expressed in a particular cell determine what that cell can do. Moreover, each step in the flow of information from DNA to RNA to protein provides the cell with a potential control point for self-regulating its functions by adjusting the amount and type of proteins it manufactures.At any given time, the amount of a particular protein in a cell reflects the balance between that protein's synthetic and degradative biochemical pathways. On the synthetic side of this balance, recall that protein production starts at transcription (DNA to RNA) and continues with translation (RNA to protein). Thus, control of these processes plays a critical role in determining what proteins are present in a cell and in what amounts. In addition, the way in which a cell processes its RNA transcripts and newly made proteins also greatly influences protein levels. Regulation of gene expression at transcription and translation level occurs in both prokaryotes and eukaryotes, but it is far more complex in eukaryotes. Interestingly, in the most complex organism, that means human there is only ~ 2.5% gene (48 Mega base pairs out of 3200 Mega base pairs) is of protein coding in nature and the rest of it is protein non-coding means junk. It means the regulatory checkpoints are constrained only to 2.5% of the whole genome and the rest 97.5% remains unregulated. This statistics remains un-noticed for a long until and unless small non-coding RNAs (miRNAs and siRNAs) were identified.

To survive, cells must be able to respond to changes in their surrounding environment. Genes interact via transcription factors or a regulatory protein with different types of receptors available inside or on the membrane surface and respond to the organism's environment. There are generally two categories of genes-constitutive and non-constitutive (regulated). Constitutive genes are always expressed, that means remains "on" regardless of environmental conditions for example, housekeeping genes. Such genes are the most important element of a genome, and they control the ability of self-replication (formation of new DNA from old one), expression (formation of proteins), and damage control (through DNA damage repair mechanism). Therefore, these genes are responsible for the panoramic view of an organism's central metabolism and protein diversity. In contrast, most of the genes are regulated genes whose expression product (protein) is needed only occasionally. It means, these genes work as an electrical safety switch which is turned "on" and "off" at the demand of the situation. But how do these genes get turned "on" and "off"?

Every eukaryotic cell (except gametes) has the same DNA with the same information, then what is the reason behind the diversity of different cell types? This diversity of cell types is due to the expression of different sets of genes. For instance, an undifferentiated fertilized egg looks and acts quite different from a lungs cell, a neuron, or a liver cell because of differences in the genes each cell expresses. Interestingly, in eukaryotes, the default state of gene expression is "off" rather than "on," as in prokaryotes. The first secret is almost all eukaryotic genes must be shut "off" in order to allow for normal cell function. The second secret lies in chromatin, or the complex of DNA and histone proteins found within the cell's nucleus. These histone proteins remodel the chromatin into two forms, famously known as heterochromatin and euchromatin. The euchromatin is a loose packing state of DNA in chromosome, which permits the replication and transcription of DNA or gene. While, in the heterochromatin stage DNA is very tightly packed into the chromosome, which resists both the replication and transcription of DNA or gene. When a specific gene is tightly bound in heterochromatin with histone, that gene is "off" . Because, histone protein majorly consists of positively charged amino acids lysine and arginine, causing them to interact electrostaticly with the negatively charged phosphate groups of the DNA nucleotides. How do eukaryotic genes manage to escape this silencing? This happens by modifications of the positively charged aminoacids of histone to create some domains in which DNA is more loosely bound and others in which it is very tightly bound. DNA methylation is one mechanism which is coordinated with histone modifications, leading to silencing of gene expression. Sometimes, histone molecules are acetylated at specific locations, causing less interaction with DNA, thereby loosely packing of DNA in euchromatic state. The regulation of the loose packing of DNA or genes in euchomatic state is a hot cake of research for exploring other ways of molecular regulation at replication, transcription and translational level. In the coming section we will discuss about the specific molecules control and their regulatory check and bounds of gene expression at transcriptional and translational level.

Molecule Regulation of Gene Expression.

Since, only some of the genes (constitutive genes) are active throughout the growth and development of an organism and the rest of the genes (non-constitutive or regulated genes) within a cell's genome get activated (turned "on") at the demand of the situation. The gene expression is regulated at various stages, but most famously at: Transcription, Post-transcription, Translation and Post-translation levels.

In prokaryotes, due to non-compartmentalization and compactness of genome, several genes are arrangend in a tandem array, famously known as operon. These "operons" are the unit of bacterial gene expression and regulation, e.g. *lac* operon (Jacob et al. 1960, Jacob and Monod 1961) , *trp* operon (Morse et al. 1969, Squires et al. 1975, Lee and Yanofsky 1977). These operon contains structural gene and regulator elements like promoter, operator and repressor arranged in a very disciplined manner from 5' end to 3' end in a contiguous fashion. In prokaryotes, most regulatory proteins are specific to one gene and one promoter, which is enough for expression of cluster of genes. For instance, some repressors bind near the start of mRNA production for an entire operon, or cluster of co-regulated genes. The regulation of such genes differs between prokaryotes and eukaryotes. In prokaryotes, mostly the presence of regulatory proteins like repressor protein prevents transcription, therefore operon is under negative control and therefore turn off the genes. Here, the prokaryotic cells rely on protein–small molecule binding for example, repressor protein-lactose binding in *lac* operon, repressor protein-tryptophan binding in *trp* operon. These small molecule like, lactose and tryptophan acts as ligands which signals the cell whether gene expression is needed or not. The repressor or activator protein binds near its regulatory target- the promoter region or promoter-operator overlapped region of a gene. Some regulatory proteins must have a ligand attached to them which make compatible to bind, whereas others are non-compatible to bind when attached to a ligand.

In eukaryotes, regulation of gene expression is much more complex than prokaryotes, simply because there is no such existence of a single promoter for a cluster of genes. We can say that a set of eukaryotic genes is not arranged under the control of only one promoter, rather enhancer and silencers are also nearly equally responsible for the regulation of a gene. The complete upstream regulatory region of a gene is split into core promoter [-30 to -40 bases away from the transcription start position (+1)], proximal promoter [-40 to -200 bases away from the transcription start position (+1)], and enhancer sequences [-200 bases to -50 kilo bases away from the transcription start position (+1)]. Also the enhancer sequences are prevalently present on far upstream [about -50 kb away from core promoter in 5' flanking region] and far downstream [about +50 kb away from core promoter in 3' flanking region] of a gene. The promoters and enhancers are collectively known as *cis*-acting elements.

Parallel to such vastly spanned upstream regulatory regions (+1 to -50 kb in 5' flanking region) or regulatory binding sites, eukaryotic genes are also regulated by several transcription factors (TFs) or trans-acting factors. Since, TFs are regulatory proteins which interact with DNA as well as surrounding proteins for activation and the functioning of the whole DNA-protein complex. Therefore, TFs have very unique modular structure one part of which is responsible for DNA binding, another for dimer formation and another for transcriptional activation. Also, the eukaryotic gene expression is usually regulated by a combination of several regulatory proteins or transcription factors (eg. TAFs, TBP, TFIIA, TFIIB, etc.) acting together, in a very coordinated manner.

2.2 Transcriptional and Post-transcriptional Gene Regulation

Gene expression is generally known as the formation of protein from the DNA gene through an intermediate molecule, messenger RNA (mRNA). Therefore, eukaryotic gene expression is combination of two parallel processes active at different locations- transcription in the nucleus and translation in the cytoplasm. Regulation of these two main steps of protein production, transcription and translation is critical to this adaptability. Cells can control which genes get transcribed and which transcripts get translated. Further, they can biochemically process transcripts and proteins in order to affect their activity. Since, thousands of mRNA transcripts are produced every second in every cell. Therefore, the amounts and isoforms of mRNA molecule in a cell decides the function of that cell. In fact, the primary control point for gene expression is usually at the very beginning of transcription and this type of regulation continues on every step of post-transcription, then on the beginning of translational and finally continues to post-translational modifications (more or less). RNA transcription makes an efficient control point because many proteins can be made from a single precursor mRNA molecule after RNA processing. Transcript processing by the means of alternative splicing provides an additional level of regulation for eukaryotes, which is a type of post-transcriptional regulation. These precursor RNA transcripts are processed in the nucleus before they are exported to the cytoplasm in the mature form for translation.

Transcriptional Regulation in Prokaryotes.

In prokaryotes, the nutrient availability allows bacteria to rapidly adjust their transcription patterns (switching "on" and "off" of genes) in response to environmental conditions. Similarly, DNA binding protein or regulatory proteins (repressors, co-repressors and activators) are also under control of nutrient availability. In addition, regulatory sites on prokaryotic DNA are typically located close to a promoter site which plays an important role in gene transcription. Repressor protein that prevent transcription, binds to an operator sequence upstream of the gene. It regulates the operon by negative control mechanism. Activator protein that promotes transcription, binds to promoter site or enhancer sites of the operon. It regulates the operon by positive control mechanism. The promoters sequences found at the 5' upstream region of a gene, which signals to regulate initiation of transcription. Most common promoter sequences in prokaryotes are TATAAT or Pribnow Box which lies 10-base pair (-10) upstream of start of transcription initiation and TTGACA–35-bp (-35) upstream of start of transcription (start codon). On the contrary, in eukaryotes the most prevalent promoter sequences are TATA Box (TATAAA) also known as Hogness-Goldberg box which lies 19-27 base pair upstream of start of transcription. Another consensus promoter sequence is CAAT Box (CAAT) upstream of TATA Box in either orientation and similarly GC Box (GGGCGG) in either orientation.

The lac Operon and its Regulation

The *lac* operon is responsible for the regulation of normal lactose (*lac*) metabolism in *E. coli* bacteria which resides in the mammalian gut. The *lac* operon of *E. coli* has four major components, (i) a common promoter (*lacP*)–site to which RNA polymerase binds (ii) operator (*lacO*)–site to which repressor protein binds (iii) three structural genes—(a) *Lac Z*-codes for beta-galactosidase, (b) *Lac Y*-codes for permease and (c) *Lac A*-codes for thiogalactosidase transacetylase and (iv) a regulator gene (*lacI*)–codes for repressor protein. A repressor protein consists of four identical polypeptide chains having two binding sites; one binds to allolactose and another binds to DNA.

First, enzyme permease actively transports extracellular lactose into the cell, which cannot be easily diffused across the *E. coli* cell membrane. Finally, enzyme beta-galactosidase breaks intracellular lactose into glucose and galactose to utilize the reserve energy of lactose. This enzyme also converts lactose into a related compound allolactose and similarly converts allactose into glucose and galactose. Allolactose acts as a key player (as an inducer) in regulating lactose metabolism. Although the function of the thiogalactoside transacetylase is still not known, it is hypothesized that this enzyme might involve in detoxification and in lactose utilization. Since, the structure and enzymatic activity of this enzyme have been found conserved in many bacterial species.

Regulation of lac Operon in the Presence of Lactose

In the presence of lactose, allolactose is formed, which binds to one of the active site of repressor protein. This binding changes the conformation of another binding site of repressor protein which binds to DNA, therefore inactivates the repressor protein to bind the DNA. Due to this phenomenon, the *lac* operator (*lacO*) site remains unoccupied so RNA polymerase binds to the promoter site (*lacP*), the transcription of all structural genes (*lacZ*, *lacY* and *lacA*) takes place back to back. Therefore, these three enzymes responsible for lactose metabolism are synthesized simultaneously, by a specific molecule allolactose known as an inducer. The *lac* operon is therefore under inducible control, it means – the presence of lactose / allolactose induces the transcription of the structural genes (turned on).

Regulation of lac Operon in the Absence of Lactose

In the absence of lactose, there is no formation of allolactose, the active repressor protein transcribed from *lacI*, binds to the lac operator (*lacO*). Since, *lac* operator partially overlaps the 3'end of the *lac* promoter (*lacP*) and 5' end of first structural gene *lacZ*. Therefore, binding of repressor protein to the operator site, prevents the binding of RNA polymerase at operator site and transcription of the structural genes (*lacZ*, *lacY* and *lacA*) are prevented. The *lac* operon is therefore under negative control because of the presence of active repressor protein that prevents transcription of the structural genes (turned off).

The trp Operon and its Regulation

The tryptophan (*trp*) operon in *E. coli*, controls the biosynthesis of the amino acid tryptophan, which has a bulky aromatic side chain and an indole group. The *trp* operon is regulated at a transcriptional level by two mechanisms—one by co-repressor and another by attenuation process. This operon is under negative repressible control, where transcription of structural genes are normally "on" and must be repressed ("off") by the binding of tryptophan to the repressor protein. The *trp* operon consists of (i) a common promoter (*trpP*)–site to which RNA polymerase binds (ii) an operator (*trpO*)–site to which repressor protein binds, (iii) a 5' untranslated region (UTR) leader sequence (162 nucleotide long) containing four regions (region 1, region 2, region 3 and region 4) (iv) five structural genes- (a) *trpE*, (b) *trpD*, (c) *trpC*, (d) *trpB* and (e) *trpA*, (v) a regulatory gene (*trpR*) –codes for repressor protein, and (vi) attenuator (A)–regulates mRNA transcription. Similar to the lac repressor, the *trp* repressor protein has also two binding sites, one that binds to DNA at operator site and another that binds to the tryptophan (as an activator).

Regulation of trp Operon in the Absence or Low Level of Tryptophan

When cellular level of tryptophan is in low level or absent, transcription of all the five structural genes forms five enzymes which synthesize tryptophan from chorismate. The *trp* operator (*trp O*) partially overlaps the 3'end of the *trp* promoter (*trp P*) and 5' end of UTR, therefore inactive repressor protein alone cannot bind to the operator site. Since, RNA polymerase easily binds to the *trp* promoter (*trp P*) and begins transcribing the DNA, producing the region1 of 5' UTR leader sequence. Then a lagging ribosome attaches to the 5' end of the 5'UTR and begin to translate region1 while region 2 is already being transcribed by leading RNA polymerase. The ribosome stalls at the Trp codons in region1 because low level of tryptophan. Because the lagging ribosome is stalled, region2 is not covered by the ribosome when region 3 is transcribed by leading RNA polymerase. When region3 is transcribed it spontaneously pairs with the complementary region 2 forming a secondary structure "antiterminator". Finally, the region 4 is transcribed by leading RNA polymerase, it cannot pair with region 3, because region 3 is already paired with region2; therefore the secondary structure "attenuator" never forms and transcription of all the five structural genes continues.

Regulation of trp Operon in the Presence of Tryptophan

When cellular level of tryptophan is high, it binds to one of the binding site of repressor protein, and conformationally makes another binding site of repressor protein compatible to bind to the DNA at operator site. Therefore, transcription of all the five structural genes is inhibited and the synthesis of tryptophan from chorismate is inhibited and the synthesis of more tryptophan does not take place. Similarly to the case of low level of trptophan, RNA polymerase easily binds to the *trp* promoter (*trp P*) and begins transcribing the DNA, producing the region1 of 5' UTR leader sequence. Then a lagging ribosome attaches to the 5' end of the 5'UTR and begin to translate region1 while region 2 is already being transcribed by leading RNA polymerase. The ribosome translates region1 while the leading RNA polymerase fastly transcribes region 3. Most importantly, the ribosome does not stall at the Trp codons in region1 because tryptophan is abundantly present. Since, the leading ribosome covers part of region 2 preventing it from complementary pairing with region 3. Finally, the region 4 is transcribed by leading RNA polymerase and pairs with region 3. This pairing of regions 3 and 4 produces a secondary structure "attenuator" that terminates transcription of all the five structural genes. This fine-tuning system of four leader regions 1, 2, 3 and 4 is known as attenuation, which uses complementary mRNA structures to stop both transcription and translation depending on the concentration of an operon's end-product. The trp operon is therefore under negative repressible control, it means –the presence of tryptophan represses the transcription of the structural genes and –negative control because of presence of active repressor protein prevents RNA polymerase that prevents transcription. In eukaryotes, there is no exact equivalent of attenuation, because transcription and translation is temporally and spatially separated, making this sort of coordinated effect impossible.

Transcriptional Regulation in Eukaryotes.

As we have discussed earlier, eukaryotic genes are not arranged under the control of only one promoter rather enhancer sequence, silencer sequence, insulator sequence, and several transcription factors are equally responsible for the expression of a gene. Normally, transcription begins when an RNA polymerase binds to a core promoter sequence on the DNA molecule. This core promoter (TATA box) sequence is almost always located just upstream from the transcriptional start (AUG / +1), though it can be located downstream of the mRNA (3' end).

Genes are also Regulated by Enhancers and Transcription Factors (TFs)

As enhancer sequences are located thousands of bases away either upstream or downstream, bound by an enhancer-binding protein or activator protein, causes the binding of TFs with the promoter of the gene. This activator protein binding causes DNA to loop out, bringing the TFs into physical proximity with RNA polymerase and other proteins in the complex which increases or enhances the initiation of transcription. The promoters and enhancers in combination act as a *cis*-acting elements. Each TF has a specific DNA binding domain that recognizes a 6-10 base-pair motif in the DNA, as well as an effector domain. Transcription factors have commonly three types of DNA binding domains: (a) Helix turn helix motif, (b)Zinc finger motif and (c) leucine zipper motif. Other examples of transcription factors are–steroid hormone receptors which bind to hormone responsive elements (HRE). For an activating TF, the effector domain recruits RNA polymerase II, the eukaryotic mRNA-producing polymerase, to begin transcription of the corresponding gene. Some activating TFs even turn on multiple genes at once. All TFs bind at the promoters just upstream of eukaryotic genes, similar to bacterial regulatory proteins.

Genes are also Regulated by Silencers and Insulators

Silencers are short control regions of DNA that, like enhancers, but may be located nearer (-20bp) or thousands of base pairs farther (-2000bp) in upstream regions from where they control the gene. However, when transcription factors bind to them, expression of the gene they control is repressed by inhibiting transcription initiation. Silencers sequence can repress promoter activity in any orientation and at any position.

Post-transcriptional gene Regulation in Prokaryotes and Eukaryotes.

Post transcriptional gene regulation is not limited only to mRNA, rather rRNA, tRNA and hnRNA (heterologous nuclear RNA) are also processed and modified for further specific functional activities and stability. Both the prokaryotic and eukaryotic mature ribosomal RNAs (rRNA) are originated and processed from longer precursors called preribosomal RNAs (pre-rRNAs) synthesized by Polymerase I. In prokaryotes, especially bacterial, mature 16S rRNAs, 23S rRNAs, 5S rRNAs and some tRNAs, are originated from a single 30S rRNA precursor sequence by base specific methylation and enzymetic cleavage carried out by the enzymes RNase III, RNase P, and RNase E. Similarly, in eukaryotes, mature 18S rRNAs, 28S rRNAs, and 5.8S rRNAs are originated from a single 45S pre-rRNA transcript by excessive methylation and a series of enzymatic cleavages processed in the nucleolus. Most eukaryotic cytoplasm have at least a pool of the 20 distinct mature tRNAs charged with 20 different amino acids, for example - tRNA^{Tyr}, - tRNA^{Ala} ,- tRNA^{Met}. Some abundantly present amino acids have multiple copies of the respective tRNA genes. These mature tRNAs are derived from longer RNA precursors by base modification and enzymatic removal of nucleotides from the 5' and 3' ends and addition of CCA to the 3' end. Therefore, any precursor RNA (pre-mRNA, pre-rRNA, pre-tRNA) which are the immediate product of transcription are modified and edited before they are translated. The most important RNA processing and modification steps are involved in the pre-mRNAs found in nucleus of eukaryotes, which are edited before they are translated into the cytoplasm.

Post-transcriptional Regulation of mRNA by 5' Capping, PolyAdenylation, RNA splicing and RNA Editing

The 5'-capping is a unique feature of eukaryotic mRNA processing in which a residue of 7-methylguanosine (7mGppp cap) is linked to the 5'-terminal nucleotide through an unusual 5', 5'-triphosphate linkage providing protection and stability to 5' end. The synthesis of 7mGppp cap is carried out by enzymes tethered to the CTD of Pol II. The cap remains tethered to the CTD through an association with the cap-binding complex (CBC). Similarly, the 3' end of primary transcript or precursor mRNA is cleaved, and 80-250 adenine (A) residues are added to create a poly-A tail. This 3' polyadenylation protects the 3' end from enzymetic destruction due to exonucleases. But there is a just opposite function of 3' polyadenylation in many prokaryotic mRNAs where, this poly-A tails stimulate decay of mRNA rather than protecting it from degradation. The third most important modification and processing of primary mRNA transcript is splicing, by which introns are removed and exons are joined to form a continuous sequence that specifies a functional polypeptide. There are many splicing methods performed by specific processing machinery like – by spliceosomes, by ribozymes, and by ribonucleasesand ligases. In eukaryotic system, there is a complex pattern of exons and intron boundaries, as well as different categories of introns (type-I, type-II and type-III), so alternative splicing acts which is a –alternative ways of splicing of introns, producing different mRNAs transcript variants or isoforms. Another uncommon mode of RNA processing is RNA editing, where large regions of mRNA are synthesized without any uridylate, and the U residues are inserted later by RNA editing, occurring in a parasitic protozoa *Trypanosoma brucei*. Here, RNA editing occurs is mitochondrial cytochrome oxidase subunit II gene (5'-----AAA GTA GAG AAC CTG GTA--------3'). Where, insertion of four U residues produces a revised/edited mRNA (5'-----AAA GUA GAU UGU AUA CCU GGU-----3'). This edition is performed by a special class of guide RNAs, complementary to the edited mRNA product, which act as templates for the editing process. This editing produces a completely new protein -----Lys Val Asp Cys Ile Pro Gly---- instead of -----Lys Val Glu Asn Leu Val-----. Therefore, these alternative splicing patterns and RNA editing, determines the proteins and their amounts to be produced from an mRNA transcript.

2.3   Translational and Post-translational Gene Regulation

Translational regulation plays a much more prominent role in eukaryotes than in prokaryotes, due to the cellular compartmentalization and complexity of metabolic activities. It means, the transcription and translation processes in prokaryotes are coupled, which provides vital proteins and enzymes for the survival of bacteria at the time of necessity. Since the transcription in eukaryotes occurs in the nucleus, while on the other hand, most mRNAs are translated in the cytoplasm. This spatial distribution of two vital processes of gene expression also imposes a temporal barrier which imposes a significant delay in the demand and supply of a protein. These barriers are further enhanced by the necessity of alternative splicing and RNA editing of intervening sequences from most eukaryotic mRNA transcripts before they can be translated. As a result, the primary mRNA transcripts which generally encode only single protein, is processed and translocated across the nuclear membrane prior to translation and therefore have relatively long half-lives. Therefore, translational regulation may play an important role in expressing protein product of certain very long eukaryotic genes, for which transcription and mRNA processing can require many hours.

Therefore, translational regulations are achieved by altering the half-life or stability of the mRNA for translation -e.g. removal of poly-A binding protein (PABP) reduces the half-life of mature mRNAs. It is also achieved by controlling the initiation and rate of translation -e.g. iron-responsive element (IRE) of the 5' UTR of human ferritin gene. There are some major points of differences in prokaryotic and eukaryotic translational initiation. In Prokaryotes, translation begins at an AUG codon (Methionine) by pairing of a special "initiator" tRNA charged with formyl-methionine fMet-tRNA$^{Met}_i$. Initiation of protein synthesis in prokaryotes involves formation of a complex between the 30S ribosomal subunit, mRNA, GTP, fMet-tRNA^{fMet}, three initiation factors (IF-1, IF-2 and IF-3), and the 50S subunit. Similarly, Shine-Dalgarno sequence (5'—AGGAGGU—3' 10 bases upstream of AUG codon) in prokaryotes is the site of action where 3' end of 16S rRNA of the 30S ribosomal subunit binds. While, in eukaryote 7-methylguanosine cap (7mGppp) at 5' end of mature mRNA is the site of action where 18S rRNA of 40S ribosomal subunit binds. Here Initiation of protein synthesis involves formation of a complex between the 40S ribosomal subunit, mRNA, GTP, Met-tRNA^{Met}i, nine initiation factors (eIF2, eIF2B, eIF3, eIF4A, eIF4B, eIF4E, eIF4G, eIF5 and eIF6), and the 60S subunit.

Post-translational Gene Regulation in Prokaryotes and Eukaryotes.

At the end of the translation, the nascent polypeptide chain is folded and processed, and then they are modified, packed and transported to their designated place as active protein for biological activity. During its synthesis, the polypeptide chain progressively assumes energetically and structurally most stable state known as "native conformation". This native conformation is attained due to the formation of appropriate hydrogen bonds, van der Waals, ionic, and hydrophobic interactions among different amino acids and its secondary folds. Correct folding is not always energetically favorable in the cytoplasm, therefore molecular chaperones (including GroE chaperonins) bind to nascent peptides and facilitate correct folding. Some newly made proteins, both prokaryotic and eukaryotic, do not attain their final biologically active conformation and specific functionality until they have been altered by one or more processing reactions called posttranslational modifications (PTMs). The most common modifications are acetylation, phosphorylation, glycosylation, Ubiquitination and enzymetic cleavage. PTMs play a key role in many cellular processes such as cellular differentiation (Grotenbreg and Ploegh 2007), protein degradation (Geiss-Friedlander and Melchior 2007), signaling and regulatory processes (Morrison et al. 2002), regulation of gene expression, and protein-protein interactions. Protein phosphorylation is the most commonly studied area of post-translational modification since it plays a vital role in intracellular signal transduction and is involved in regulating cell cycle progression, differentiation, transformation, development, peptide hormone response, and adaptation (Hubbard and Cohen 1993, Pawson and Scott 1997, Hunter 2000, Cohen 2002). It has been estimated that one third of mammalian proteins may be phosphorylated and this modification often plays a key role in modulating protein function.

The formyl group on the initiating Met residue of poly-peptides that are synthesized in prokaryotes is almost always removed by a deformylase enzyme. Only rarely is N-formyl Met found at the N-terminus of a mature protein. In about half the proteins of both prokaryotic and eukaryotic cells, the initiating Met residue is removed from the nascent chain by a ribosome-associated Met-aminopeptidase. Whether it is removed depends primarily on the second amino acid residue. Small residues (Gly, Ala, Ser, Cys, and Thr) favor removal of the Met residue in prokaryotes; large, hydrophobic, and charged residues seem to prevent removal.

Phosphorylation

Phosphorylation is the addition of a phosphate group (PO_4 or PO_3^{-2}) to certain amino acids (serine, tyrosine or threonine residue) in a peptide chain reversibly. Other amino acid residues in prokaryotes like - Asp, His, and Lys residues may also be phosphorylated. The phosphoryl groups (PO_3^{-2}) are added by specific protein kinases, using ATP as the phosphoryl donor:

$$\text{protein} + \text{ATP} \longrightarrow \text{protein}—PO_3^{-2} + \text{ADP}$$

The phosphoryl groups are removed by specific phosphatases:

$$\text{protein}—PO_3^{-2} + \text{H2O} \longrightarrow \text{Protein} + HPO_3^{-2}$$

These two reactions are catalyzed by kinases and phosphatases enzymes, and their activities of the phosphorylated proteins are regulated and are strictly controlled. The sites of phosphorylation are usually the hydroxyl groups of specific Ser, Thr, or Tyr residues. The addition or removal of a phosphate group can alter protein conformation (and therefore function) by locally altering the charge and hydrophobicity where it is added. It plays an important role in regulating many important cellular processes such as cell cycle, growth, apoptosis (programmed cell death) and signal transduction pathways. For example, in signalling, kinase cascades are turned on or off by reversible phosphorylation either by addition or removal of a phosphate group.

N-Acetylation

This process involves the transfer of an acetyl group to nitrogen and it occurs almost in all eukaryotic proteins. It has both reversible and irreversible mechanisms. Methionine aminopeptidase (MAP) is an enzyme responsible for N-terminal acetylation which results in the cleavage of N-terminal methionine before replacing the amino acid with an acetyl group from acetyl-coA by the enzyme N-acetyltransferase. Acetylation helps in protein stability, protection of the N-terminus and the regulation of protein-DNA interactions in the case of histones. N-terminus acetylation can occur whether or not the initiating Met residue is still present. Whether acetylation occurs depends to some extent on the nature of the N-terminal residue. In a survey using mutagenesis of the N-terminus of one particular protein, those forms acetylated had N-terminal Gly, Ala, Ser, and Thr residues. Although acetylation is the main covalent modification made to the amino ends of proteins, a great variety of other modifications have been observed in particular cases, which include addition of formyl, pyruvoyl, fatty acyl, a-keto acyl, glucuronyl, and methyl groups.

Glycosylation

Attachment of carbohydrate group is one of the most prevalent PTM of eukaryotic proteins, especially of secreted and membrane proteins. This addition of a carbohydrate or sugar moiety to proteins ranges from simple monosaccharide modifications of nuclear transcription factors to the complex branched polysaccharide chains of cell surface receptors. The most relevant properties of glycosyl groups attached to proteins are - their variable structures, which permit specificity in their interactions with other molecules; their hydrophilic natures, which keep them in aqueous solution; and their bulk, which markedly affects the surface properties of the protein to which they are attached. There are two types of glycosylation, called N-linked and O-linked depending on the atom of the protein to which the carbohydrate is attached. N-linked glycosylation occurs exclusively on the nitrogen atom of Asparagine side chains, whereas O-linked glycosylation occurs on the oxygen atoms of hydroxyls, particularly those of serine/threonine residues. N-glycosylation occurs co-translationally soon after the Asn residue emerges into the ER. Whereas, O-glycosylation was thought to be confined to proteins that pass through the ER and Golgi apparatus, but recently it has been found to occur in a surprising number of cytoplasmic and nuclear proteins. In this case, N-acetyl glucosamine (GlcNAc) groups are attached to the side chains of Ser and Thr residues, but little is known about the process. Glycosylations are often required for correct peptide folding and can increase protein stability and solubility and protect against degradation.

Ubiquitination and Targeted Protein Degradation

Proteins are in a continual state of flux, being synthesized and degraded. In addition, when proteins become damaged they must be degraded to prevent aberrant activities of the defective proteins and/or other proteins associated with those that have been damaged. The attachment of protein ubiquitin (an 8kDa polypeptide consisting of 76 amino acid residues) linked to an amine group of lysine in target protein via its C-terminal glycine. Poly-ubiquitinated proteins are targeted for destruction which leads to component recycling and the release of ubiquitin. An example of this is in the cell cycle where ubiquitination marks cyclins for destruction at defined time points. After ubiquitination protein becomes target of destruction by a complex structure referred to as the proteosome. In eukayotic cells the proteasome is found in the cytosol and the nucleus and has a large mass such that it has a sedimentation coefficient of $26S$. The $26S$ proteasome comprises a $20S$ barrel-shaped catalytic core as well as $19S$ regulatory complexes at both ends. Degradation of proteins in the proteasome occurs via an ATP-dependent mechanism.

Enzymatic cleavage (Proteolysis)

Most proteins undergo proteolytic cleavage following translation. The simplest form of this is the removal of the initiation methionine. Many proteins are synthesized as inactive precursors that are activated under proper physiological conditions by limited proteolysis. Pancreatic enzymes and enzymes involved in clotting are examples of the latter. Inactive precursor proteins that are activated by removal of polypeptides are termed "proproteins." Breakdown of peptide bond in a protein into smaller functional units which can happen anywhere in a protein, is known as "proteolysis." There are several number of protease enzymes with a varied array of specificity, localization, length of activity and mechanism of peptide bond cleavage. Proteolysis is a thermodynamically favourable and irreversible reaction and is therefore under tight regulatory control. The control mechanisms include regulation by cleavage in either cis or trans and compartmentalization. Degradative proteolysis is important as it removes unassembled protein subunits and misfolded proteins and also maintains protein concentration at homeostatic concentrations. Some proteases are classified based on their site of action like the aminopeptidases which act on amino terminus and carboxipeptidases which act on carboxy terminus of a protein respectively. Others are classified based on the active site group of a protease that are involved in proteolysis. These proteases include; serine proteases, cysteine proteases, aspartic acid proteases and zinc metalloproteases. Proteolysis can also release useful cleavage fragments and remove autoinhibitory domains from proteins. This regulation can prevent fibrous and polymer proteins from assembling in inappropriate locations and keep proteins which could otherwise have damaging effects like enzymes and growth factors inactive until they reach their target location. It is also used to remove features of a protein which are not needed in the mature form, particularly targeting signal sequences and the N-terminal methionine. A complex example of post-translational processing of a preproprotein is the cleavage of prepro-opiomelanocortin (POMC) synthesized in the pituitary. This preproprotein undergoes complex cleavages, the pathway of which differs depending upon the cellular location of POMC synthesis. Another is example of a preproprotein is insulin. Since insulin is secreted from the pancreas it has a prepeptide. Following cleavage of the 24 amino acid signal peptide the protein folds into proinsulin. Proinsulin is further cleaved yielding active insulin which is composed of two peptide chains linked togehter through disulfide bonds.

Lipidation

Lipidation attaches a lipid group, such as a fatty acid, covalently to a protein. In general, lipidation helps in cellular localization and targeting signals, membrane tethering and as mediator of protein-protein interactions. Important types include palmitoylation which creates a thioester link between long-chain fatty acids and cysteine residues, N-myristorlation of glycine residues which plays a role in membrane targeting and GPI-anchor addition which links a glycosyl-phosphatidylinositol (GPI) to and extracellular protein to mediate its attachment to the plasma membrane.

Acylation

Many proteins are modified at their N-termini following synthesis. In most cases the initiator methionine is hydrolyzed and an acetyl group is added to the new N-terminal amino acid. Acetyl-CoA is the acetyl donor for these reactions. Some proteins have the 14 carbon myristoyl group added to their N-termini. The donor for this modification is myristoyl-CoA. This latter modification allows association of the modified protein with membranes. The catalytic subunit of cyclicAMP-dependent protein kinase (PKA) is myristoylated.

Methylation

Post-translational methylation of proteins occurs on nitrogens and oxygens. The activated methyl donor is S-adenosylmethionine (SAM). The most common methylations are on the ε-amine of lysine residues. Methylation of lysine residues in histones in DNA is an important regulator of chromatin structure and consequently of transcriptional activity. Lysine methylation was originally thought to be a permanent covalent mark, providing long-term signaling, including the histone-dependent mechanism for transcriptional memory. However, recent evidence has shown that lysine methylation, similar to other covalent modifications, can be transient and dynamically regulated by an opposing de-methylation activity. Recent findings indicate that methylation of lysine residues affects gene expression not only at the level of chromatin, but also by modifying transcription factors.

Sulfation

Sulfate modification of proteins occurs at tyrosine residues. As many as 1% of all tyrosine residues present in the eukaryotic proteome are modified by sulfate addition making this the most common tyrosine modification. Tyrosine sulfation is accomplished via the activity of tyrosylprotein sulfotransferases (TPST) which are membrane-associated enzymes of the *trans*-Golgi network. There are two known TPSTs identified as TPST-1 and TPST-2. The universal sulfate donor for these TPST enzymes is 3'-phosphoadenosyl-5'-phosphosulphate (PAPS). Addition of sulfate occurs almost exclusively on secreted and trans-membrane spanning proteins. Since sulfate is added permanently it is necessary for the biological activity and not used as a regulatory modification like that of tyrosine phosphorylation.

2.4 Gene Regulation by small noncoding RNA (especially miRNA and siRNA)

Most eukaryotes also make use of small regulatory noncoding RNAs (ncRNA) like miRNAs and siRNAs to regulate their gene expression. There is a booming research in recently discovered small regulatory RNAs in animals and plants which largely falls into two categories, one is snoRNAs and another is miRNAs/siRNAs. These small ncRNAs are very diverse set of tiny RNAs that does not encode a protein, but contains information for diverse nature of functions during all stages of growth and development. The human genome consists of at least 30% introns (group I, group II, nucler pre-mRNA introns, transfer RNA introns etc.) which may be the major source of regulatory ncRNAs produced in parallel with protein-coding transcripts and others genes transcripts (Mattick 2001, Mattick and Gagen 2001, Mattick 2003). Almost all snoRNAs and a large proportion of miRNAs in animals are encoded in introns of either protein-coding genes or non-protein-coding genes (Mattick and Makunin 2005, Rodriguez et al. 2004, Cai et al. 2004, Baskerville and Bartel 2005, Ying and Lin 2005). Recently, research showed that a number of mammalian miRNAs are derived from tandem repeats, mainly transposons (Smalheiser and Torvik 2005). The presence of methylated guanosine caps at their 5' ends of some human snoRNAs shows that they are independently transcribed (Tycowski et al. 2004).

Generally, snoRNAs are 60-300 nucleotides long ncRNAs which guide the site-specific modification of nucleotides in target RNAs (may be rRNA, mRNA or snRNA) via short regions of base-pairing. There are two major classes, the box C/D snoRNAs which guide 2'-O-ribose-methylation, and the box H/ACA snoRNAs which guide pseudouridylation of target RNAs (Meier 2005, Bachellerie et al. 2002, Henras et al. 2004, Kiss et al. 2004). The most noted regulatory functions of some snoRNAs are tissue-specific and developmental regulation, and imprinting. One such snoRNAs is linked to the aberrant splicing of the serotonin receptor 5-HT(2C)R gene in Prader–Willi syndrome patients (Kishore and Stamm 2006, Cavaille et al. 2000).

miRNAs and siRNAs are short, ~22-30 nucleotides long RNA molecules derived either from hairpin or double-stranded RNA precursors. For example, the enzyme Dicer finds double-stranded regions of RNA and cuts out short pieces that can serve in a regulatory role. Argonaute is another enzyme that is important in regulation of small noncoding RNA–dependent systems. miRNAs suppress translation via nonperfect pairing with target mRNAs, involving a seed pairing in 3' region (second to eighth nucleotide) or similar, to siRNAs cause degradation of target RNAs by the siRNA induced silencing complex (siRISC) in the case of perfect complementarity with the target site. The mature miRNA guides the miRNA induced silencing complex (miRISC) to partially complementary sequences, termed miRNA recognition elements (MREs), in target mRNAs to repress mRNA translation, promote transcript decay or both (Baig and Khan 2013). This phenomenon of repressing mRNA translation or promoting transcript decay or both is known as RNA interference (RNAi). It is estimated that about 33% of human protein coding genes are regulated by miRNAs (Du and Zamore 2005). Over 800 known functional ncRNAs genes have been identified and catalogued in mammals in recent years, excluding tRNAs, rRNAs and snRNAs genes (Pang et al. 2005 and Liu et al. 2005). There are a number of well-characterized antisense transcripts which appear to play a regulatory roles in relation to their sense gene, including those opposite FGF-2 (fibroblast growth factor-2), HIF-1 (hypoxia inducible factor-1) and myosin heavy chain in mammals (Werner 2005). ncRNAs have also been implicated in many diseases, including various cancers and neurological diseases (Mattick J.S. 2003, Pang et al. 2005).

2.5 Concluding Remarks

In conclusion, the gene expression and regulation at transcriptional, post-transcriptional, translational and post-translational level has brought about great insights into the fields of molecular biology, molecular genetics, genomics, proteomics, and protein functions in relation to gene functions. We have discussed only two most simplified models (*lac* and *trp* operon) of gene expression regulation in prokaryotes, and discussed the eukaryotic gene regulatory system in brief, but in a very effective way to illuminate the readers. In this section, the prokaryotic transcriptional regulatory mechanisms by co-repression, attenuation and feedback inhibition are considered in *lac* and *trp* operon which are under inducible control and negative control. While, eukaryotic transcriptions are regulation by enhancers, transcription factors, silencers and Insulators are discussed. The positive control elements are recognized by activator or enhancer proteins that stimulate transcription. Post-transcriptional regulation of tRNA, rRNA, snRNA and mRNA by 5' capping, polyAdenylation, RNA splicing and RNA editing are highlighted. However, some other facts and features are ignored or simplified. Conclusively, the comparative analysis of transcription, post-transcription, translation and post-translational modifications are sufficient enough to assert that the present knowledge of prokaryotic gene expression is very much simpler but fundamentally different than eukaryotic gene expression. These fundamental differences are essential for eukaryotic organisms to express genes in the incredibly diverse patterns that are necessary for complex biological processes. As far as gene regulation is concerned, it is very much complex in eukaryotes at every level, from the very beginning of chromatin modification up to the post-translational modifications. Most importantly, a eukaryotic promoter can exist in a variety of stable and intermediate states that are transcriptionally inactive. These states correspond to the chromatin modifying activities recruited by particular activators or repressors and hence can be regulated by cell cycle and developmental signals.

Also, eukaryotic gene expression is extremely complex and every regulatory check points are very finely tuned by the assembly of a large number of protein complexes, acting in a very well coordinated manner to control the rate of gene transcription and translation. We can also say that, present research on eukaryotic gene regulation is still in infancy and not modeled at the molecular scale to accurately explain all the details of its components. Here we have discussed only introductory information on transcription, posttranscription, translation, posttranslation, expression and regulation as well as ncRNAs based gene regulation in prokaryotic and eukaryotic system due to the limitations of this book.

2.6 Future Dimensions

We have tried to highlight fundamental points of differences and similarities of gene expression and regulation in eukaryotes and prokaryotes at every level. These descriptions want to illustrate the versatility and economy of all types of regulatory check and bounds at every level, whether by attenuation mechanisms in prokaryotes or by insulators in eukaryotes. A variety of biomolecules, cell organelles, processes, and events provide the solid base for decisions whether or not to terminate transcription ahead of the structural genes of operons. Similarly, DNA-protein interactions, protein-protein interactions, and other molecular interactions, are responsible for the decision of continuation or termination of translation for synthesizing the functionally active protein or truncated proteins in eukaryotes. We should have enough courage to explore other dimensions of eukaryotic gene expression and regulation by prompting the interdisciplinary researchers to correlate these biological information to the mathematical and computational resources to model their processivity and dynamism. Future, an interactive cooperation between experimental data and computational theory should be established to obtain best parameters to test the dynamic response of these models under different circumstances, and to improve the model formulation. The human health care system can be improved if the determination and quantification of posttranscriptionl modified proteins and misfolded proteins are identified on right time. Therefore, molecular detection of diseases due to protein misfolding, heart ailments, cancer, neurodegenerative disorders and diabetes is possible before they become invasive.

References

Adams, M.D., J.M. Kelley, J.D. Gocayne, M. Dubnick, M.H. Polymeropoulos. et al. 1991. Complementary DNA sequencing: Expressed Sequence Tags and Human Genome Project. *Science.* 252:1651–56.

Adams, M.D., S.E. Celniker, R.A. Holt, CA. Evans, J.D. Gocayne, P.G. Amanatides, S.E. Scherer, P.W. Li, R.A. Hoskins, R.F. Galle. et al. 2000. The Genome Sequence of *Drosophila melanogaster. Science.* 287 (5461): 2185-95.

Akiva, P., A. Toporik, S. Edelheit, Y. Peretz, A. Diber, R. Shemesh, A. Novik, and R. Sorek., 2006. Transcription-mediated gene fusion in the human genome. Genome Res. 16:30–36.

Al-Hajj, M. 2007. Cancer Stem Cells and Oncology Therapeutics. Curr. Opin. Oncol. 19: 61–64.

Allen, E., Z. Xie, A.M. Gustafson, and J.C. Carrington, 2005. MicroRNA-Directed Phasing During Trans-Acing Sirna Biogenesis in Plants. Cell. 121: 207-21.

Aravin, A.A., R. Sachidanandam, A. Girard, K. Fejes-Toth, and G.J. Hannon. 2007. Developmentally regulated piRNA clusters implicate MILI in transposon control. Science. 316: 744-47.

Avery, O., T. MacLeod, C. M. and M. McCarty, 1944. Studies on the Chemical Nature of the Substance Inducing Transformation of Pneumococcal Types. Induction of Transformation by a Deoxyribonucleic Acid Fraction Isolated from Pneumococcus Type III. J. Exp. Med. 79:137.

Bachellerie, J., P. Cavaille, J. and A. Huttenhofer, 2002. The expanding snoRNA world. Biochimie. 84: 775–90.

Balakirev, E.S., and F.J. Ayala. 2003. Pseudogenes: Are they "junk" or functional DNA? Annu. Rev. Genet. 37:123–51.

Baralle, F. E. 1977. Complete nucleotide sequence of the 5' noncoding region of rabbit B-globin mRNA, Cell. 10: 549-58.

Baskerville, S. and D.P. Bartel, 2005. Microarray profiling of microRNAs reveals frequent coexpression with neighboring miRNAs and host genes. RNA. 11: 241–47.

Brent, M.R. 2008. Steady Progress and Recent Breakthroughs in the Accuracy of Automated Genome Annotation. Nat. Rev. Genet. 9: 62–73

Brownlee, G.G. and E.M. Cartwright, 1977. Rapid gel sequencing of RNA by primed synthesis with reverse transcriptase. J. Mol. Biol. 114: 93-117.

Cai, X., C.H. Hagedorn, and B.R. Cullen, 2004. Human microRNAs are processed from capped, polyadenylated transcripts that can also function as mRNAs. RNA. 10: 1957–1966.

Carrington, J.C., and V. Ambros. 2003. Role of microRNAs in Plant and Animal Development. Science. 301: 336-38.

Cavaille, J. Buiting, K. Kiefmann, M. Lalande, M. Brannan, C., I. Horsthemke, B. Bachellerie, J.P. Brosius, J. and A. Huttenhofer, 2000. Identification of brain-specific and imprinted small nucleolar RNA genes exhibiting an unusual genomic organization. Proc. Natl Acad. Sci. USA. 97: 14311–14316.

Cavaille, J., Seitz, H. Paulsen, M. Ferguson-Smith, A.C. and J.P. Bachellerie, 2002. Identification of tandemly repeated C/D snoRNA genes at the imprinted human 14q32 domain reminiscent of those at the Prader-Willi/Angelman syndrome region. Hum. Mol. Genet. 11: 1527–38.

Cavaille, J. Vitali, P. Basyuk, E. Huttenhofer, A. and J.P. Bachellerie, 2001. A novel brain-specific box C/D small nucleolar RNA processed from tandemly repeated introns of a non-coding RNA gene in rats. J. Biol. Chem. 276, 26374–83.

Chang, J.C., G.F. Temple, R. Poon, K. H. Neumann, and Y.W. Kan. 1977. The nucleotide sequences of the untranslated 5' regions of human alpha- and beta-globin mRNAs. Proc. Nati. Acad. Sci. USA. 74: 5145-49.

Cohen, P.T.W. 2002. Protein phosphatase 1- targeted in many directions. J Cell Sci. 115:241-56.

Collins, F.S., et al. 2003. A Vision for the Future of Genomics Research. Nature. 422(6934): 835-47.

Collins, F.S., M. Morgan, and A. Patrinos. 2003. The Human Genome Project: Lessons from Large-scale Biology. Science. 300 (5617): 286–90.

Crick, Francis. 1954. Structure and Function of DNA. Discovery. 15: 12-17.

Crick, Francis. 1954. The Complementary Structure of DNA. *Proceedings of the National Academy of Sciences of the United States of America.* 40: 756-58.

Crick, Francis. 1956. "The Structure of DNA." The Chemical Basis of Heredity. Johns Hopkins University Press Pp. 532-39.

Crick, Francis. and James D. Watson. 1954. The Complementary Structure of Deoxyribonucleic Acid. Proceedings of the Royal Society of London 223, Series A: 80-96.

Doench, J.G., C.P. Petersen, and P.A. Sharp. 2003. siRNAs Can Function as miRNAs, Genes Dev. 17: 438-442.

Donis-Keller, H., A. Maxam, and W. Gilbert. 1977. Mapping Adenines, Guanines, and Pyrimidines in RNA. Nucleic Acids Res. 4: 2527-38.

Dunham, I., N. Shimizu, B.A. Roe, S. Chissoe, A.R. Hunt, J.E. Collins, R. Bruskiewich, D.M. Beare, M. Clamp, L.J. Smink, R. Ainscough, J.P. Almeida, A. Babbage, C. Bagguley, J. Bailey, K. Barlow, K.N. Bates, O. Beasley, C.P. Bird, S. Blakey, A.M. Bridgeman, D. Buck, J. Burgess, W.D. Burrill, and K.P. O'Brien. 1999. The DNA sequence of human chromosome 22. Nature. 402(6761): 489-95.

Eichler, E.E. 2001. Recent duplication, domain accretion and the dynamic mutation of the human genome. Trends Genet. 17:661–69

Filipowicz, W., S.N. Bhattacharyya, and N. Sonenberg. 2008. Mechanisms of Post-Transcriptional Regulation by microRNAs: Are the Answers in Sight? Nat. Rev. Genet. 9: 102–14

Fleischmann, R.D., M.D. Adams, O. White, R.A. Clayton, E.F. Kirkness, A.R. Kerlavage, C.J. Bult, J.F. Tomb, BA. Dougherty, J.M. Merrick. et al. 1995. Whole-genome random sequencing and assembly of Haemophilus influenzae Rd. Science. 269(5223): 496-512.

Francis H. Crick 1958. Symp. Soc. Exp. Biol., The Biological Replication of Macromolecules. XII, 138.

Francis H. Crick 1970. Central Dogma of Molecular Biology. Nature. Vol 227.

François Jacob and Jacques Monod. 1961. Genetic Regulatory Mechanisms in the Synthesis of Proteins. Journal of Molecular Biology. 3: 318-56.

Geiss-Friedlander, R. and F. Melchior. 2007. Concepts in sumoylation: a decade on. Nat. Rev. Mol Cell Biol. 8: 947-56.

Gerald, F.J. 1989. RNA evolution and the origins of life. Nature. 338: 217-24.

Gilbert, W., A.M. Maxam, R. Tizard, and K.G. Skryabin. 1977. In Eukaryotic Genetics System: ICN-UCLA Symposia on Molecular and Cellular Biology, eds. J. Abelson and G.W. Wilcox. Academic Press: New York, 8: 15-23.

Goffeau, A., B.G. Barrell, H. Bussey, R.W. Davis, B. Dujon, H. Feldmann, F. Galibert, J. D. Hoheisel, C. Jacq, M. Johnston, E.J. Louis, H.W. Mewes, Y. Murakami, P. Philippsen, H. Tettelin, and S.G. Oliver. 1996. Life with 6000 genes. Science. 274(5287): 546, 563-67.

Greenman, C., P. Stephens R. Smith G.L. Dalgliesh, C. Hunter. et al. 2007. Patterns of Somatic Mutation in Human Cancer Genomes. Nature. 446: 153–58.

Grotenbreg, G. and Ploegh, H. 2007. Chemical Biology: Dressed-up Proteins. Nature. 446:993-95.

Han, J., D. Kim, and K.V. Morris. 2007. Promoter-associated RNA is Required for RNA-Directed Transcriptional Gene Silencing in Human Cells. Proc. Natl. Acad. Sci. USA. 104:12422-27.

Hanahan, D., and R.A. Weinberg. 2000. The Hallmarks of Cancer. Cell. 100: 57–70.

Hattori, M., A. Fujiyama, T.D. Taylor, H. Watanabe, T. Yada, H.S. Park, A. Toyoda, K. Ishii, Y. Totoki, D.K. Choi, E. Soeda, M. Ohki, T. Takagi, Y. Sakaki, S. Taudien, K. Blechschmidt, A. Polley, U. Menzel, J. Delabar, K. Kumpf, R. Lehmann, D. Patterson, K. Reichwald, A. Rump, M. Schillhabel, and A. Schudy. 2000. The DNA Sequence of Human Chromosome 21. The Chromosome 21 Mapping and Sequencing Consortium. Nature. 405(6784): 311-19.

Henras, A., K. Dez, C. and Henry, Y. 2004. RNA structure and function in C/D and H/ACA s(no)RNPs. Curr. Opin. Struct. Biol. 14: 335–43.

Hillier, L.D., G. Lennon, M. Becker, M.F. Bonaldo, B. Chiapelli. et al. 1996. Generation and Analysis of 280,000 Human Expressed Sequence Tags. Genome Res. 6:807–28.

Hodnett, J.L. and H. Busch. 1968. Isolation and Characterization of Uridylic Acid-rich 7 S Ribonucleic Acid of Rat Liver Nuclei. J. Biol. Chem. 243 (24): 6334-42.

Holley, R.W., J. Apgar, G.A. Eerett, J.T. Madison, M. Marquisce, S.H. Merrill, J.R. Penswick, and A. Zamir. 1965. Science. 147: 1462–65.

Hoopes, L. 2008. Introduction to The Gene Expression and Regulation Topic Room. Nature Education. 1(1): 160.

Horiuchi, T., and T. Aigaki, 2006. Alternative trans-splicing: A novel mode of pre-mRNA processing. Biol. Cell. 98:135–40.

Houwing, S., L.M. Kamminga, E. Berezikov, D. Cronembold, A. Girard, H. Vanden-Elst, D.V. Filippov, H. Blaser, E. Raz, C.B. Moens, R.H. Plasterk, G.J. Hannon, B.W. Draper, and R.F. Ketting. 2007. A Role for Piwi and piRNAs in Germ Cell Maintenance and Transposon Silencing in Zebrafish. Cell. 129: 69-82.

Hubbard, M.J. and P. Cohen. 1993. On target with a new mechanism for the regulation of protein phosphorylation. Trends Biochem. Sci. 18:172-77.

Hunter, T. 2000. Signaling - 2000 and beyond. Cell. 100(1):113-27.

Jacob, F., D. Perrin, C. Sanchez, and J. Monod. 1960. C. R. Seances Acad. Sci. 250: 1727–29.

Jones-Rhoades, M.W. and D.P. Bartel. 2004. Computational identification of plant microR-NAs and their targets, including a stress-induced miRNA. Mol. Cell. 14: 787-99.

Kim, V.N. 2005. MicroRNA biogenesis: Coordinated cropping and dicing. Nat. Rev. Mol. Cell Biol. 6: 376-85.

Kishore, S. and Stamm, S. 2006. The snoRNA HBII-52 regulates alternative splicing of the serotonin receptor 2C. Science. 311: 230–32.

Kiss, A., M. Jady, B., E. Bertrand, E. and T. Kiss, 2004. Human box H/ACA pseudouri-dylation guide RNA machinery. Mol. Cell. Biol. 24: 5797–807.

Kramer, F.R. and D.R. Mills. 1978. RNA sequencing with radioactive chain-terminating ribonucleotides. Proc. Natl. Acad. Sci. USA. 75,5334-38.

Kuchino, Y. and S. Nishimura. 1989. Enzymatic RNA sequencing Methods. Enzymol., 180: 154–63.

Lee, F. and C. Yanofsky 1977. Transcription termination at the trp operon attenuators of *Escherichia coli* and *Salmonella typhimurium:* RNA secondary structure and regula-tion of termination. Proc. Natl. Acad. Sci. USA. 74(10): 4365–69.

Lewis, B.P., C.B. Burge, and D.P. Bartel. 2005. Conserved seed pairing, often flanked by adenosines, indicates that thousands of human genes are microRNA targets. Cell. 120: 15–20.

Liu, W.Y., X.R. Gu, and J.E. Cao. 1980. A New Fluorescent Hydrazide for Sequencing Ribonucleic-Acids, Scientia Sinica. 23: 1296–308.

Lockard, R.E., B. Alzner-Deweerd, J.E. Heckman, J. MacGee, M.W. Tabor, and U.L. RajBhandary. 1978. Nucleic Acids Res. 5: 37-56.

M. Sarwar Baig and Jawaid A. Khan 2013. Identification of Gossypium hirsutum miRNA targets in the genome of Cotton leaf curl Multan virus and betasatellite- Ind. J. Biotech. 12: 336-42.

Mattaj, I.W., D. Tollervey, and B. Seraphin. 1993. Small nuclear RNAs in messenger RNA and ribosomal RNA processing. Federation of American Societies for Experimental Biology Journal. 7: 47-53.

Mattick, J.S. 2001. Non-coding RNAs: The Architects of Eukaryotic Complexity. EMBO Rep. 2, 986–91.

Mattick, J.S. 2003. Challenging the dogma: the hidden layer of non-protein-coding RNAs in complex organisms. Bioessays. 25: 930–39.

Mattick, J.S. and M.J. Gagen. 2001. The evolution of controlled multitasked gene net-works: the role of introns and other non-coding RNAs in the development of com-plex organisms. Mol. Biol. Evol. 18: 1611–30.

Mattick, J.S. and I.V. Makunin. 2005. Small regulatory RNAs in mammals. Hum. Mol. Genet. 14: R121–32.

Mattick, J.S. and I.V. Makunin. 2006. Non-coding RNA. Human Molecular Genetics. 15 (1): R17-R29.

Maxam, A.M. and W. Gilbert. 1977. A New Method for Sequencing DNA, Proc. Natl. Acad. Sci. USA. 74: 560-64.

Mayer, M.G., and L.M. Floeter-Winter. 2005. Pre-mRNA Trans-splicing: From Kinetoplastids to Mammals, An Easy Language for Life Diversity. Mem. Inst. Oswaldo Cruz. 100: 501–13.

McGeoch, D.J. and N.T. Turnball, 1978. Analysis of the 3'-terminal nucleotide sequence of vesicular stomatitis virus N protein mRNA. Nucleic Acids Res. 5 (11): 4007-24.

Meier, U.T. 2005. The many facets of H/ACA ribonucleoproteins. *Chromosoma*, 114: 1–14.

Meinke, D.W., J.M. Cherry, C. Dean, S.D. Rounsley, and M. Koornneef. 1998. *Arabidopsis thaliana:* a model plant for genome analysis. Nature, 282(5389): 662, 679-82.

Michelle, A.C. and J.H. Gregory. 2004, RNase III Enzymes and The Initiation of Gene Silencing. Nature Structural & Molecular Biology. 11: 214 - 18.

Miyazaki, M. 1974. Studies on the Nucleotide Sequence of Pseudouridine-containing 5 S RNA from *Saccharomyces cerevisiae. J. Biochem.* 75 (6): 1407-10.

Morrison, R.S., Kinoshita, Y., Johnson M.D., Uo, T., Ho, J.T., McBee, J.K., Conrads, T.P., and Veenstra, T.D. 2002. Proteomic Analysis in the Neurosciences. Mol Cell Proteomics. 1: 553–560.

Morse, D., E. Mosteller, R., D. and Yanofsky C. 1969. Dynamics of synthesis, translation, and degradation of trp operon messenger RNA in *E. coli.* Cold Spring Harb Symp Quant Biol. 34: 725–40.

Pal, S., R. Gupta, H. Kim, P. Wickramasinghe, V. Baubet, L.C. Showe, N. Dahmane, and R.V. Davulur. 2011. Alternative transcription exceeds alternative splicing in generating the transcriptome diversity of cerebellar development. Genome Res. 21:1260–72.

Parra, G., A. Reymond. N. Dabbouseh, E.T. Dermitzakis, R. Castelo, T.M. Thomson, S.E. Antonarakis, and R. Guigo. 2006. Tandem chimerism as a means to increase protein complexity in the human genome. Genome Res. 16:37–44.

Pawson, T. and Scott, J.D. 1997. Signaling through scaffold, anchoring, and adaptor proteins. Science. 278:2075-80.

Peattie, D.A. 1979. Direct chemical method for sequencing RNA. Proc. Natl. Acad. Sci. USA. 76: 1760–64.

Prober, J.M., G.L. Trainor, R.J. Dam, F.W. Hobbs, C.W. Robertson., R.J. Zagursky, A.J. Cocuzza, M.A. Jensen, and K. Baumeister. 1987. A system for rapid DNA sequencing with fluorescent chain-terminating dideoxynucleotides. Science. 238: 336–41.

Rodriguez, A., Griffiths-Jones, S., Ashurst, J.L. and Bradley, A. 2004. Identification of mammalian microRNA host genes and transcription units. *Genome Res.* 14: 1902–10.

Rogelj, B. and Giese, K.P. 2004. Expression and function of brain specific small RNAs. Rev. *Neurosci.* 15: 185–98.

Ross, A. and R. Brimacombe. 1978. Application of a rapid gel method to the sequencing of fragments of 16S ribosomal RNA from *Escherichia coli. Nucleic Acids Res.* 5: 241- 56.

Ryan J.T. D.K. Craig, S. Cas, and J.S. Mattick. 2009. Evolution, biogenesis and function of promoter-associated RNAs. *Cell Cycle.* 8(15): 2332-38.

Sanger, F. and A.R. Coulson. 1978. The use of thin acrylamide gels for DNA sequencing. *FEBS Lett.* 87: 107-110.

Sanger, F., G.G. Brownlee, and B.G. Barrell, 1965. A Two-dimensional Fractionation Procedure for Radioactive Nucleotides. J. *Mol. Biol.* 13: 373–98.

Seki, M., M. Narusaka, A. Kamiya, J. Ishida, M. Satou. et al. 2002. Functional annotation of a full-length Arabidopsis cDNA collection. *Science.* 296: 141–45.

Simard, M.J. and G. Hutvagner. 2005. RNA silencing. *Science.* 309: 1518.

Simoncsits, A., G.G. Brownlee, R.S. Brown, J.R. Rubin, and H. Guilley. 1977. Nature (London) 269: 833-36.

Smalheiser, N.R. and Torvik, V.I. 2005. Mammalian microRNAs derived from genomic repeats. Trends Genet. 21: 322–26.

Smith, L.M., J.Z. Sanders, R.J. Kaiser, P. Hughes, C. Dodd, C.R. Connell, C. Heiner, S.B.H. Kent, and L.E. Hood. 1986. Fluorescence detection in automated DNA sequence analysis. Nature. 321: 674–679.

Squires C.L., Lee F.D., and Yanofsky C. 1975. Interaction of the trp repressor and RNA polymerase with the trp operon. J Mol Biol. 92(1): 93–111.

Stanley, J. and S. Vassilenko. 1978. A different approach to RNA sequencing, Nature (London). 274: 87- 89.

Tanaka,Y., T.A. Dyer. and G.G. Brownlee. 1980. An improved direct RNA sequence method; its application to Vida faba 5.8S ribosomal RNA. Nucleic Acids Res. 8: 1259–72.

The *C. elegans* Sequencing Consortium. Genome sequence of the nematode *C. elegans*: a platform for investigating biology. *Science*, 282(5396): 2012-18, Dec 1998.

Thomas, R.K., A.C. Baker, R.M. Debiasi, W. Winckler, T. Laframboise. et al. 2007. High-Throughput Oncogene Mutation Profiling in Human Cancer. *Nat. Genet*. 39: 347–51.

Tycowski, K., T. Aab, A. and Steitz, J.A. 2004. Guide RNAs with 5'caps and novel box C/D snoRNA-like domains for modification of snRNAs in metazoa. Curr. Biol. 14: 1985–95.

Velculescu, V.E., L. Zhang, B. Vogelstein, and K.W. Kinzler. 1995. Serial Analysis of Gene Expression. Science. 270: 484–87.

Volpe, T.A., C. Kidner, I.M. Hall, G. Teng, S.I. Grewal. and R.A. Martiensson. 2002, Regulation of Heterochromatic Silencing and Histone H3 lysine-9 methylation by RNAi, *Science*. 297: 1833-37.

Walter, Gilbert. 1987. The Exon Theory of Genes, *Cold Spring Harb Symp Quant Biol.* 52: 901-905.

Wang, J., W. Wang, R. Li, Y. R, Li G. Tian. et al. 2008. The Diploid Genome Sequence of an Asian Individual. Nature. 456: 60–65.

Wang, S.M. 2007. Understanding SAGE data. Trends Genet. 23: 42–50.

Watanabe, T., Y. Totoki, A. Toyoda, M. Kaneda, S. Kuramochi-Miyagawa, et al. 2008. Endogenous siRNAs from naturally formed dsRNAs Regulate Transcripts in Mouse Oocytes. *Nature*. 453: 539–43.

Watson, James D. and Francis Crick. 1953. Molecular Structure of Nucleic Acids: A Structure for Deoxyribose Nucleic Acid. *Nature*. 171: 4356.

Watson, James D. and Francis Crick. 1953. The Structure of DNA. Cold *Spring Harbor Symposia on Quantitative Biology.* 18.

Werner, A. 2005. Natural antisense transcripts. *RNA Biol*. 2: 53–62.

Wheeler, D.A., M. Srinivasan, M. Egholm, Y. Shen, L. Chen, et al. 2008. The Complete Genome of an Individual by Massively Parallel DNA Sequencing. *Nature*. 452: 872–76.

Wood, L.D., D.W. Parsons, S. Jones, J. Lin, T. Sjoblom. et al. 2007. The Genomic Landscapes of Human Breast and Colorectal Cancers. *Science*. 318: 1108–13.

Ying, S.Y. and Lin, S.L. 2005. Intronic microRNAs. Biochem. Biophys. *Res. Commun*. 326: 515–20.

Zelent, A., M. Greaves, and T. Enver. 2004. Role of the TEL-AML1 fusion gene in the molecular pathogenesis of childhood acute lymphoblastic leukaemia. *Oncogene*. 23:4275–4283

Zheng, D., Z. Zhang, P.M. Harrison, J. Karro, N. Carriero, and M. Gerstein. 2005. Integrated pseudogene annotation for human chromosome 22: Evidence for transcription. *J. Mol. Biol*. 349: 27–45.

Zimmern, D. and P. Kaesberg. 1978. 3'-terminal Nucleotide Sequence of Encephalomyocarditis Virus RNA Determined By Reverse Transcriptase and Chain-Terminating Inhibitors. *Proc. Nati. Acad. Sci. USA*. 75: 4257-61.

5

Modern Technologies and Approaches for Decoding Non-Coding RNA-Mediated Biological Networks in Systems Biology and Their Applications

Devyani Samantarrai, Mousumi Sahu, Garima Singh, Jyoti Roy, Chandra Bhushan and Bibekanand Mallick[1]*

Abstract

Cellular specialization and its functionality are determined by organization of diverse types of interacting elements forming biological networks. Of these interacting elements, non-coding RNAs (ncRNAs) have emerged as key regulators of biological networks by modulating gene expression at various levels. The ncRNAs such as microRNAs (miRNAs), piwi-interacting RNAs (piRNAs), long ncRNAs (lncRNAs), small interfering RNAs (siRNAs) and transcription initiation RNAs (tiRNAs) have been observed to regulate cellular development, differentiation and many more biological processes. The deregulation of these ncRNAs that form regulatory circuits by interacting with other groups of ncRNAs and coding transcripts affect biological networks leading to alteration and abnormalities in cellular activity causing diseases, such as cancers. Thus, conceptualizing ncRNA-mediated regulatory networks incorporating inferences from high-throughput data and interactomics is crucial to understand cellular behavior in different conditions. Here, we have provided snapshots of ncRNA-mediated regulatory networks operating at transcriptional and post-transcriptional level and have discussed modern technologies and available resources needed for ncRNA-mediated interactomics and network study. Moreover, different types of network properties such as topology, cluster, module, motif, dynamics and evolution as well as their significance in illustrating and interpreting regulatory pathways have been discussed in this chapter.

1. Introduction

The amazing complexity of living organisms, starting from the simplest bacteria to the most complex humans has always intrigued scientist for generations. Uncovering

RNAi and Functional Genomics Laboratory, Department of Life Science, National Institute of Technology, Rourkela - 769008, Odisha, India
* Corresponding author : vivek.iitian@gmail.com

the reasons behind this complexity of living systems has been one of the primary goals of scientists. The 21st century has witnessed an upsurge in the use of different approaches to study this complexity wherein there has been a marked development in discovery of new technologies. These technologies encompass different high throughout methodologies to unearth information at genomic, transcriptomic, and proteomic level to explore the intricacies of living systems. These developments have made possible to combine systems level and molecular biology based approaches to address the missing link between them, thus giving rise to the interdisciplinary field of science called '*systems biology*' (Diaz-Beltran et al. 2013). Systems biology focuses on the identification and characterization of complex and dynamic interactions between biological molecules such as DNA, RNA, proteins, and metabolic intermediates. These diverse types of interacting molecules form complex molecular networks within a living cell. The precise interaction as well as regulation among these molecules is a requisite for the proper execution of complex biological processes and normal functioning of a biological system. Hence, to understand the mystery of how the biological processes, such as cellular proliferation, differentiation, cell death, etc. occur with such precision at the molecular level requires unraveling the regulatory switches and networks that modulate these processes.

The study of network analysis of biological systems is promising and gaining increasing acceptance as an useful method to decipher underlying mechanisms of various biological processes, pathways and their regulatory modes controlling fate of the cells. The biological information represented as networks classically considers the interacting molecules as nodes and their interactions as edges. These networks provide information about the nodes and how they regulate their interacting partners, thereby organizing them into regulatory maps. These regulatory maps or biological networks can be represented in various ways such as directed graphs, directed acyclic graphs (DAGs), undirected graphs, trees, minimum spanning trees, steiner trees, boolean networks, etc. (Dittrich et al. 2008). The biological networks can include transcription factor and gene regulatory network, protein–protein interactions, protein phosphorylation networks, metabolic interactions (Karp et al. 1996), genetic interaction networks, cell signaling networks, kinase-substrate networks, epistasis interaction networks (Segre et al. 2005), disease gene interaction networks (Goh et al. 2007) and drug interaction networks (Yildirim et al. 2007) etc.

Among the interacting molecules forming various types of networks, non-coding RNAs (ncRNAs) are the most intriguing molecules that are known to regulate gene expression at various levels and have emerged recently as key regulators of biological networks. The role of these ncRNAs, such as microRNAs (miRNAs), piwi-interacting RNAs (piRNAs), long ncRNAs (lncRNAs), long intergenic ncRNAs (lincRNA), small interfering RNAs (siRNAs) and transcription initiation RNAs (tiRNAs) have been extensively studied in regulating cellular development, differentiation, and oncogenesis, etc. (Taft et al. 2010). The study of these regulatory ncRNAs is presently expanding at an unprecedented rate and new exciting developments have emerged over the years. The Encyclopedia of DNA elements consortium (ENCODE) transcriptome project revealed that only ~1.5% - 2% of the human genome codes for proteins, whereas 80% of the genome in eukaryotes is transcribed, thus suggesting a vast number of transcripts are non-protein coding RNAs (ncRNAs) (Wilhelm et al. 2008). With the discovery of a plethora of ncRNAs, the biological complexity of living organisms has been correlated to them and not to protein coding genes as thought earlier (Taft et al. 2007). Further, control of cellular functions may be corroborated with the interactions between ncRNAs and proteins because the majority of chromatin modifying enzymes lack DNA binding capacity. Thus, they use a mediator for binding to DNA which studies indicate to be ncRNAs, thus acting as guide for chromatin modifying enzymes (Knowling and Morris, 2011). The ncRNAs are often involved in intricate regulatory circuits among themselves and other protein-coding transcripts, the best example being the recently reported ceRNAs (competitive endogenous RNA) (Salmena et al. 2011). The ceRNAs regulate other RNA transcripts by competing for common miRNAs, thus forming a large-scale regulatory network across the transcriptome. The co-regulatory interaction as well as ceRNA crosstalk mainly depends on the MREs (miRNA Response Elements) located on the interacting transcripts thus making it crucial to be identified (Salmena et al. 2011). The key challenge lies in deciphering the meticulousness with which they interact among each other and understanding their behaviour in disease systems. Of all the ncRNAs explored till now, miRNAs are the most extensively investigated species with several resources documenting their roles in diseases (Ambros, 2008; Bartel, 2009; Stefani and Slack, 2008; Xiao et al. 2009). Other ncRNAs such as piRNAs, lncRNAs, snoRNAs have also been demonstrated to be associated with diseases (Esteller, 2011; Sana et al. 2012). To investigate ncRNA-mediated complex regulatory networks, systems biology is the appropriate approach which considers that the regulatory networks differ between normal and disease conditions (Hood et al. 2004). Thus, understanding the cellular behavior in different disease conditions require conceptualizing ncRNA-mediated regulatory networks by incorporating inference from high-throughput data and interactomics study.

With the increasing discovery of significance of ncRNAs in different biological processes via involvement in the complex regulatory networks and affecting gene expression, their regulatory mechanism remains the prime focus. The deregulation of these network components affects biological processes leading to alteration and abnormalities in cellular activity. Unraveling the complexity of ncRNA-mediated networks will help in identifying the critical regulators modulating the ncRNAs expression as well as those that are modulated by the ncRNAs. With a deeper understanding of the ncRNAs cross-talks in the pathogenesis of a disease, it may be used as a target for therapeutic intervention for the disease. Moreover, the advances in understanding of the complexity of ncRNA-mediated regulatory networks may transform the current diagnostic and therapeutic approaches towards development of personalized medicines.

In this chapter, we have focused on the ncRNA-mediated regulatory networks operating at transcriptional and post-transcriptional level. We have discussed the various approaches for ncRNA-mediated regulatory network generation and visualization. This includes interactomics study using modern high-throughput technologies and methods, such as ChIP-seq, CLIP-seq, etc; data mining to construct ncRNA-mediated regulatory network and various network visualization tools. The biological network architecture which includes the different types of network properties such as topology, cluster, module, motif, dynamics and evolution and their substantial role in illustrating regulatory pathways have been discussed. Furthermore, we have covered different ways to analyze the ncRNA-mediated regulatory networks as well as their applications and challenges.

2. ncRNA-mediated regulatory networks generation and visualization

Construction of ncRNA-mediated regulatory networks essentiates a deep understanding of ncRNAs functionality and their interacting partners at various level of gene expression. ncRNAs influence the expression of a gene by various mechanisms like transcriptional regulations, post-transcriptional regulation, transposon silencing and epigenetic modifications etc. Before focusing on how the ncRNA mediated regulatory network functions at various levels, brief understanding of some of the important regulatory ncRNAs and the levels at which these imparts their function is essential (*see* Table 1).

TABLE 1 ncRNAs and their regulatory functional levels

Type	Full Name	Length	Regulatory levels	References
miRNA	MicroRNA	18-25	Post-transcriptional gene regulation Transcriptional gene regulation	(Kaikkonen et al. 2011; Winter et al. 2009)
piRNA	PIWI-interacting RNA	24-30	Post-transcriptional gene regulation Transcriptional gene regulation	(Le Thomas et al. 2013; Watanabe and Lin, 2014)
Endo-siRNA	Endogenous-small interfering RNA	21-22	Post-transcriptional gene regulation Transcriptional gene regulation	(Kaikkonen et al. 2011; Taft et al. 2010)
lncRNA	Long non-coding RNA	>200	Post-transcriptional gene regulation Transcriptional gene regulation	(Kaikkonen et al. 2011; Taft et al. 2010)
snoRNA	Small nucleolar RNA	80-200	Post-transcriptional gene regulation	(Matera et al. 2007; Taft et al. 2010)
eRNA	Enhancer RNA	100-9000	Transcriptional gene regulation	(Kaikkonen et al. 2011)
PAR *	Promoter-associated RNA	16-200	Transcriptional gene regulation	(Kaikkonen et al. 2011)
circRNA	Circular RNA	>200	Transcriptional gene regulation by acting as miRNA sponge	(Memczak et al. 2013)

(*PARs include- PASR,TSSa-RNA,tiRNA, PROMPT)

Of the above listed ncRNAs, miRNAs are well known to regulate post-transcriptional gene expression in the cytoplasm, but studies have also suggested that miRNAs may alternatively have nuclear roles where they modulate gene expression by interacting with the promoters at transcriptional level. Place et al. provided the first evidence in 2008 that miR-373 induced the expression of CDH1 and CSDC2 by targeting the promoter (Place et al. 2008). A cis-regulatory role of miR-320 was reported by Kim et al. where miR-320, which is encoded within the promoter of POLR3D, targeted its own genomic location and led to transcriptional silencing of POLR3D (Kim et al. 2008). Further studies have also shown a significant enrichment of putative miRNA target sites within the promoter region of genes from genome-wide evaluations of promoters (Younger et al. 2009).

piRNAs are the largest class of small ncRNAs whose primary function is to silence the 'genomic parasites'- transposable elements (TEs) in the germ line cells and maintain the genomic integrity. Besides transposon silencing, piRNAs have also been found to be involved in repression of protein coding genes (Weick and Miska, 2014), both at the transcriptional and post-transcriptional level (Faulkner et al. 2009; Siomi et al. 2011; Watanabe and Lin, 2014). Intriguingly, recent investigations on piRNAs with predominant nuclear localization and robust sensitivity to serotonin, required for memory, have revealed their role in regulating memory storage with abundant expression in brain (Rajasethupathy et al. 2012). This has added new dimensions to the functional diversity of piRNAs and many similar reports are gradually emerging.

Endo-siRNA was discovered some seven years ago (Ghildiyal et al. 2008) as compared to exogenous siRNA that was discovered almost two decades ago. These endo-siRNAs are known to be involved in transposon silencing, anti-viral defence, transcriptional regulation by chromatin remodelling and post-transcriptional gene regulation through argonaute (Ago) mediated cleavage of target transcripts (Carthew and Sontheimer, 2009; Ghildiyal and Zamore, 2009).

lncRNAs are transcribed by RNA polymerase II (RNAPII) that lack an open reading frame (ORF) and are longer than 200 nucleotides in length (Bonasio and Shiekhattar, 2014; Cabili et al. 2011; Derrien et al. 2012). Their size distinguishes them from other classes of small ncRNAs. These regulate the expression of protein-coding genes at both transcriptional and post-transcriptional level. At transcriptional level, they either work in *cis* or *trans* and regulate gene expression either positively or negatively (Kornienko et al. 2013). In the *cis* mode, they can act either via their product or by the process of transcription, where only the process of transcription of lncRNAs is able to affect target gene expression through RNAPII that either traverses a regulatory element or alters general chromatin organization of the locus. A good example of *cis* mode regulation where they act via their product is the human HOTTIP lncRNA which is expressed in the HOXA cluster and activates transcription of flanking genes by causing histone H3 lysine-4 trimethylation (H3K4me3) at their promoters (Wang et al. 2011). In the *trans* mode, lncRNAs act via their product to inhibit protein coding gene expression, one example being HOTAIR lncRNA expressed from HOXC cluster repressing transcription across 40kb of the HOXD cluster (Kornienko et al. 2013). HOTAIR lncRNA is required for repressive H3K27me3 of HOXD cluster by interacting with Polycomb repressive complex 2 (PRC2). Post-transcriptional regulation by lncRNA involves either modulation of mRNA stability by homologous base pairing or by acting as ceRNAs to regulate the level of miRNAs (Sen et al. 2014).

snoRNAs are the family of ncRNAs that are primarily known to guide chemical modification of rRNA, tRNA and snRNA nucleotides (Bachellerie et al. 2002). Their cellular function has continued to expand and are seen to mediate mRNA splicing and play a role in post-transcriptional regulation of mRNA by functioning as a miRNA. ACA45, a human snoRNA has been shown to be processed into a 21-nts-long mature miRNA by the RNAse III family endoribonuclease Dicer (Ender et al. 2008). Another snoRNA, HBII-52 (SNORD115) has been reported to regulate alternative splicing of the trans gene transcript (Kishore and Stamm, 2006).

eRNAs are mostly found to be expressed from the extragenic DNA sequence at the enhancer region (Fedoseeva et al. 2012). These ncRNAs have a specific histone methylation signature typical of enhancers. Their exclusive features include generation from regions defined by high mono-methylation on lysine 4 of histone 3 (H3K4me1) and low tri-methylation on lysine 4 of histone 3 (H3K4me3) (De Santa et al. 2010; Kim et al. 2010). These regions are enriched for RNA polymerase II (PolII) and transcriptional co-regulators, such as the p300 co-activator. Though their function is mostly unknown, many studies have proposed their role in transcriptional activation (Kaikkonen et al. 2011).

PARs class of ncRNAs were discovered through various genome tiling and high-throughput sequencing methods and were found to be associated with the promoters. PARs, characterized according to their locations include promoter-associated small RNA (PASR), transcription start site-associated RNA (TSSa-RNA), transcription initiation RNA (tiRNA) and promoter upstream transcript (PROMPT). PARs are found to be associated with transcriptional activation and repression (Han et al. 2007; Wang et al. 2008).

circRNAs are a group of highly stable ncRNAs that affect transcript regulation by acting as miRNA sponges. miRNA sponges are transcripts with multiple miRNA binding sequences, acting as competitive inhibitor of miRNAs that sequester the miRNAs from its target thereby de-repressing the target gene expression (Kluiver et al. 2012). The circRNAs are advantageous to act as a miRNA sponge as it lacks poly(A) tails that help in escaping from deadenylation decapping and degradation caused by miRNAs (Hansen et al. 2013; Memczak et al. 2013).

Apart from regulating genes at the transcriptional level by either promoting or inhibiting the target gene transcription ncRNAs are also known to be regulated at the transcriptional level by different TFs. Different TFs bind to the promoter of ncRNAs and modulate their activation or repression. The transcriptional regulation of lncRNAs, miRNAs, and other ncRNAs is being widely explored with the recent advances in the chromatin immunoprecipitation with next-generation DNA sequencing (ChIP-seq) technology that has enabled the detection of transcription factor binding sites (TFBSs) with unprecedented sensitivity.

3. Interactomics

For the ncRNA-mediated regulatory networks generation, the first step is to study the whole set of molecular interactions among the ncRNAs and other biological molecules taking place in the cellular environment. These interactions include ncRNAs → target mRNAs, TFs → ncRNA genes and ncRNAs → ncRNAs. To study these interactions taking place within the cells, many modern technologies have been developed, such as cross-linking immunoprecipitation (CLIP), (ChIP-seq), Cross-linking ligation and sequencing of hybrids (CLASH), RNA-Binding Protein Immunoprecipitation-Microarray Profiling (RIP-Chip), RNA-Binding Protein

Immunoprecipitation-sequencing (RIP-seq) and Systematic evolution of ligands by exponential enrichment (SELEX). The high-throughput data generated using these techniques are primarily available at NCBI-Sequence Read Archive (SRA) and DNAnexus-SRA+. A brief description of these modern technologies are given below.

(i) **CLIP-seq**-CLIP-based technique integrates UV crosslinking with immuno-precipitation to analyze interactions of proteins with RNAs (Darnell, 2010). The different CLIP based techniques include HITS-CLIP, PAR-CLIP and iCLIP. These various CLIP techniques are the state-of-art high-throughput methods that promise to provide an in-depth understanding of where and how protein-RNA complexes interact to regulate gene expression at the post-transcriptional level by mapping RNA binding sites for a protein of interest on a genome-wide scale. Integration of these data will strengthen the perspective of the ncRNA regulatory networks.

(a) **HITS-CLIP** (High-throughput Sequencing-Crosslinking and Immuno Precipitation) is the combination of high-throughput sequencing tech-niques with CLIP for identification of RNA binding proteins (RBPs) at a resolution of 30-60 nucleotides (Zhang and Darnell, 2011). RBPs are key players in the post-transcriptional regulation of gene expression like polyadenylation, splicing, mRNA stabilization and localization, etc, (Lunde et al. 2007). HITS-CLIP was first developed to get insight into the alternative processing of mouse brain by generating genome wide binding of the neuron specific splicing factor Nova (Licatalosi et al. 2008). Subsequently, this technique has been applied to various other factors like SFRS1(Sanford et al. 2009), PTB (Xue et al. 2009), and the argonaute (AGO) for finding miRNA binding sites (Thomson et al. 2011).

(b) **PAR-CLIP** (Photoactivatable Ribonucleoside Enhanced Crosslinking Immunoprecipitation) identifies the binding sites of cellular RBPs and miRNA- containing ribonucleoprotein complexes (miRNPs) (Hafner et al. 2010). This method is based on incorporation of photoactivatable nucleoside analogs into nascent RNAs. These analogs facilitate highly efficient crosslinking of proteins to RNA and overcome the crosslink-ing efficiency problems faced in HITS-CLIP.

(c) **iCLIP** (Individual-Nucleotide Resolution Crosslinking and Immuno Precipitation) similar to other CLIP methods allows identification of protein-RNA interactions. Unlike other CLIP methods, this technique enables PCR amplification of truncated cDNAs, which is achieved by an intramolecular cDNA circularization step and thereby identifies pro-tein–RNA crosslink sites with single nucleotide resolution (Huppertz et al. 2014).

(ii) **ChIP-seq** – It is a technique to study protein-DNA interactions on a genome scale (Johnson et al. 2007). In this method, the regions of DNA in contact with the TFs or other proteins are isolated by chromatin immunoprecipita-tion followed by massively parallel sequencing. These regions are then ana-lyzed to determine the interaction of the protein with its direct target DNA.

This technique is primarily used to determine the binding regions/motifs of TFs with DNA, called the TF binding sites (TFBSs) that regulates the gene expression. It is also used to determine the binding of other chromatin associated proteins with the DNA. Moreover, ChIP-seq data also provides information about the TF-ncRNAs interations (Marson et al. 2008).

(iii) CLASH (Cross-linking, ligation and sequencing of hybrids) – This cross-linking, ligation and sequencing of hybrids technique allows direct high-throughput mapping of RNA-RNA interactions (Kudla et al. 2011). It allows to determine direct miRNA-target pairs as chimeric reads from next generation sequencing data. With emerging evidences of miRNA targets not being limited to only protein-coding transcripts, but also to other ncRNAs (Poliseno et al. 2010) or lncRNAs (Jalali et al. 2013), this technique will help in transcriptome wide identifications of these interactions.

(iv) RIP-Chip – This technique couples RNA-Binding Protein Immunoprecipitation with reverse transcription, followed by microarray chip analysis to study interactions among one protein and many different RNA species (Khalil et al. 2009).

(v) RIP-Seq – This is an alternative to the RIP-Chip method where the RNA pulled down is analyzed using high-throughput sequencing rather than performing microarray.

(vi) SELEX – The Systematic Evolution of Ligands by Exponential Enrichment method is a powerful technique to detect protein binding site on RNA. It depends on the ability of RBPs to select and bind to high-affinity RNA ligands from a pool of RNAs (Manley, 2013).

Apart from the above mentioned techniques, there are other techniques that provide information on genome-wide transcriptional activity, which include RNA-sequencing (RNA-seq), intron RNA-seq (iRNA-seq), small RNA sequencing (sRNA-seq), genomic run-on-sequencing (GRO-seq) and DNA microarray. The data generated by adopting these methods are often submitted at NCBI- Gene Expression Omnibus (GEO) repository, NCBI-Sequence Read Archive (SRA), DNAnexus-SRA, Oncomine, TCGA etc, which can be analyzed to unravel expression profiles of genes and ncRNAs. These expression data can be further correlated to find regulatory relationships among ncRNAs and genes. Then, extracting the relationships among different transcripts in a particular cellular condition will be challenging which needs a comprehensive knowledge of gene regulatory mechanisms. For example, in a system, genes negatively correlated in expression with miRNAs are more likely predicted as targets (Tsang et al. 2007). These co-relations can be further linked to generate ncRNA regulatory network. We have outlined some of the techniques used now a days for studying genome-wide transcriptional activity which forms the foundation for a generation of regulatory networks:

(i) **RNA-seq** – It is the most revolutionary technique developed for transcriptome profiling that uses next generation sequencing to reveal a snap-shot of the transcriptome. It provides accurate measurement of the levels of transcripts and all their isoforms (Wang et al. 2009). For interpreting the functional element of the genome and understanding their role in disease and development, the study of the transcriptome of a cell is essential which includes all species of transcripts like mRNA, small RNAs and ncRNAs like lncRNAs. More recently, Xiao et al used RNA-seq data to generate lncRNA and protein coding gene transcript profiles and generated a regulatory network which revealed regulatory relationships between lncRNAs and protein coding genes (Xiao et al. 2015).

(ii) **iRNA-seq** – This technique based on the analysis of intron coverage from total RNA-seq provides a genome-wide assessment of transcriptional activity. It provides a more accurate determination of the sensitive transcriptional activity as compared to RNA-seq, that provides the steady-state levels of RNA including existing levels of RNA (Madsen et al. 2015).

(iii) **Small RNA sequencing** – It is a type of next generation sequencing used to sequence all the small RNAs of length ~17-35 nts with unprecedented sensitivity. Unlike RNA-seq, this method uses starting material enriched with small RNAs. It also helps in the discovery of novel small ncRNAs.

(iv) **GRO-seq** – This technique maps the distribution of short transcripts of transcriptionally engaged RNA polymerase II that are allowed to transcribe (run-on) a short distance and incorporate the nascent RNAs with an affinity tag like BrU-tag. Sequencing of the RNAs and alignment to the genome helps in mapping and quantifying transcriptionally engaged polymerase density genome-wide (Core et al. 2008).

(v) **Microarray** – In microarray for expression analysis, the cDNA derived from mRNA of known genes is immobilized on a solid support called chips as probes. Which contain known genes. These chips are then used to determine complementary binding of the unknown sequences from a sample, thus allowing simultaneous analysis for gene expression and gene discovery. Microarray provides efficient and quick analysis of the expression of many genes in a single reaction. The microarray study of various ncRNAs have also been emerged in recent days to know their expression patterns in different disease systems (Maskos and Southern, 1992).

As regulatory ncRNAs unfold their action via interactions, we can study these interactions using various tools. Table 2 briefs about the major tools to study different types RNA associated interactions.

TABLE 2 Tools for ncRNA associated interaction predictions

Name	Website	References
PicTar	http://pictar.mdc-berlin.de	(Krek et al. 2005)
miRanda	http://www.microrna.org	(Enright et al. 2003)
miRTarCLIP	http://mirtarclip.mbc.nctu.edu.tw	(Chou et al. 2013)
miRTrail	http://mirtrail.bioinf.uni-sb.de	(Backes et al. 2007)
miRDB	mirdb.org	(Wong and Wang, 2015)
TargetScan	http://www.targetscan.org	(Lewis et al. 2005)
RNAhybrid	http://bibiserv.techfak.uni-bielefeld.de/rnahybrid	(Rehmsmeier et al. 2004)
microT-CDS	http://diana.imis.athena-innovation.gr/DianaTools/index.php?r=microT_CDS/index	(Reczko et al. 2012)
miRNA – Target Gene Prediction at EMBL	http://www.russelllab.org/miRNAs	(Stark et al. 2003)
RNA22	https://cm.jefferson.edu/rna22	(Miranda et al. 2006)
psRNATarget: A Plant Small RNA Target Analysis	plantgrn.noble.org/psRNATarget	(Dai and Zhao, 2011)
RegRNA	http://regrna2.mbc.nctu.edu.tw	(Chang et al. 2013)
STarMir	http://sfold.wadsworth.org/cgi-bin/starmirtest2.pl	(Ding et al. 2004)
Magia	http://gencomp.bio.unipd.it/magia2/start	(Sales et al. 2010)
miRGator	http://mirgator.kobic.re.kr	(Nam et al. 2008)
mirDIP	http://ophid.utoronto.ca/mirDIP	(Shirdel et al. 2011)
MicroInspector	http://bioinfo.uni-plovdiv.bg/microinspector	(Rusinov et al. 2005)
PITA	http://genie.weizmann.ac.il/pubs/mir07/mir07_data.html	(Kertesz et al. 2007)
LncTar	http://www.cuilab.cn/lnctar	(Li et al. 2014a)
RNAplex	http://www.bioinf.uni-leipzig.de/Software/RNAplex	(Tafer and Hofacker, 2008)
IntaRNA	http://rna.informatik.uni-freiburg.de/IntaRNA/Input.jsp	(Busch et al. 2008)
RactIP	http://www.ncrna.org/software/ractip	(Kato et al. 2010)
RIsearch	http://rth.dk/resources/risearch	(Wenzel et al. 2012)
PETcofold	http://rth.dk/resources/petcofold	(Seemann et al. 2011)
catRAPID omics	http://s.tartaglialab.com/page/catrapid_omics_group	(Agostini et al. 2013)
RBP map	http://rbpmap.technion.ac.il	(Paz et al. 2014)

In addition to the above interaction prediction tools, several databases are publicaly available those house both validated and predicted interctions among RNA-RNA, TF-ncRNA and TF-gene. Some of these databases are listed in Table 3.

TABLE 3 Databases for mining ncRNA-mediated regulatory interactions

Database	Description	References
starBase	RNA–RNA and protein–RNA (both coding and non coding RNA)	(Li et al. 2014b).

ChIPBase	TF-lncRNA, TF-miRNA, TF-other ncRNA, TF-mRNA and TF-miRNA-mRNA	(Yang et al. 2013).
StarmiRDB	miRNA-target genes	(Rennie et al. 2014)
miRecords	Both validated and predicted miRNA-target interactions	(Xiao et al. 2009)
miRTarBase	Experimentally curated miRNA-target interactions	(Hsu et al. 2011)
TarBase	Experimentally validated miRNA-target interactions	(Vlachos et al. 2015)
TransmiR	Experimentally validated TF-miRNA regulations along with mechanisms	(Wang et al. 2010)
doRiNA	miRNA-target interactions and RBP binding sites	(Blin et al. 2015)
Micro-PIR	miRNA target sites within human promoter sequences	(Piriyapongsa et al. 2012)

4. Network visualization tools

Biological network visualization is known to be very crucial that helps in complete illustration of network topology and other associated properties.The various interactions obtained either directly from the experiments or databases and those obtained using interaction prediction tools can be combined to generate the ncRNA-mediated regulatory network. The interactions can be visualized as a network by using various network visualization tools. Some of the widely used network visualization tools are listed in Table 4.

TABLE 4 Various network visualization tools

Visualization tools	Websites
Cytoscape	www.cytoscape.org
Gephi	http://gephi.github.io
Graphviz	http://www.graphviz.org
Pajek	http://vlado.fmf.uni-lj.si/pub/networks/pajek
GUESS	http://graphexploration.cond.org
GVF	http://gvf.sourceforge.net
JUNG	http://jung.sourceforge.net
BioFabric	http://www.biofabric.org
Walrus	http://www.caida.org/tools/visualization/walrus
VisANT	http://visant.bu.edu
SNAVI	https://code.google.com/p/snavi
AVIS	http://actin.pharm.mssm.edu/AVIS2
yED	http://www.yworks.com/en/products/yfiles/yed

5. Biological Network Architecture

Understanding the complex molecular systems in a cell often requires myriads of components like ncRNAs, genes and TFs and their interactions to be presented as networks. Unraveling these interwoven networks requires understanding of the different structures and associated functional properties of the biological network. The basic network features includes the following:

5.1 Network topology

Network topology encompasses information about specific properties of a network like nodes (interacting entities) and edges (interactions). It also includes general properties of the entire network called global topological properties. Network topology plays an essential role in understanding network architecture and performance. The topological features of a node include-

(i) **Degree** It is the number of edges/link a node forms. In a directed network, the number of head endpoints on a node is termed "in-degree", and the number of tail points on a node is termed "out-degree" of the node. Higher the degree of a node, the better it is connected to a network and plays a critical role maintaining the structural integrity of the network.

(ii) **Node betweenness centrality** It is the number of shortest paths that pass through a node among all other shortest paths between all possible pairs of nodes. It is an indicator of a nodes centrality in a network.

(iii) **Closeness centrality** It is the average of the shortest path from one node to all other nodes.

(iv) **Eigenvector centrality** It measures the closeness to highly connected nodes.

Apart from these features of a node, classifying nodes with specific functional groups is essential which assigns its function, cellular location, etc. Several approaches have been extensively used for this purpose like annotating genes with specific Gene Ontology (GO) terms, signaling pathways, etc. The topological features of an edge, which may represent activating or inhibiting relationships between a pair of nodes, in a directed graph, includes-

(i) **Betweenness centrality** It is the sum of the fraction of all-pairs shortest paths that pass through it.

(ii) **Edge directionality** which specifies the upstream and downstream nodes that are connected by a particular edge.

Global topological characteristics of networks consist of -

(i) **Degree distribution** It is the probability distribution of the degrees of nodes over the whole network and is usually represented by a histogram.

(ii) **Distance/Characteristic path length** It represents the average shortest path between all pairs of nodes

(iii) **Clustering coefficient** It is a measure of the degree of interactions by determining the connectivity of a node with its neighbors averaged over the entire network. A high clustering coefficient for a network is an indicator of a smaller network.

(iv) **Grid coefficient** It looks beyond the first neighbors to the second neighbors of a node when maintaining the connectivity.

(v) **Network diameter-** It is the maximum distance between any two nodes and is also termed as the graph diameter. Depending on it, a graph may be a small network or a large network.

All the topological parameters discussed broadly fall into six categories: degree, betweenness, edge directionality, characteristic path length, network diameter, and clustering coefficient. The six topological parameters are depicted in Fig. 1.

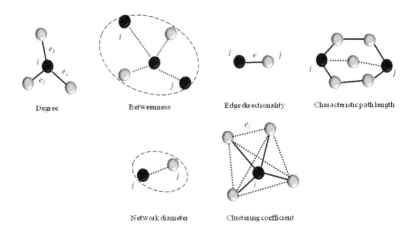

FIGURE 1   Six network topology parameters. i, j : node, e : Edge.

Importance of network topology

Congregation of interactions into networks has revealed that a recurrent property of regulatory biological networks is that they have a "scale-free" topology, wherein a number of nodes have only few connections and only a few nodes called "hubs" that are furnished with large number of connections which bind the network together (Barabasi and Oltvai, 2004). Hubs are extremely important as they play an essential role in biological systems. Much of the regulations in a network in mediated through hubs. Apart from the information about important regulatory ncRNA hubs which can be obtained from a ncRNA regulatory network, assigning functions to un-annotated ncRNAs or genes can also be performed. The network hub-based method for prediction of ncRNA function is one of the most direct methods. It can assign a function to the un-annotated ncRNAs according to the functional enrichment of its immediate neighboring genes. Liao and group were the first to apply this method to predict the function of lncRNAs (Liao et al. 2011).

Similarly, identifying key components that form the up-stream molecules called as master regulators is essential. These critical master regulators modulate multiple downstream genes either directly or via a cascade of gene expression changes, and when misexpressed, they re-orient the fate of cells to become diseased cells. A good example is the gene Twist, which is a master regulator controlling embryonic morphogenesis and its de-regulation is responsible for tumor invasion and metastasis (Yang et al. 2004).

Furthermore, interactions between lncRNA–mRNA, lncRNA–miRNA, and miRNA–mRNA in ncRNA-mediated network can be modulated by introducing sponges and masks that change the stochiometry of the binding site which effectively affects the topology and function of these networks. In a ceRNA network, the accurate topology leads to the identification of bonafide ceRNA regulations that would have otherwise seem implausible based only on miRNA and mRNA ratio (Sanchez-Mejias and Tay, 2015).

5.2 Network module/cluster

It is the local unit of the network which is defined as sub-graph. It is a dense area of connectivity with a loose link, separated from regions of low connectivity (Newman, 2006). Liao and group have also predicted the function of lncRNAs by network modules using Markov cluster algorithm (MCL) (Liao et al. 2011). This opens up the avenues for determining the function of ncRNAs in a ncRNA mediated regulatory network using network modules. However, with the ubiquitous nature of the networks modules in a biological network, it may help to understand the interplay between network structure and function well.

5.3 Network motif

A common pattern or a recurring circuit in a network is called a network motif. Network motifs are composed of a few nodes and edges and are found to occur more frequently in the topology of a biological network. ncRNAs such as miRNAs are usually involved in many regulatory pathways or networks and are enriched with motifs, such as positive or negative feed forward loops (FFLs) (Mangan et al. 2003) and feedback loops (FBLs) (Angeli et al. 2004). This involvement of miRNAs in complex motifs makes the use of systems biology necessary for understanding their regulation. Identification of putative regulatory motifs which is associated with inherent complexity is ideal for designing experiments to analyze their features and regulations. The role of miRNAs as complex regulatory motifs is well elucidated by Vera and colleagues (Vera et al. 2013). The FFLs are found to be overrepresented in transcriptional network and is a three gene pattern composed of one TF interacting with one miRNA and both jointly regulating a target gene. In mammals, typically four patterns of FFLs are extensively studied where the TF can either induce or repress expression of miRNA and target gene while miRNA always inhibits expression of the target gene (Tsang et al. 2007). The FBL occurs when output of a system is feed back as input to the system, forming a loop. These can be either positive, where it accelerates a process or negative, where it slows down a process. A FBL is particularly important because it directly influences a systems dynamics. For finding enriched three-element and four-element motifs in a network, the tool mfinder can be used (Milo et al. 2002). Apart from FFLs and FBLs, another important motif identified in ncRNA mediated networks is bifans (Ingram et al. 2006; Lipshtat et al. 2008). These network motifs can be identified in directed or undirected networks called graphlets (Przulj et al. 2006). Bifan motif is a four gene pattern, where two source nodes directly cross-regulate two target nodes. These are overrepresented motifs in transcriptional and mammalian cell signaling network.

5.4   Network Dynamics

The dynamics of complex biological networks allows cells to respond to various conditions or cell states like proliferation, differentiation and apoptosis. Network dynamics study has shown that in an interaction network, the interacting molecules vary their partners according to time and locations. This led to the discovery of two types of hubs by Han et al. called "party hubs" and "date hubs" (Han et al. 2004). They studied the network dynamics and revealed that the party hubs interact with all their partners at the same time at spatial locations and have a higher probability of function within the same cellular process. The date hubs varied their interacting partners from time to time and thus linked various biological processes. Further in-silico analysis of networks reveals that the party hubs are more likely to be the module organizers and the date hubs act as module connectors. Studies by Luscombe and colleagues on dynamics of the transcriptional network for five different cellular conditions revealed sub-networks with different topologies at global and local level (Luscombe et al. 2004). Furthermore, roles of lncRNAs in ncRNA regulatory network topologies and dynamics have been well studied by different groups (Kung et al. 2013; Mercer and Mattick, 2013).

5.5   Network Evolution

Almost all real-world networks evolve with time by addition or removal of nodes and edges. Various models have been proposed to study networks during evolution. One model of network growth described by Barabasi and Albert assumes that as most biological networks follow a scale-free topology, it is considered that the networks expand by the continuous addition of new nodes with fixed degree (Barabasi and Oltvai, 2004). These nodes attach preferentially to well-connected nodes. The addition of new nodes in biological networks is most likely due to gene duplication. Understanding network evolution of ncRNA-mediated regulatory networks will affect network functionality by impacting the connectivity among its components. However, till date no significant work has been reported on this compelling prospective of ncRNA regulatory network alteration.

---

6.   Analyzing ncRNA-mediated regulatory network

The ncRNA regulatory networks generated using various tools can then be analyzed for generation of new hypothesis which can further be validated by designing new experimental procedures. For analyzing a functional network associated with a par-ticular biological process, the networks can be built taking seed nodes. For example, differentially regulated genes and proteins in a disease or treatment condition as compared to normal or untreated could also be used as seed for building disease-associated networks (Berger et al. 2007). These functional networks or disease networks are built by mapping the seed nodes onto a bigger regulatory interaction network obtained using all the individual regulatory interactions. Figure 2 provides an idea of the generation of a disease specific regulatory hub from a larger regulatory network done by our group (Figure 2) (Samantarrai et al. 2015).

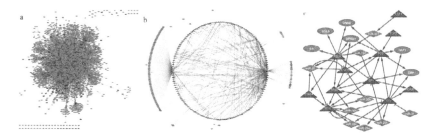

FIGURE 2 Generation of disease-specific regulatory hub from a larger regulatory network (a) Curated TF-miRNA regulatory network (b) Active sub-network in STS metastasis (c) Notch signaling specific network generated from active sub-network.

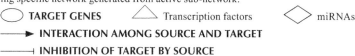

⬭ **TARGET GENES**   △ Transcription factors   ◇ miRNAs

———▶ **INTERACTION AMONG SOURCE AND TARGET**

———⊣ **INHIBITION OF TARGET BY SOURCE**

6.1   Integration of experimental resources

Integration of networks with different experimental resources or any type of back-ground knowledge is helpful to predict functional interactions, and to improve the accuracy of the predicted interactions. The experimental resources and background knowledge includes validated interaction or validated information on the involve-ment of genes, ncRNAs or proteins in a disease system. For example, Jiang et al integrated known target interactions of ncRNAs and regulation of ncRNAs to pre-dict novel interactions among them in Alzheimer's disease (Jiang et al. 2013). Apart from this, similar data sets which are generated using different methods can also be integrated to improve data quality and get information on missing data. For exam-ple, Han and colleagues removed flawed interactions from the yeast protein-protein interaction network and constructed a "filtered yeast interactome"- FYI using high quality interactions obtained from two experiments (Han et al. 2004). Similar type of studies can be performed to explore ncRNA-mediated regulatory network analysis. Apart from this, integrating expression data from microarray or high-throughput sequencing like RNA-seq, small RNA seq, etc can add weightage to the network for better understanding. Integrating expression data in the transcriptional network has also proven to improve the quality of data and disclose cis-regulatory modules (Bar-Joseph et al. 2003). Our group has also recently published on unraveling TF-miRNA regulatory crosstalk in metastasis of soft tissue sarcomas, where data from experimentally validated resources and microarray expression was integrated to generate novel regulatory interactions (Samantarrai et al. 2015). Thus, network integration involves combining various data from different resources that assists in uncovering critical regulators and present a more comprehensive view on their cellular functions.

6.2 Algorithms and tools to infer regulatory network

The generated functional networks or disease network can be analyzed further using different algorithms like mean-first-pass-time (MFPT) (Noh and Rieger, 2004), Steiner trees (Dittrich et al. 2008; Huang and Fraenkel, 2009) nearest neighbor expansion and shortest path search algorithm to find connections between genes, ncRNAs, etc. The steiner tree was used by White and colleagues to minimally connect up-regulated genes in breast and colorectal cancer (White et al. 2007). It has also been used to connect signaling pathways to TFs (Huang and Fraenkel, 2009). The shortest path algorithm has been used to predict key regulators of neurite outgrowth triggered by cannabinoid receptor signaling network (Bromberg et al. 2008). Implication of these algorithms in assimilating ncRNA-mediated regulatory networks are believed to be useful in finding functional relations between nodes in addition to predicting nodes that have not been detected experimentally. Apart from these algorithms, many tools can be used to infer valuable regulatory and functional information from the network (Table 5).

TABLE 5 Tools for inferring ncRNA-mediated regulatory network

Tools	Descriptions	Websites
Cytoscape	Many useful plug-ins provided by it helps in inferring network architecture, functional annotations, pathway enrichment, pathway detection, cluster generation, etc	http://apps.cytoscape.org
RiNAcyc (RNA-interacting Nodes in Acyclic path)	Generates acyclic paths from unidirectional binary interactions among various biological entities, especially genes, miRNAs and transcription factors (TFs)	http://vvekslab.in/tools.html
DAVID	It identifies enriched GO terms, functional related gene groups, links disease-related genes, etc.	http://david.abcc.ncifcrf.gov
MetaCoreTM	A commercial software suite for performing functional analysis of NGS data, microarray, metabolic, proteomics, siRNA, miRNA, etc,.	https://portal.genego.com
IPA	A commercial software suite that provides curated information on genes, miRNAs, cellular and disease processes, signaling and metabolic pathways, biomarkers, etc. It also generates information on key relationships and novel interactions.	http://www.ingenuity.com
DIANA-mirPath	It provides miRNA pathway analysis with accurate statistics.	http://diana.imis.athena-innovation.gr/DianaTools

7.   Application of non-coding RNA-mediated biological network

The study of association of ncRNAs with gene regulatory networks and their deregulations in complex diseases is one of the hot topics in genome research now-a-days. With the discovery that the genome of complex organisms have an enormous portion representing ncRNAs, many transcriptome studies have revealed that these ncRNA are the key regulators of biological processes and form a veiled layer of molecular genetics signals. Hence, construction of ncRNA-mediated regulatory network using different algorithms would help in better understanding of the ncRNA based biological functions playing possible critical roles in disease etiology. Various ncRNA-mediated networks have been generated and studied for several purposes like highlighting the role of ncRNAs in cancer (Li et al. 2011; Xiao et al. 2015), studying effect of infections on the networks (Wu et al. 2011), analyzing transcriptional regulation network in T-cells (Sun et al. 2013) etc. These network studies have provided many critical information regarding their regulations and involvement in diseases.

Diversity and complexity found among mammals and other complex organisms despite the relative commonality in their protein coding genes is one of the paradoxes of molecular biology which is explained by ncRNA-mediated regulatory networks. These enormous numbers of ncRNAs were hidden because of lack of development of techniques for detecting and analyzing them. The development of high-throughput technologies, different bioinformatics approaches and analytical techniques like RT-PCR, etc for analyzing the genome has revealed the true complexity of the organisms. With the application of modern bioinformatics techniques, it is possible to develop a generic approach, which will be applicable to any regulatory RNA molecule, thus revolutionizing the study of ncRNAs.

Till date, various studies have been performed by different groups of scientist with an aim to provide the best generic approach for generating and unraveling ncRNA-mediated regulatory network. The use of different curated and experimental data in generation of these networks will help deciphering novel enriched interacting molecules that might form the master regulatory component called hub. Further enrichment analysis will lead to identification of the most enriched hubs and their association to key biological pathways. The deregulation of these hubs are know to deregulate the regulatory network. These critical regulatory hubs can be considered as biomarker for a specific system and can act as a potential therapeutic target for a disease or stages of a disease. For example, Sun et al in 2012 identified critical miRNAs in the Notch signaling pathway from miRNA-TF regulatory network for glioblastoma, based on identification of hubs (Sun et al. 2012).

In modern systems biology study, various efficient network generation and analysis approaches have been developed. For example, Xiao et al generated a Bayesian gene regulatory network of lncRNAs and protein coding genes (Xiao et al. 2015). The Bayesian algorithm represents a set of random variables and their conditional dependencies via a directed acyclic graph. It revealed the regulatory relationship between the lncRNAs and protein coding genes in a prostate cancer model and predicted the function of lncRNA linked to highly connected coding genes through functional enrichment. They found 762 lncRNAs in the constructed network which were assigned function. Recently, our group worked on unraveling novel TF-miRNA regulatory network in sarcoma metastasis, and developed a depth first search algorithm (DFS) algorithm-based tool called RiNAcyc for deciphering acyclic path which are potentially active in soft tissue sarcoma metastasis (Samantarrai et al. 2015). Depth first search algorithm is an algorithm which searches by traversing the depth of any particular path in a graph before exploring its breadth. We found 12 significantly active paths from a sub-network which was potentially active in STS metastasis.

Analyzing the network motifs in a biological network, Herranz and Cohen showed that let-7 and IL6 in oncogenic transformation are seen to be in a FFL in which NF-κB activation leads to depletion of let-7 and alleviates repression of IL6 (Herranz and Cohen, 2010). The feedback regulatory mechanism, another dominant network motif of the biological network can also lead to understanding of the complexity of biological function. A study by Aguda and colleagues revealed a feedback loop formed by Myc, E2F and miRNA cluster miR-17-92 (Aguda et al. 2008).

Another crucial ncRNA-mediated regulatory network is the genome-scale ceRNA network (ceRNET) study. The prediction of interaction topology of ceRNAs are now being extensively expanded with ample availability of genome-wide ncRNA-mRNA expression data over a wide range of disease systems. A recent work studied breast and thyroid cancer specific ceRNET and identified potential hubs discriminating respective cancer, as well as metastatic risk (Zhou et al. 2014). Similarly, ceRNET has been revealed in papillary thyroid carcinoma, gastric cancer and breast cancer exploring the cross regulation amongs lncRNA, miRNAs and mRNA (Huang et al. 2014; Paci et al. 2014; Xia et al. 2014). These network obtained in one system can have implication in other system with some novel or similar kind of functional regulations.

The emerging science of systems biology demands handling of large-scale high throughput data for the analysis of ncRNA-mediated regulations, which is simplified by generation and interpretation of these networks using different bioinformatics approaches. These networks will provide insight into transcriptome and its regulation by ncRNAs, which form the major chunk of the transcriptome.

8.   Challenges and Future Directions

A complete and thorough understanding of the ncRNA-mediated regulatory networks is dream as of now. The key challenge remains in understanding the precise role of ncRNAs whose functions and regulations are still unexplored, particularly in relation to diseases. The attention that ncRNAs are receiving has greatly increased after the ENCODE project which revealed that 80% of the genome in eukaryotes is transcribed of which only 2% codes for proteins and the remaining are RNAs with no coding ability. Also, there is a constant increase in the detection of new types of ncRNAs. Inspite of an increase in interest among scientists in deciphering the role of ncRNAs, and the remarkable development of new technologies and resources as evident from the increasing number of scientific article related to this area, there still remains a long way to go to completely understand and decipher it.

The characterization of ncRNA transcripts that form major chunk of human genome and their biosynthesis pathways are still in nascent stage and will remain as the primary focus for molecular biologist for many years to come. Surprisingly, all types of ncRNAs identified till now form only a small fraction of the possible ncRNAs to be deciphered. Many issues make their identification and functional annotation difficult, like incomplete understanding of the functional motifs present in ncRNAs, identification of their regulatory region and their low expression. Furthermore, despite the development of modern technologies, the analysis of the transcriptome by different methods and technology is primitive, as these methods do not allow full-length sequencing of all the transcripts. The identification of new ncRNAs, their interactions with other transcripts forming networks, assigning biological roles to all ncRNAs coded by a genome and studying their modes of regulatory networks is a daunting task and a significant challenge faced today. With the advent of technologies, it is expected that the rising genomics and bioinformatics approaches will help in uncovering and deciphering their functionality.

The construction of a disease-specific ncRNA-mediated regulatory network requires proper understanding of biological meaning of the data to be used, associating appropriate regulatory interactions and their accurate data mining. Apart from these, an extensive functional annotation of the generated network as well as appropriate integration of inter-disciplinary approaches will also significantly add to the understanding of network biology.

References

Agostini, F., Zanzoni, A., Klus, P., Marchese, D., Cirillo, D., and G.G. Tartaglia, 2013. catRAPID omics: a web server for large-scale prediction of protein-RNA interactions. *Bioinformatics* 29, 2928-2930.

Aguda, B.D., Kim, Y., Piper-Hunter, M.G., Friedman, A., and C.B. Marsh, 2008. MicroRNA regulation of a cancer network: consequences of the feedback loops involving miR-17-92, E2F, and *Myc. Proc Natl Acad Sci U S A* 105, 19678-19683.

Ambros, V. 2008. The evolution of our thinking about microRNAs. *Nat Med* 14, 1036-1040.

Angeli, D., Ferrell, J.E., Jr., and E.D. Sontag, 2004. Detection of multistability, bifurcations, and hysteresis in a large class of biological positive-feedback systems. *Proc Natl Acad Sci U S A* 101, 1822-1827.

Bachellerie, J.P., Cavaille, J., and A. Huttenhofer, 2002. The expanding snoRNA world. *Biochimie* 84, 775-790.

Backes, C., Keller, A., Kuentzer, J., Kneissl, B., Comtesse, N., Elnakady, Y.A., Muller, R., Meese, E., and H.P. Lenhof, 2007. GeneTrail--advanced gene set enrichment analysis. *Nucleic Acids Res* 35, W186-192.

Bar-Joseph, Z., Gerber, G.K., Lee, T.I., Rinaldi, N.J., Yoo, J.Y., Robert, F., Gordon, D.B., Fraenkel, E., Jaakkola, T.S., Young, R.A., et al. 2003. Computational discovery of gene modules and regulatory networks. *Nat Biotechnol* 21, 1337-1342.

Barabasi, A.L., and Z.N. Oltvai, 2004. Network biology: understanding the cell's functional organization. *Nat Rev Genet* 5, 101-113.

Bartel, D.P. 2009. MicroRNAs: target recognition and regulatory functions. *Cell* 136, 215-233.

Berger, S.I., Posner, J.M., and Ma'ayan, A. 2007b. Genes2Networks: connecting lists of gene symbols using mammalian protein interactions databases. *BMC Bioinformatics*. 8: 372.

Blin, K., Dieterich, C., Wurmus, R., Rajewsky, N., Landthaler, M., and A. Akalin, 2015. DoRiNA 2.0--upgrading the doRiNA database of RNA interactions in post-transcriptional regulation. *Nucleic Acids Res* 43, D160-167.

Bonasio, R., and R. Shiekhattar, 2014. Regulation of transcription by long noncoding RNAs. *Annu Rev Genet* 48, 433-455.

Bromberg, K.D., Ma'ayan, A., Neves, S.R., and R. Iyengar, (2008). Design logic of a cannabinoid receptor signaling network that triggers neurite outgrowth. *Science* 320, 903-909.

Busch, A., Richter, A.S., and R. Backofen, 2008. IntaRNA: efficient prediction of bacterial sRNA targets incorporating target site accessibility and seed regions. *Bioinformatics* 24, 2849-56.

Cabili, M.N., Trapnell, C., Goff, L., Koziol, M., Tazon-Vega, B., Regev, A., and Rinn, J.L. (2011). Integrative annotation of human large intergenic noncoding RNAs reveals global properties and specific subclasses. *Genes Dev* 25, 1915-27.

Carthew, R.W. and E.J. Sontheimer, 2009. Origins and Mechanisms of miRNAs and siRNAs. Cell 136, 642-55.

Chang, T.H., Huang, H.Y., Hsu, J.B., Weng, S.L., Horng, J.T. and H.D. Huang, 2013. An enhanced computational platform for investigating the roles of regulatory RNA and for identifying functional RNA motifs. *BMC Bioinformatics* 14 Suppl 2: S4.

Chou, C.H., Lin, F.M., Chou, M.T., Hsu, S.D., Chang, T.H., Weng, S.L., Shrestha, S., Hsiao, C.C., Hung, J.H., and H.D. Huang, 2013. A computational approach for identifying microRNA-target interactions using high-throughput CLIP and PAR-CLIP sequencing. *BMC Genomics* 14 Suppl 1: S2.

Core, L.J., Waterfall, J.J., and J.T. Lis, 2008. Nascent RNA sequencing reveals widespread pausing and divergent initiation at human promoters. *Science* 322: 1845-48.

Dai, X., and P.X. Zhao, 2011. psRNATarget: a plant small RNA target analysis server. *Nucleic Acids Res* 39: W155-159.

Darnell, R.B. 2010. HITS-CLIP: panoramic views of protein-RNA regulation in living cells. *Wiley Interdisc Rev RNA* 1: 266-86.

De Santa, F., Barozzi, I., Mietton, F., Ghisletti, S., Polletti, S., Tusi, B.K., Muller, H., Ragoussis, J., Wei, C.L., and G. Natoli, 2010. A large fraction of extragenic RNA pol II transcription sites overlap enhancers. *PLoS Biol* 8: e1000384.

Derrien, T., Johnson, R., Bussotti, G., Tanzer, A., Djebali, S., Tilgner, H., Guernec, G., Martin, D., Merkel, A., Knowles, D.G., et al. 2012. The GENCODE v7 catalog of human long noncoding RNAs: analysis of their gene structure, evolution, and expression. *Genome Res* 22: 1775-1789.

Diaz-Beltran, L., Cano, C., Wall, D.P., and F.J. Esteban, 2013. Systems biology as a comparative approach to understand complex gene expression in neurological diseases. Behav Sci (Basel) 3, 253-272.

Ding, Y., Chan, C.Y., and Lawrence, C.E. 2004. Sfold web server for statistical folding and rational design of nucleic acids. *Nucleic Acids Res* 32: W135-141.

Dittrich, M.T., Klau, G.W., Rosenwald, A., Dandekar, T., and T. Muller, 2008. Identifying functional modules in protein-protein interaction networks: an integrated exact approach. *Bioinformatics* 24: i223-31.

Ender, C., Krek, A., Friedlander, M.R., Beitzinger, M., Weinmann, L., Chen, W., Pfeffer, S., Rajewsky, N., and G. Meister, 2008. A human snoRNA with microRNA-like functions. Mol Cell 32, 519-28.

Enright, A.J., John, B., Gaul, U., Tuschl, T., Sander, C., and Marks, D.S. 2003. MicroRNA targets in Drosophila. *Genome* Biol 5: R1.

Esteller, M. 2011. Non-coding RNAs in human disease. *Nat Rev Genet* 12, 861-74.

Faulkner, G.J., Kimura, Y., Daub, C.O., Wani, S., Plessy, C., Irvine, K.M., Schroder, K., Cloonan, N., Steptoe, A.L., Lassmann, T., et al. 2009. The regulated retrotransposon transcriptome of mammalian cells. *Nat Genet* 41: 563-71.

Fedoseeva, D.M., Kretova, O.V., and N.A. Tchurikov, 2012. Molecular analysis of enhancer RNAs and chromatin modifications in the region of their synthesis in Drosophila cells possessing genetic constructs. *Dokl Biochem Biophys* 442: 7-11.

Ghildiyal, M., Seitz, H., Horwich, M.D., Li, C., Du, T., Lee, S., Xu, J., Kittler, E.L., Zapp, M.L., Weng, Z., et al. 2008. Endogenous siRNAs derived from transposons and mRNAs in Drosophila somatic cells. *Science* 320: 1077-81.

Ghildiyal, M., and P.D. Zamore, 2009. Small silencing RNAs: an expanding universe. *Nat Rev Genet* 10: 94-108.

Goh, K.I., Cusick, M.E., Valle, D., Childs, B., Vidal, M., and A.L. Barabasi, 2007. The human disease network. *Proc Natl Acad Sci U S A* 104, 8685-90.

Hafner, M., Landthaler, M., Burger, L., Khorshid, M., Hausser, J., Berninger, P., Rothballer, A., Ascano, M., Jr., Jungkamp, A.C., Munschauer, M., et al. 2010. Transcriptome-wide identification of RNA-binding protein and microRNA target sites by PAR-CLIP. *Cell* 141: 129-41.

Han, J., Kim, D., and K.V. Morris, 2007. Promoter-associated RNA is required for RNA-directed transcriptional gene silencing in human cells. *Proc Natl Acad Sci U S A* 104: 12422-427.

Han, J.D., Bertin, N., Hao, T., Goldberg, D.S., Berriz, G.F., Zhang, L.V., Dupuy, D., Walhout, A.J., Cusick, M.E., Roth, F.P., et al. 2004. Evidence for dynamically organized modularity in the yeast protein-protein interaction network. *Nature* 430: 88-93.

Hansen, T.B., Jensen, T.I., Clausen, B.H., Bramsen, J.B., Finsen, B., Damgaard, C.K., and J. Kjems, 2013. Natural RNA circles function as efficient microRNA sponges. *Nature* 495: 384-88.

Herranz, H., and S.M. Cohen, 2010. MicroRNAs and gene regulatory networks: managing the impact of noise in biological systems. *Genes Dev* 24: 1339-44.

Hood, L., Heath, J.R., Phelps, M.E., and B. Lin, 2004. Systems biology and new technologies enable predictive and preventative medicine. *Science* 306, 640-43.

Hsu, S.D., Lin, F.M., Wu, W.Y., Liang, C., Huang, W.C., Chan, W.L., Tsai, W.T., Chen, G.Z., Lee, C.J., Chiu, C.M., et al. 2011. miRTarBase: a database curates experimentally validated microRNA-target interactions. *Nucleic Acids Res* 39: D163-69.

Huang, C.T., Oyang, Y.J., Huang, H.C., and H.F. Juan, 2014. MicroRNA-mediated networks underlie immune response regulation in papillary thyroid carcinoma. *Sci Rep* 4: 6495.

Huang, S.S., and E. Fraenkel, 2009. Integrating proteomic, transcriptional, and interactome data reveals hidden components of signaling and regulatory networks. *Sci Signal* 2: ra40.

Huppertz, I., Attig, J., D'Ambrogio, A., Easton, L.E., Sibley, C.R., Sugimoto, Y., Tajnik, M., Konig, J., and J. Ule, 2014. iCLIP: protein-RNA interactions at nucleotide resolution. *Methods* 65: 274-87.

Ingram, P.J., Stumpf, M.P., and Stark, J. 2006. Network motifs: structure does not determine function. *BMC Genomics* 7: 108.

Jalali, S., Bhartiya, D., Lalwani, M.K., Sivasubbu, S., and V. Scaria, 2013. Systematic transcriptome wide analysis of lncRNA-miRNA interactions. *PLoS One* 8: e53823.

Jiang, W., Zhang, Y., Meng, F., Lian, B., Chen, X., Yu, X., Dai, E., Wang, S., Liu, X., Li, X., et al. 2013. Identification of active transcription factor and miRNA regulatory pathways in Alzheimer's disease. *Bioinformatics* 29: 2596-602.

Johnson, D.S., Mortazavi, A., Myers, R.M., and B. Wold, 2007. Genome-wide mapping of *in vivo* protein-DNA interactions. *Science* 316, 1497-502.

Kaikkonen, M.U., Lam, M.T., and Glass, C.K. 2011. Non-coding RNAs as regulators of gene expression and epigenetics. *Cardiovasc Res* 90: 430-40.

Karp, P.D., Riley, M., Paley, S.M., and A. Pelligrini-Toole, 1996. EcoCyc: an encyclopedia of Escherichia coli genes and metabolism. *Nucleic Acids Res* 24: 32-39.

Kato, Y., Sato, K., Hamada, M., Watanabe, Y., Asai, K., and T. Akutsu, 2010. RactIP: fast and accurate prediction of RNA-RNA interaction using integer programming. *Bioinformatics* 26: i460-66.

Kertesz, M., Iovino, N., Unnerstall, U., Gaul, U., and E. Segal, 2007. The role of site accessibility in microRNA target recognition. *Nat Genet* 39: 1278-84.

Khalil, A.M., Guttman, M., Huarte, M., Garber, M., Raj, A., Rivea Morales, D., Thomas, K., Presser, A., Bernstein, B.E., van Oudenaarden, A., et al. 2009. Many human large intergenic noncoding RNAs associate with chromatin-modifying complexes and affect gene expression. *Proc Natl Acad Sci U S A* 106: 11667-72.

Kim, D.H., Saetrom, P., Snove, O., Jr., and Rossi, J.J. 2008. MicroRNA-directed transcriptional gene silencing in mammalian cells. Proc Natl Acad Sci U S A 105, 16230-16235.

Kim, T.K., Hemberg, M., Gray, J.M., Costa, A.M., Bear, D.M., Wu, J., Harmin, D.A., Laptewicz, M., Barbara-Haley, K., Kuersten, S., et al. 2010. Widespread transcription at neuronal activity-regulated enhancers. Nature 465, 182-187.

Kishore, S., and Stamm, S. 2006. The snoRNA HBII-52 regulates alternative splicing of the serotonin receptor 2C. Science 311, 230-232.

Kluiver, J., Slezak-Prochazka, I., Smigielska-Czepiel, K., Halsema, N., Kroesen, B.J., and van den Berg, A. 2012. Generation of miRNA sponge constructs. Methods 58, 113-117.

Knowling, S., and Morris, K.V. 2011. Non-coding RNA and antisense RNA. Nature's trash or treasure? Biochimie 93, 1922-1927.

Kornienko, A.E., Guenzl, P.M., Barlow, D.P., and Pauler, F.M. 2013. Gene regulation by the act of long non-coding RNA transcription. BMC Biol 11, 59.

Krek, A., Grun, D., Poy, M.N., Wolf, R., Rosenberg, L., Epstein, E.J., MacMenamin, P., da Piedade, I., Gunsalus, K.C., Stoffel, M., et al. 2005. Combinatorial microRNA target predictions. Nat Genet 37, 495-500.

Kudla, G., Granneman, S., Hahn, D., Beggs, J.D., and Tollervey, D. 2011. Cross-linking, ligation, and sequencing of hybrids reveals RNA-RNA interactions in yeast. Proc Natl Acad Sci U S A 108, 10010-10015.

Kung, J.T., Colognori, D., and Lee, J.T. 2013. Long noncoding RNAs: past, present, and future. Genetics 193, 651-669.

Le Thomas, A., Rogers, A.K., Webster, A., Marinov, G.K., Liao, S.E., Perkins, E.M., Hur, J.K., Aravin, A.A., and Toth, K.F. 2013. Piwi induces piRNA-guided transcriptional silencing and establishment of a repressive chromatin state. Genes Dev 27, 390-399.

Lewis, B.P., Burge, C.B., and Bartel, D.P. 2005. Conserved seed pairing, often flanked by adenosines, indicates that thousands of human genes are microRNA targets. Cell 120, 15-20.

Li, J., Ma, W., Zeng, P., Wang, J., Geng, B., Yang, J., and Cui, Q. 2014a. LncTar: a tool for predicting the RNA targets of long noncoding RNAs. Brief Bioinform.

Li, J.H., Liu, S., Zhou, H., Qu, L.H., and Yang, J.H. 2014b. starBase v2.0: decoding miRNA-ceRNA, miRNA-ncRNA and protein-RNA interaction networks from large-scale CLIP-Seq data. Nucleic Acids Res 42, D92-97.

Li, X., Gill, R., Cooper, N.G., Yoo, J.K., and Datta, S. 2011. Modeling microRNA-mRNA interactions using PLS regression in human colon cancer. BMC Med Genomics 4, 44.

Liao, Q., Liu, C., Yuan, X., Kang, S., Miao, R., Xiao, H., Zhao, G., Luo, H., Bu, D., Zhao, H., et al. 2011. Large-scale prediction of long non-coding RNA functions in a coding-non-coding gene co-expression network. Nucleic Acids Res 39, 3864-3878.

Licatalosi, D.D., Mele, A., Fak, J.J., Ule, J., Kayikci, M., Chi, S.W., Clark, T.A., Schweitzer, A.C., Blume, J.E., Wang, X., et al. 2008. HITS-CLIP yields genome-wide insights into brain alternative RNA processing. Nature 456, 464-469.

Lipshtat, A., Purushothaman, S.P., Iyengar, R., and Ma'ayan, A. 2008. Functions of bifans in context of multiple regulatory motifs in signaling networks. Biophys J 94, 2566-2579.

Lunde, B.M., Moore, C., and Varani, G. 2007. RNA-binding proteins: modular design for efficient function. Nat Rev Mol Cell Biol 8, 479-490.

Luscombe, N.M., Babu, M.M., Yu, H., Snyder, M., Teichmann, S.A., and Gerstein, M. 2004. Genomic analysis of regulatory network dynamics reveals large topological changes. Nature 431, 308-312.

Madsen, J.G., Schmidt, S.F., Larsen, B.D., Loft, A., Nielsen, R., and S. Mandrup, 2015. iRNA-seq: computational method for genome-wide assessment of acute transcriptional regulation from total RNA-seq data. *Nucleic Acids Res* 43: e40.

Mangan, S., Zaslaver, A., and U. Alon, 2003. The coherent feedforward loop serves as a sign-sensitive delay element in transcription networks. *J Mol Biol* 334: 197-204.

Manley, J.L. 2013. SELEX to identify protein-binding sites on RNA. Cold Spring *Harb Protoc* 2013: 156-63.

Marson, A., Levine, S.S., Cole, M.F., Frampton, G.M., Brambrink, T., Johnstone, S., Guenther, M.G., Johnston, W.K., Wernig, M., Newman, J., et al. 2008. Connecting microRNA genes to the core transcriptional regulatory circuitry of embryonic stem cells. *Cell* 134: 521-33.

Maskos, U., and E.M. Southern, 1992. Oligonucleotide hybridizations on glass supports: a novel linker for oligonucleotide synthesis and hybridization properties of oligonucleotides synthesised in situ. *Nucleic Acids Res* 20: 1679-84.

Matera, A.G., Terns, R.M., and M.P. Terns, 2007. Non-coding RNAs: lessons from the small nuclear and small nucleolar RNAs. *Nat Rev Mol Cell Biol* 8: 209-20.

Memczak, S., Jens, M., Elefsinioti, A., Torti, F., Krueger, J., Rybak, A., Maier, L., Mackowiak, S.D., Gregersen, L.H., Munschauer, M., et al. 2013. Circular RNAs are a large class of animal RNAs with regulatory potency. *Nature* 495: 333-38.

Mercer, T.R., and Mattick, J.S. 2013. Structure and function of long noncoding RNAs in epigenetic regulation. *Nat Struct Mol Biol* 20: 300-307.

Milo, R., Shen-Orr, S., Itzkovitz, S., Kashtan, N., Chklovskii, D., and U. Alon, 2002. Network motifs: simple building blocks of complex networks. *Science* 298: 824-27.

Miranda, K.C., Huynh, T., Tay, Y., Ang, Y.S., Tam, W.L., Thomson, A.M., Lim, B., and Rigoutsos, I. 2006. A pattern-based method for the identification of MicroRNA binding sites and their corresponding heteroduplexes. *Cell* 126: 1203-17.

Nam, S., Kim, B., Shin, S., and S. Lee, 2008. miRGator: an integrated system for functional annotation of microRNAs. *Nucleic Acids Res* 36: D159-164.

Newman, M.E. 2006. Modularity and community structure in networks. *Proc Natl Acad Sci U S A* 103: 8577-82.

Noh, J.D., and H. Rieger, 2004. Random walks on complex networks. *Phys Rev Lett.* 92: 118701.

Paci, P., Colombo, T., and L. Farina, 2014. Computational analysis identifies a sponge interaction network between long non-coding RNAs and messenger RNAs in human breast cancer. *BMC Syst Biol* 8: 83.

Paz, I., Kosti, I., Ares, M., Jr., Cline, M., and Y. Mandel-Gutfreund, 2014. RBPmap: a web server for mapping binding sites of RNA-binding proteins. *Nucleic Acids Res* 42: W361-67.

Piriyapongsa, J., Bootchai, C., Ngamphiw, C., and S. Tongsima, 2012. microPIR: an integrated database of microRNA target sites within human promoter sequences. *PLoS One* 7: e33888.

Place, R.F., Li, L.C., Pookot, D., Noonan, E.J., and R. Dahiya, 2008. MicroRNA-373 induces expression of genes with complementary promoter sequences. *Proc Natl Acad Sci U S A*. 105: 1608-1613.

Poliseno, L., Salmena, L., Zhang, J., Carver, B., Haveman, W.J., and P.P. Pandolfi, 2010. A coding-independent function of gene and pseudogene mRNAs regulates tumour biology. *Nature* 465: 1033-38.

Przulj, N., Corneil, D.G., and I. Jurisica, 2006. Efficient estimation of graphlet frequency distributions in protein-protein interaction networks. *Bioinformatics* 22: 974-80.

Rajasethupathy, P., Antonov, I., Sheridan, R., Frey, S., Sander, C., Tuschl, T., and E.R. Kandel, 2012. A role for neuronal piRNAs in the epigenetic control of memory-related synaptic plasticity. *Cell* 149: 693-707.

Reczko, M., Maragkakis, M., Alexiou, P., Grosse, I., and A.G. Hatzigeorgiou, 2012. Functional microRNA targets in protein coding sequences. *Bioinformatics* 28: 771-76.

Rehmsmeier, M., Steffen, P., Hochsmann, M., and R. Giegerich, 2004. Fast and effective prediction of microRNA/target duplexes. *RNA* 10: 1507-17.

Rennie, W., Liu, C., Carmack, C.S., Wolenc, A., Kanoria, S., Lu, J., Long, D., and Y. Ding, 2014. STarMir: a web server for prediction of microRNA binding sites. *Nucleic Acids Res*. 42: W114-118.

Rusinov, V., Baev, V., Minkov, I.N., and M. Tabler, 2005. MicroInspector: a web tool for detection of miRNA binding sites in an RNA sequence. *Nucleic Acids Res* 33: W696-700.

Sales, G., Coppe, A., Bisognin, A., Biasiolo, M., Bortoluzzi, S., and C. Romualdi, 2010. MAGIA, a web-based tool for miRNA and Genes Integrated Analysis. *Nucleic Acids Res*. 38: W352-359.

Salmena, L., Poliseno, L., Tay, Y., Kats, L., and Pandolfi, P.P. 2011. A ceRNA hypothesis: the Rosetta Stone of a hidden RNA language? *Cell* 146: 353-58.

Samantarrai, D., Sahu, M., Roy, J., Mohanty, B.B., Singh, G., Bhushan, C., and B. Mallick, 2015. Unraveling novel TF-miRNA regulatory crosstalk in metastasis of Soft Tissue Sarcoma, *Sci Rep*. 5: 9742.

Sana, J., Faltejskova, P., Svoboda, M., and O. Slaby, 2012. Novel classes of non-coding RNAs and cancer. *J Transl Med* 10: 103.

Sanchez-Mejias, A., and Y. Tay, 2015. Competing endogenous RNA networks: tying the essential knots for cancer biology and therapeutics. *J Hematol Oncol* 8: 30.

Sanford, J.R., Wang, X., Mort, M., Vanduyn, N., Cooper, D.N., Mooney, S.D., Edenberg, H.J., and Liu, Y. 2009. Splicing factor SFRS1 recognizes a functionally diverse landscape of RNA transcripts. *Genome Res*. 19: 381-94.

Seemann, S.E., Richter, A.S., Gesell, T., Backofen, R., and J. Gorodkin, 2011. PETcofold: predicting conserved interactions and structures of two multiple alignments of RNA sequences. *Bioinformatics* 27, 211-19.

Segre, D., Deluna, A., Church, G.M., and R. Kishony, 2005. Modular epistasis in yeast metabolism. *Nat Genet.* 37: 77-83.

Sen, R., Ghosal, S., Das, S., Balti, S., and J. Chakrabarti, 2014. Competing endogenous RNA: the key to posttranscriptional regulation. *Scientific World Journal.* 2014: 896206.

Shirdel, E.A., Xie, W., Mak, T.W., and I. Jurisica, 2011. NAViGaTing the micronome--using multiple microRNA prediction databases to identify signalling pathway-associated microRNAs. *PLoS One* 6: e17429.

Siomi, M.C., Sato, K., Pezic, D., and A.A. Aravin, 2011. PIWI-interacting small RNAs: the vanguard of genome defence. *Nat Rev Mol Cell Biol.* 12: 246-58.

Stark, A., Brennecke, J., Russell, R.B., and S.M. Cohen, 2003. Identification of Drosophila MicroRNA targets. *PLoS Biol* 1: E60.

Stefani, G., and F.J. Slack, 2008. Small non-coding RNAs in animal development. *Nat Rev Mol Cell Biol* 9: 219-30.

Sun, J., Gong, X., Purow, B., and Z. Zhao, 2012. Uncovering MicroRNA and Transcription Factor Mediated Regulatory Networks in Glioblastoma. *PLoS Comput Biol* 8: e1002488.

Sun, Y., Tawara, I., Zhao, M., Qin, Z.S., Toubai, T., Mathewson, N., Tamaki, H., Nieves, E., Chinnaiyan, A.M., and P. Reddy, 2013. Allogeneic T cell responses are regulated by a specific miRNA-mRNA network. *J. Clin Invest* 123, 4739-54.

Tafer, H., and I.L. Hofacker, 2008. RNAplex: a fast tool for RNA-RNA interaction search. *Bioinformatics.* 24, 2657-63.

Taft, R.J., Pang, K.C., Mercer, T.R., Dinger, M., and J.S. Mattick, 2010. Non-coding RNAs: regulators of disease. *J Pathol.* 220: 126-39.

Taft, R.J., Pheasant, M., and J.S. Mattick, 2007. The relationship between non-protein-coding DNA and eukaryotic complexity. *Bioessays.* 29: 288-99.

Thomson, D.W., Bracken, C.P., and G.J. Goodall, 2011. Experimental strategies for microRNA target identification. *Nucleic Acids Res* 39: 6845-53.

Tsang, J., Zhu, J., and A. van Oudenaarden, 2007. MicroRNA-mediated feedback and feed-forward loops are recurrent network motifs in mammals. *Mol Cell* 26: 753-67.

Vera, J., Lai, X., Schmitz, U., and O. Wolkenhauer, 2013. MicroRNA-regulated networks: the perfect storm for classical molecular biology, the ideal scenario for systems biology. *Adv Exp Med Biol* 774, 55-76.

Vlachos, I.S., Paraskevopoulou, M.D., Karagkouni, D., Georgakilas, G., Vergoulis, T., Kanellos, I., Anastasopoulos, I.L., Maniou, S., Karathanou, K., Kalfakakou, D., et al. 2015. DIANA-TarBase v7.0: indexing more than half a million experimentally supported miRNA:mRNA interactions. *Nucleic Acids Res* 43: D153-59.

Wang, J., Lu, M., Qiu, C., and Q. Cui, 2010. TransmiR: a transcription factor-microRNA regulation database. *Nucleic Acids Res.* 38: D119-22.

Wang, K.C., Yang, Y.W., Liu, B., Sanyal, A., Corces-Zimmerman, R., Chen, Y., Lajoie, B.R., Protacio, A., Flynn, R.A., Gupta, R.A., et al. 2011. A long noncoding RNA maintains active chromatin to coordinate homeotic gene expression. *Nature.* 472: 120-24.

Wang, X., Arai, S., Song, X., Reichart, D., Du, K., Pascual, G., Tempst, P., Rosenfeld, M.G., Glass, C.K., and R. Kurokawa, 2008. Induced ncRNAs allosterically modify RNA-binding proteins in cis to inhibit transcription. *Nature.* 454: 126-30.

Wang, Z., Gerstein, M., and M. Snyder, 2009. RNA-Seq: a revolutionary tool for transcriptomics. *Nat Rev Genet.* 10: 57-63.

Watanabe, T., and H. Lin, 2014. Posttranscriptional regulation of gene expression by Piwi proteins and piRNAs. *Mol Cell* 56, 18-27.

Weick, E.M., and E.A. Miska, 2014. piRNAs: from biogenesis to function. Development 141, 3458-71.

Wenzel, A., Akbasli, E., and J. Gorodkin, 2012. RIsearch: fast RNA-RNA interaction search using a simplified nearest-neighbor energy model. *Bioinformatics* 28, 2738-46.

White, AG.; A. Ma'ayan, 2007. Connecting seed lists of mammalian proteins using Steiner Trees. Signals, Systems and Computers; ACSSC Conference Record of the Forty-First Asilomar Conference; 4–7: *IEEE*, 155-59

Wilhelm, B.T., Marguerat, S., Watt, S., Schubert, F., Wood, V., Goodhead, I., Penkett, C.J., Rogers, J., and J. Bahler, 2008. Dynamic repertoire of a eukaryotic transcriptome surveyed at single-nucleotide resolution. *Nature* 453: 1239-43.

Winter, J., Jung, S., Keller, S., Gregory, R.I., and S. Diederichs, 2009. Many roads to maturity: microRNA biogenesis pathways and their regulation. *Nat Cell Biol* 11, 228-234.

Wong, N., and X. Wang, 2015. miRDB: an online resource for microRNA target prediction and functional annotations. *Nucleic Acids Res* 43, D146-52.

Wu, Y.H., T.F., Hu, Y.C., Chen, Y.C., Y.N., Tsai, Y.H., Tsai, C.C., Cheng, and Wang, H.W. 2011. The manipulation of miRNA-gene regulatory networks by KSHV induces endothelial cell motility. *Blood* 118, 2896-905.

Xia, T., Q., Liao, X., Jiang, Y., Shao, B., Xiao, Y., Xi, and J. Guo. 2014. Long noncoding RNA associated-competing endogenous RNAs in gastric cancer. *Sci Rep* 4, 6088.

Xiao, F., Z., Zuo, G., Cai, S., Kang, X., Gao, and T. Li. 2009. miRecords: an integrated resource for microRNA-target interactions. *Nucleic Acids Res.* 37, D105-110.

Xiao, Y., Y., Lv, H., Zhao, Y., Gong, J., Hu, F., Li, J., Xu, J., Bai, F., Yu, and X. Li. 2015. Predicting the functions of long noncoding RNAs using RNA-seq based on Bayesian network. *Biomed Res Int*: 839590.

Xue, Y., Zhou, Y., Wu, T., Zhu, T., Ji, X., Kwon, Y.S., Zhang, C., Yeo, G., Black, D.L., Sun, H., et al. 2009. Genome-wide analysis of PTB-RNA interactions reveals a strategy used by the general splicing repressor to modulate exon inclusion or skipping. *Mol Cell* 36: 996-06.

Yang, J., Mani, S.A., Donaher, J.L., Ramaswamy, S., Itzykson, R.A., Come, C., Savagner, P., Gitelman, I., Richardson, A., and R.A. Weinberg, 2004. Twist, a master regulator of morphogenesis, plays an essential role in tumor metastasis. *Cell* 117: 927-39.

Yang, J.H., Li, J.H., Jiang, S., Zhou, H., and L.H. Qu, 2013. ChIPBase: a database for decoding the transcriptional regulation of long non-coding RNA and microRNA genes from ChIP-Seq data. *Nucleic Acids Res* 41: D177-87.

Yildirim, M.A., Goh, K.I., Cusick, M.E., Barabasi, A.L., and M. Vidal. 2007. Drug-target network. *Nat Biotechnol* 25: 1119-26.

Younger, S.T., Pertsemlidis, A., and D.R. Corey. 2009. Predicting potential miRNA target sites within gene promoters. *Bioorg Med Chem Lett* 19: 3791-794.

Zhang, C., and R.B. Darnell, 2011. Mapping *in vivo* protein-RNA interactions at single-nucleotide resolution from HITS-CLIP data. *Nat Biotechnol* 29: 607-14.

Zhou, X., J., Liu, and W. Wang. 2014. Construction and investigation of breast-cancer-specific ceRNA network based on the mRNA and miRNA expression data. *IET Syst Biol* 8: 96-103.

Section 3
Proteins

6

Annotation of Hypothetical Proteins- a Functional Genomics Approach

Sonali Mishra[1], Utkarsh Raj[2], Pritish Kumar Varadwaj[3]*

Abstract

Proteins are the most abundant bio-molecules found in living cells. The role of individual proteins and their interaction with other proteins are crucial for several cellular processes. Protein interacting with protein, DNA and ligand often drives the fate of various biochemical processes. There have been several efforts to annotate the individual protein functionality and its role in performing cellular processes. The impairment of protein functions can lead to an array of diseases and disorders in biosystems. With the advent of the post-genomic era there exists a large class of proteins with their functions still unknown. Around 150 genomes that are recently sequenced are yet to be assigned their physiological function and almost 30-50% of the genes are still having undefined functional characterization for their encoded proteins. Such class of proteins is termed as hypothetical proteins. A hypothetical protein is generally predicted to be expressed from ORFs (Open Reading Frame) with no experimental evidence of translation. These proteins constitute a considerable fraction of the proteomes of most of the organisms. Through the domain extraction of these proteins, it can be helpful to search and reveal many gene coding proteins, their structure and function. In order to understand the biological system of these species, their functional characterization of proteins is very important.

There are several prediction techniques available for the annotation of hypothetical proteins. The methods are most prevalent in both wet and dry laboratory. Although the traditional methods involving extraction of proteins from whole cell or their subcellular fraction can assign functions to these genes accurately, the process is expensive and time consuming at the same time. Computational approach for annotation of such hypothetical proteins utilizing the databases available on the basis of genome and proteome of certain organism are proved to be more precise, fast and accurate method. In this chapter, proteins with their relevance in organisms and the methods

---

[1] 965/200 – A, Sohbatiyabagh, Allahabad 211006, India. Tel: +91-9452262943
   E-mail: sonalimishra013@gmail.com
[2] RSA-304, IIIT-Allahabad, Allahabad 211012, India. Tel: +91-9936100030
   E-mail: utkarsh.iiita@gmail.com
[3.] CC2-4203, IIIT-Allahabad, Allahabad 211012, India. Tel: +91-9415264599
* Corresponding author : pritish@iiita.ac.in

for annotation of hypothetical proteins have been described. The various methods for annotation of hypothetical proteins such as protein-protein interaction-based method-comparative genomics-based method, genome context method, clustering method and in silico structural modeling based method, etc., have been described in brief with a detailed step-wise methodology for annotating gene or protein of any genus.

We have shown all those steps such as retrieval of sequence, prediction of gene ontology, homology search/structural modeling and finally the validation of various physicochemical and functional parameters through various softwares and databases. A comparative outlook of the methods and their relevance has been presented. Further we have shown the process in detail on a model organism *Candida dubliniensis* which contains a wide range of uncharacterized proteins. The various steps of its functional and structural characterization along with results have been described. As the field of bioinformatics is progressing day-by-day, the motive of research should also be shifted towards characterization rather than just discovery of new genomic sequences. The purpose of this chapter is to demonstrate the screening of hypothetical proteins by the application of various in silico methods which are intended to predict and validate various protein functions.

1. Introduction

Proteins are biological macromolecules capable of performing various functions in all living organisms. Each protein consists of multiple amino acids, which are the basic unit of proteins. There are 20 amino acids with specific functional group and distinct three-dimensional structure and conformation. The diversity of the proteins is basically maintained by the difference in the sequence of occurrence of amino acids. According to the amino acids composition and structure of these proteins, they perform various functions within the bio-system. The sequence of arrangement of the amino acids in a protein is decided by the genes responsible for their encoding. There exist 64 genetic codes which generally encode around 20 amino acids, although two new amino acids, namely, selenocysteine and pyrrolysine have also been recently added to this group (Aravind and Koonin 1999, Backert and Meyer 2006).

The amino acids composition of a protein sequence is crucial to decide the fate of protein folding in three-dimensional space and this event is often associated with coupling of chemo-mechanical parameters which allows proteins to act as various key bio-molecules, viz. enzymes, receptors for signal transduction, bio-motors and switches (Arigoni et al. 1998).

The Central Dogma (as shown in Figure 1) is based on the conversion of the genetic message embedded in DNA to a functional mRNA moiety through the process called transcription and subsequent conversion of the copied genotype to a phenotype in the form of proteins, through translation process (Crick 1970).

The understanding of this framework can describe the transfer of sequence information between biopolymers. Initially, the genetic information is copied from DNA to DNA through the replication and process further the information is propagated to m-RNA via transcription. Lastly, the proteins are synthesized in translation where m-RNA acts as a template carrying the information regarding the role and fate of the protein (Crick 1956).

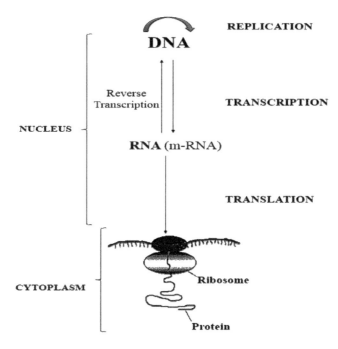

FIGURE 1   Central dogma of biology.

The translated proteins have diverse physiological functions in the organisms. As the huge protein data lies in the range of HPs whose functional annotation might get a better understanding of the organism's molecular biology. The functions of these proteins can be classified in different categories as mentioned in Table 1.

TABLE 1 Examples of Protein Functions

Proteins	Functions in nature	Instances
Structural	Provide structural basis	Collagen found in cartilage; keratin found in hair and nails.
Contractile	Move muscles	Myosin and actin contracting muscles fiber.
Transport	Transportation of essential substances in the entire body	Haemoglobin caries oxygen, lLipoprotein caries lipids.
Storage	Serves for nutrients storage	Casein for milk, ferritin for iron
Hormones	Regulate many metabolic reactions and the nervous system.	Insulin for blood glucose level, growth hormone for body growth.
Enzymes	Works as a catalyst in many biochemical reactions.	Sucrose and trypsin catalyze the hydrolysis of sucrose and protein respectively.
Protection	Protection against foreign substances.	Immunoglobulins stimulate immune responses in the body.

1.1   Protein structure

Protein structures can be classified at three levels— primary, secondary and tertiary (Figure 2). Primary protein structure is the linear sequence of amino acids through a series of amino peptide bonds. The planar structure exhibits resonance stabilization due to the presence of hydrogen bonds (Lodish et al. 2000) between neighbouring residues.

Protein secondary structures are result from the folding in primary structure due to the presence of hydrogen bonds between amide planes. These structural forms are broadly classified as alpha-helix, beta-sheets and loops. The alpha helix structure gets its typical shape due to intercalating hydrogen bonds. The beta-sheet is further classified into parallel and anti-parallel beta sheet depending on the patterns of hydrogen bond. The loop often facilitates the folding of protein complex structures by providing flexibility to rigid planar structure (Anfinsen 1973).

The tertiary structure of protein can be considered as the spatial arrangements a containing combination of several secondary structural moieties. For example, alpha-keratins are the fibrous proteins with two alpha right-handed alpha-helix intertwined to form an alpha-coiled coil, usually cross-linked by weak interactions. The structural intricacy of each protein, with specific combinations of alpha-helices, beta-sheets and turns is responsible for its associated function (Anfinsen 1973). Quaternary structure consists of multiple subunits of tertiary structure arranged in specific stereo-spatial order to make the native structure of protein.

1.2   Protein Synthesis

Proteins are synthesized through a process called *translation* which occurs in the cytoplasm where genetic codes are translated into respective proteins (Baran et al. 2006). Ribosomes mediate translation of genetic codes into polypeptide chains, which further undergo several post-translational modifications to become functional proteins.

Steps of translation

1. **Initiation**- mRNA enters into the cytoplasm and gets associated with ribosome. Base pairing (A-U, G-C) between m-RNA codons and t-RNA anticodons determines the order of amino acids in a protein.
2. **Elongation**- Ribosomes carry information along with m-RNA; the t-RNA transfer its amino acid to the growing protein chain and produces protein.
3. **Termination**- When the ribosomes encounter a stop codon, i.e., UAA, UGA, or UAG, the ribosome falls and the process is terminated.

1.3   Hypothetical proteins

Over the last few years, more than 3000 genomes of diverse bacteria, archaea and eukaryotes have been successfully sequenced. Rapid progress has been made towards defining the structures and functions of HPs, identifying the translocated effector molecules and elucidating the mechanisms by which the effectors subvert eukaryotic cellular processes during infection (Aravind and Koonin 1999). Although many organisms' genomes have been sequenced completely but still the potential functions of many genes are yet to be discovered. It is reported that in newly sequenced bacterial genome, around 30-40% of the genes have not been assigned definite function. We still have very little idea of what proteins they code and what are their potential target sites (Backert and Meyer2006). Many of these proteins are found in conserved domains with unknown functions and their functional annotation is very important for the acute knowledge of functional genomics and general biology as well.

It is noteworthy that crucial elements of various central pathways of information processing and metabolism could still be lurking among the 'conserved hypothetical's' because of which important processes and mechanisms, such as signaling and stress response in the respective organisms, remain unraveled. It can be inferenced from the fact that when a complete open reading frame (ORF) is annotated as a 'conserved hypothetical protein', it generally means that the gene is conserved across many organisms but its function is still unknown (Arigoni et al. 1998). In most of the cases, the general prediction of a gene's function can be made on the basis of a conserved sequence motif or subtle sequence similarity either in comparison with previously characterized proteins or in the presence of characteristic structural features. Looking at the solution, we see that many of these conserved proteins can be easily predicted and characterized as ATPases, GTPases, methyltranferases, metalloproteases, DNA and RNA binding proteins and membrane transporters, etc.

The major region of the genomes reported is still under suspicion regarding their function and comes under the range of hypothetical proteins (Fig 2).

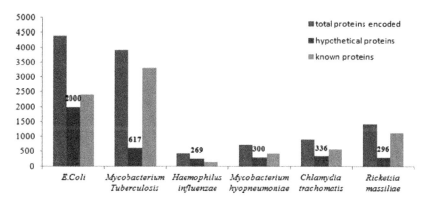

FIGURE 2 Histogram showing the comparative representation of the total number of encoded proteins, HPs and known proteins in various organisms.

There are many proteins for which general biochemical functions have been predicted but still we are unaware of their exact biological function. In order to gather information about such proteins, we can access certain databases such as COG for phylogenetic classification of proteins encoded in complete genomes and their functional group R. COG is a genome database which contains phylogenetic classification of proteins encoded in complete genomes. Out of all the uncharacterized proteins, there are two categories of proteins which are found in closely related organisms while others are found in distantly related organisms. There are many parasitic pathogens which have to reduce their genome size for adapting to a parasitic lifestyle which leads to a drastic loss of genes which might be encoding for many metabolic enzymes, transcriptional regulators and membrane permeases. So the genes that have been retained, are considered essential for cell survival and become an important part of hypothetical's classification which also give them an attracting significant attention as potential therapeutic targets.

Some *in-silico* techniques have been initiated to characterize a hypothetical protein and uncover their structural and functional information. Several earlier experimentations were being conducted to study the presence and role of hypothetical proteins in various eukaryotic and prokaryotic organisms. It has been reported that the flowering plant *Arabidopsis thaliana* which contains 25,498 genes, encodes proteins that covers 11,000 families and thus several HPs could be in annotated by its genome level comparison with other organisms like *Drosophila* and *Caenorhabditis elegans* (Galperin and Koonin 2004, Rounsley et al. 2000).

As the inference of an *in-silico* study on *Arabidopsis thaliana, it* was concluded that beta-hairpin-alpha-hairpin repeats are the characteristics of the ankyrin family. The presence of ankyrin repeats indicated its role in protein-protein interaction which is essential for various metabolic processes in organisms (Heynh et al. 2005). The structural and functional features were predicted by the folds that were recognized as beta-hairpin-alpha-hairpin repeat with ankyrin repeat super family and the dominance of such coiled regions indicated the high level of conservation and stability of the protein structure (Li et al. 2006).

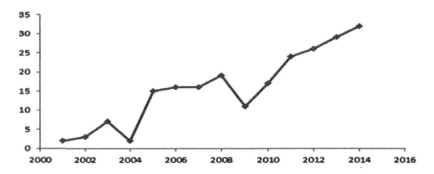

FIGURE 3   Work done on annotation of hypothetical proteins in last 15 years

From the reference of PubMed articles, there are 6360 research articles available on hypothetical proteins and its predictions. From this entire record, around 197 of them are entirely on functional annotation of these proteins in which 20 and more are recent works published in 2013-14 (Fig 3).

In 2014, the ribosomal phosphoprotein P_0 of the human malaria parasite *Plasmodium falciparum* (P_fP_0) was identified as a surface protein working for protection. Its function is also found in the nucleus of Drosophila and further could be classified into many classes such as ribosomal proteins, nucleotide binding and hypothetical integral membrane proteins (Aruna et al. 2004).

In 2006, a study on *Mycobacterium tuberculosis* uncovered an important function of proteins through biochemical studies and presented their role in hydrating short trans-2-enoyl-coenzyme pathway genomes. Another HP of *Mycobacterium tuberculosis* Rv0130 protein was found to be highly conserved R-specific hydrolase motif buried deeply between the two monomers (Castell et al. 2005).

Mycoplasma hyopneumoniae, a pathogenic organism which causes enzootic pneumonia, was studied for the annotation of HPs in 2012. It has been reported that about 42% of the 716 coding sequence are annotated as hypothetical proteins. In the study, it was found that three proteins were involved in various metabolic processes and two proteins were found to be crucial in the transcription process (da Fonsêca et al. 2012).

Arabidopsis thaliana is a model plant that contains many genes that are conserved in other plants as well and thus is very useful in comparative genomic analysis to explore hypothetical protein. One such instance can be presented as in case of *Oryza sativa* in 2012 where SA-JA signaling pathway was considered to screen 13 known gene sequences similar in both plants and the results suggested the conserved nature of these proteins (Indra et al. 2012).

Again in 2013, an important experimentation in the direction of annotation was successful in reporting *Rv* hypothetical proteins of a virulent strain *of Mycobacterium tuberculosisH37Rv*. Further, its comparative genomics with a model of *M tuberculosis H37Rv* lead to uncover the pathway with mechanism of virulence (Zahra 2013).

In 2013, *Mycobacterium leprae* was studied for annotation of hypothetical proteins in order to find new therapeutic drug targets. In this study, active site detection was performed by modeling and simulating an important protein Acyl-CoA synthetase which was also found to be playing an important role in fatty acid metabolism in *M leprae* (Anjum 2013).

In the same year, a research article was published describing the annotation of genes from *Staphylococcus aureus*. The study has revealed the functions of several HPs as ATP binding proteins, multiple antibiotic resistance (MAR) export proteins, helix turn helix domains, arsenate reductase, elongation factors, ribosomal proteins, etc. (Ramadevi and Subhashree 2012).

The study carried out in 2014, has suggested the presence of 2.2% of proteins of *Candida dubliniensis* as conserved hypothetical protein (HPs). Further, 27 HPs were found with well-defined functions and were characterized as enzymes, nucleic acid binding proteins, transport proteins, etc. Five of them showed adhesion character which is likely to be essential for the survival of yeast and pathogenesis. This study was found to be very helpful in understanding the mechanism of virulence, drug resistance, pathogenesis, adaptability to host and most importantly drug discovery for the treatment of *C dubliniensis* infections (Kumar et al. 2014).

Further, in 2014, *Pseudomonas aeruginosa* (a bacterium resistant to a large number of antibiotics and disinfectants) was being studied to find the structure and function of PAZ481 protein to determine its role in antibiotic resistance. Through its HPs functional annotation, it was found that many proteins are involved in the physiology and electron transport and protein pumping to generate ATP of this bacterium (David et al. 2014).

Another work published in 2014 on *Bacillus lehensis*, suggested the role of HPs in metal binding. An HP termed as Bleg1_2507 was found to contain Thioredoxin (Trx) domain and possess highly conserved metal binding residues Cys69, Cys73 and His159 which are conserved in Sco proteins of all prokaryotes and eukaryotes (Soo et al. 2014).

In the same case, Bleg1_2507 was found to have low sequence identity (47%) with BsSco but interestingly its metal binding site residues were located at flexible active loops which corresponds to other Sco proteins. This shows another peculiar nature of HPs and importance of their study (Noor et al. 2014).

In the same year, a research article was published on functional annotation of an important pathogenic organism *Chlamydea trachomatis*. This research could successfully annotate the functions of HPs as protease, ligase, synthase, translocase and zinc finger domain (Mishra et al. 2013). The data results from this study facilitate identification of potential therapeutic targets and enable the search for new inhibitors or vaccines.

In 2014, Biochem Pharmacology published a study on the functional annotation of hypothetical proteins of *Leptospiro interrogans*. Looking at the pathogenic nature of *Leptospiro interrogans* it was very important to annotate the genes and proteins of this organism and through *in-silico* study, it was found that proteins from families Ado Met DC, LRR and PilZ are mostly conserved in many microorganisms and thus can be targeted to develop many efficacious drug molecules (Bidkar et al. 2014).

Through *in-silico* based methods, around 35 proteins out of 114 conserved hypothetical proteins of *Rickettsia massiliae* (MTU5), have been annotated. As *Rickettsia massiliae* is the cause of Rocky Mountain spotted fever (RMSF) and is the prototype bacterium in the spotted fever group of *Rickettsia* so its functional annotation becomes an important aspect to find the potential target sites for the cure of many diseases (Hoskeri et al. 2010).

Neisseria gonorrhoeae causes gonorrhoea. This particular family of proteins has increasing prevalence of strains with resistance to antibiotics. Annotation of these proteins finds way to development of drugs targeted specifically at these proteins and thus it still remains a challenge to researchers (Neeraj et al. 2013).

Haemophilus influenzae is a multi-drug resistance strain and it demands development of better/new drugs against its pathogen proteins (HPs). Through the help of *in- silico*-based methods the amino acid sequences of all 429 HPs have been extensively annotated and function is allotted to around 296. HPs were found to belong to various classes of proteins such as enzymes, transporters, carriers, receptors, signal transducers, binding proteins, virulence and other proteins (Shahbaaz et al. 2013).

2. Methodology for functional annotation of hypothetical proteins

The hypothetical proteins may contain many biologically important functional properties which could be annotated by various methods. In the following section, it has been tried to explain the brief methodology of such methods with required description of the softwares and databases used on the basis of the results obtained through various researches done on the organism that belong to different families, genera and species. There are some general wet laboratory-based and some function and structure-based methods such as Rosetta stone method, genome context method, comparative genomics, clustering and *in-silico* based approach.

2.1 Methods based on protein-protein interactions

Proteins interact with each other in order to carry out certain functions such as transcriptional factors that also interact among themselves in order to perform transcription. So, somehow this interaction pattern can reveal the function of the protein. Rosetta stone method can predict the interaction pathway of such proteins very easily (Enault et al. 2005). It can be illustrated as if we have two proteins A and B and their analog AB is found in some other organism that indicates that they have interaction. This method can be considered effective because a biochemical function in many cases depends on the action of a multi-meric complex demonstrating a correlation between co-interacting proteins and their functions (Shailesh 2008).

2.2 Methods based on comparative genomics

Comparative genomics method is based on the assumption that the proteins that function together either in metabolic pathway or in structural complex could be expected to have been originated from the same evolution. Phylogenetic profiling can tell if the proteins are functionally linked. Moreover, phylogenetic profile is like a string with one bit and '*n*' entries, here *n* is the number of genomes under consideration. So, if the *n*th genome contains a homolog then the *n*th entry can be represented as unity in the Phylogenetic profile (Ranea et al. 2007). Further, clustering of these profiles can determine the proteins having common profile.

2.3 Genome context methods

In this method, we predict functional associations between protein-coding genes by analyzing gene fusion events, the conservation of gene neighborhood, or the significant co-occurrence of genes across different species (Fan et al. 2012). It is different from an homology-based approach and predicts functional associations between proteins, such as physical interactions, or co-membership in pathways, regulators or other cellular processes (Fiser and Sali 2003).

2.4 Clustering approaches

Clustering of genes is done by several methods and it can reveal which proteins come under the same cluster. It's a process of grouping on the basis that genes of the same cluster are involved in similar functions. Hence, the protein that is coded by this gene will also have the same function (Catell 1943).

2.5 *In-silico* Structural Modeling-based Method

This method has been the most feasible, sophisticated and relevant for annotation of hypothetical proteins from the last many years. In this section, we have tried to collect all the possible strategies for *in-silico* based annotation of hypothetical proteins in simplified steps such as retrieval of sequence/data collection, structural modeling/structural analysis and functional characterization/annotation. Although several approaches have been utilized in the recent past for annotation of HPs, a generalized flowchart summarizing all such efforts is given in **Annexure I**.

The steps involved in identification and annotations as shown in Figure 4 and also been summarized as follows:

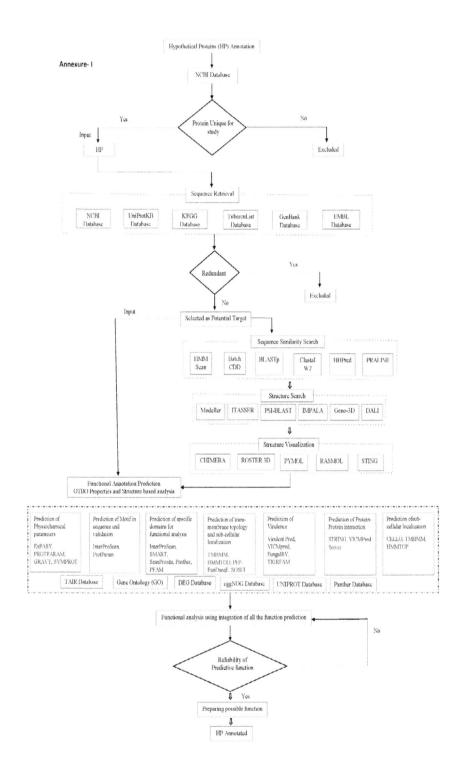

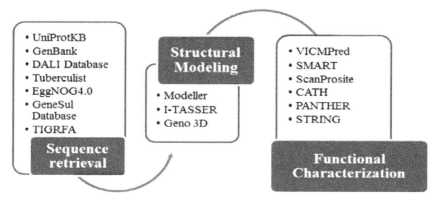

FIGURE 4   Flow of annotation with related databases and software's

A. Retrieval of Sequence

Hypothetical proteins sequence is extracted from the database for prediction of genes. The primary sequence of the desired organism could be obtained from the available genomic database (Table-2). Further, the primary sequence is compared to homologous sequences found in database using similarity search tools to find conserved domains. After this search, we can analyze the secondary structure of the protein (Galperin and Koonin 2004). BLAST or any similarity search tool will find the conserved region. The structure could be modeled if not present in the Protein Data Bank (PDB) in order to deduce the 3D structure of proteins and analyze. This could be done by taking the templates from PDB and further homology modeling could be done by SWISS PDB, it's a fully automated protein structure homology-modelling server and easily accessible via the ExPASy web server (Christophe et al. 2002). For a better understanding of the procedure, we have added the mechanism, results and specific software details of the model organism *Candida dubliniensis.*

TABLE 2   Predicted function of HPs from Candida dubliniensis

S.N	Gene ID	UniProt ID	Protein Function
1	8045310	B9W9J1	Peroxiredoxin activity
2	8047341	B9WFD2	Phosphoinositide binding
3	8047346	B9WFD7	Structural protein
4	8047351	B9WFE4	ATP binding
5	8047353	B9WFE6	RNA binding
6	8047358	B9WFF1	Protein binding
7	8047371	B9WFG4	Phosphatase
8	8047376	B9WFG9	RNA binding
9	8047379	B9WFH2	Transporter activity
10	8047381	B9WFH4	Kinase activity
11	8047605	B9WFM3	Protein binding
12	8047460	B9WFR1	Transferase activity

S.N	Gene ID	UniProt ID	Protein Function
13	8047467	B9WFR8	DNA binding
14	8047468	B9WFR9	Transferase activity
15	8047470	B9WFS1	Hydrolase activity
16	8047471	B9WFS2	DNA binding
17	8047600	B9WFS4	Protein binding
18	8047474	B9WFS6	RNA binding
19	8047491	B9WFU3	Oxidoreductase activity
20	8047495	B9WFU7	DNA binding
21	8047497	B9WFU9	Hydrolase
22	8047654	B9WFV3	DNA binding
23	8047663	B9WFW2	Kinase activity
24	8047669	B9WFW8	DNA binding
25	8048708	B9WIB2	Protein binding
26	8048742	B9WIF0	Oxidoreductase activity
27	8048763	B9WIH5	Transport activity

As in the case of *Candida dubliniensis,* the sequences were retrieved from NCBI. A sequence similarity search was carried out using BLAST. ClustalW2 is a multiple sequence alignment tool for aligning more than two sequences. Further, the physical and biochemical properties of the sequences were calculated by ProtParam. The Gene IDs and UniProt ID of *Candida dubliensis* and conserved domains were identified as shown in Table 3 (Kumar et al. 2014).

TABLE 3 List of domains identified in the HPs from Candida dubliniensis

S.N	UniProt ID	Conserved Domain (super family)
1.	B9W9J1	Carboxymuconolactone decarboxylase (CMD)
2.	B9WBA5	Hypothetical protein FLILHELTA
3.	B9WFD2	ANTH domain family
4.	B9WFE4	Archaeal ATPasea
5.	B9WFG4	PP2Cc super family
6.	B9WFG8	lipoprotein A (RlpA)-like double-psi beta-barrel
7.	B9WFG9	PIN domain
8.	B9WFH2	Major facilitator superfamily (MFS) Sugar (and other) transportera
9.	B9WFH4	Diacylglycerol kinase catalytic domain (DAG) LCB5; Sphingosine kinase and enzymesa
10.	B9WFM3	CUE domain
11.	B9WFP3	Oxidoreductase-like protein, N-terminal
12.	B9WFR8	Oxidoreductase-like protein, N-terminal, GAL4-like Zn2Cys6 binuclear cluster DNA-binding domain GAL4-like Zn(II)2Cys6 (C6 zinc) binuclear cluster DNA-binding Domain
13.	B9WFR9	CoA-transferase family III Predicted acyl-CoA transferases/carnitine dehydratasea
14.	B9WFS1	Putative lysophospholipase, Alpha/beta hydrolase family

S.N	UniProt ID	Conserved Domain (super family)
15.	B9WFS2	fungal transcription factor regulatory middle homology region, GAL4-like Zn2Cys6 binuclear cluster DNA-binding domain, GAL4-like Zn(II)2Cys6 (C6 zinc) binuclear cluster DNA-binding domain
16.	B9WFS4	Chaperone for protein-folding within the ER, fungal
17.	B9WFS6	Putative RNA methyltransferase
18.	B9WFU3	Protein disulfide isomerase (PDIa) family, Protein disulfide oxidoreductases and proteins with a thioredoxin fold
19.	B9WFU7	Fungal transcription factor regulatory middle homology region, GAL4-like Zn2Cys6 binuclear cluster DNA-binding domain, GAL4-like Zn(II)2Cys6 (C6 zinc) binuclear cluster DNA-binding Domain
20.	B9WFV3	GAL4-like Zn2Cys6 binuclear cluster DNA-binding domain, GAL4-like Zn(II)2Cys6 (C6 zinc) binuclear cluster DNA-binding domain
21.	B9WFW2	Yersinia pseudotuberculosis carbohydrate kinase-like subgroup, Nucleotide-binding domain of the sugar kinase/HSP70/actin superfamily FGGY-family pentulose kinase
22.	B9WFW8	Rad17 cell cycle checkpoint protein

B. Structural Modeling

The structure of protein is also helpful in function prediction as it reveals many physicochemical characteristics of proteins. Therefore, in order to predict the 3D structure, the templates of proteins could be used. Further, many tools could be used such as Modeller for homology and comparative modeling of protein 3D structure (Fiser and Sali 2003), I-TASSER which is the number one server for protein structure and function prediction and uses the hierarchical method for Threading (Roy et al. 2010), Geno 3D, it's a comparative molecular modeling software (Christophe 2002, Mayrose et al. 2004) (Table-1). The comparison of structural information of proteins, such as arrangement of alkyrin repeats (ANK), alpha helix and beta sheets, with other organisms can reveal their functional properties (Huaiyu et al. 2013). The structure prediction methods can be classified in two ways—*ab initio methods* which predicts a protein structure based on physico-chemical principles directly, and *template-based methods*, which uses known protein structures as templates. Template based methods involve homology or comparative modeling, and fold recognition via threading. An online Protein Structure Prediction Server (PS2-v2 - PS Square version 2) used for template based method which is considered as the most user-friendly and generally uses the principle of pairwise and multiple sequence alignment (Aloy et al. 2005).There are various factors of protein' functions which could be depicted from its structural information such as dominance of coiled regions indicate high levels of conservation and stability of protein. Alkyrin repeats exclusively function to mediate protein-protein interaction some of which can be directly involved in human cancer and other diseases. The folds were recognized using PFP-FunDSeqE as mentioned in Table 4.

TABLE 4 Different types of folds identified in HPs from Candida dubliniensis

S.N	Fold type	UniProt ID
1.	Beta-trefoil	B9W9J1, B9WFG9
2.	Small inhibitors, toxins, lectins	B9WBA5, B9WFS4
3.	Immunoglobulin-like	B9WFD2, B9WFE4,B9WFF7, B9WFG4, B9WFH2, B9WFH4, B9WFP3, B9WFR1, B9WFS0, B9WFS6, B9WFT3, B9WFT7, B9WFT8, B9WFV7, B9WIB2, B9WIC3, B9WIF4, B9WIG2, B9WIH4
4.	DNA binding 3-helical	B9WFD7, B9WFR8
5.	4-helical cytokines	B9WFE6, B9WFS2
6.	Ob-fold	B9WFF1
7	Viral coat and capsid proteins	B9WFG8, B9WFU7, B9WFX1, B9WIG1, B9WIH5
8.	4-helical up and down bundle	B9WFM3
9.	TIM-barrel	B9WFR9, B9WFW8
10.	Hydrolases	B9WFS1
11.	Thioredoxin like	B9WFU3, B9WIF0
12.	Belta-grasp	B9WFU9
13.	Cupredoxins	B9WFV3
14.	Ribonuclease h-like motif	B9WFW2
15.	Cona-like lectin/glucanases	B9WIA6

C. Functional Characterization

Functional characterization of proteins classifies them into categories based on their functions and other basic properties which could be useful for the studies related to protein-protein interaction, target sites, protein functions, etc. The functions could be predicted from structural features as well as from the physicochemical properties. So, modeled structures could be further applied for both structural and functional classification and characterization of HP's.

There are various tools that can identify functions of HP's such as the conserve domain search. **Conserved Domain** could be used for the identification of motifs and domains in proteins and it is an important aspect as per the classification of protein sequences and functional annotation (Marchler et al. 2015). The specific domains and particular amino acid sequences **InterProScan** (for the classification of proteins into families for their functional analysis), **Interpro**, **ScanProsite** (PROSITE consists of documentation entries describing protein domains, families and functional sites as well as associated patterns and profiles to identify them) and **SMART**. The evolutionary relation and biological process can also reveal the functions (**Panther** which is a library of protein families and subfamilies indexed by function, **Pfam** is a large collection of protein families, each represented by multiple sequence alignments and hidden Markov models (HMMs)). A novel genome-wide domain prediction method, SECOM, is also in practice which first indexes all the proteins in the genome by using a hash seed function (Fan et al. 2012). The physical properties also play a very important role in functional classification (SOSUI is a classification and secondary structure prediction system for membrane proteins, **TMHMM** is for prediction of transmembrane helices in proteins and it has been rated as the best for such prediction, **Psort-2** is a bioinformatics tool used for the prediction of protein locali-sation sites in cells. It receives the information of an amino acid sequence and its taxon, **SignalP** server predicts the presence and location of signal peptide cleavage sites in amino acid sequences from different organisms: Gram-positive prokaryotes, Gram-negative prokaryotes, and eukaryotes **and HMMTOP** are for the prediction of transmembrane helices and topology of proteins). Further, other factors such as virulence factors can be predicted by **VICMPred** for prediction of Virulence factors, Information molecule, Cellular process and Metabolism molecule in the Bacterial proteins, cellular localization of proteins can be checked by **Cello software** which-has been working very successfully with the NHS.

In order to assign a precise function to HPs, all the sequences could be analyzed from different databases such as Conserved Domain Database (CDD) (Marchler-Bauer et al., 2011), SMART (Letunic et al., 2012), ScanProsite (Sigrist et al. 2012), CATH (Cuff et al. 2011) and PANTHER (Paul et al. 2003) etc. CDD contains curated domain model based on the tertiary structure of the protein to provide sequence/ structure/function relationship in an organized hierarchy of family and superfamily. The function of SMART is to compare glutamate or aspartate and a histidine, in their catalytic domain.

Certain characters of these proteins such as pathogenicity can be understood by phosphate and transferase levels. The decrease in phosphate concentration of the host may lead to increase in virulence of pathogens like *C. albicans, Candida glabrata* and *Saccharomyces cerevisiae.* On the other hand, in yeast, some tranferases have been found to play significant role against oxidative stress. Many HPs possess enzy-matic activities and are categorized as hydrolases, phosphatase, tranferases, kinase, oxidoreductases, and peroxiredoxin. In case of our model organism, *Candida dub-liniensis,* the functional categories of the HPs is mentioned in Table 5 and the list of softwares which are used in functional annotation has also been listed (Table 6). Table 7 comprises a list of databases required for functional annotation of HPs.

TABLE 5 Functional categories of HPs

Predicted functions	**(Hypothetical Proteins)HPs**
Enzymatic activity	
Hydrolase activity	B9WFS1, B9WFU9
Phosphatase activity	B9WFG4
Transferase activity	B9WFR1, B9WFR9
Kinase activity	B9WFH4, B9WFW2
Oxidoreductase activity	B9WIF0, B9WFU3
Peroxiredoxin activity	B9W9J1
Binding protein	
DNA binding	B9WFR8, B9WFS2, B9WFU7, B9WFV3, B9WFW8
RNA binding	B9WFE6, B9WFS6, B9WFG9
Protein binding	B9WFF1, B9WFM3, B9WFS4, B9WIB2
ATP binding	B9WFE4
Phosphoinositide-binding	B9WFD2
Other proteins	
Transport activity	B9WFH2, B9WIH5
Structural protein	B9WFD7

TABLE 6 List of softwares used in functional annotation

S. no.	Softwares Used	Description	Links
1.	I-TASSER	For protein structure and function predictions	http://zhanglab.ccmb.med. umich.edu/I-TASSER/ (Yang et al. 2015)
2.	Geno3D	It's a web server to generate protein 3D model.	https://geno3dprabi.ibcp.fr/ (Roshni et al. 2014)
3.	InterPro	Protein sequence analysis & classification.	http://www.ebi.ac.uk/interpro/ (Mulder et al. 2008)
4.	BLASTcds	Finds regions of similarity between biological sequences.	http://blast.ncbi.nlm.nih.gov/ Blast.cgi (Altschul et al. 1990)
5.	COGs	It's a sequence alignment for each domain is constructed, which allows a novel sequence to be matched rapidly to domains already in the library.	http://www.ncbi.nlm.nih.gov/ COG (Neidhardt et al. 1996)
6.	CDART	Domain Architecture Retrieval Tool	http://www.ncbi.nlm.nih.gov/ Structure/lexington/lexington. cgi (Geer et al. 2002)
7.	SCOP	Its a tool for browsing and analyzing structural classification of proteins (SCOP) data.	http://supfam.org/ SUPERFAMILY/hmm.html (Meier and Söding 2015)
8.	Gene Ontology Consortium	Structural classification of proteins from Gene Ontology Database.	http://geneontology.org/ (Carbon et al. 2009)
9.	Gene Threader	It is a software to compute gene structure predictions	http://genomethreader.org/ (Gremme et al. 2005)
10.	PROSPECT-PSPP	2D and 3D structure prediction.	http://csbl.bmb.uga.edu/ protein_pipeline (Moult et al. 2003)
11.	InterProScan	Matches against the InterPro collection of protein signature databases.	http://www.ebi.ac.uk/Tools/pfa/ iprscan5/ (Bruce and Reid 2015)
12.	TrEMBL	It is a computer-annotated supplement of SWISS-PROT that contains all the translations of EMBL nucleotide sequence entries.	http://www.ebi.ac.uk/uniprot/ TrEMBLdocs/trembl_release_ notes_13.html (Emmert et al. 1994)
13.	DISULFIND	It checks the disulfide bonding state and the cysteine connectivity prediction server.	http://disulfind.dsi.unifi.it/ (Fariselli et al. 1999)
14.	CDHIT	Clustering analysis of various types of DNAs and RNAs.	http://weizhong-lab.ucsd.edu/ cdhit_suite/cgibin/index.cgi (Weizhong et al. 2001)
15.	ProtoNet	ProtoNet provides automatic hierarchical classification of protein sequences.	http://www.protonet.cs.huji. ac.il/introduct.php? global=prot onet|no|6|61|lifetime (Sasson et al. 2003)
16.	MED-SUMO	They find similar binding sites and thus may perform similar functions.	http://www.medit.fr/ (Doppelt et al. 2009)

17.	SCANPROSITE	It is a new and improved version of the web-based tool for detecting PROSITE signature matches in protein sequences.	http://prosite.expasy.org/scanprosite/ (Gattiker et al. 2002)
18.	VAST(Vector Alignment Search Tool)	Stands for Vector Alignment Search Tool. Is a computer algorithm developed at NCBI and used to identify similar protein 3-dimensional structures	http://structure.ncbi.nlm.nih.gov/VAST/vast.shtml (Madej et al. 2014)
19.	DaliLite(EMBL_EBI)	DaliLite computes optimal and suboptimal structural alignments between two protein structures.	http://www.ebi.ac.uk/Tools/structure/dalilite/ (Brooksbank et al. 2014)
20.	GOR	Its a Protein Secondary Structure Prediction Server	http://gor.bb.iastate.edu/
21.	STRING	Search Tool for the Retrieval of Interacting Genes/Proteins. The interactions include direct (physical) and indirect (functional)	http://string-db.org/
22.	SOSUI(MAFET SERVER)	For secondary structure prediction of membrane protein.	http://nhjy.hzau.edu.cn/kech/swxxx/jakj/dianzi/Bioinf7/Expasy/Expasy10.files/sosuimenu0.htm (Hirokawa et al. 1998)
23.	IMPALA	Integrated Molecular Pathway Level Analysis. It does the pathway over-representation and enrichment analysis with expression and / or metabolite data	http://impala.molgen.mpg.de/ (Kamburov et al. 2011)
24.	T-COFFEE	A collection of tools for Computing, Evaluating and Manipulating Multiple Alignments of DNA, RNA, protein sequences and structures.	http://tcoffee.vitalit.ch/apps/tcoffee/index.html(Notredame et al. 2000)
25.	Q Site Finder	An energy-based method for the prediction of protein–ligand binding sites	http://www.ebi.ac.uk/pdbe-site/pdbemotif/(Aloy 2001)
26.	ProFunc	Is a popular web server composed of a compendium of structure–based and sequence–based methods.	http://www.ebi.ac.uk/thornton-srv/databases/profunc/
27.	SuMo	Used for 3D search for protein functional sites.	http://sumo-pbil.ibcp.fr/cgi-bin/sumo-welcome
28.	Site Engine	Recognizes regions on the surface of one protein that resembles a specific binding site of another.	http://bioinfo3d.cs.tau.ac.il/SiteEngine/
29.	PatchFinder	Identification of functional regions in proteins.	http://patchfinder.tau.ac.il/ (Mandel et al. 1995)

Computational Biology and Bioinformatics

TABLE 7 List of databases required for functional annotation

S. no.	Databases	Description	Links
1.	UniProtKB	The mission of UniProt is to provide the scientific community with a comprehensive, high-quality and freely accessible resource of protein sequence and functional information.	http://www.uniprot.org/ (Jain et al. 2009)
2.	GenBank(NCBI)	Nucleic acid sequences provide the fundamental starting point for describing and understanding the structure, function, and development of genetically diverse organisms.	http://www.insdc.org/ files/feature_table.html#1 (Benson et al. 2004)
3.	DALI Database	The Dali Database is based on all-against-all 3D structure comparison of protein structures in the Protein Data Bank (PDB).	http://ekhidna.biocenter. helsinki.fi/dali/start (Nykyri et al. 2012)
4.	Tuberculist	The TubercuList knowledge base integrates genome details, protein information, drug and transcriptome data, mutant and operon annotation, bibliography, structural views and comparative genomics, in a structured manner required for the rational development of new diagnostic, therapeutic and prophylactic measures against tuberculosis.	http://tuberculist.epfl.ch/ (Rosenkrands et al. 2000)
5.	EggNOG4.0	eggNOG (evolutionary genealogy of genes: Non-supervised Orthologous Groups) is a database of orthologous groups of genes.	http://eggnog.embl.de/ version_4.0.beta/ (Powell et al. 2012)
6.	QuikGO(EBI)	GO slims are lists of GO terms that have been selected from the full set of terms available from the Gene Ontology project.	http://www.ebi.ac.uk/ QuickGO/
7.	DEG(Database of Essential Genes)	Essential genes are those indispensable for the survival of an organism, and therefore are considered a foundation of life. DEG hosts records of currently available essential genomic elements, such as protein-coding genes and non-coding RNAs, among bacteria, archaea and eukaryotes.	http://tubic.tju.edu.cn/ deg/ (Kobayashi et al. 2003)
8.	GeneSul Database	The Gene Ontology (GO) project is a collaborative effort to address the need for consistent descriptions of gene products across databases.	http://geneontology.org/
9.	TAIR	Data available from TAIR includes the complete genome sequence along with gene structure, gene product information, gene expression, DNA and seed stocks, genome maps, genetic and physical markers, publications, and information about the Arabidopsis research community.	http://www.arabidopsis. org/ (Rhee et al. 2006)

10.	CD Search	To find Conserved Domains and Protein Classification	http://www.ncbi.nlm.nih. gov/Structure/cdd/docs/ cdd_search.html
11.	Genome Cube	Source BioScience LifeSciences are European leaders in DNA sequencing, genomic services, bioinformatic analyses and offer a comprehensive portfolio of clones, genomic reagents and antibodies.	http://www.lifesciences. sourcebioscience.com/ about-us/
12.	Joint Center	Joint Center for Computational Biology and Bioinformatics	http://www.jcbi.ru/EN/
13.	Gene Ontology(GO)	The Gene Ontology (GO) project is a collaborative effort to address the need for consistent descriptions of gene products across databases.	http://geneontology.org/ (Carbon et al. 2009)
14.	CDD BLAST	Its collection of domain models includes a set curated by NCBI, which utilizes 3D structure to provide insights into sequence/structure/function relationships.	http://www.ncbi.nlm.nih. gov/cdd (Finn et al. 2010)
15.	INTEPROSCAN	This form allows to scan sequence for matches against the InterPro collection of protein signature databases.	http://www.ebi.ac.uk/ Tools/pfa/iprscan5/ (Altschul et al. 1997)
16.	PFAM	PFAM database is a large collection of protein families, each represented by multiple sequence alignments and hidden Markov models (HMMs).	http://pfam.xfam.org/ (Sonnhammer et al. 1997)
17.	TIGRFAMs	TIGRFAMs is a resource consisting of curated multiple sequence alignments, Hidden Markov Models (HMMs) for protein sequence classification, and associated information designed to support automated annotation of (mostly prokaryotic) proteins.	http://www.jcvi.org/ cgibin/tigrfams/index.cgi (Haft et al. 2001)
18.	KEGG	Database of metabolic pathway (Kyoto Encyclopedia of Genes and Genomes).	http://www.genome.jp/ kegg/ (Kanehisa et al. 2014)

3. Conclusion

Annotation of hypothetical proteins is an essential requirement in the present era. Looking at the large number of genomic data and protein sequence data availability, there exists a requirement of further attempts to study their physiology and functions. The functions of these proteins can reveal many new crucial proteins which can act as potential therapeutic targets and can also identify their role in pathways of many biochemical cycles. A systematic process of functional annotation of these hypothetical proteins through bioinformatics tools has proven to be very useful in recent years. *In-silico* based approach throws better choice over the other traditional methods, as it provides a less time-consuming and more sophisticated approach with quality control on each intermediate steps. The step-wise procedure of functional annotation of hypothetical proteins can further utilize the essence of machine learning algorithms to classify the proteins on the basis of their functional properties.

The proper utilization of the tools and databases that are already available, facilitate knowledge discovery. The identification of novel protein functions might reveal many functional properties of the organism and can be described on the basis of the molecular features associated with the organism. An in-depth study of the hypothetical proteins could open several new functional pathways and will help the current molecular biology, drug designing and proteomics studies.

References

Aloy, P. 2001. Automated structure-based prediction of functional sites in proteins: applications to assessing the validity of inheriting protein function from homology in genome annotation and to protein docking. J. Mol. Biol. 311: 395–408.

Aloy, P., Pichaud, M., and R.B. Russell, 2005. **Protein complexes: structure prediction challenges for the 21(st) centuries.** *Curr Opin Struct Biol.* **15**(1):15-22.

Altschul, S.F., Gish, W., Miller, W., Myers, E.W., and D.J. Lipman, 1990. Basic local alignment search tool. J. Mol. Biol. 215, 403–410.

Altschul, S.F., Madden, T.L., Schaffer, A.A., Zhang, J., Zhang, Z., Miller, W., Lipman, D.J. 1997. Gapped Blast and Psi-Blast: a new generation of protein database search programs. Nucleic Acids Res., 25, 3389–3402.

Anfinsen, C.B., 1973. The principles that governs the folding of protein chains. Science 181:223-30.

Anjum, S. 2013. Computational Genome analysis of Hypothetical Protein in *Mycobacterium lapraeTN* For Therapeutic Drug Target Identification, Int. Res. J. of Science & Engineering, 1(3):90-91, 2322-0015.

Aravind, L. and E.V. Koonin, 1999. Gleaning non-trivial structural, functional and evolutionary information about proteins by iterative database searches. J Mol Bioi 287: 1023-1040.

Arigoni, F., Talabot, F., and M. Peitsch, 1998. A genome-based approach for the identification of essential bacterial genes. Nature Biotechnology 16: 851-856.

Aruna, K., Tritha, C., Savitri, N., Abdul, M., S. Alfica, and S. Shubhona, 2004. Identification of a hypothetical membrane protein interactor of ribosomal phosphoprotein P0, J. Biosci., Vol. 29, No. 1, 33–43.

Backert, S. and T.F. Meyer, 2006. Type IV secretion systems and their effectors in bacterial pathogenesis. Curr Opin Microbial.9: 207-217.

Baran, M.C., Moseley, H.N., Aramini, J.M., Bayro, M.J., Monleon, D., Locke, J.Y., and G.T. Montelione, 2006. SPINS: a laboratory information management system for organizing and archiving intermediate and final results from NMR protein structure determinations. Proteins: Structure, Function, and Bioinformatics, 62(4), 843-851.

Benson, D.A., Karsch, M.I., Lipman, D.J., Ostell, J. and D.L. Wheeler, 2004. GenBank: update. Nucleic Acids Res. 32: 23–26.

Bidkar, A.P., Thakur, K.K., Bolshette, N.B., Dutta, J., and R. Gogoi, 2014. *In-silico* Structural and Functional Analysis of Hypothetical Proteins of *Leptospira Interrogans*. Biochem Pharmacol 3:136.

Brooksbank, C., Bergman, M.T., Apweiler, R., Birney, E., and J. Thornton, 2014. The European Bioinformatics Institute's data resources. Nucleic Acids Res., 42, 18–25.

Bruce, A.R., and T. Reid, 2015. Functional and Phylogenetic Characterization of Proteins Detected in Various Nematodes Intestinal Compartments MCP. 14: 812-827.

Carbon, S., Ireland, A., Mungall, C.J., Shu, S., Marshall, B., and S. Lewis, 2009. AmiGO Hub, Web Presence Working Group. AmiGO: online access to ontology and annotation data. Bioinformatics. 25: 288-89.

Carbon, S., Ireland, A., Mungall, C.J., Shu, S., Marshall, B., and S. Lewis, 2009. AmiGO Hub, Web Presence Working Group. AmiGO: online access to ontology and annotation data. Bioinformatics. 25, 288-9.

Castell, A., Johansson, P., Unge, T., Jones, T.A., and K. Bäckbro, 2005. *Rv0216*, a conserved hypothetical protein from *Mycobacterium tuberculosis* that is essential for bacterial survival during infection, has a double hotdog-fold. Protein Sci. 14: 1850–1862.

Cattell, R. B. 1943. The description of personality: Basic traits resolved into clusters. Journal of Abnormal and Social Psychology. 38: 476–506.

Christophe, C., Martin, J., Gilbert, D., and G. Christophe, 2002. Geno3D: automatic comparative molecular modelling of protein, Volume 18, Issue 1, Oxford University Press.

Crick, F. 1970. Central dogma of molecular biology. Nature, 227, 561–63.

Crick, F.H.C. 1956. On Protein Synthesis. Symp. Soc. Exp. Biol. XII: 139-63.

Cuff, A.L., Sillitoe, I., Lewis, T., Clegg A.B., Rentzsch, R., Furnham, N., Jones, D., Thornton, J., and C.A. Orengo, 2011. "Extending CATH: increasing coverage of the protein structure universe and linking structure with function." Nucleic Acids Res. 39.

da Fonsêca, M. M., Zaha, A., Caffarena, E. R., and A. T. R. Vasconcelos, 2012. Structure-based functional inference of hypothetical proteins from Mycoplasma hyopneumoniae. Journal of Molecular Modeling, 18(5): 1917-1925.

David, A.D., George, E.B., and G.S. Janneth, 2014. Structural and Functional Prediction of the Hypothetical Protein Pa2481 in *Pseudomonas Aeruginosa Pao,* Advances in Intelligent Systems and Computing. 232: 47-55.

Doppelt, A.O., Moriaud, F., and F. Delfaud, 2009. Analysis of HSP90 related folds with MED-SuMo classification approach. Drug Des Dev Therapy.3: 59–72.

Emmert, D.B., Stoehr, P.J., Stoesser, G., and G.N. Cameron, 1994. The Ribosomal Database Project (RDP). Nucleic Acids Res. 22: 3445-49.

Enault, F., Suhre, K., and J. Claverie, 2005. Gene Function Predictor: A gene annotation tool based on genomic context analysis, BMCBioinformatics. 6: 247.

Fan, M., Wong, K. C., Ryu, T., Ravasi, T., and X. Gao, 2012. Secom: A novel hash seed and community detection based-approach for genome-scale protein domain identification. PLoS ONE 2012 7(6): e39475.doi:10.1371/journal.pone.0039475.

Fariselli P., Riccobelli P., and R. Casadio 1999. Role of evolutionary information in predicting the disulfide-bonding state of cysteine in proteins. Proteins. 36: 340–346.

Finn, R.D, Mistry, J., Tate, J., Coggill, P., Heger, A., Pollington, J.E., Gavin, O.L., Gunasekaran, P., Ceric, G., Forslund, K., 2010. The Pfam protein family's database. Nucleic Acids Res. 38, 211-222.

Fiser, A. and A. Sali, 2003. Modeller: generation and refinement of homology-based protein structure models. Meth. Enzymol. 374: 461–91.

Galperin, M.Y. and E.V. Koonin, 2004. Conserved hypothetical' proteins: prioritization of targets for experimental study, Nucleic Acids Res. 12: 5452-63.

Gattiker, A., Gasteiger, E., and A. Bairoch, 2002. ScanProsite: a reference implementation of a PROSITE scanning tool. Appl. Bioinformatics. 1, 107–108.

Geer, L.Y., Domrachev, M., Lipman, D.J., and S.H. Bryant, 2002. CDART: protein homology by domain architecture. Genome Res. Oct. 12: 1619-23.

Gremme, G., Brendel, V., Sparks, M.E., and S. Kurtz, 2005. Engineering a software tool for gene structure prediction in higher organisms. Information and Software Technology, 47: 965-978.

Haft, D.H., Loftus, B.J., Richardson, D.L., Yang, F., Eisen, J.A., Paulsen, I.T. White, O. 2001. TIGRFAMs: a protein family resource for the functional identification of proteins. Nucleic Acids Res., 29, 41–43.

Heynh, A., Bhattacharjee, H., Choudhury, U., Maheswari, S, R., and B, D. Joshi, 2005. *In-silico* prediction of structural and functional aspects of a hypothetical protein of *Arabidopsis thaliana*. Drought and Salt Tolerance in Plants. Crit. Rev. Plant Sci. 24: 2358.

Hirokawa, M., and C. Boon, 1998. SOSUI: Classification and secondary structure prediction for membrane proteins, Bioinformatics. 14: 378-79.

Hoskeri, J.H., Krishna, V., and C. Amruthavalli, 2010. Functional Annotation of Conserved Hypothetical Proteins in *Rickettsia Massiliae* MTU5. J Comput Sci Syst Biol. 3: 050-052.

Huaiyu, M., Anushya, M., and D.T. Paul, 2013. PANTHER in 2013: modeling the evolution of gene function, and other gene attributes, in the context of phylogenetic trees. Nucleic acids research, 41(D1), D377-D386.

Indra, S., Pragati, A., and S. Kavita, 2012. In search of function for hypothetical proteins encoded by genes of SA-JA pathways in *Oryza sativa* by *in-silico* comparison and structural modeling, ISSN 0973-2063, Bioinformation 8(1): 001-005.

Jain, E., Bairoch, A., Duvaud, S., Phan, I., Redaschi, N., Suzek, B.E., Martin, M.J., McGarvey, P., and E. Gasteiger, 2009. Infrastructure for the life sciences: design and implementation of the UniProt website. BMC Bioinformatics, 10: 136.

Kamburov, A., Cavill, R., Ebbels, T.M., Herwig, R., and H.C. Keun, 2011. Integrated pathway-level analysis of transcriptomics and metabolomics data with IMPaLA. Bioinformatics. 27: 2917-2918.

Kanehisa, M., Goto, S., Sato, Y., Kawashima, M., Furumichi, M., Tanabe, M. 2014. Data, information, knowledge and principle: back to metabolism in KEGG. Nucleic Acids Res. 42, 199–205.

Kobayashi, K., Ehrlich, S.D., Albertini, A., Amati, G., Andersen, K.K., Arnaud, M., Asai, K., Ashikaga, S., Aymerich, S., and P. Bessieres, 2003. Essential *Bacillus subtilis* genes. Proc. Natl Acad. Sci. USA, 100, 4678–83.

Kumar, K., Prakash, A., Tasleem, M., Islam, A., Ahmad, F., and M. I. Hassan, 2014. Functional annotation of putative hypothetical proteins from *Candida dubliniensis*. Gene, 543(1): 93-100.

Li. J., Mahajan, A., and M, D. Tsai, 2006. Alkyrin repeat: A unique motif mediating protein-protein interactions, Biochemistry. 45: 5168-78.

Lodish, H., Berk, A., Zipursky, S.L., Matsudaira, P., Baltimore, D., and J. Darnell, 2000. Section 3.1, Hierarchical Structure of Proteins. Molecular Cell Biology. 4th edition. W. H. Freeman & Co.

Madej, T., Lanczycki, C.J., Zhang, D., and P.A. Thiessen, 2014 Geer RC, Marchler-Bauer A, Bryant SH. MMDB and VAST+: tracking structural similarities between macromolecular complexes. Nucleic Acids Res.42: 297-303.

Mandel, G.Y., Schueler, O., and H. Margalit, 1995. Comprehensive analysis of hydrogen bonds in regulatory protein DNA-complexes: in search of common principles. J. Mol. Biol. 253: 370–82.

Marchler, B.A., Derbyshire, M.K., Gonzales, N.R., Lu, S., Chitsaz, F., Geer, L.Y., Geer, R.C., He, J., Gwadz, M., Hurwitz, D.I., Lanczycki, C.J., Lu, F., and G.H. Marchler, 2015. CDD: NCBI's conserved domain database. Nucleic Acids Res. 43.

Mayrose, I., Graur, D., Ben-Tal, N., and Pupko, T. 2004. Comparison of site specific rate-inference methods for protein sequences: empirical Bayesian methods are superior. Mol. Biol. Evol. 21: 1781–91.

Meier, A. and J. Söding, 2015. Context similarity scoring improves protein sequence alignments in the midnight zone Bioinformatics. 31 (5): 674-81.

Mishra, P.K., Sonkar, S.C., Raj, S.R., Chaudhry, U., and D. Saluja, 2013. Functional Analysis of Hypothetical Proteins of *Chlamydia Trachomatis*: A Bioinformatics Based Approach for Prioritizing the Targets. *J Comput Sci Syst Biol*, 7:1.

Moult, J., Fidelis, K., Zemla, A. and T. Hubbard, 2003. Critical assessment of methods of protein structure prediction (CASP)-round V. Proteins. 53: 334–39

Mulder, N.J., Apweiler, R., Attwood, T.K., Bairoch, A., Bateman, A., Binns, B., Bork, P., Buillard, V., Cerutti, L., Copley, R., 2008. New developments in the InterPro database. Nucleic Acids Res. 35, 224–228.

Neeraj, N., Neera, M., and C. Sayan, 2013. Analysis of Annotation Strategies for Hypothetical Proteins: A Case Study of Neisseria, Journal of Natural Science, Biology and Medicine. 02(03), 2319-1163.

Neidhardt, F.C., Curtiss, R., Ingraham, J.L., Lin, E.C.C., Low, K.B., Magasanik, B., Reznikoff, W.S., Riley, M., Schaechter, M., and H.E. Umbarger, 1996. Escherichia coli and Salmonella. Cellular and Molecular Biology, 2nd edn.

Noor, Y.M., Samsulrizal, N.H., Jema'on, N.A., Low, K.O., Ramli, A.N., Alias ,N.I., Damis, S.I., Fuzi, S.F., Isa, M.N., Murad, A.M., Raih, M.F., Bakar, F.D., Najimudin, N., Mahadi N.M., Illias and R.M. 2014. A comparative genomic analysis of the alkali tolerant soil bacterium *Bacillus lehensis* G1. Gene 545: 253-261.

Notredame, C., Higgins, D.G., and J. Heringa, 2000. "T-Coffee: A novel method for fast and accurate multiple sequence alignment". J Mol Biol. 302, 205–17.

Nykyri, J., Niemi, O., Koskinen, P., Nokso, K.J., Pasanen, M., Broberg, M. 2012. Revised Phylogeny and Novel Horizontally Acquired Virulence Determinants of the Model Soft Rot Phytopathogen *Pectobacterium wasabiae* SCC3193 PLoS Pathogens. 8: e1003013.

Paul, D., Thomas, M.J., Campbell, A.K., Huaiyu, Mi., Brian, K., Robin, D., Karen, D., Anushya, M., and N. Apurva, 2003. PANTHER: a library of protein families and subfamilies indexed by function. Genome Res., 13: 2129-141.

Powell, S., Szklarczyk, D., Trachana, K., Roth, A., Kuhn, M., Muller, J., Arnold, R., Rattei, T., Letunic, I., Doerks, T., Jensen, L.J., and P. Bork, 2012. eggNOG v3.0: orthologous groups covering 1133 organisms at 41 different taxonomic ranges. Nucleic Acids Research. 40: 284–289.

Ramadevi, M. and V. Subhashree, 2012. Computational structural and functional analysis of hypothetical proteins of *Staphylococcus aureus*, Bioinformation 8(15): 722-28.

Ranea, J.A.G., Yeats, C., Grant, A., and Orengo, C.A. 2007. Predicting Protein Function with Hierarchical Phylogenetic Profiles: The Gene3D Phylo-Tuner Method Applied to Eukaryotic Genomes. PLoS Comput Biol. 3(11): e237.

Rhee, S.Y., Dickerson, J., and D. Xu, 2006. Bioinformatics and its Applications in Plant Biology. Annual Review of Plant Biology. 57: 335-60.

Rosenkrands, I., Weldingh, K., Jacobsen, S., Hansen, C.V., Florio, W., Gianetri, I., and P. Andersen, 2000. Mapping and identification of Mycobacterium tuberculosis proteins by two-dimensional gel electrophoresis, microsequencing and immunodetection. Electrophoresis. 21(5): 935-48.

Roshni, P., Susan, S.A., and P.K. Cleave, 2014. *In silico* predictive studies of mAHR congener binding using homology modelling and molecular docking. Toxicol Ind Health 30 (8): 765-76.

Rounsley, S., Bush, D., Subramaniam, S., Levin, I., S. Norris, 2000. Analysis of the genome sequence of the flowering plant *Arabidopsis thaliana,* Nature 408: 796-815.

Roy, A., Kucukural, A., and Y. Zhang, 2010. I-TASSER: a unified platform for automated protein structure and function prediction. Nature Protocols, 5: 725-38.

Sasson, O., Vaaknin, A., Fleischer, H., Portugaly, E., Bilu, Y., Linial, N. and M. Linial, 2003. ProtoNet: hierarchical classification of the protein space. Nucleic Acids Res. 31: 348–352.

Shahbaaz, M., Hassan, M.I., and Ahmad, F. 2013. Functional Annotation of Conserved Hypothetical Proteins from *Haemophilus influenzae* Rd KW20. PLoS ONE 8(12): e84263.

Shailesh, V. D. 2008. The Rosetta Stone Method. Bioinformatics Methods in Molecular Biology. 453: 169-180.

Sigrist, C., Cerutti, L., Cuche, B.A., Hulo, N., Bridge, A., Bougueleret, L., and I. Xenarios, 2012. New and continuing developments at PROSITE, Nucleic Acids Research, 2012, 1–4. doi:10.1093/nar/gks1067.

Sonnhammer, E.L.L., Eddy, S.R., Durbin, R. 1997. Pfam: A comprehensive database of protein domain families based on seed alignment. Proteins, 28, 405-420.

Soo, H.T., Yahaya, M.N., Adam, T.C.L., Abu, B.S., and A.K. Roghayeh, 2014. A Sco protein among the hypothetical proteins of *Bacillus lehensis* G1: Its 3D macromolecular structure and association with Cytochrome C Oxidase.BMC Struct Biol. 19: 14-11.

Weizhong, L., Lukasz, J., Adam, G., 2001. "Clustering of highly homologous sequences to reduce the size of large protein database", Bioinformatics, 17, 282-283.

Yang, J., Yan, R., Roy, A., Poisson, J., and Y. Zhang, 2015. The I-TASSER Suite: Protein structure and function prediction. Nature Methods. 12: 7-8.

Zahra, M. 2013.*In Silico* Investigation of *Rv* Hypothetical Proteins of Virulent Strain *Mycobacterium tuberculosis H37Rv*, Indian J. Pharm. Biol. Res. 4: 81-88.

7

Protein-Protein Functional Linkage Predictions: Bringing Regulation to Context

Anne Hahn[1], and Vijaykumar Yogesh Muley[2]*

Abstract

The cell brings various proteins in specific contexts with respect to internal and external perturbations to invoke appropriate responses. The context of a protein could be immune response against invading pathogens or a metabolic pathway in which a set of proteins breaks down glucose molecules to provide energy. Therefore, the contextual information of a protein helps us in understanding its function at local and global level. The last two decades have witnessed a significant progress in identifying protein context and understanding protein organization at systems level. The context of proteins is conceptualized in the form of a network or a graph where proteins that participate in related functions are connected by edges. The global network can be derived by assembling contextual information of all proteins encoded by a cell and provides a perspective on the functioning of proteins in context of others. Several methods have been proposed to infer contextual information of a protein. Genomic context based methods assume that the co-evolutionary signals and neighborhood of protein encoding genes within the genome sequences reflect their functional dependence. Co-expression of protein coding genes also turned out to be a hallmark of their linked roles in specific functions. These methods have been providing a wealth of information on the organization of pathways at the global level, on functional clues for uncharacterized proteins based on their connectivity with known proteins, and identification of disease related proteins and selectively targeting drugs. This chapter reviews the computational protein-protein functional linkage prediction methods developed in the post-genomic era. The chapter concludes with the proposal of a gene co-regulation based, novel method for functional linkage prediction.

[1] Friedrich Schiller University, Department of Molecular Medicine, Medical faculty, Bachstraße 18, 07743, Jena, Germany. Tel: +4936417939072. Email: anne.hahn@uni-jena.de

[2] Network Modeling, Leibniz Institute for Natural Product Research and Infection Biology - Hans Knöll Institute, Beutenbergstraße 11, 07745, Jena, Germany. Tel: +4936415321389. Email: vijay@iscb.org

* Corresponding author : vijay.muley@gmail.com

1.   Protein Functions in the Post-Genomic Era

The information needed for the response to external influences and for the development of a living organism is written in its genome in the form of arrays of genes. Proteins are gene products involved in almost every physiological function of an organism, and their various functional aspects have been studied using model organisms. The completely sequenced genomes meanwhile available offer possibilities to grow in-depth knowledge of the genomic organization of the encoded proteins to define their biological functions at different scales (Galperin and Koonin 2001).

The basis of any protein's architecture is a unique sequence of covalently bound amino acids encoded by the nucleotide sequence of the respective gene. Sequence modules of independent evolutionary origin, called domains, fold independently and also do not need the context of the whole protein to perform their specific function (Chothia, et al. 2003). For example, receptor tyrosine kinases usually have multiple domains, for ligand binding, dimerization, autophosphorylation and providing a platform for the assembly of downstream signaling complexes. Basically, a combination of domains in a protein allows it to perform multiple functions (Vogel, et al. 2004). Conventionally, the molecular role of a protein has been investigated by various experimental methods and often its three-dimensional structure later has been used as a framework to explain known functional properties. But it's also possible to vice versa predict an uncharacterized protein's functions based on its primary and secondary structure.

The majority of the computational approaches available uses the sequences or structural information of already characterized proteins related to the protein of interest to solve this problem (Loewenstein, et al. 2009, Procter, et al. 2010). These approaches, commonly referred to as homology-based protein function prediction methods, imply that a high sequence similarity of two proteins is likely to occur due to a common origin of those proteins that are *homologous* (Fitch 1995). Different species can acquire a gene from their last common ancestor in two ways (reviewed in Koonin 2005): Two homologous genes are *paralogues* if they diverged from the same ancestral gene by a duplication event and are retained in the same species. There they may execute the same or very different functions. On the other hand, *orthologues* also diverged from the same ancestral gene, however were retained in two species following a speciation event and perform the same function. Consequently, in order to predict a protein's function it would be more useful to focus on orthologues with similar functions rather than on paralogues. However, defining orthologues and paralogues of a gene is not an easy task considering that gene gains and losses are very common during evolution (Koonin 2005, Reeck, et al. 1987). Homology-based approaches are at the base of functional annotation of uncharacterized proteins and newly sequenced genomes (Gabaldon and Koonin 2013, Goldsmith-Fischman and Honig 2003, Koonin, et al. 1996, Tatusov, et al. 1997).

However, not every protein's function can be predicted by homology-based approaches, and even if light is shed on the molecular function or localization of a protein, this information might not suffice to understand its cellular role. Furthermore, there are challenges like *moonlight proteins,* a class of multifunctional proteins, in which a single polypeptide chain performs multiple physiologically relevant biochemical or biophysical functions (Jeffery 2014). Moonlighting proteins are expressed throughout the evolutionary tree and participate in many different biochemical pathways. Currently, there is no reliable method available to predict which proteins exhibit such behavior and what cellular processes they contribute to. For that, the knowledge about a protein's context of action is crucial. In the past 15 years, non-homology based protein function prediction methods have been developed which elucidate protein function at cellular level (reviewed in Petrey and Honig 2014, Shoemaker and Panchenko 2007, Yamada and Bork 2009). These methods assemble networks from proteins that are under the same functional constraints. Networks alone do not actually provide the molecular function of a protein but its cellular context, which is a firsthand clue about its function.

2. Computational Methods for Predicting Functional Linkages between Proteins

The prediction of a protein's function using homology-based methods involves finding homologous proteins based on sequence similarity in a structure or sequence database of characterized proteins (Pearson 1995). By studying the Multiple Sequence Alignment (MSA) of homologous proteins, observing the presence of conserved motifs of known function and structural properties, the molecular function of a protein of interest can be inferred (McClure, et al. 1994, Procter, et al. 2010). Even one characterized homologous protein's sequence is sufficient for this task. However, after continuous efforts in experimental determination of protein function over several years, even in model organisms still more than half of the proteins are uncharacterized yet (Hu, et al. 2009, Jaroszewski, et al. 2009) (Fig. 1). This limits the use of homology-based approaches for proteins of unknown functions.

Genome sequencing projects faced this problem at the end of the 20th century, which led to the development of methods for protein function prediction based on the *guilt-by-association* principle (Aravind 2000). This principle assumes that two interacting proteins usually participate in the same or related cellular functions. For instance, if a protein of previously unknown function has direct or indirect links to many proteins that participate in a specific biological process, then it is very likely that the uncharacterized protein is also involved in that process. The guilt-by-association approaches provide information about the context of the unknown protein's action which is inferred from its characterized interacting proteins, that do not necessarily have sequence similarity to it. Many methods have been proposed to decipher functional associations of the protein of interest (Marcotte, et al. 1999, Petrey and Honig 2014, Yamada and Bork 2009). By principle, these methods work at genome scale, taking into consideration both genomic context and expression data and hence provide a snap-shot of all proteins that are linked to one another in the global operational network of an organism. Every method has its own premise to use the guilt-by-associations approach to predict functions of a protein of interest. In the following, an overview shall be given concerning the basic assumptions underlying the most commonly used methods.

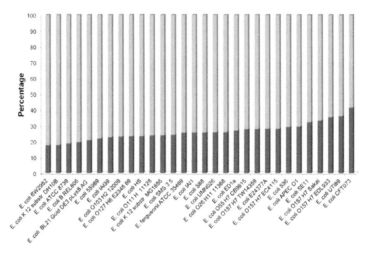

FIGURE 1   The extent of annotation available for species belonging to the *Escherichia* genus. Percentages of proteins of annotated (light gray) and of putative, hypothetical and unknown functions (dark gray) in the completely sequenced genome of the model organism *Escherichia coli* and its close relatives are shown. On an average, 26% of proteins of various species are still uncharacterized. Protein table files (ptt) for *Escherichia* genomes were downloaded from ftp://ftp.ncbi.nlm.nih.gov/genomes/Bacteria/. The proteins with functions termed "putative", "hypothetical" and "unknown" in protein table files were treated as uncharacterized and the remaining as annotated, either computationally or experimentally. Robust evaluation may further increase the number of uncharacterized proteins.

---

3.   Phylogenetic Profiling

The *Phylogenetic Profiling* (PP) method assumes that if two proteins are dependent on one another in performing a specific function, they would both be maintained in the organisms requiring that function and absent otherwise (Pellegrini, et al. 1999). In other words, these proteins co-evolve due to the constraints imposed by their linked functions. *Phylogenetic profiles* represent the presence and absence of all proteins of an organism of interest (providing the query genome) in a set of reference genomes: The reference genomes are searched for orthologues for all proteins encoded in the query genome and a phylogenetic profile (or phyletic pattern or co-occurrence profile) can be set up (Fig. 2). The information on the presence of an orthologue of the query protein in a certain reference genome can be given simply as binary digits; but common are also e-values and bit scores of the alignment between the sequence of the query protein and its orthologue's sequence in the reference genome (Date and Marcotte 2003, Enault, et al. 2003). After having set up the phylogenetic profile, it is analyzed which proteins of the query genome show similar distribution patterns across the reference genomes.

Similarity between phylogenetic profiles of two proteins can be computed using various metrics. The most common similarity measures are Pearson correlation coefficient and mutual information (Kensche, et al. 2008). Every query genome protein's profile is compared against every other protein profile and the pairs are ranked according to their scores.

	Gene	G1	G2	G3	G4	G5	G6	G7	G8	G9	G10	G11	G12	G13	G14	G15
Histidine biosynthesis	hisA	0	0	0	0.25	0.25	0.24	0.28	0.27	0.63	0.69	0.7	0.8	0.97	0.96	0.93
	hisB	0	0	0	0.22	0.23	0.22	0.25	0.24	0.65	0.59	0.79	0.81	0.97	0.97	0.95
	hisD	0	0	0	0.36	0.33	0.28	0.32	0.33	0.59	0.57	0.68	0.73	0.92	0.93	0.93
	hisF	0	0	0	0.46	0.45	0.38	0.39	0.39	0.81	0.76	0.84	0.87	0.97	0.98	0.97
SoxR reducing system	rsxA	0	0	0	0	0	0	0	0	0.79	0.83	0.88	0.89	0.98	0.98	0.99
	rsxB	0	0	0.39	0	0	0	0	0	0.58	0.59	0.79	0.76	1	1	0.92
	rsxC	0	0	0	0	0	0	0	0	0.54	0.54	0.62	0.63	0.95	0.95	0.89
	rsxD	0	0	0	0	0	0	0	0	0.49	0.6	0.67	0.73	0.99	1	0.91
	rsxE	0	0	0	0	0	0	0	0	0.59	0.63	0.77	0.8	0.97	0.97	0.94
	rsxG	0	0	0	0	0	0	0	0	0.41	0.46	0.68	0.72	0.97	0.97	0.9

FIGURE 2 A schematic representation of the phylogenetic profiling method for predicting protein-protein linkages. Each row and column represent a protein or a reference genome respectively. The presence of an *Escherichia coli* protein in a genome is represented by the normalized bit score of the alignment with its orthologue in that corresponding genome; otherwise the matrix element is zero. Phylogenetic profiling considers co-occurrence of proteins across various genomes to link proteins. The figure depicts co-occurrence of four histidine biosynthesis pathway proteins in genome G4 to G15, and in G9 to G15 genes proposed in the SoxR reducing system to protect *Escherichia coli* cells against superoxide and nitric oxide. The phylogenetic profile method distinguishes these two pathway's proteins based on their unique pattern and sequence similarity scores across genomes.

The *Mutual Information* (MI) describes the reduction of information entropy (as a measure of uncertainty) of a system or variable, when information about a new variable is introduced. It can be calculated as follows for the phylogenetic profiles of protein X and Y (Date and Marcotte 2003),

$$MI(X, Y) = H(X)+H(Y) - H(X,Y) \tag{1}$$

The MI will show whether a reduction of information entropy can be observed if the two proteins are examined together (equation 3), compared to considering them separately (equation 2). That would be the case if the proteins showed co-occurrence; because only from the phylogenetic profile of one of the proteins predictions could be made on the occurrence of the other. The empirical information entropies are given by,

$$H(X) = \sum_{x} \frac{n_x}{N} \tag{2}$$

where x is the number of possible states for X (i.e. "present" and "absent"); n_x is the frequency of a certain state in all reference genomes and N the number of reference genomes; and

$$H(X, Y) = \sum_{x,y} \left(\frac{n_{(x,y)}}{N} \log \frac{n_{(x,y)}}{N} \right) \tag{3}$$

Again, (x,y) are different states, now the different possible combinations of the presence or absence of X and Y. In a binary system ("present" and "absent") there are four combinations possible: both proteins present, both proteins absent and first or second protein present alone, respectively. If instead of binary values real values like bit scores are used in the phylogenetic profile, they should be binned in intervals of 0.1 (Date and Marcotte 2003). If, instead of log, $\log_2$ is used, $H(X,Y)$ will be in bits, and in nats if natural logarithm is used.

The *Pearson Correlation Coefficient* (PCC) has the great advantage of providing information on anti-correlation. Anti-correlation can be observed in the profiles of pairs of proteins that never occur in the same genome together and are likely to perform *analogous* functions. The functional replacement of a protein by another protein of non-related sequence is often termed *non-orthologous displacement* in literature (Koonin, et al. 1996). Correlated and anti-correlated phylogenetic profiles from more than 200 genomes have been analyzed and modules of highly correlated profiles have been identified (Kim and Price 2011, Slonim, et al. 2006). Each of these modules often is enriched with proteins performing related functions. For example, motility and sporulation related proteins form unique modules due to their respective presence in and restriction to motile and sporulating organisms (Slonim, et al. 2006).

A major challenge of phylogenetic profiling analysis is the selection of reference genomes. If reference genomes of species closely related to the query organism are chosen, then many phylogenetic profiles would be similar because speciation has taken place quite recently, and not due to functional relatedness of the proteins. On the other hand, reference genomes of species that have diverged from the query genome very early, yield high similarity among phylogenetic profiles mainly of proteins that are conserved in diverse organisms. Briefly, a reference set of closely related genomes may lead to several false predictions due to high scoring of phylogenetic profiles of proteins conserved among closely related species; whereas a reference set of distant relatives is likely to present high scores only for phylogenetic profiles of universal proteins, but not for proteins that engage in specialized processes occurring in a subset of organisms only. Normalization of bit scores and e-values has been recommended to overcome these problems and gain significant accuracy in comparison to the standard implementation of the method (Ferrer, et al. 2010, Muley and Ranjan 2012). Phylogenetic profiling can provide high quality results by selecting an appropriate set of reference genomes according to the specific need of the study and proper normalization of bit scores (Muley 2012, Muley and Acharya 2013).

4. Analysis of Correlated Mutations in Protein Families by Mirrortree Approach: Indicator of Protein-Protein Interaction

Phylogenetic profiling relies entirely on the co-occurrence of proteins to infer their co-evolutionary behavior. Intuitively, if two proteins physically interact and mutation takes place in one of the proteins' sequences at the interface of the two, this does not necessarily impair their interaction if there is a corresponding change in the protein sequence of the interacting partner (reviewed in Pazos and Valencia 2008). These amino acid mutations are called *correlated mutations* and often are observed at proteins' ligand-binding or functional sites, that adapt to their ligand or substrate, or protein interaction sites which co-evolve if a protein contributes to a larger complex (Gobel, et al. 1994). Identifying correlated mutations between two protein families is a computationally demanding task, especially at genome-level. The *Mirrortree* method provides an elegant solution to track this problem in a reasonable computational time. The underlying idea is that phylogenetic trees of two interacting protein families would have comparable topology due to correlated mutations in them (Fig. 3).

The simplest form of mirrortree method uses MSA of two protein families to quantify the extent of correlated mutation between them as hint at interaction (Pazos and Valencia 2001). First, MSAs are constructed for two proteins of interest and their orthologues from a set of *n* reference genomes. A distance matrix of the dimensions *n* by *n* is computed for each protein, using the MSA results. This matrix contains the sequence differences among all orthologues from the reference genomes. It's a necessary precondition to allow only those reference genomes encoding for orthologues of both proteins, so the distance matrices have the same dimensions and are comparable. The degree of correlation of the distance matrices quantifies the extent of correlated mutations in the two proteins in question and thus indirectly rates the degree of similarity between the phylogenetic trees. It is also possible to derive distance matrices from actual phylogenetic trees, however, the tree reconstruction process is time consuming for genome-scale applications of the method.

One of the major problems is that sequence distances between orthologues from closely related species are small, compared to those between distantly related species. This is called *background similarity*, and occurs due to the fact that similarities of phylogenetic trees often represent common speciation events. In other words, if two proteins have gone through similar speciation events, there would be a similarity of their phylogenetic trees which does not reflect the ability of those proteins to interact

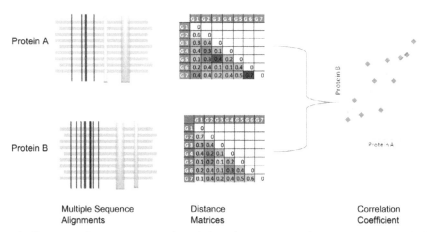

FIGURE 3   A schematic representation of the mirrortree approach. The mirrortree method compares distance matrices derived from the alignment of the query protein and its orthologues. Various approaches can be used to correct these matrices to exclude speciation information prior to comparison. Subsequently, the correlation coefficient between matrices can be calculated which reflects the degree of co-evolution of the proteins A and B, and hence is an indicator of interaction.

with each other. The background similarity has to be removed from the distance matrix in order to use only information that is fairly unaffected by speciation events. The number of false positives can be reduced by normalization of the distance matrices or by setting up a distance matrix as described above for 16S rRNA and subtracting its values from the protein distance matrices (Pazos, et al. 2005). That way, every distance between orthologues from genomes A and B in all matrices is corrected by the distance between the 16S rRNA genes of A and B. 16S rRNA is present in most prokaryotes and diverse enough to be used for species identification. That is, its sequence can be used to identify and evaluate evolutionary relationships. For distantly related A and B where high distances are found, this is compensated by a considerably high correction value. That way, distance values are brought to scale. Still, the sequence distance of 16S rRNA can not be transferred directly to protein distances but should be scaled prior to corrections (Pazos, et al. 2005). We have also proposed an alternative approach to correct distances by using distances between species based on the number of shared orthologues between them (Muley and Ranjan 2012). This approach outperforms 16S rRNA normalization method but at the cost of more computational time.

Conceptually, the mirrortree method, also referred to as *in-silico two-hybrid*, is the only available method with the ability to detect physical interactions. Therefore, it can be used to discover protein complexes.

5.   Chromosomal Proximity of Genes Reflects their Functional Links

The methods previously described to predict protein-protein linkages are based on the co-evolution of two proteins or the amino acids within their sequence. There are several reasons to believe that not just the co-evolution of two proteins is a clue for functional links between them, but also the presence of two genes encoding them on the same stretch of DNA within reasonable distance. The accessibility of the DNA is a prerequisite for gene expression. The higher order DNA structure has to be partially unpacked to enable binding of the transcription machinery at a particular locus. This DNA unwinding cannot be confined to just the gene that is to be transcribed, and eventually the expression of neighboring genes might be affected, too. Therefore, it would be more efficient to place genes adjacent to each other that are needed in the same context and co-regulate them.

The *operon*, a group of co-transcribed and co-regulated adjacent genes on the same genomic strand, is one of the earliest and most central concepts of bacterial genetics (Jacob and Monod 1961). The tendency of genes to be organized in operons across the genome is far more noticeable in prokaryotes, though not completely absent from eukaryotic organisms (Lathe, et al. 2000). Co-transcription and co-regulation define the simultaneous expression of operonic genes in a cell. Thus, gene products encoded in an operon usually perform related functions, but operonic genes with seemingly unrelated functions are also common (Lathe, et al. 2000, Tamames, et al. 1997). The latter is often the result of genomic rearrangements that happened during evolution. Comparative genomics revealed that the joining of small operons or breaking down of large operons is a common phenomenon during evolution. There are several lines of evidence on the conservative nature of these rearrangements that invariably maintained individual genes in very specific functional and regulatory contexts (Korbel, et al. 2004, Lathe, et al. 2000). These observations suggest both that operons can be used to identify functionally linked genes and that fragmented operons in a query genome can be assembled by evidence of their chromosomal proximity in other genomes. This led to the development of many protein-protein linkage prediction methods (Ferrer, et al. 2010), two of which shall be described here to demonstrate their potential to reconstruct operons that are diffused in the query genome.

5.1   Gene Cluster Method

On the basis of short intergenic distances (below 100 nucleotides) between adjacent genes that are encoded on the same genomic strand (same orientation), it is possible to assign them to operons with a maximum accuracy of 88% in the *Escherichia coli* genome (Overbeek, et al. 1999, Salgado, et al., 2000). It was observed that in average 35% of the genes that occur in clusters are involved in the same pathways as their neighbors (Tamames, et al. 1997), and they typically interact with each other physically (Dandekar et al. 1998); a trend which reflects selection against the deleterious effects of protein complex subunits being co-regulated but not localized in close vicinity. However, the inference of functional linkage between genes on the basis of their operon or gene cluster organization is limited to the genes that are adjacent on the query genome sequence. That is, for a query genome with n genes, the *Gene Cluster* (GC) method can generate scores for at most n (in case of circular chromosomes, otherwise $n-1$) gene pairs from the $(n^2-n)/2$ possible pairs. Thereby, the genes that are co-regulated and also functionally linked would not be detected by a simple GC approach, if they have been placed away from each other on the chromosome in the process of rearrangement during evolution.

Comparative genomic approaches have become a powerful tool to deduce the rearrangements of genes based on chromosomal proximity of orthologous genes. This yields a better version of the GC method (Fig. 4A), where gene clusters in reference genomes are defined as sets of co-directional genes within a certain maximum intergenic distance, generally below 200 base pairs (Muley and Ranjan 2012). Then, the GC algorithm takes into consideration every possible pair of genes of the query genome and calculates the frequency of the orthologous proteins in the reference genomes being encoded in the same gene cluster. GC scores above zero indicate

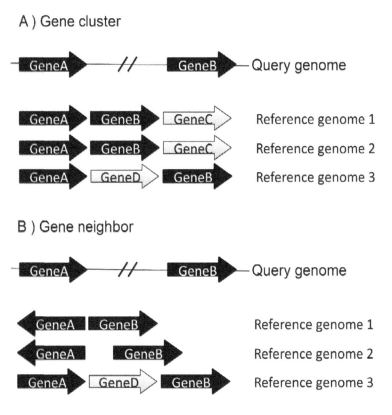

FIGURE 4 A schematic representation of the gene cluster and gene neighbor methods for predicting protein-protein functional links. Methods are exemplified using the hypothetical example of gene A and gene B which are, in the query organism, not located adjacent to each other on the chromosome. A) The gene cluster method calculates the probability of orthologues of the query proteins to co-occur in the same gene cluster in the reference genomes. A gene cluster is defined as a set of unidirectional genes adjacent to each other within an intergenic distance of 200 nucleotide bases. In the given example, genes encoding orthologues of the query proteins A and B co-occur in all three reference genomes, hence 3/3 is the interaction score between them. B) The gene neighbor method calculates interaction scores for query protein pairs based on the minimum chromosomal distance between their orthologue encoding genes in any one reference genome. In the given example, the minimum distance is obtained from the 1st reference genome, so that distance would be used to calculate the interaction score for query gene A and B. Both methods identify re-arranged operons. The gene neighbor method has an advantage in detecting divergently transcribed genes pairs as well.

that orthologues of the query gene pair are within the same gene cluster in at least one of the reference genomes and thus likely to be co-transcribed there. Therefore, GC reveals operons that have been re-arranged in the query genome, based on the evidence of their intact operon structure in multiple reference genomes. Muley and Ranjan have shown that this version of GC method is one of the more accurate methods in terms of predicting linkages with high specificity (Muley and Ranjan 2012).

5.2 Gene Neighbor Method

The GC method can detect re-arranged operons but not the divergently transcribed, conserved gene pairs that have been shown to be often co-regulated and functionally linked (Korbel, et al. 2004). Unlike the GC algorithm, the *Gene Neighbor* (GN) method assumes chromosomal proximity of orthologous genes across a set of reference genomes as an indicator of functional linkage, without constraint on the relative gene orientations. Thereby, it not only predicts re-arranged operons but also divergently transcribed gene pairs.

Over the years, the GN method has been modified into several forms (Ferrer, et al. 2010). A recent version of the method focuses on the minimum genomic distance between orthologous genes that can be found in one of the reference genomes (Janga, et al. 2005). Bacterial chromosomes usually are circular, so the intergenic distance has to be calculated both in clockwise and anti-clockwise direction, but only the minimum distance is considered for further steps (Fig. 4B). On a very small chromosome, an intergenic distance of 200 nucleotides might be quite a large distance whereas on a large chromosome genes with the same intergenic distance would be considered to be localized very close to each other. Therefore, to obtain the GN score, the minimum intergenic distance d_{min} has to be normalized by the chromosome size n and multiplied by two, to scale this relative distance to a value between 0 and 1,

$$D_t = 2\frac{d_{min}}{n} \tag{4}$$

A slight modification gives n as number of genes and d_{min} as number of number of genes interposed between the two genes under consideration, plus one, so the intergenic distance between two adjacent genes were 1. This would put into perspective long intergenic distances that occur because of one very long gene between the two genes in question.

As opposed to GC, GN predicts a high number of false positives and should be used with caution, but it also offers the possibility to infer metabolic gene linkages, which often is a difficult task for other methods (Muley and Ranjan 2013).

6. Expression Similarity of Genes as an Indicator of Functional Linkage

Genes maintained in specific functional and regulatory contexts within genome sequences may be co-expressed. It has been shown in yeast that interacting pairs of proteins show high correlation of gene expression compared to non-interacting pairs (Jansen, et al. 2002). Therefore, it is possible to predict interactions between two proteins on the basis of similar expression patterns of their coding genes. Genome-scale expression data obtained from various physiological conditions can be represented as a matrix, so that each row would represent expression values for one gene and each column of the matrix would be the condition in which this expression has been measured. There are several metrics one can use to compute the similarity among expression profiles of genes (Song, et al. 2012). PCC is a common measure since it distinguishes positively and negatively associated gene pairs. Horvath and colleagues brought together the network theory and the field of microarray data analysis in Weighted Gene Correlation Network Analysis (WGCNA) to infer high quality co-expression gene modules (Horvath and Dong 2008). WGCNA is available as R package which offers several cutting-edge approaches for understanding gene expression data. These techniques recently have been applied to understand embryonic development. It is observed that each developmental stage can be delineated concisely by a small number of functional modules of co-expressed genes, indicating a sequential order of transcriptional changes in pathways of cell cycle and gene regulation, translation and metabolism, acting in a step-wise fashion from oocyte cleavage up to the morula stage of development (Xue, et al. 2013).

Expression Similarity (ES) based methods differ from the aforementioned prediction methods since they don't require comparative genomics analysis. Therefore, they are easy to implement and can be applied effectively using hundreds of expression datasets readily available for model organisms. They are also computationally inexpensive compared to the methods mentioned above and perform exceptionally well in predicting functional linkages, compared to the aforementioned genomic context based methods (Fig. 5).

7. From Transcriptional Regulation to Predicting Protein-Protein Functional Linkages: A Novel Approach

Co-regulation has been shown to be an indicator of functional coupling of genes. However, so far direct relationships of the expression of transcription factors and of genes have never been investigated for functional linkage prediction studies. For instance, the expression similarity between all transcription factors and a gene gives a vector in which each value represents the relation of a corresponding transcription factor to the expression of that gene. Likewise, similar vectors can be calculated for all query genes. The high correlation between vectors of two genes could result from the same set of transcription factors influencing them. It suggests that these genes are co-regulated and their expression is likely to be required at the same time for related functions. There are several ways to estimate the influence of a particular transcription factor on the expression of a particular gene (Marbach, et al. 2012). Here we demonstrate the power of the proposed method using PCC as a measure of similarity between genes and all transcription factors, and then for similarity between gene profile pairs. We compared the performance of our method with four of the existing methods reviewed above and studied in (Muley and Ranjan 2012) (Fig. 5).

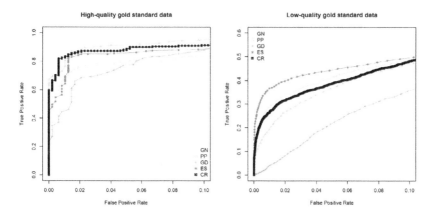

FIGURE 5 Performance comparison of four existing methods with the novel approach based on co-regulation of genes. All five methods were tested on high quality (HQG) and low quality (LQG) gold standards (Muley and Ranjan 2012). The co-regulation (CR) based method outperforms others in terms of low false prediction rate on both datasets. The methods compared are Gene Neighbor (GN), Phylogenetic profiling (PP), a Mirrortree method variant called Genome Distance (GD), Expression Similarity (ES) and the proposed method, CR. Interaction scores for the existing methods were computed as described in (Muley and Ranjan 2012). The ES method shows much better performance than the genomic context methods. CR does not outperform, but at high interaction scores works better than the existing methods.

Performance of the methods was determined as described in the following section; on two different datasets. The High Quality Gold Standard dataset (HQG) was acquired from three different databases (DIP, a Database of Interacting Partners, EcoCyc protein complexes, and KEGG, the Kyoto Encyclopedia of Genes and Genomes pathway annotations) and contains interacting protein pairs that are conserved among at least 200 genomes (Muley and Ranjan 2012). They belong to the same functional category and reportedly interact on physical and/or complex-associated level. The Low Quality Gold Standard dataset (LQG) was obtained from the same databases, but without the constraint on the proteins' phylogenetic distribution. The new method ranks second best predictor of functional links between proteins. Furthermore, it returns a low number of false positive predictions at a high interaction score cutoff (Fig. 5). The time expenditure is, like for the ES method, negligible. Interaction scoring can be done within few minutes even for an expression matrix of more than 10,000 genes measured across 2,000 samples.

8. Measuring Prediction Performance of Methods

In the sections above, different methods for protein-protein functional linkage inference were introduced. Consequently, to give a full overview of a protein's function and role in an organism (Hu, et al. 2009), or to elucidate new mechanisms (Li, et al. 2005, Menche, et al. 2015), the prediction of functional pathways needs to be as accurate as possible. So how do these methods perform in predicting pathways ?

To investigate the prediction performance of methods, a gold standard dataset of interacting and non-interacting protein pairs is required, commonly referred to as 'positives' and 'negatives', unify quotation marks (compare to "present" and "absent" on p.5/157) respectively. Interacting pairs detected by low-throughput experiments are few, even in model organisms and compared to all possible protein pairs encoded by any organism. Some of the high-throughput methods (such as yeast two-hybrid) are noisy and it is recommendable to avoid using these datasets as gold standards whenever possible. High quality information on the pathway memberships of proteins as provided in databases such as KEGG and BioCyc is also suitable as gold standard for the validation of predicted links.

Obtaining non-interacting pairs is not an easy task because such information is rarely reported in literature. The simplest solution is to generate all possible pairs among the proteins of an organism and then remove all reported interacting pairs from that dataset. The remaining pairs can then be considered as non-interacting pairs and further filtering can be done to only include pairs in which the partners neither are present in the same pathway nor the same sub-cellular compartment. It is assumed that proteins that belong to different sub-cellular localizations and pathways will not interact with each other (Jansen and Gerstein 2004).

Each prediction method returns a likelihood score of interaction, with the exception of the GN method, which returns a distance measure. High values here mean, in contrary to other methods, high unlikeliness of interaction, so this has to be adjusted by calculating the difference to a fixed value. An interaction score cutoff has to be defined, above which functional links are assumed. This cutoff should be set with respect to the best performance possible for a particular method. Having a gold standard dataset one can estimate the predictive power of a method at several cutoffs. For a chosen cutoff, positives with a score greater than or equal to the threshold are classified as *true positives (TP)* and those with a score below the cutoff are classified as *false negatives (FN)*. Similarly, negatives with a score greater than or equal to the threshold are classified as *false positives (FP)* and those with a score below the threshold are classified as *true negatives (TN)*. Using these four parameters, several performance details can be deduced. The most common prediction accuracy measures are the Receiver Operator Characteristics (ROC) curve and its integral, the Area Under ROC Curve (AUC). In ROC curves, the True Positive Rate (TPR) is plotted against the False Positive Rate (FPR) which are calculated as below, at a series of interaction score cutoffs,

$$TPR = \frac{TP}{TP + FN} \tag{5}$$

$$FPR = 1 - \frac{TN}{TN + FP} \tag{6}$$

For a random predictor these rates would always be equal: Without respect to the true relationship of two proteins, in 50% of all cases an interaction would be predicted. The ROC curve of a random predictor is a diagonal and AUC equals to 0.5 (dashed lines in Fig. 5). A good predictor would have an ROC curve above this diagonal. In contrast, an ideal predictor would produce a TPR of 100% irrespective of the FPR, which would result in a rectangular ROC curve with AUC being 1. The cutoff where TPR is highest and the FPR is lowest can be chosen as proxy to define function links.

The prediction accuracy determined with a certain gold standard dataset should be treated cautiously though; as different methods for protein-protein linkage prediction show noticeable variance in their performance, depending on the specific functional context the investigated proteins are annotated to (Muley and Ranjan 2013). The true performance in a certain task can be very different from the power of the method as estimated using the gold standard dataset.

9. Perspectives

With the development of sequencing technology, biological databases were flooded with several hundred genomes in the hope that the availability of the complete genome sequence would lead to understanding an organism in detail. However, it was soon realized that we do not have any functional clues for more than 50% of all genes encoded from these genomes (Doerks, et al. 2004, Galperin and Koonin 2004). At the same time, comparative genomics revealed several aspects of gene evolution such as non-conserved gene order, co-occurrences of functionally linked genes, and that if genes are present in close vicinity on the chromosome often they perform related functions (Tamames, et al. 1997). Upon these observations the efforts to identify uncharacterized proteins' functions gained more momentum. This led to the development of the methods discussed here and several variants. Muley studied the aforementioned methods in detail to understand the factors affecting their performances, evaluation of various gold standard datasets, ability to predict functional pathways, and protein function predictions using machine learning derived protein-protein interaction networks (Muley 2012). It has been shown that a high performance for predicting protein-protein linkages can be achieved even with 100–150 reference genomes (Muley and Ranjan 2012). In a subsequent study, it was observed that the prediction of metabolic pathway protein interactions is a challenging task for all methods mentioned here; possibly due to the flexible/independent evolutionary histories of these proteins (Muley and Ranjan 2013). These methods perform better than a random predictor in functional associations of proteins involved in amino acid, nucleotide, glycan and vitamin & co-factor pathways but random on carbohydrate, lipid and energy metabolism. On the contrary, genetic information processing and specialized processes such as motility related protein-protein linkages, that occur in a subset of organisms are predicted with comparably good accuracy. In general, the ES method outperforms the others in most of the functional pathways. These observations reflect the shortcomings of the existing methods. Here we proposed a novel method for functional linkage prediction based on co-regulatory gene pairs. It performs as good as the existing top performing methods and can be combined with other methods to further increase accuracy by reducing the number of false predictions.

In spite of the challenges in predicting functional linkages using individual methods, the integrative application of these methods and the resulting networks have provided a wealth of information about biological systems. Network-based studies support a hierarchical modular organization of biological systems (Barabasi and Oltvai 2004, Li, et al. 2005). Protein-protein interaction network analysis also can be used to propose relations between disease genes (Sahni, et al. 2015) and among diseases and use this for therapeutic drug target identification (Menche, et al. 2015, Vidal, et al. 2011). These methods have also been employed for large-scale protein annotations and relating genes to phenotypic traits, followed by experimental validation (Doerks, et al. 2004, Slonim, et al. 2006, Yamada, et al. 2012). We believe that our proposed method will be of immense help in inferring networks in the era of next generation sequencing.

Acknowledgements

We acknowledge the anonymous reviewers for constructive comments on the manuscript.

References

Aravind, L. 2000 Guilt by association: contextual information in genome analysis, *Genome Res*, **10**, 1074-77.

Barabasi, A.L. and Z.N. Oltvai, 2004 Network biology: understanding the cell's functional organization, *Nat Rev Genet*. **5**: 101-13.

Chothia, C., Gough, J., Vogel, C. and Teichmann, S.A. 2003 Evolution of the protein repertoire, *Science*. **300**: 1701-703.

Dandekar, T., Snel, B., Huynen, M. and P. Bork, 1998 Conservation of gene order: a fingerprint of proteins that physically interact, *Trends Biochem Sci*. **23**: 324-28.

Date, S.V. and E.M. Marcotte, 2003 Discovery of uncharacterized cellular systems by genome-wide analysis of functional linkages, *Nat Biotechnol*. **21**: 1055-62.

Doerks, T., von Mering, C. and P. Bork, 2004 Functional clues for hypothetical proteins based on genomic context analysis in prokaryotes, *Nucleic Acids Res*. **32**: 6321-26.

Enault, F., Suhre, K., Abergel, C., Poirot, O. and J.M. Claverie, 2003 Annotation of bacterial genomes using improved phylogenomic profiles, *Bioinformatics*. **19 Suppl 1**: i105-107.

Ferrer, L., Dale, J.M. and P.D. Karp, 2010 A systematic study of genome context methods: calibration, normalization and combination, *BMC Bioinformatics*. **11**: 493.

Fitch, W.M. 1995 Uses for evolutionary trees, *Philos Trans R Soc Lond B Biol Sci*. **349**: 93-102.

Gabaldon, T. and E.V. Koonin, 2013 Functional and evolutionary implications of gene orthology, *Nat Rev Genet*. **14**: 360-66.

Galperin, M.Y. and E.V. Koonin, 2001 Comparative genome analysis, *Methods Biochem Anal*. **43**: 359-92.

Galperin, M.Y. and E.V. Koonin, 2004 'Conserved hypothetical' proteins: prioritization of targets for experimental study, *Nucleic Acids Res*. **32**: 5452-63.

Gobel, U., Sander, C., Schneider, R. and A. Valencia, 1994 Correlated mutations and residue contacts in proteins, *Proteins*. **18**: 309-17.

Goldsmith-Fischman, S. and B. Honig, 2003 Structural genomics: computational methods for structure analysis, *Protein Sci*. **12**: 1813-21.

Horvath, S. and J. Dong, 2008 Geometric interpretation of gene coexpression network analysis, *PLoS Comput Biol*. **4**: e1000117.

Hu, P., Janga, S.C., Babu, M., Diaz-Mejia, J.J., Butland, G., Yang, W., Pogoutse, O., Guo, X., Phanse, S., Wong, P., Chandran, S., Christopoulos, C., Nazarians-Armavil, A., Nasseri, N.K., Musso, G., Ali, M., Nazemof, N., Eroukova, V., Golshani, A., Paccanaro, A., Greenblatt, J.F., Moreno-Hagelsieb, G. and A. Emili, 2009 Global functional atlas of *Escherichia coli* encompassing previously uncharacterized proteins, *PLoS Biol.* **7**: e96.

Jacob, F. and J. Monod, 1961 Genetic regulatory mechanisms in the synthesis of proteins, *J Mol Biol.* **3**, 318-56.

Janga, S.C., Collado-Vides, J. and G. Moreno-Hagelsieb, 2005 Nebulon: a system for the inference of functional relationships of gene products from the rearrangement of predicted operons, *Nucleic Acids Res.* **33**: 2521-30.

Jansen, R. and M. Gerstein, 2004 Analyzing protein function on a genomic scale: the importance of gold-standard positives and negatives for network prediction, *Curr Opin Microbiol.* **7**: 535-545.

Jansen, R., Greenbaum, D. and M. Gerstein, (2002) Relating whole-genome expression data with protein-protein interactions, *Genome Res.* **12**: 37-46.

Jaroszewski, L., Li, Z., Krishna, S.S., Bakolitsa, C., Wooley, J., Deacon, A.M., Wilson, I.A. and A. Godzik, 2009 Exploration of uncharted regions of the protein universe, *PLoS Biol.* **7**, e1000205.

Jeffery, C.J. 2014 An introduction to protein moonlighting, *Biochem Soc Trans.* **42**: 1679-83.

Kensche, P.R., van Noort, V., Dutilh, B.E., and M.A. Huynen, 2008 Practical and theoretical advances in predicting the function of a protein by its phylogenetic distribution, *J R Soc Interface.* **5**: 151-70.

Kim, P.J. and N.D. Price, 2011 Genetic co-occurrence network across sequenced microbes, *PLoS Comput Biol.* **7**: e1002340.

Koonin, E.V. 2005 Orthologs, paralogs, and evolutionary genomics, *Annu Rev Genet.* **39**: 309-38.

Koonin, E.V., Mushegian, A.R. and P. Bork, 1996 Non-orthologous gene displacement, *Trends Genet*, **12**: 334-36.

Koonin, E.V., Tatusov, R.L. and K.E. Rudd, 1996 Protein sequence comparison at genome scale, *Methods Enzymol.* **266**: 295-322.

Korbel, J.O., Jensen, L.J., von Mering, C. and P. Bork, 2004 Analysis of genomic context: prediction of functional associations from conserved bidirectionally transcribed gene pairs, *Nat Biotechnol.* **22**: 911-17.

Lathe, W.C., 3rd, Snel, B. and P. Bork, 2000 Gene context conservation of a higher order than operons, *Trends Biochem Sci.* **25**: 474-79.

Li, H., Pellegrini, M. and D. Eisenberg, 2005 Detection of parallel functional modules by comparative analysis of genome sequences, *Nat Biotechnol.* **23**: 253-60.

Loewenstein, Y., Raimondo, D., Redfern, O.C., Watson, J., Frishman, D., Linial, M., Orengo, C., Thornton, J. and A. Tramontano, 2009 Protein function annotation by homology-based inference, *Genome Biol.* **10**: 207.

Marbach, D., Costello, J.C., Kuffner, R., Vega, N.M., Prill, R.J., Camacho, D.M., Allison, K.R., Kellis, M., Collins, J.J. and G. Stolovitzky, 2012 Wisdom of crowds for robust gene network inference, *Nat Methods.* **9**: 796-804.

Marcotte, E.M., Pellegrini, M., Ng, H.L., Rice, D.W., Yeates, T.O. and Eisenberg, D. (1999) Detecting protein function and protein-protein interactions from genome sequences, *Science.* **285**: 751-53.

McClure, M.A., Vasi, T.K. and W.M. Fitch, 1994 Comparative analysis of multiple protein-sequence alignment methods, *Mol Biol Evol*. **11**: 571-92.

Menche, J., Sharma, A., Kitsak, M., Ghiassian, S.D., Vidal, M., Loscalzo, J. and A.L. Barabasi, 2015 Disease networks. Uncovering disease-disease relationships through the incomplete interactome, *Science*. **347**: 1257601.

Muley, V.Y. 2012 Improved computational prediction and analysis of protein-protein interaction networks. *Ph.D. Thesis*. Manipal Universtity, Manipal, India, http://hdl.handle.net/10603/5399

Muley, V.Y. and Acharya, V. 2013 Co-Evolutionary Signals Within Genome Sequences Reflect Functional Dependence of Proteins, in Tucker, M. (ed), *Genome-wide prediction and analysis of protein-protein functional linkages in bacteria*. Springers Briefs in Systems Biology, 19-32.

Muley, V.Y. and A. Ranjan, 2012 Effect of reference genome selection on the performance of computational methods for genome-wide protein-protein interaction prediction, *PLoS One*, **7**, e42057.

Muley, V.Y. and A. Ranjan, 2013 Evaluation of physical and functional protein-protein interaction prediction methods for detecting biological pathways, *PLoS One*. **8**: e54325.

Overbeek, R., Fonstein, M., D'Souza, M., Pusch, G.D. and N. Maltsev, 1999 The use of gene clusters to infer functional coupling, *Proc Natl Acad Sci U S A*. **96**: 2896-901.

Pazos, F., Ranea, J.A., Juan, D. and , M.J. Sternberg 2005 Assessing protein co-evolution in the context of the tree of life assists in the prediction of the interactome, *J Mol Biol*: **352**, 1002-15.

Pazos, F. and A. Valencia, 2001 Similarity of phylogenetic trees as indicator of protein-protein interaction, *Protein Eng*. **14**, 609-14.

Pazos, F. and A. Valencia, 2008 Protein co-evolution, co-adaptation and interactions, *EMBO J*, **27**, 2648-55.

Pearson, W.R. 1995 Comparison of methods for searching protein sequence databases, *Protein Sci*, **4**, 1145-60.

Pellegrini, M., Marcotte, E.M., Thompson, M.J., Eisenberg, D. and T.O. Yeates, 1999 Assigning protein functions by comparative genome analysis: protein phylogenetic profiles, *Proc Natl Acad Sci U S A*. **96**, 4285-88.

Petrey, D. and B. Honig, 2014 Structural bioinformatics of the interactome, *Annu Rev Biophys*. **43**, 193-210.

Procter, J.B., Thompson, J., Letunic, I., Creevey, C., Jossinet, F. and G.J. Barton, (2010) Visualization of multiple alignments, phylogenies and gene family evolution, *Nat Methods*. **7**, S16-25.

Reeck, G.R., de Haen, C., Teller, D.C., Doolittle, R.F., Fitch, W.M., Dickerson, R.E., Chambon, P., McLachlan, A.D., Margoliash, E., T.H. Jukes, and et al. 1987 "Homology" in proteins and nucleic acids: a terminology muddle and a way out of it, *Cell*. **50**, 667.

Sahni, N., Yi, S., Taipale, M., Fuxman Bass, J.I., Coulombe-Huntington, J., Yang, F., Peng, J., Weile, J., Karras, G.I., Wang, Y., Kovacs, I.A., Kamburov, A., Krykbaeva, I., Lam, M.H., Tucker, G., Khurana, V., Sharma, A., Liu, Y.Y., Yachie, N., Zhong, Q., Shen, Y., Palagi, A., San-Miguel, A., Fan, C., Balcha, D., Dricot, A., Jordan, D.M., Walsh, J.M., Shah, A.A., Yang, X., Stoyanova, A.K., Leighton, A., Calderwood, M.A., Jacob, Y., Cusick, M.E., Salehi-Ashtiani, K., Whitesell, L.J., Sunyaev, S., Berger, B., Barabasi, A.L., Charloteaux, B., Hill, D.E., Hao, T., Roth, F.P., Xia, Y., Walhout, A.J., Lindquist, S. and M. Vidal, 2015 Widespread macromolecular interaction perturbations in human genetic disorders, *Cell*. **161**: 647-60.

Salgado, H., Moreno-Hagelsieb, G., Smith, T.F. and J. Collado-Vides, 2000 Operons in Escherichia coli: genomic analyses and predictions, *Proc Natl Acad Sci U S A*. **97**: 6652-57.

Shoemaker, B.A. and A.R. Panchenko, 2007 Deciphering protein-protein interactions. Part II. Computational methods to predict protein and domain interaction partners, *PLoS Comput Biol*, **3**. e43.

Slonim, N., Elemento, O. and S. Tavazoie, 2006 Ab initio genotype-phenotype association reveals intrinsic modularity in genetic networks, *Mol Syst Biol*. **2**: *2006 0005*.

Song, L., Langfelder, P. and S. Horvath, 2012 Comparison of co-expression measures: mutual information, correlation, and model based indices, *BMC Bioinformatics*. **13**: 328.

Tamames, J., Casari, G., Ouzounis, C. and A. Valencia, 1997 Conserved clusters of functionally related genes in two bacterial genomes, *J Mol Evol*. **44**: 66-73.

Tatusov, R.L., Koonin, E.V. and D.J. Lipman, 1997 A genomic perspective on protein families, *Science*. **278**: 631-37.

Vidal, M., Cusick, M.E. and A.L. Barabasi, 2011 Interactome networks and human disease, *Cell*, **144**, 986-98.

Vogel, C., Bashton, M., Kerrison, N.D., Chothia, C. and S.A. Teichmann, 2004 Structure, function and evolution of multidomain proteins, *Curr Opin Struct Biol*. **14**: 208-16.

Xue, Z., Huang, K., Cai, C., Cai, L., Jiang, C.Y., Feng, Y., Liu, Z., Zeng, Q., Cheng, L., Sun, Y.E., Liu, J.Y., Horvath, S. and G. Fan, 2013 Genetic programs in human and mouse early embryos revealed by single-cell RNA sequencing, *Nature*. **500**: 593-97.

Yamada, T. and Bork, P. 2009 Evolution of biomolecular networks: lessons from metabolic and protein interactions, *Nat Rev Mol Cell Biol*, **10**: 791-803.

Yamada, T., Waller, A.S., Raes, J., Zelezniak, A., Perchat, N., Perret, A., Salanoubat, M., Patil, K.R., Weissenbach, J. and P. Bork, 2012 Prediction and identification of sequences coding for orphan enzymes using genomic and metagenomic neighbours, *Mol Syst Biol*. **8**: 581.

Section 4
Epigenetics

8

Epigenomic Analysis of Chromatin Organization and DNA Methylation

Xin Wang¹, Helen M. McCormick¹, Djordje Djordjevic¹,
Eleni Giannoulatou¹, Catherine M. Suter¹ and Joshua W. K. Ho¹*

Abstract

One of the most fascinating aspects of biology is that the same genome can give rise to a diversity of cell types and cellular functions in a multicellular organism. This suggests that in addition to the DNA sequence there must be additional "epigenetic" features that are associated with cell type or condition specific gene regulation. In eukaryotic organisms, two important mechanisms for epigenetic regulation of gene expression are the dynamic regulation of chromatin organization and DNA methylation. A large number of sequencing-based assays, such as ChIP-seq, DNase-seq, Hi-C, whole genome bisulfite sequencing and Reduced Representation Bisulfite Sequencing (RRBS), have been developed to survey the epigenomes in a genome-wide fashion. In this chapter, we survey the state-of-the-art bioinformatics methodologies and tools for performing these epigenomic analyses. We also review the application of these epigenomic analyses, especially in terms of identifying the location and possible function of non-coding regulatory regions in a genome-wide fashion.

1. Introduction

There are two important mechanisms of epigenetic regulation: chromatin organization and DNA methylation. Both mechanisms have been extensively studied and individually exhibit strong effects on gene expression (Allis et al. 2008). Many different experimental procedures have been developed to generate genome-wide profiles of chromatin organization and DNA methylation. In this chapter, we survey and discuss state-of-the-art bioinformatic methods for analyzing these profiles. In the following sections, we will describe the analysis of chromatin and DNA methylation

¹· Victor Chang Cardiac Research Institute, Darlinghurst, NSW 2010, Australia
   The University of New South Wales, Sydney, NSW 2052, Australia
 * Corresponding author : j.ho@victorchang.edu.au

data separately, but it is important to note that chromatin organization and DNA methylation act in unison to regulate the genome.

2. Chromatin organization

In eukaryotic organisms, the genomic DNA is packaged by histone proteins to form chromatin. Studying the organization and chemical composition of chromatin is important in understanding gene regulation (Zhou et al 2011). The primary functions of chromatin are to stabilize and protect DNA, particularly during mitosis or meiosis, to package DNA into a smaller volume and to control DNA replication and gene expression.

The basic unit of chromatin in eukaryotes is called a "nucleosome". A nucleosome consists of 147 bp of DNA wrapped around an octamer of histone proteins (Luger et al. 1997). The adjacent nucleosomes are separated by ~80 bp of linker DNA. The nucleosome occupancy is dynamic, and is associated with the accessibility of regulatory proteins binding to a local chromatin region. At specific regulatory regions, such as active promoters, a histone protein (e.g. H2A, H2B, H3 and H4) can be substituted with a histone variant (e.g., histone variant H2A.Z), or can be chemically modified (e.g., tri-methylation of histone H3 at lysine 4; H3K4me3) (Zhou et al. 2011). These chemical modifications of histones usually involve the methylation and acetylation of one or more N-terminal lysine (K) and arginine (R) amino acids. Different histone modifications are correlated with different genomic elements and regulatory states (Table 1). As such, the combinations of histone modifications, histone variants and chromatin accessibility are important markers of the potential regulatory function of a genomic locus (Ren et al. 2000, Ernst and Kellis 2010).

TABLE 1 Histone modifications and the regulatory region they are enriched in

Type of modification	me1 (monomethylation)	me2 (dimethylation)	me3 (trimethylation)	ac (acetylation)
H3K4	Enhancer	Promoter and enhancer	Promoter	Promoter
H3K9	Hetero-chromatin	Hetero-chromatin	Hetero-chromatin	Promoter and enhancer
H3K27	In transcribed genes, and active transcription	Ubiquitously enriched across the genome	Repressed chromatin	Active promoter and enhancer
H3K36	Transcribed gene body	Transcribed gene body	Transcribed gene body	Promoter
H3K79	Transcribed gene body	Transcribed gene body	Transcribed gene body	

Beside histones, there is another class of regulatory proteins called "transcription factors" (TFs). They often bind to specific DNA sequences and affect gene expression (Brivanlou et al. 2002). Transcription factors can act alone or interact with other proteins in a complex way, by inhibiting or promoting the recruitment of RNA polymerase to specific genes (Fig. 1). There are ~2,000 proteins that contain DNA-binding domains and are considered to function as transcription factors in human (Babu et al. 2004, Ravasi et al. 2010). In addition, combinations of several different transcription factors in binding sites often play a vital role in expression of nearby genes. Thus, different combinations of human transcription factors could create different gene expression patterns in human cells (Brivanlou et al. 2002). DNA can also be bound to other non-sequence-specific chromatin binding proteins, called chromatin regulators (CRs) (Ram et al 2011). One of the most commonly studied chromatin regulators is EP300, a histone acetyltransferase which is responsible for acetylation of H3K27 in active enhancers. The binding of EP300 is commonly used as a marker of enhancer (Visel et al. 2009). They bind in a combinatorial manner to the DNA, presumably through interaction with transcription factors and the chromatin environment. Together, the study of chromatin organization involves the study of histone modification landscape, and the binding of transcription factors and chromatin regulators. All of these can be measured in a genome-wide fashion using ChIP-seq, which is the main topic of discussion in the next section.

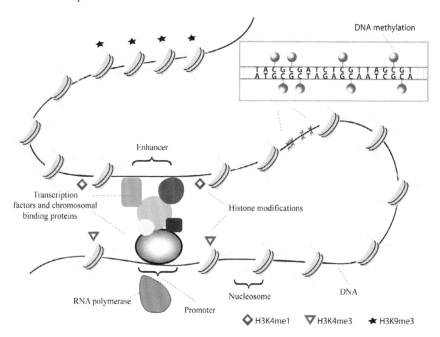

FIGURE 1   The major structures of chromatin and DNA methylation.

2.1   ChIP-seq

Chromatin immuno-precipitation followed by high throughput sequencing (ChIP-seq) is a method used to analyze protein-DNA interactions and identify the binding sites of DNA-associated proteins *in vivo*. This state-of-the-art technique can be used to map genome-wide binding sites for any TF or chromatin regulators of interest (e.g., CTCF or EP300) as well as chemically modified histones (e.g., H3K4me1 or H3K9mc3). The detection sensitivity of ChIP-seq is much better than its microarray predecessor, ChIP-chip (Ho et al. 2011). An overview of the ChIP-seq experiment is shown in Fig. 2. It involves protein-DNA cross-linking of chromatin, extraction of chromatin, fragmentation of the chromatin by sonication or enzymatic digestion, enrichment of DNA that is bound to a specific protein by an antibody "pull-down", removal of cross-link, and sequencing of one or both of the enriched DNA fragments using a next generation sequencing (NGS) platform.

2.2   ChIP-seq experimental design

There are several important experimental design issues that should be considered carefully before carrying out ChIP-seq:

1.   Availability of high quality antibodies

One important issue to consider is whether good quality ChIP-grade antibody specific to your protein of interest is available. One should consult public repositories, such as the Antibody Validation Database (http://compbio.med.harvard.edu/antibodies) when choosing an antibody (Egelhofer et al. 2011).

2.   Read depth

Whether or not enough reads have been generated by the next-generation sequencing experiment is always a crucial issue for researchers to consider. Jung et al. (2014) studied this question by analyzing a deeply sequenced histone modifications ChIP-seq data for *D. melanogaster* (fly) and human. They found that for fly, the sufficient read quantity was 20 million reads. For human, although they found no obvious saturation point in their study, they recommended at least 40 million reads for most marks (Jung et al. 2014). Also, the quantity of reads should be adjusted depending on the expected coverage of that mark in the genome, e.g., histone modifications that are expected to exhibit broad enrichment patterns should be sequenced in a higher depth.

3.   Control sample

There are three major types of control samples for ChIP-seq experiments:

(a) *"Input"*. Chromatin can be cross-linked and fragmented with the same treatment as the ChIP experiment but without the antibody pulldown enrichment step.

(b) *"IgG"*. Using a non-specific antibody that reacts with an irrelevant, non-nuclear antigen for the pulldown experiment.

(c) *"H3"*. The genome-wide profile of histone H3 provides a baseline distribution of nucleosomes in the genome.

All of these control samples should approximately get the same read depth as the ChIP-seq sample, and have good quality (Ho et al. 2011, Chen et al. 2012, Landt et al. 2012).

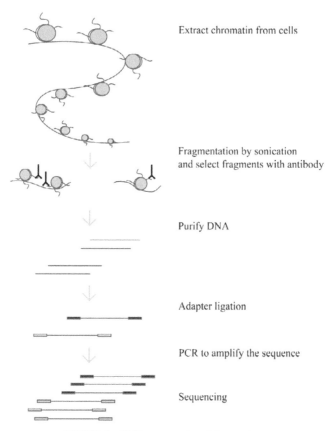

Extract chromatin from cells

Fragmentation by sonication
and select fragments with antibody

Purify DNA

Adapter ligation

PCR to amplify the sequence

Sequencing

FIGURE 2   ChIP-seq experimental procedure.

Replication

It is important to consider inclusion of biological or technical replicates in ChIP-seq experiments. Technical replicates will enable a fair assessment of reproducibility of the binding peaks observed in an individual replicate. Once the reproducibility is established, one can merge the reads from the replicates to produce a single ChIP-seq data set with a much higher depth for downstream analysis. Reproducibility between two replicates can be measured by the Irreproducible Discovery Rate (IDR) (Li et al. 2011). The benefit of IDR is that it can be used to consistently combine data across samples from different laboratories (Landt et al. 2012). Biological replicates may also be important if individual-to-individual variation is an important aspect of the experiment.

Single-end vs. paired-end sequencing

One important aspect of ChIP-seq analysis is to estimate the average fragment size. Obtaining a good estimate of fragment size is important for obtaining a high resolution ChIP-seq profile and peak call. This is not a problem if we have pair-end reads as fragment size can be directly calculated based on the coordinates of the mate pair.

Nonetheless, we can obtain a fairly reasonable estimation of fragment size based on methods such as cross-correlation analysis of the positive and negative strand reads in a single-end sequencing data set (Kharchenko et al. 2008, Heinz et al. 2010). Given the additional cost of doing paired-end sequencing, doing paired-end analysis may not be justified.

2.3   ChIP-seq data analysis

After a ChIP-seq experiment, we obtain FASTQ files as the output of a NGS machine. A typical FASTQ file has tens of millions of reads and their quality scores. The quality of the reads can be checked by a program called FastQC (http://www.bioinformatics.babraham.ac.uk/projects/fastqc/). If the quality is reasonable, the reads are then mapped to a reference genome. This step can be processed by tools such as Bowtie (Langmead et al. 2009) and Bowtie2 (Langmead and Salzberg 2012). For good quality ChIP-seq data, we should expect 70-90% of the reads to be mappable to the genome. This alignment and quality control step is similar to the analysis of other DNA-based NGS applications.

Once the reads are mapped, ChIP-seq specific analysis begins. The main goal of this analysis is to identify the binding of a specific TF or CR (detectable as short ChIP-seq signal peaks) or presence of histone modifications or variants (usually detectable as broad enrichment ChIP-seq regions). Different analysis programs are optimized for different tasks. Furthermore, some tools require a control sample (i.e., Input, IgG, or H3) to normalize the data. Many ChIP-seq analysis programs have been developed (Table 2), and here we summarize the key steps that are common among these programs:

Removal of bad reads

The first step is typically to remove low quality reads (i.e., reads with low quality score), and local read clusters that are most likely PCR amplification artefacts.

Fragment size estimation

For single-end sequencing data, one important step is to estimate the average fragment size. One way to estimate fragment size from single-end read data is the cross-correlation analysis of the positive and negative strand reads (Kharchenko et al. 2008). Once the fragment size is estimated, this information can be used to either extend or shift the reads at their 3' end. Accurate fragment size estimation and read extension or shifting is important for obtaining accurate peak calling results.

Peak calling and broad enrichment identification

This is perhaps one of the most important tasks in ChIP-seq analysis. The idea is to identify local regions in which the number of reads in the ChIP sample is significantly more than what is expected by chance based on the known background read distribution. The background distribution can be modeled by Poisson or a negative binomial distribution, and the parameters of these models can be estimated from the control sample. Most programs perform a sliding window scan of the genome to identify local enrichment regions, and use heuristics to merge neighbouring local

TABLE 2 Common ChIP-seq software and their features

Name	Use for	Control	Remarks	Reference
MACS	Both	Opt.	Employs a dynamic Poisson distribution to adjust the biases of the input file.	(Zhang et al. 2008)
S2P	Both	Opt.	An asymmetric distribution of tags on positive and negative strands is considered to improve the spatial precision.	(Kharchenko et al. 2008)
PeakSeq	Both	Man.	Employs a compensatory algorithm which considers the input files to adjust the bias of the signal caused by open chromatin.	(Rozowsky et al. 2009)
SICER	HM	Opt.	Merges together enrichment information from neighboring nucleosomes to increase true positive discovery rate. It calls spatial clusters of signals that are unlikely to arise randomly.	(Zang et al. 2009)
HOMER	Both	Opt.	By default, HOMER outputs peaks with a fixed peak size, which is determined by autocorrelation analysis. Focuses on identifying sites where the ChIP antibody makes a single contact with the DNA.	(Heinz et al. 2010)
CCAT	HM	Man.	Uses a library-swapping program to calculate the false discovery rate.	(Xu et al. 2010)
ZINBA	Both	Opt.	Uses a sliding-window approach to detect enriched regions. It adjusts chromosome-wide bias via an automated model selection algorithm.	(Rashid et al. 2011)
GEM	TF	Opt.	Improves motif discovery using positional priors from a generative probabilistic model of ChIP data. It further ameliorates false positive identification of binding events by using motif information.	(Guo et al. 2012)

TF: transcription factor or chromatin regulator; HM: histone modification; Man.: mandatory; Opt: optional.

enrichment regions. For identification of TF binding sites, it is also possible to make use of motif[1] information to refine the location of the binding sites (Guo et al. 2012).

De novo motif discovery

For the analysis of TF ChIP-seq data, one can perform *de novo* motif discovery using the sequences under the called peaks. HOMER (Heinz et al. 2010) and MEME-ChIP (Machanick and Bailey, 2011) are good examples of this type of analysis.

Other analysis

Other common tasks include visualization of the genome-wide ChIP-seq signal (with or without normalization with control data) in a genome browser, such as the Integrative Genomic Viewer (IGV) (James et al. 2011). Many ChIP-seq analysis programs can produce "wiggle files" that facilitate such visualization. Another common task is to analyze the local ChIP-seq signals in specific genomic regions such as enhancers, promoters, gene bodies, or TF binding sites. The signal of these regions may result in an average signal profile or a meta-gene profile (e.g., see Box 3 of Ferrari et al. 2014)

2.4   Chromatin states analysis

A chromatin state is defined as a combination of different chromatin features, usually focusing on histone modifications. With n distinct histone modification profiles, in theory there can be up to 2^n histone modification co-occurrence patterns. Nonetheless, in practice a much smaller number of distinct co-occurrence patterns are observed, and these chromatin states correlate with distinct regulatory regions or activity status (Ernst et al. 2011, Kharchenko et al. 2011). Therefore, it is of great interest to identify all the chromatin states in a genome. If we have a sufficient number and diversity of histone modification profiles, this approach allows us to annotate the genome with potential regulatory function (Ernst and Kellis 2010, Filion et al. 2010, Ernst et al. 2011, Kharchenko et al. 2011, Julienne et al. 2013). Furthermore, chromatin state dynamics between samples reveal cell-type specific regulatory regions (Sohn et al. 2015). A number of tools have been developed for chromatin state analysis (Table 3).

TABLE 3 Common chromatin state discovery software and their features

Name	Remarks	Reference
ChromHMM	Based on a multivariate Hidden Markov Model.	(Ernst and Kellis 2012)
Segway	Employs a Dynamic Bayesian Network (DBN) model for analyzing multiple tracks of genomics data at a high genomic resolution.	(Hoffman et al. 2012)
hiHMM	Based on a hierarchically-linked infinite hidden Markov Model (hiHMM). It enables cross-species, and cross-cell type inference of chromatin states.	(Sohn et al. 2015)

---

[1] Here motif is a pattern of nucleotide sequence that has, or is speculated to have a biological significance of TF binding affinity.

2.5 DNase-seq and FAIRE-seq

DNase I hypersensitive sites sequencing (DNase-seq) is a method that identifies the location of "open" or accessible chromatin, often representing active regulatory regions, via sequencing of regions of DNA sensitive to DNase I endonuclease cleavage (Crawford et al. 2006). Although the DNase I endonuclease digests DNA sequences without bias, in the complex chromatin environment it is prone to digest DNA sequences from unbound open chromatin (Furey 2012). After mapping the digested DNA fragments to the reference genome, clear "footprints" emerge consisting of enriched peaks flanking an unmapped gap in the signal. These footprints represent locations where proteins were bound to DNA, which prevents the DNA being digested by DNase I while leaving the neighboring DNA accessible to digestion. Sequenced genomic DNA or naked DNase-I digestion can be used as control for the DNase-seq experiment.

Formaldehyde-Assisted Isolation of Regulatory Elements (FAIRE-seq) is another method used for determining the locations of open chromatin genome-wide (Giresi et al 2007). Comparison studies revealed that FAIRE-seq and DNase-seq showed high rate of cross-validation (Song et al 2011). As the formaldehyde cross-linking is more efficient in nucleosome-bound DNA compared with that in nucleosome-depleted regions of the genome, the method isolates the non cross-linked DNA found in open chromatin, and then sequences it. Compared to DNase-seq, the FAIRE-seq experimental protocol is easier to conduct in the wet lab. FAIRE-seq data were generated as part of the ENCODE project and examples of these can be visualized at the UCSC Genome Browser. The analysis of DNase-seq and FAIRE-seq data is largely similar to ChIP-seq analysis. Nonetheless, specific programs have been developed to deal with specific features of DNase-seq and FAIRE-seq experiments. A summary of several common analysis programs is shown in Table 4.

TABLE 4 Common DNase-seq software and their features

Name	Remarks	Reference
F-seq	Produces a continuous tag sequence density estimation to call DNase footprint by detecting regions of open chromatin, or to call TF binding sites by finding peaks.	(Boyle et al. 2008)
pyDNase	Focuses on the sites in which the cleavage happens. By using the Wellington and Wellington 1D footprinting algorithms, pyDNase can specifically count cuts on the positive or negative reference strand.	(Piper et al. 2013)
PIQ	Applies a machine learning algorithm (expectation propagation) to find the genomic binding sites of transcription factors (TFs) at corresponding motifs from DNase-Seq data. Also uses machine learning in Input data normalization.	(Sherwood et al. 2014)

By using DNase-seq alone, researchers have obtained some exciting results. For example, Neph et. al. (2012) built an extensive core human regulatory network comprising connections among 475 sequence-specific TFs and analyzed the dynamics of these connections across 41 diverse cell and tissue types. Two years later, the same lab constructed a regulatory network from mouse tissues and compared it to the human network (Stergachis et al. 2014). Nonetheless, DNase-seq also has limitations for transcription factor detection, as many TFs exchange with specific binding sites in living cells on a timescale of seconds, instead of remaining stably bound for a long period (Sung et al 2014). Hence, the combination of ChIP-seq and DNase-seq would greatly improve the resolution as well as the confidence of TF detection.

2.6   Hi-C

How distal regulatory elements in a genome, such as enhancers and insulators, affect the expression of genes that are far apart on the linear chromosome is an interesting question in functional genomics. A novel method named "Hi-C", which combines DNA proximity ligation with high-throughput sequencing genome-wide, was invented to comprehensively detect chromatin interactions (Lieberman-Aiden et al. 2009). Hi-C was developed based on Chromosome Conformation Capture (3C), an older technique, in which chromatin is cross-linked with formaldehyde, then digested and re-ligated such that only DNA fragments that are covalently linked together form ligation products. The ligation complex contains multiple strands of DNA that reside physically close to each other in the three dimensional organization of the genome. Combining this technique with next generation sequencing makes Hi-C a powerful method to find chromatin interactions genome-wide (Belton et al. 2012).

Rao et al. (2014) used Hi-C to construct haploid[1] and diploid connectivity maps of nine human cell lines with as small as 1 kb resolution. They found that genomes could be divided into topologically associated domains with a median length of 185 kb, which are related with different histone marks and could be categorized into six sub-compartments. They also identified about 10,000 loops linking gene promoters and enhancers, which showed both conservation and novelty across species and cell types (Rao et al. 2014). Their huge dataset, which is browsable online, is a useful resource for understanding the complex chromatin environment.

3.   DNA Methylation Analysis

DNA methylation, the reversible addition of methyl groups to the fifth carbon of cytosine nucleotides, is an important regulator of gene expression and thus phenotypic variation and disease states (Fig. 1). DNA methylation occurs primarily at CG dinucleotides in vertebrates, but is also found in a range of other contexts; CHH and CHG methylation occurs commonly in plants (Cokus et al. 2008) and can also be found in certain vertebrate cell types, such as mammalian pluripotent cells (Laurent et al. 2010). In mammals, DNA methylation is predominantly established by the methyltransferase enzymes Dnmt3a and Dnmt3b, and subsequently maintained through semiconservative replication by the action of Dnmt1.

DNA methylation is generally associated with gene silencing and is thought to, as part of a larger system of epigenetic regulators including nucleosome phasing

---

[1] They used KBM-7 cells, which is a chronic myelogenous leukemia (CML) cell line. A unique aspect of the KBM-7 cell line is that it is near-haploid, meaning it contains only one copy for most of its chromosomes.

and histone modification, lead to inaccessibility of the DNA to the transcriptional machinery (Ehlrlich et al. 1993, Nan et al. 1998, Ng et al. 1999, Wade et al. 1999). However, the association with gene expression is not wholly understood and appears to be very context dependent. For example, methylation within promoters and first exons is highly correlated with transcriptional repression, but dense methylation within downstream exons and introns does not necessarily lead to repression and may even be associated with active transcription (Lister et al. 2009, Brenet et al. 2011, Jones 2012). The transcriptional machinery thus appears to interpret the occurrence of cytosine methylation differently depending on its exact location within the coding sequence.

DNA methylation, as one modification in the wider system of epigenetic regulation, ultimately contributes to the formation of many different phenotypes from the same genome, as is the case with the process of cellular differentiation. Being a covalent modification, DNA methylation is more stable than other highly dynamic epigenetic marks, such as histone modification, and is generally maintained over the life cycle of the cell; some genomic regions, such as pericentromeres and other repeat elements are maintained as methylated for the entire life cycle. The stability of DNA methylation makes it an attractive marker for studies that seek epigenetic causation or association in complex human disease e.g. epigenome-wide association studies (EWAS). Over the last decade or so, multiple methods have been developed to allow profiling of DNA methylation on a genome-wide scale, and state-of-the-art approaches involve the coupling of next-generation sequencing with the gold standard technique of bisulfite modification (Clark et al., 1994). Such methods have led to the identification of areas of differential methylation between samples or cohorts in EWAS across a wide variety of systems.

3.1 Whole genome bisulfite sequencing (WGBS)

Bisulfite sequencing allows DNA methylation state detection on standard sequencing platforms, and has led to significant advances in our understanding of DNA methylation and epigenetic regulation (Fig. 3). Sample preparation for bisulfite sequencing involves treatment of genomic DNA with sodium bisulfite, which deaminates cytosine residues, so converting them to uracil, while 5-methylcytosine is unaffected. During subsequent PCR, uracil nucleotides produced from the bisulfite conversion are converted to thymine (Clark et al., 1994). Following DNA sequencing the resulting reads are aligned and compared to a reference DNA sequence, and C/T mismatches of aligned reads are used to determine the methylation states in the original DNA molecules. In this way, single-base resolution maps of DNA methylation state are obtained for cytosines across the whole genome.

Generally a researcher will be interested in detecting differentially methylated sites where one sample group, such as samples taken from cancers, will display a significant difference in methylation at particular cytosines compared to a control sample group, such as a healthy sample from the same tissue type. Differentially methylated regions (DMRs) may also be found, which are regions of the genome that are abundant in individual differentially methylated cytosines. DMRs may be obtained from tiling approaches which are often based on a set length or may be empirically formed based on distance to the next differentially methylated cytosine (DMC) and the presence of a set number DMCs within each defined tile. Standard bisulfite sequencing cannot distinguish between 5-methylcytosine and other forms of the modification, such as hydroxymethylation, an important epigenetic mark in mammalian brain tissue, but additional steps can be added to the sample preparation in order to achieve this.

Whole genome bisulfite sequencing (WGBS) is currently the state-of-the-art method for obtaining high resolution DNA methylation data and has the advantage that almost the entire genome is interrogated. For the most common library preparation method (used for Illumina sequencing) genomic DNA is sheared during bisulfite conversion, this bisulfite-treated single-stranded DNA is then random-primed using a polymerase capable of reading uracil nucleotides. This polymerase synthesizes DNA containing a sequence tag. 3' ends of the new strands are then selectively tagged with a second fragment of specific sequence, resulting in di-tagged DNA molecules with known sequence at their 5' and 3'ends. As with standard Illumina sequencing, samples can be barcoded and sequenced in multiplex, but for minimally sufficient coverage for the purposes of DNA methylation analysis, a single library from a human sample is often run across two lanes of an 8-lane flow cell, giving around 120 GB sequence data.

The cost of WGBS may be prohibitive for medium to large scale experiments or small laboratories, and this has led to the development of methods that reduce breadth of coverage of the genome, allowing more samples to be included in the same sequencing run or lane and maximising coverage at areas of interest.

3.2   Reduced representation bisulfite sequencing

Reduced representation bisulfite sequencing (RRBS) (Smith et al 2009) is a method that allows enrichment of the genome for regions that are likely to be of most interest for epigenetic regulation, such as CpG islands, promoters and enhancer sequences. This reduced representation lowers the cost of sequencing and also gives increased depth of coverage, facilitating the resolution of more subtle changes in methylation levels at a given site.

RRBS library preparation utilizes restriction enzymes that cut genomic DNA at specific sequences, independently of methylation status. The enzymes Msp I and Taqα1 are commonly used, which have restriction sites of 5'...C|CGG...3' and 5'... T|CGA...3' respectively; these enzymes are insensitive to CpG methylation. The resulting fragments are then size-selected, most commonly by gel electrophoresis, for fragments of between 40 and 220 bp in length. Selection of this fragment size gives information for the majority of CpG islands in the mammalian genome and also for a range of other regions, including genes, enhancers and repeat sequences. This step is followed by fragment end-repair, A-tailing, sequencing adapter ligation and finally bisulfite conversion.

3.3   Special considerations for WGBS and RRBS

Many unique technical considerations must be taken into account for bisulfite sequencing methods. As some genomic material is lost over the course of the process, particularly during bisulfite conversion, the library preparation can call for large amounts of genomic DNA, ranging from hundreds of nanograms and up to 2 micrograms or more, depending on the protocol. Recently several methods have been described that allow RRBS libraries to be created from relatively small amounts of starting material, in the order of around 1ng of DNA, or DNA from only several hundred cells (Smallwood and Kelsey 2012, Schillebeeckx et al. 2013). The accuracy of the size selection process is also an important factor for RRBS, as it is hard to avoid contamination with fragments smaller than the ideal 40 bp size cut-off. In the case of smaller fragments, if the read length is larger than the DNA fragment, the adapter sequences at either end may be sequenced, leading to contamination of the sequencing read library. These contaminating sequences then need to be trimmed from the resulting reads prior to sequence alignment or mapping efficiency can be greatly reduced. A number of tools have been developed for this and are described in the next section. The most common read lengths used for DNA methylation sequencing, particularly on Illumina platforms, is 50-100 bp. As with other sequencing applications performed on Illumina platforms, the quality of data deteriorates as the sequencing cycles progress, and PHRED scores tend to drop significantly after a read length of about 70 bp. Thus, longer reads lead to a higher probability of errors in base calls, and thus a higher error rate in methylation state calling. The choice of single (directional) or paired-end (non-directional) sequencing is also an important consideration. Single-ended sequencing is the most common strategy as paired-end sequencing does not yield information for both the top and bottom strands, but rather creates redundant information for the same DNA strand. It can however give a marginal increase in mapping efficiency. Bias in directionality produced from paired-end sequencing is another potential problem. In single-ended sequencing experiments, the reads resulting from Msp I restricted libraries will all begin with either CGG or TGG, depending on the genomic methylation state at that site. Paired-end sequencing leads to an increase in the number of starting trinucleotide possibilities, as instead of just coming from the original top or bottom strand, some reads will yield information from the complementary sequences of the original top or bottom stands. Thus as well as the two possibilities of CGG and TGG, we now will also have CAA and CGA to take into account.

A special protocol is also required during the actual sequencing process for RRBS and related library types. Because the MspI recognition site yields fragments where the first three bases are non-random, substantial data loss can occur during the cluster calling phase of sequencing due to high apparent cluster density and poor cluster localization. To minimize these problems, imaging and cluster localization are delayed until the fourth sequencing cycle, in a process termed 'dark sequencing', which requires that the sequencing camera is turned off during the first three cycles of sequencing chemistry, and cluster calling to be based upon cycles 4-7. Minor adjustments in sequencing software are required to achieve this.

3.4 Data pre-processing, quality control and alignment methods

As previously mentioned, adapter contamination in sequencing reads can lead to significant decreases in mapping efficiency and thus the loss of large numbers of sequencing reads. It is therefore important to remove as much of this adapter contamination as possible from the reads before the step of alignment to the reference genome. Several methods have been developed for this purpose, such as cutadapt (Martin 2011) and trim_galore (http://www.bioinformatics.babraham.ac.uk/projects/

trim_galore/), which is a wrapper script combining the functions of both adapter trimming with cutadapt and quality control measurements using the FastQC program (http://www.bioinformatics.babraham.ac.uk/projects/fastqc/).

General quality control metrics can be gained by the use of FastQC. This is the most widely-used tool for this purpose for both standard and bisulfite sequencing experiments, though several other methods are available. Some particular considerations of the quality control process for bisulfite sequencing applications such as WGBS and RRBS are:

- Poor qualities (as deduced from PHRED score) can lead to incorrect methylation calls or mis-mapping.

- Poor bisulfite conversion efficiency can lead to erroneous calling of methylation state. Generally accepted minimum bisulfite conversion efficiency is 98%.

- Adapter contamination leading to incorrect methylation calls or poor alignment and data loss.

- Redundant information created in the case of paired-end sequencing. This extra information will generally need to be removed from datasets prior to any downstream differential methylation analyses.

- Positions filled in during the end-repair step may give different methylation state calls than the original genomic DNA. This filled-in position can however be used to calculate the efficiency of bisulfite conversion.

The most widely-used alignment and mapping tool for bisulfite sequencing data is Bismark (Krueger and Andrews 2011). This software combines the short-read alignment tool Bowtie2 (Langmead and Salzberg 2012) with methylation calling functionality, and creates output for methylation in CpG, CHG and CHH contexts. Other available tools for alignment of base-space WGBS/RRBS data include BS-Seeker (Chen et al 2010) and RRBSMAP (Xi et al 2012), but as with all fields in high-throughput sequencing, new software is continually being developed.

The presence of batch effects is another important consideration for all high-throughput sequencing experiments. Batch effects may result from a number of sources, including samples within a study being sequenced in separate run or the use of a different technician for library preparation or sequencing. Such non-biological variation needs to be accounted for or it may impede downstream statistical analysis. The program RnBeads (Yassen Assenov, et. al. 2014) applies the method of surrogate variable analysis (SVA) (Leek and Storey 2007) in order to detect and correct for batch effects, and is able to take a variety of different file formats, including those produced from Bismark and common methylation microarray software.

In genome-wide methylation studies the researcher may be interested in the genomic methylation states of a particular cell type, such as a cancer cell or a particular type of white blood cell. However, many samples are obtained from pre-curated tissue or DNA banks, or the cost of cell-sorting many samples may be prohibitive, and samples will therefore contain genomic DNA from several different cell populations. This is a potential confounder in the analyses (Jaffe and Irizarry 2014). One way to deal with the problem of cellular heterogeneity is by the use of corrective

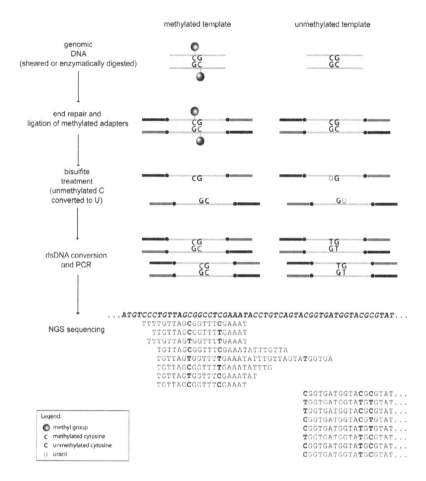

FIGURE 3 Bisulfite sequencing: Sodium bisulfite treatment deaminates cytosine residues, converting them to uracil, while 5-methylcytosine is unaffected. Following PCR, at which point uracils are converted to thymines, the fragments are sequenced in high-throughput and methylated and unmethylated cytosines can be distinguished from one another upon alignment with a reference genome sequence.

algorithms such as FaST-LMM-EWASher (Zou et al. 2014) and CETS (Guintivano et al. 2013) for neuronal cell samples, but due to a lack of empirical testing the performance of these methods is not well established.

3.5 Currently available methods for statistical analysis of genomic DNA methylation

The choice of statistical analysis methods is a crucial one for genome-wide DNA methylation studies, as different methods can lead to detection of quite different sets of differentially methylated cytosines. The field of statistical analysis of DNA methylation data is a relatively immature one, as we are only just coming to understand some of the complex correlation structures and mechanisms of DNA methylation patterning within different experimental designs. No systematic comparison or benchmarking of statistical methods has yet been performed, and as new packages and pipelines are developed at such a rapid pace it is a complex field to navigate. Most methods developed for statistical analysis of methylation data fall into two general types: binomial test-based and beta-binomial model-based methods. The distribution of DNA methylation percentage (or methylation ratio) is bimodal and display either near complete (100%) methylation or near zero, giving two modes. The number of methylated reads can be modelled by a binomial distribution. Beta-binomial model-based methods are therefore a more natural choice for this type of data, and many software packages such as BiSeq, MOABS and DSS use this approach. Most of the available software for DNA methylation data analysis is implemented in Bioconductor, a site for open-source bioinformatics tools written in R, though some may be in other languages. All of the programs are detailed below (Table 5) are free or open-source.

TABLE 5 Common DNA methylation analysis software and their features

Name	Implementation	Remarks	References
BSmooth and Bs-seq	R	Combines tools for alignment, quality control and identification of DMRS. A local likelihood smoothing algorithm is implemented, allowing it to perform particularly well with low coverage data. Can be used to identify regions of high methylation variability, so called "hypervariable" regions. After the calculation of sample-specific methylation profiles, DMRs are identified using a signal-to-noise statistic similar to t-test.	Hansen, Langmead and Irizarry, 2012.
Methylkit	R	Performs a variety of general, basic statistics and plots, including sample correlation matrices, hierarchical clustering and identification of DMCs and DMRs. DMCs are identified by a Fisher's Exact Test (FET) in the case of a one by one comparison (or pooled treatment groups) or logistic regression if there are biological replicates. Incorporates a basic tiling function.	Akalin et al. 2012
eDMR	**R**	Used as an extension to Methylkit (Akalin et al. 2012). Takes a myDiff object from methylkit as input and constructs DMRs using a weighted cost function and empirical tiling method. The significance of the inferred DMRs is calculated by combining the p-values of DMCs within that region. The user can input parameters required for a DMR to be called, such as distance to the next DMC and number of CpGs and DMCs per region.	Li et al. 2013.

DMAP	C	For differential methylation analysis of both WGBS and RRBS data. Takes SAM (Sequence Alignment/Map format) file or methylation extractor output from Bismark. Uses a fragment-based approach, where MspI fragment sizes of 40-220 bp are used as the unit of analysis, rather than fixed-length tiles or windows. Includes a variety of basic statistical analyses, such as ANOVA, Chi-squared test and Fisher's exact test and, as with many of the other programs, also performs annotation with proximal genic and CpG features.	Stockwell et al. 2014
BiSeq	R	Combines functions for differential methylation detection within target regions by incorporating spatial dependencies. A hierarchical testing procedure is applied where following a traditional DMR detection approach, localization of DMRs in rejected regions is also considered. Takes input from targeted bisulfite sequencing experiments such as RRBS and enhanced RRBS (eRRBS).	Hebestreit et al. 2013
MethylSig	R	Analysis of both WGBS and RRBS data. Uses a beta-binomial model and factors in coverage and variation within groups at each CpG or DMR, and uses calibrated Type I error rate. A statistic based on the likelihood ratio test is used to evaluate the significance level of the difference in methylation. Performs site-specific or sliding window tests and also includes functions for data annotation and visualization as well as transcription factor binding site enrichment.	Park et al. 2014

4.   Bioinformatics databases and resources

Researchers can download a vast amount of epigenomic data from databases such as NCBI's GEO (Barrett et al. 2013), ENCODE (ENCODE Project Consortium 2004) and Roadmap Epigenomics (Bernstein et al. 2010). There are also many organ or disease specific databases such as BloodChIP (Chacon et al. 2014) and HAEMCODE (Ruan et al. 2013). In these databases, users can usually find the mapped read files or even the final output files in various formats, including signal tracks, bed files and called peak files from chromatin experiments and also raw FASTQ files from experiment. Here we review several major public repositories.

4.1   NCBI's Gene Expression Omnibus (GEO) and Sequence Read Archive (SRA)

GEO (http://www.ncbi.nlm.nih.gov/gds) and SRA (http://www.ncbi.nlm.nih.gov/sra) are the largest central repositories for genomics data globally and due to the excellent search functionality they form a crucial resource for many bioinformatics analyses. Much of the data from the databases below and the results of many other experiments from researchers worldwide are accessible through GEO and SRA searches.

4.2   ENCODE: Encyclopedia of DNA Elements project

ENCODE is a major research project (https://www.encodeproject.org) started by the US National Human Genome Research Institute (NHGRI) in September 2003 (ENCODE Project Consortium 2004). As a successor to the Human Genome Project, ENCODE aims to discover all regulatory elements in the human genome. ENCODE involves many research groups worldwide, and has already produced a large amount of data that can be downloaded from the UCSC server (http://genome.ucsc.edu/ENCODE), including ChIP-seq, DNase-seq, FAIRE-seq, RNA-seq, Methyl RRBS, Methyl Array data and other data such as from CAGE and ChIA-PET (Fullwood et al. 2009) for many human cell types. Be aware that more data sets are available for older genome assemblies such as hg18.

4.3   Mouse ENCODE and modENCODE projects

The Mouse ENCODE (http://www.mouseencode.org) and modENCODE (model organism ENCODE, http://www.modencode.org) projects extend the ENCODE approach and assay data to three model organisms: *Mus musculus* (mouse), *Drosophila melanogaster* (fly) and *Caenorhabditis elegans* (worm) (Stamatoyannopoulos et al. 2012, Gerstein et al. 2010, Roy et al. 2010, Boyle et al. 2014, Gerstein et al. 2014, Ho et al. 2014).

4.4   FANTOM project

Functional Annotation of The Mammalian Genome (FANTOM) is an inter-national project (http://fantom.gsc.riken.jp) established in 2000 to function-ally annotate the full-length cDNAs collected during the Mouse Encyclopedia Project at RIKEN (Kawai et al. 2001). Although experimental techniques have changed throughout FANTOM 1 to FANTOM 5, the project aims to reveal the function of transcripts and the transcriptional regulatory network. FANTOM 5 now focuses on examining how our unique genome controls the variety of cell types in a human body. Users can also find human enhancer annotation files from FANTOM 5 (Andersson et al. 2014).

4.5   The Roadmap Epigenomics project

NIH (National Institute of Health) launched the Roadmap Epigenomics Project in 2007 (http://www.roadmapepigenomics.org). As a complement to the ENCODE Project, the Roadmap Epigenomics Project aims to comprehensively understand the epigenetic events of a variety of cell types during development, differentiation and during human diseases (Bernstein et al. 2010). The project has also built a large reference epigenome database available for public download, containing whole genome bisulfite sequencing data, RRBS data and a wealth of histone modification data across different organs and tissues.

---

5. Applications of epigenomics

With technological advances in ChIP-seq, DNase-seq, FAIRE-seq, WGBS, RRBS and chromatin state discovery tools, the pace of genome annotation and discovery of functional genomic elements has accelerated rapidly since 2008, fuelled by research from the large consortium projects described above.

5.1 Major functional regulatory insights gained through epigenomic profiling

Meissner et al. (2008) revealed the widespread dynamics of DNA methylation during cellular differentiation and development using RBBS. Their description of developmentally regulated demethylation of highly conserved non-coding regulatory elements provided a mechanism for Waddington's epigenetic landscape model.

The worm modENCODE analysis discovered that elements in HOT-regions (High-Occupancy Target regions bound by many TFs) in L1 larvae cells drive expression in most adult cell types, in contrast to other genes with a largely tissue-specific expression (Gerstein et al. 2010) . They also found differences between the hierarchies of TFs at each developmental level of worms - TFs acting at the lower levels tended to be more uniformly expressed across multiple tissues. Kharchenko et al. (2011) presented a genome-wide chromatin landscape of fly based on eighteen histone modifications, which they summarized into nine chromatin states, demonstrating that transcriptionally active genes showed distinct chromatin patterns. They also found several histone modification signatures among Polycomb targets, providing clues to how regulatory elements interact with Polycomb group proteins (Kharchenko et al. 2011).

A breakthrough in cancer epigenomics came when Hansen et al. (2011) discovered that cancer genomes display significant methylation instability including large blocks of hypomethylation, sometimes affecting more than half of the genome. Genes in this region showed extreme gene expression variability and are involved in mitosis and matrix remodelling, hinting at an epigenetic mechanism for tumor heterogeneity (Hansen et al. 2011).

Using the ChromHMM algorithm, Ernst et al. (2011) discriminated six classes of human chromatin states in multiple cell lines. They characterized promoters with active, weak and poised expression activity, strong and weak enhancers, insulator regions, transcribed regions, repressed and inactive states. They found that strong enhancer and polycomb repressed states are missing in embryonic stem cells, consistent with the nature of embryonic cells. They also found that promoter states appeared more stable across different cell-types than enhancers, which appear in cell-type specific enhancer clusters that drive tissue specific gene expression. Furthermore, highly disease associated single nucleotide polymorphisms were often located within the active enhancers of related cell types, and in some cases affect a motif site for TF binding, suggesting a mechanism for the genetic etiology of the disease (Ernst et al. 2011).

In 2012, the human ENCODE paper (Dunham et al. 2012), revealed that a somatic mutation is less likely to occur within ENCODE annotated regulatory regions and that GWAS SNPs presented more often within these regulatory regions. Amongst other findings was that mRNA expression levels could be predicted with high accuracy using histone modification and TF binding data and that most TFs have a non-random association between each other. In order to quantify the conservation of regulatory mechanisms between human and mouse, researchers built a TF-TF cross-regulatory network for each species (Yue et al 2014). These networks included ~500 transcription factors and were based on DNase-seq footprints located on binding motifs in the promoter and enhancer regions of TF genes. They found that while only 22% of apparently active transcription factor motifs are strictly conserved between species, in total, around 50% of connections in the TF-TF regulatory circuitry are conserved through the same or different binding sites in regulatory regions (Yue et al 2014). Ho et al. (2014) studied cross-species chromatin organization between human, fly and worm using 1,400 ENCODE and modENCODE data sets. They discovered that although the three organisms shared many conserved features in the organization of the chromatin, the organization and composition of repressive chromatin are quite different in worm compared to fly and human.In early 2015, the Roadmap Epigenomics Project published a series of articles with many interesting discoveries and resources (Romanoski et al. 2015, http://www.roadmapepigenomics.org/publications/). In one study, Kundaje et al. (2015) performed an integrative study of 111 reference human cell-type epigenomes with gene expression data. This allowed them to globally map regulatory elements, define the makeup of regulatory modules, confirm that disease and trait associated variants occur in tissue specific epigenomic marks and calculate a similarity distance between cell types in a multi-dimensional scaling analysis. Another study focused on investigating higher-order chromosomal structure and allele-specificity in genetic regulation (Dixon et al. 2015). They found that while local interaction domain borders were stable during development, a dramatic 36% of the genome transitioned between compartment types A and B in at least one lineage, with significant changes in functionality. Change in H3K4me1 density was found to be the most predictive feature of changes in long-range chromatin interactions, reflecting the important role of enhancer looping in defining chromatin structure (Dixon et al. 2015). The consortium also released an efficient epigenomics browser for visualising hundreds of datasets at once in the context of disease causing variation (http://epigenomegateway.wustl.edu/browser/roadmap, Zhou et al. 2015).

5.2   Epigenome wide association studies (EWAS)

During the last decade many genetic variants have been discovered that influence the occurrence of complex human diseases such as cancer, obesity and hypertension. This has been made possible partly by the genome-wide association study (GWAS) approach, where SNPs are compared on a genome-wide scale between, for example, healthy and diseased samples. However, useful findings from GWAS have nowhere near reached the expectations of the scientific community and as the significance of epigenetic variation has come more into light, an *epi*genome-wide approach has become popular. Alterations in epigenetic states can occur spontaneously and potentially lead to heritable disease predisposition (Suter et al. 2004); epigenetic states can also respond to environmental stimuli and parental influence (Cropley et al. 2006 and 2012). Because epigenetic changes can occur in multiple individuals at the same genes but on multiple haplotypes, they are invisible to genome-wide association studies (GWAS), which rely on defining common sequence variation. A systematic study of epigenetic marks on a genome-wide scale can provide an unbiased assessment of epigenetic variants that may associate with a trait. Much like GWAS, epigenome-wide association studies (EWAS) involve interrogation of variants using high-throughput platforms; the stability of DNA methylation, and its amenability to study, renders this epigenetic mark the most suitable for EWAS.

EWAS are in their infancy, and most of the early studies copied the key design and analytical principles of GWAS, which are now known to be problematic in the context of EWAS. Standards in study design, sample size selection, and statistical analysis for EWAS are still emerging (Bock, 2012, Michels et al. 2013). The complex nature of epigenetic states renders the optimal design and analysis of human EWAS conceptually distinct and potentially more difficult than GWAS. *De novo* study of human cohorts runs a high risk of missing relevant epigenetic changes, since these may be obscured by high backgrounds of genetic variation, epigenetic variation unrelated to the disease, and epigenetic variation consequential to the disease, its treatment, or co-morbid conditions. Such problems are not easily overcome, and may require EWAS with many thousands of individuals. Despite these challenges, the EWAS approach has led to some new insight into possible epigenetic contributions to disease and into how environmental influences, such as those experienced in utero, may affect offspring via epigenetic mechanisms. One example of such a discovery is that of Joubert et al. (2012) who identified a number of genes with significantly altered methylation in children whose mothers smoked while pregnant. Several of these methylation changes were later confirmed in additional studies (Markunas et al., 2014), and could very well be involved in the development of disease arising from maternal smoking. Technological advancement and decreasing costs of high-throughput sequencing will hopefully lead to significant improvements in our understanding of common, complex diseases in the coming years.

6.   Emerging fields in epigenetics

Genome-wide epigenomic profiling has already made a significant impact on our understanding of gene regulation using the current technologies described in earlier sections. New experimental and computational technologies, and new applications of these technologies, are pushing the boundary of this field. In this section, we discuss three leading-edge research areas that we believe will drive the next waves of breakthroughs: regulatory grammar of cis-regulatory elements (new application of epigenomic data); single cell epigenetics (new methodology); and comparative epigenomics of evolutionarily diverse organisms (new area of epigenomic research).

6.1   Understanding the hidden grammar of cis-regulatory elements

One application of epigenomic profiling is the identification of regulatory regions at a genome-wide scale. Researchers around the world have contributed to the generation of epigenomic profiles, and used them to discover consensus DNA motifs of transcription factors via ChIP-seq and DNase-seq. We can now easily browse this knowledge using an online open-access database of transcription factor binding profiles, such as JASPAR (Sandelin et al. 2004, Mathelier et al. 2014). Using these databases, Kheradpour et al. (2013) performed a systematic study on regulatory motifs by synthesizing plasmids with different selections of activating and repressing motifs. They found that different forms of motif disruption, such as scrambling, total removal, 1bp insertion or deletion, and so on, showed a variety of effects on related gene expression. Smith et al. (2013) went further in their research, synthesizing 4,970 regulatory element sequences within three groups: homotypic / same motif (class I), simple heterotypic / 2+ motifs (class II) and more complex heterotypic motif combinations (class III). With this experimental design, they firstly observed a significant correlation between the size of the homotypic cluster and gene expression for around half of their class I motifs. They also found that in general, heterotypic motifs drive stronger expression than homotypic motifs (Smith et al 2013). These findings represent pioneering work in using massively parallel assays to quantitatively test the effect of sequence composition in regulatory elements on gene expression. We believe these types of studies will unravel the combinatorial grammar behind transcription factor activity.

6.2   Single cell analysis

Single-cell genomics is leading to major advances in our ability to confidently answer questions about cellular regulation and was fittingly named the Method of the Year in 2013 by the journal Nature Methods. This technology is enabling genome-wide sequencing at the level of individual cells, instead of averaging the signals from a heterogeneous cell population. This has facilitated the fine-scale identification of gene expression status during cellular differentiation (Trapnell et al. 2014) and has quantified how much intercellular variability is masked by lower experimental resolution (Wills et al. 2013). This applies directly to epigenetics, where it is difficult to say whether histone modifications and other epigenetic marks truly co-occur at a locus, or whether it is an effect of population averaging across multiple experiments.

Much work is occurring in the cutting-edge field of single-cell epigenomics, from the development of technologies to profile the epigenome of single cells, to the algorithms/methods to analyze the data and interpret the results. Many of the developing methods and associated challenges are reviewed by Hyun et al. (2015) and Bheda et al. (2014). Farlik et al. (2015) have already succeeded in creating a method for single cell whole genome bisulfite sequencing in model cell lines and used this to temporally order differentiating cells based on their individual methylomes. Investigating chromatin structure and histone modifications at the single cell level has not yet been achieved using traditional methods like ChIP-seq and DNase-seq. However, approaches such as ATAC-seq (Assay for Transposase-Accessible Chromatin with high throughput sequencing) which investigates chromatin accessibility, have reduced the number of cells required by several orders of magnitude, down to below 1000 (Buenrostro et al. 2013). Single molecule nanopore-based technologies have shown an ability to detect the slight electrical perturbations caused by chemical modifications to DNA (Carlsen et al. 2014). Non-sequencing approaches like fluorescence microscopy partnered with 3D reconstruction are also being applied to characterize single cells based on their epigenomes and single cell Hi-C technology bridges the gap between chromosomal microscopy and genomics (Tajbakhsh et al. 2015, Nagano et al. 2013).

While it will take years before we can truly profile the entire epigenome at single cell resolution, until then we can gain insights by combining our established bulk cell and low cell input technologies with true single cell genomic data that we can generate now - DNA sequencing, RNA-seq, DNA methylation and chromosomal structure through Hi-C.

6.3 Comparative epigenomics: Beyond model organisms

As genetic regulation is increasingly well studied, its central contribution to phenotypic diversity is evident. An interesting emerging idea is that heritable changes to the regulatory epigenome could be a major evolutionary driver. This is a major contributor to the complexity of the human brain (Skinner et al. 2014, Krubitzer and Stolzenberg 2014). Thus it is increasingly important to study epigenomics across the tree of life. Several studies have discovered differences in histone modification co-occurrence patterns between humans and other model organisms, including fly and worm (Ho et al. 2014, Sohn et al. 2015). These studies showed that in worm there is frequently co-occurrence of the marks H3K9me3 and H3K27me3 in repressive regions, whereas in human and fly these marks are largely mutually exclusive and define heterochromatic and polycomb-repressed regions respectively. Sohn et al. (2015) found further differences in the distribution of several less commonly profiled histone modifications within promoters and transcriptional regions between the two species. Do these changes in histone modification co-occurrence indicate another level of epigenomic dynamics we are yet to understand? The extension of cross-species epigenomics to the whole spectrum of chromatin features across many more organisms in the years to come will shed light on this question.

One area where non-model organism epigenetics can have immediate medical impact is in understanding host-pathogen interactions. Many pathogenic eukaryotes, such as fungi and their host organisms are known to switch their transcriptional programs dependent on infection status and new signalling pathways are implicated at the interface between human hosts and fungal pathogen during infection (Liu et al. 2015, Soyer et al. 2015). An emerging model of virulence evolution is that fungi may epigenetically de-repress retrotransposon activity to increase genome plasticity during infection (Gijzen et al. 2015). There is a concerted effort emerging to perform genome wide BS-seq and ChIP-seq across many fungal species and their hosts during an infection life-cycle, in order to describe the epigenetic mechanisms that control pathogenicity. This will hopefully lead to new therapeutic approaches that will have a significant impact on human health as well as many industries in which fungal species are crucial players.

Acknowledgements

This work was supported in part by funds from the New South Wales Government Ministry of Health, a Human Frontier Science Program Grant (RGY0084/2014), and a Ramaciotti Establishment Grant (ES2014/010). We thank Dr Cheryl Li for drawing part of Fig. 1.

References

Allis C. D., Jenuwein T., Reinberg D., and M.L. Caparros 2008. Epigenetics. Cold Spring Harbor Laboratory Press, New York.

Akalin, A., Kormaksson, M., Li, S., Garrett-Bakelman, F. E., Figueroa, M. E., Melnick, A., and C.E. Mason, 2012. methylKit: a comprehensive R package for the analysis of genome-wide DNA methylation profiles. *Genome Biology*, 13(10): R87.

Andersson, R., Gebhard, C., Miguel-Escalada, I., Hoof, I., Bornholdt, J., Boyd, M., and A. Sandelin, 2014. An atlas of active enhancers across human cell types and tissues. *Nature*, 507: 455–61.

Assenov, Y., Müller, A., Lutsik, P., Walter, J., Lengauer, T. and C. Bock, 2014. Comprehensive Analysis of DNA methylation Data with RnBeads, *Nature Methods*, 11(11):1138-40.

Babu, M. M., Luscombe, N. M., Aravind, L., Gerstein, M., and S. A. Teichmann, 2004. Structure and evolution of transcriptional regulatory networks. *Current Opinion in Structural Biology*. 14(3): 283-291.

Barrett, T., Wilhite, S. E., Ledoux, P., Evangelista, C., Kim, I. F., Tomashevsky, M., and A. Soboleva, 2013. NCBI GEO: archive for functional genomics data sets—update. *Nucleic Acids Research*. 41(D1): D991-D995.

Belton, J. M., McCord, R. P., Gibcus, J. H., Naumova, N., Zhan, Y., and J. Dekker, 2012. Hi-C: A comprehensive technique to capture the conformation of genomes. *Methods*. 58, 268–276.

Bernstein, B. E., Stamatoyannopoulos, J. A., Costello, J. F., Ren, B., Milosavljevic, A., Meissner, A., and J. A. Thomson, 2010. The NIH Roadmap Epigenomics Mapping Consortium. Nature *Biotechnology*. 28, 1045–1048.

Bheda, P., and R. Schneider, 2014. Epigenetics reloaded: the single-cell revolution, *Trends in Cell Biology*, 24, (11): 712-723

Bock, C. 2012. Analysing and interpreting DNA methylation data. Nature Reviews Genetics; 13: 705-19.

Boyle, A. P., Guinney, J., Crawford, G. E., and T. S. Furey, 2008. F-seq: a feature density estimator for high-throughput sequence tags. *Bioinformatics*. 24(21), 2537-38.

Boyle, A. P., Araya, C. L., Brdlik, C., Cayting, P., Cheng, C., Cheng, Y., and M. Snyder, 2014. Comparative analysis of regulatory information and circuits across distant species. *Nature*, 512(7515), 453-56.

Brenet, F., Moh, M., Funk, P., Feierstein, E., Viale, A. J., Socci, N. D., and J. M. Scandura, 2011. DNA methylation of the first exon is tightly linked to transcriptional silencing. *PLoS One*. 6(1):e14524

Brivanlou, A. H., and Jr., J. E. Darnell 2002. Signal transduction and the control of gene expression. Science, 295, 813–818.

Buenrostro, J. D., Giresi, P. G., Zaba, L. C., Chang, H. Y., and W. J. Greenleaf, 2013. Transposition of native chromatin for fast and sensitive epigenomic profiling of open chromatin, DNA-binding proteins and nucleosome position. *Nature Methods*, 10(12), 1213-18.

Carlsen, A. T., Zahid, O. K., Ruzicka, J. A., Taylor, E. W., and A. R. Hall, 2014. Selective detection and quantification of modified DNA with solid-state nanopores. *Nano Letters*. 14(10), 5488-92.

Chacon, D., Beck, D., Perera, D., Wong, J. W., and J. E. Pimanda, 2013. BloodChIP: a database of comparative genome-wide transcription factor binding profiles in human blood cells. *Nucleic Acids Research*, D172-7.

Chen, P. Y., Cokus, S. J., and Pellegrini, M. 2010. BS Seeker: precise mapping for bisulfite sequencing. *BMC Bioinformatics.* 11(1), 203.

Chen, Y., Negre, N., Li, Q., Mieczkowska, J. O., Slattery, M., Liu, T., and Liu, X. S. 2012. Systematic evaluation of factors influencing ChIP-seq fidelity. *Nature Methods.* 9(6), 609-614.

Clark, S. J., Harrison, J., Paul, C. L., and M. Frommer, 1994. High sensitivity mapping of methylated cytosines. *Nucleic Acids Research*, 22(15), 2990–97.

Cokus, S. J., Feng, S., Zhang, X., Chen, Z., Merriman, B., Haudenschild, C. D., and S. E. Jacobsen, 2008. Shotgun bisulfite sequencing of the Arabidopsis genome reveals DNA methylation patterning. *Nature.* 452(7184), 215–19.

Crawford, G. E., Holt, I. E., Whittle, J., Webb, B. D., Tai, D., Davis, S., and F. S. Collins, 2006. Genome-wide mapping of DNase hypersensitive sites using massively parallel signature sequencing (MPSS). *Genome Research.* 16, 123–31.

Cropley, J. E., Suter, C. M., Beckman, K. B., and D. I. Martin, 2006.Germ-line epigenetic modification of the murine Avy allele by nutritional supplementation. *Proceedings of the National Academy of Sciences.* 103(46), 17308-12.

Cropley, J. E., Dang, T. H., Martin, D. I., and C. M. Suter, 2012. The penetrance of an epigenetic trait in mice is progressively yet reversibly increased by selection and environment. *Proceedings of the Royal Society of London B: Biological Sciences*, 279(1737), 2347-53.

Dixon JR, Jung I, Selvaraj S, Shen Y, Antosiewicz-Bourget JE, Lee A.Y., Ye Z., Kim A., Rajagopal N., Xie W., Diao Y., Liang J., Zhao H., Lobanenkov V.V., Ecker J.R., Thomson J.A., and B.. Ren Chromatin architecture reorganization during stem cell differentiation. *Nature.* 2015 Feb 19; 518(7539): 331-36.

Dunham, I., Kundaje, A., Aldred, S. F., Collins, P. J., Davis, C. A., Doyle, F., and E. Birney, 2012. An integrated encyclopedia of DNA elements in the human genome. *Nature*, 489(7414), 57-74.

Egelhofer, T. A., Minoda, A., Klugman, S., Lee, K., Kolasinska-Zwierz, P., Alekseyenko, A. A., and J. D. Lieb, 2011. An assessment of histone-modification antibody quality. *Nature Structural & Molecular Biology*, 18(1), 91-93.

Ehrlich, M., and K. C. Ehrlich. 1993. Effect of DNA methylation on the binding of vertebrate and plant proteins to DNA, p. 145-168. In J. P. Jost and H. P. Saluz (eds.), DNA methylation: *Molecular Biology and Biological Significance.* Birkhauser Verlag, Basel, Switzerland.

ENCODE Project Consortium. 2004. The ENCODE (ENCyclopedia of DNA elements) project. *Science*,306(5696), 636-40.

Ernst, J. and M. Kellis, 2010. Discovery and characterization of chromatin states for systematic annotation of the human genome. *Nature Biotechnology*, 28, 817–25.

Ernst, J. and M. Kellis, 2012. ChromHMM: automating chromatin-state discovery and characterization. *Nature Methods*, 9(3), 215-16.

Ernst, J., Kheradpour, P., Mikkelsen, T. S., Shoresh, N., Ward, L. D., Epstein, C. B., and B. E. Bernstein, 2011. Mapping and analysis of chromatin state dynamics in nine human cell types. *Nature.* 473, 43–49.

Farlik, M., Sheffield, N. C., Nuzzo, A., Datlinger, P., Schönegger, A., Klughammer, J., and C. Bock, 2015.Single-Cell DNA Methylome Sequencing and Bioinformatic Inference of Epigenomic Cell-State Dynamics. *Cell Reports*, 10(8), 1386-97.

Feng, H., Conneely, K. N., and H. Wu, 2014. A Bayesian hierarchical model to detect differentially methylated loci from single nucleotide resolution sequencing data. *Nucleic Acids Research.* 42(8), e69.

Feng, J., Liu, T., Qin, B., Zhang, Y., and X. S. Liu, 2012.Identifying ChIP-seq enrichment using MACS. *Nature Protocols*, 7(9), 1728-40.

Ferrari, F., Alekseyenko, A. A., Park, P. J., and M. I. Kuroda, 2014. Transcriptional control of a whole chromosome: emerging models for dosage compensation. *Nature Structural and Molecular Biology*. 21(2), 118-25.

Filion, G. J., van Bemmel, J. G., Braunschweig, U., Talhout, W., Kind, J., Ward, L. D., and B. van Steensel, 2010. Systematic Protein Location Mapping Reveals Five Principal Chromatin Types in Drosophila Cells. *Cell*. 143: 212–224.

Fullwood, M. J., Liu, M. H., Pan, Y. F., Liu, J., Xu, H., Mohamed, Y. B., and Ruan, Y. 2009. An oestrogen-receptor-agr-bound human chromatin interactome. *Nature*, 462(7269), 58-64.

Furey, T. S. 2012. ChIP–seq and beyond: new and improved methodologies to detect and characterize protein–DNA interactions. *Nature Reviews Genetics*. 13(12): 840-852.

Gerstein, M. B., Lu, Z. J., Van Nostrand, E. L., Cheng, C., Arshinoff, B. I., Liu, T., and R. H. Waterston, 2010. Integrative analysis of the Caenorhabditis elegans genome by the modENCODE project. *Science*. 330: 1775–1787.

Gerstein, M. B., Rozowsky, J., Yan, K. K., Wang, D., Cheng, C., Brown, J. B., and J. Lagarde, 2014. Comparative analysis of the transcriptome across distant species. Nature, 512(7515), 445-448.

Gijzen, M., Ishmael, C., and S. D. Shrestha, 2014. Epigenetic control of effectors in plant pathogens. *Frontiers in Plant Science*, 5: 638.

Giresi, P. G., Kim, J., McDaniell, R. M., Iyer, V. R., and J. D. Lieb, 2007. FAIRE (Formaldehyde-Assisted Isolation of Regulatory Elements) isolates active regulatory elements from human chromatin. *Genome Research*, 17: 877–885.

Guintivano, J., Aryee, M. J., and Z. A. Kaminsky, 2013.A cell epigenotype specific model for the correction of brain cellular heterogeneity bias and its application to age, brain region and major depression. *Epigenetics*, 8(3): 290-302.

Guo, Y., Mahony, S., and D. K. Gifford, 2012. High resolution genome wide binding event finding and motif discovery reveals transcription factor spatial binding constraints. *PLoS Computational Biology*. 8(8): e1002638.

Hansen, K. D., Langmead, B., and A. R. Irizarry, 2012. BSmooth: from whole genome bisulfite sequencing reads to differentially methylated regions. *Genome Biology*, 13: R83.

Hansen, K. D., Timp, W., Bravo, H. C., Sabunciyan, S., Langmead, B., McDonald, O. G., and A. P. Feinberg, 2011. Increased methylation variation in epigenetic domains across cancer types. *Nature Genetics*. 43(8), 768–75.

Hebestreit, K., Dugas, M., and H. U. Klein, 2013.Detection of significantly differentially methylated regions in targeted bisulfite sequencing data. *Bioinformatics*, 29(13), 1647-53.

Heinz, S., Benner, C., Spann, N., Bertolino, E., Lin, Y. C., Laslo, P., and C. K. Glass, 2010. Simple Combinations of Lineage-Determining Transcription Factors Prime cis-Regulatory Elements Required for Macrophage and B Cell Identities. *Molecular Cell*, 38, 576–89.

Hoffman, M. M., Buske, O. J., Wang, J., Weng, Z., Bilmes, J. A., and W. S. Noble, 2012. Unsupervised pattern discovery in human chromatin structure through genomic segmentation. *Nature Methods*, 9(5), 473-76.

Ho, J. W. K., Bishop, E., Karchenko, P. V., Nègre, N., White, K. P., and P. J. Park, 2011. ChIP-chip versus ChIP-seq: lessons for experimental design and data analysis. BMC Genomics, 12(1), 134.

Ho, J. W. K., Jung, Y. L., Liu, T., Alver, B. H., Lee, S., Ikegami, K., and P. J. Park, 2014. Comparative analysis of metazoan chromatin organization. *Nature*, 512(7515), 449–52.

Hyun, B., McElwee, J. L. and P. D. Soloway 2015. Single molecule and single cell epigenomics, Methods, Vol 72, 15; 72: 41-50.

Jaffe, A. E., and R. A. Irizarry, 2014. Accounting for cellular heterogeneity is critical in epigenome-wide association studies. *Genome Biology*, 15(2), R31.

James T. Robinson, Helga Thorvaldsdóttir, Wendy Winckler, Mitchell Guttman, Eric S. Lander, Gad Getz, and Jill P. Mesirov. 2011. Integrative Genomics Viewer. Nature Biotechnology 29, 24–26

Jones PA. 2012. Functions of DNA methylation: islands, start sites, gene bodies and beyond. *Nature Reviews Genetics*; 13: 484 - 92

Joubert, B. R., Håberg, S. E., Nilsen, R. M., Wang, X., Vollset, S. E., Murphy, S. K., and S. J. London, 2012. 450K Epigenome-Wide Scan Identifies Differential DNA methylation in Newborns Related to Maternal Smoking during Pregnancy. *Environmental Health Perspectives*. 120(10), 1425–1431.

Julienne, H., Zoufir, A., Audit, B., and A. Arneodo, 2013.Human Genome Replication Proceeds through Four Chromatin States. PLoS Computational Biology, 9(10): e1003233.

Jung, Y. L., Luquette, L. J., Ho, J. W., Ferrari, F., Tolstorukov, M., Minoda, A., and P. J. Park, 2014. Impact of sequencing depth in ChIP-seq experiments.Nucleic Acids Research, 42(9), e74-e74.

Kawai, J., Shinagawa, A., Shibata, K., Yoshino, M., Itoh, M., Ishii, Y., and Hayashizaki, Y. 2001. Functional annotation of a full-length mouse cDNA collection. Nature, 409, 685–690.

Kharchenko, P. V, Alekseyenko, A. A., Schwartz, Y. B., Minoda, A., Riddle, N. C., Ernst, J., and P. J. Park, 2011. Comprehensive analysis of the chromatin landscape in Drosophila melanogaster. Nature, 471, 480–485.

Kharchenko, P. V, Tolstorukov, M. Y., and P. J. Park, 2008. Design and analysis of ChIP-seq experiments for DNA-binding proteins. Nature Biotechnology, 26, 1351–1359.

Kheradpour, P., Ernst, J., Melnikov, A., Rogov, P., Wang, L., Zhang, X., and M. Kellis, 2013. Systematic dissection of regulatory motifs in 2000 predicted human enhancers using a massively parallel reporter assay. Genome Research, 23, 800–811.

Kidder, B. L., Hu, G., and K. Zhao, 2011. ChIP-seq: technical considerations for obtaining high-quality data. Nature Immunology, 12(10), 918-922.

Krubitzer L., and D. S. Stolzenberg 2014. The evolutionary masquerade: genetic and epigenetic contributions to the neocortex, Current Opinion in Neurobiology, 24, 157-165

Krueger, F., and S. R. Andrews, 2011. Bismark: a flexible aligner and methylation caller for Bisulfite-Seq applications. Bioinformatics, 27(11), 1571-1572.

Kundaje, A., Meuleman, W., Ernst, J., Bilenky, M., Yen, A., and R. E. Consortium, 2015. Integrative analysis of 111 reference human epigenomes. Nature, 518(7539), 317–330.

Landt, S. G., Marinov, G. K., Kundaje, A., Kheradpour, P., Pauli, F., Batzoglou, S., and Snyder, M. 2012. ChIP-seq guidelines and practices of the ENCODE and modENCODE consortia. Genome Research, 22(9), 1813-1831.

Langmead, B. and Salzberg, S. L. 2012.Fast gapped-read alignment with Bowtie 2. Nature Methods, 9(4), 357-359.

Laurent L, Wong E, Li G, Huynh T, Tsirigos A, Ong CT, Low HM, Kin Sung KW, Rigoutsos I, and J. Loring 2010. Dynamic changes in the human methylome during differentiation. Genome Research. 20; 320-31

Leek, J. T., and J. D. Storey, 2007. Capturing heterogeneity in gene expression studies by surrogate variable analysis. PLoS Genetics, 3(9), e161.

Skinner, M. K., Gurerrero-Bosagna, C., Haque, M. M., Nilsson, E. E., Koop, J. A., Knutie, S. A., and Clayton, D. H. 2014. Epigenetics and the Evolution of Darwin's Finches. Genome Biology and Evolution,6(8), 1972-1989.

Shiraki, T., Kondo, S., Katayama, S., Waki, K., Kasukawa, T., Kawaji, H., and Hayashizaki, Y. 2003. Cap analysis gene expression for high-throughput analysis of transcriptional starting point and identification of promoter usage. *Proceedings of the National Academy of Sciences*, 100(26), 15776-15781.

Smallwood, S., & Kelsey, G. 2012. Genome-wide analysis of DNA methylation in low cell numbers by reduced representation bisulfite sequencing. Methods in Molecular Biology, 925, 187-197.

Smith, R. P., Taher, L., Patwardhan, R. P., Kim, M. J., Inoue, F., Shendure, J., and N. Ahituv, 2013. Massively parallel decoding of mammalian regulatory sequences supports a flexible organizational model. *Nature Genetics*, 45, 1021–28.

Smith ZD, Gu H, Bock C, Gnirke A, Meissner A. High-throughput bisulfite sequencing in mammalian genomes. *Methods*. 2009; 48(3): 226-232.

Sohn, K. A., Ho, J. W. K., Djordjevic, D., Jeong, H. H., Park, P. J., and J. H. Kim, 2015. hiHMM: Bayesian non-parametric joint inference of chromatin state maps. *Bioinformatics*. pii: btv117

Song, L., Zhang, Z., Grasfeder, L. L., Boyle, A. P., Giresi, P. G., Lee, B. K., and T. S. Furey, 2011. Open chromatin defined by DNaseI and FAIRE identifies regulatory elements that shape cell-type identity. *Genome Research*, 21, 1757–1767.

Soyer, J. L., El Ghalid, M., Glaser, N., Ollivier, B., Linglin, J., *et al*. 2014. Epigenetic Control of Effector Gene Expression in the Plant Pathogenic Fungus Leptosphaeria maculans. *PLoS Genetics*. 10(3): e1004227.

Stamatoyannopoulos, J. A., Snyder, M., Hardison, R., Ren, B., Gingeras, T., Gilbert, D. M., and L. B. Adams, 2012. An encyclopedia of mouse DNA elements (Mouse ENCODE). *Genome Biology*. 13(8), 418.

Stergachis, A. B., Neph, S., Sandstrom, R., Haugen, E., Reynolds, A. P., Zhang, M., and J. A. Stamatoyannopoulos, 2014. Conservation of trans-acting circuitry during mammalian regulatory evolution. *Nature*, 515(7527), 365–370.

Stockwell, P. A., Chatterjee, A., Rodger, E. J., and Morison, I. M. 2014. DMAP: differential methylation analysis package for RRBS and WGBS data.Bioinformatics, 30(13): 1814-22.

Sun, D., Xi, Y., Rodriguez, B., Park, H. J., Tong, P., Meong, M., and Li, W. 2014. MOABS: model based analysis of bisulfite sequencing data. Genome Biology, 15(2), 38.

Sung, M. H., Guertin, M. J., Baek, S., and Hager, G. L. 2014. DNase Footprint Signatures Are Dictated by Factor Dynamics and DNA Sequence. Molecular Cell, 56(2), 275–285.

Suter, C. M., Martin, D. I. and Ward, R. L. 2004.Germline epimutation of MLH1 in individuals with multiple cancers. Nature Genetics, 36: 497-501

Tajbakhsh, J., Stefanovski, D., Tang, G., Wawrowsky, K., Liu, N., & Fair, J. H. 2015. Dynamic heterogeneity of DNA methylation and hydroxymethylation in embryonic stem cell populations captured by single-cell 3D high-content analysis. Experimental Cell Research, 332(2), 190-201.

Trapnell, C., Cacchiarelli, D., Grimsby, J., Pokharel, P., Li, S., Morse, M., Lennon, NJ., Livak, KJ., Mikkelsen, TS., and J.L. Rinn, 2014. The dynamics and regulators of cell fate decisions are revealed by pseudotemporal ordering of single cells. Nature Biotechnology 32: 381–386

Visel, A., Blow, M. J., Li, Z., Zhang, T., Akiyama, J. A., Holt, A., and L. A. Pennacchio, 2009. ChIP-seq accurately predicts tissue-specific activity of enhancers. Nature, 457(7231), 854-858.

Wade, P. A., A. Gegonne, P. L. Jones, E. Ballestar, F. Aubry, and A. P. Wolffe. 1999. Mi-2 complex couples to DNA methylation chromatin remodeling and histone deacetylation. Nature Genetics 23: 62-66.

Wills, Q. F., Livak, K. J., Tipping, A. J., Enver, T., Goldson, A. J., Sexton, D. W., Holmes, C. 2013. Single-cell gene expression analysis reveals genetic associations masked in whole-tissue experiments. Nature Biotechnology 31: 748–752

Xi, Y., Bock, C., Müller, F., Sun, D., Meissner, A., & Li, W. 2012. RRBSMAP: a fast, accurate and user-friendly alignment tool for reduced representation bisulfite sequencing. Bioinformatics, 28(3), 430-432.

Xu, H., Handoko, L., Wei, X., Ye, C., Sheng, J., Wei, C. L., ... Sung, W. K. 2010. A signal-noise model for significance analysis of ChIP-seq with negative control. Bioinformatics, 26, 1199–1204.

Yue, F., Cheng, Y., Breschi, A., Vierstra, J., Wu, W., Ryba, T., ... Consortium, T. M. E. 2014. A comparative encyclopedia of DNA elements in the mouse genome. Nature, 515(7527), 355–364.

Zang, C., Schones, D. E., Zeng, C., Cui, K., Zhao, K., & Peng, W. 2009. A clustering approach for identification of enriched domains from histone modification ChIP-seq data. Bioinformatics, 25, 1952–1958.

Zhang, Y., Liu, T., Meyer, C. A., Eeckhoute, J., Johnson, D. S., Bernstein, B. E., ... Liu, X. S. 2008. Model-based analysis of ChIP-seq (MACS). Genome Biology, 9, R137.

Zhou, V. W., Goren, A., & Bernstein, B. E. 2011. Charting histone modifications and the functional organization of mammalian genomes. Nature Reviews Genetics, 12, 7–18.

Zhou, X., Li, D., Zhang, B., Lowdon, R. F., Rockweiler, N. B., Sears, R. L., ...& Wang, T. (2015). Epigenomic annotation of genetic variants using the Roadmap Epigenome Browser. Nature biotechnology, 33(4), 345-346.

Zou, J., Lippert, C., Heckerman, D., Aryee, M., & Listgarten, J. 2014. Epigenome-wide association studies without the need for cell-type composition. Nature Methods, 11(3) 309-311.

9

Gene Body Methylation and Transcriptional Regulation: Statistical Modelling and More

Shaoke LOU[1]*

Abstract

DNA methylation is an important type of epigenetic modification involved in gene regulation. Although strong DNA methylation at promoters is widely recognized to be associated with transcriptional repression, many aspects of DNA methylation remain not fully understood, including the quantitative relationships between DNA methylation and expression levels, and the individual roles of promoter and gene body methylation. Integrated analysis of whole-genome bisulfite sequencing and RNA sequencing data from human samples and cell lines find that while promoter methylation inversely correlates with gene expression as generally observed, the repressive effect is clear only on genes with a very high DNA methylation level. By means of statistical modeling, we find that DNA methylation is indicative of the expression class of a gene in general, but gene body methylation is a better indicator than promoter methylation. These findings are general in that a model constructed from a sample or cell line could accurately fit the unseen data from another. We will review the latest study on gene body methylation and discuss the possible mechanism how gene body methylation regulates gene expression. We also suggest that future studies on gene regulatory mechanisms and disease-associated differential methylation should pay more attention to DNA methylation at gene bodies and other non-promoter regions.

1. Introduction

Epigenetics is the study of heritable changes in gene expression that are not due to changes in DNA sequence. It refers to functionally relevant modifications to the genome, such as DNA methylation and histone modification, both of which serve to regulate gene expression without altering the underlying DNA sequence. Specific epigenetics of biological processes involve many aspects, such as imprinting, gene silencing, X-chromosome inactivation, reprogramming, the progress of carcinogenesis, regulation of histone modification and heterochromatin (Lister et al. 2009).

[1] 266 Whitney Ave BASS 437, New Haven, CT 06511, USA. Tel: 203-432-5405, loushaoke@gmail.com
* Corresponding author : loushaoke@gmail.com

The molecular basis of epigenetics is complex, which includes chemical modification to the DNA (DNA methylations) or to the protein that are closely associated with chromatin structure (Histone modifications).

DNA methylation is a typical characteristic of higher organisms and some of its features are conserved across many species. It refers to methylation at the carbon 5 position of cytosine (5me C) which mostly happens within CpG, CHG and CHH DNA patterns ('H' denotes non-cytosine bases: A, G, T). DNA methylation as an important epigenetics layer contributes to the regulation of transcription and formation of chromatin structures. DNA methylation is stable as well as dynamic, where 'stable' means cytosine methylation pattern is stable modification of the genomic DNA that can be inherited, and 'dynamics' denotes that the modifications change during the lifespan of certain cells of tissues and it is also susceptible to environmental conditions, such as diet, drinks, toxin and air quality etc.

The basis for understanding the function of DNA methylation is the knowledge of its distribution in the genome. The spectrum of methylation levels and patterns is very broad and 60% ~ 90% of all CpGs are methylated in mammals (Ehrlich et al. 1982; Lister et al. 2009). While non-CpG cytosine (CHH and CHG) methylation mainly occurs in human somatic tissue, and is particularly prevalent in brain tissue, which is reproducible across many individuals (Varley et al. 2013).

As early as 1983, Busslinger found that DNA methylation around 5' terminal (-760 ~ 1000bp) of the genes plays a direct role in the regulation of gene expression (Busslinger, Hurst, and Flavell 1983). The inverse relationship of promoter methylation with mRNA expression is consistent with the role of DNA methylation in modulating the spatial pattern of gene expression (Petkova, Seigel, and Otteson 2011). The DNA methylation on promoter region can affect the recognition and binding of a transcription factor to activate or repress gene transcription (Bird 2002). Similarly, microRNA can also be regulated by the methylation of CpG island at the start site (Vrba et al. 2010).

To systematically study DNA methylation at the genomic scale, it is necessary to identify many, ideally all, methylated sites in a genome. Various high-throughput methods have been invented for large-scale detection of methylation events (Suzuki and Bird 2008; Beck and Rakyan 2008; Jones 2012; Laird 2010). These methods differ in the way genomic regions enriched for methylated or unmethylated DNA are identified, and how genomic locations of these regions or their sequences are determined. The former includes the use of methylation-sensitive restriction enzyme digestion (Khulan et al. 2006; Lippman et al. 2005), immunoprecipitation (Weber et al. 2005; Weber et al. 2007; Zhang et al. 2006), affinity capture (Brinkman et al. 2010; Illingworth et al. 2008), and bisulfite conversion of unmethylated cytosines to uracils (Cokus et al. 2008; Lister et al. 2008; Lister et al. 2009; Li et al. 2010). The identities of the collected regions are determined by microarray (Khulan et al. 2006; Lippman et al. 2005; Weber et al. 2005; Weber et al. 2007; Zhang et al. 2006) or sequencing (Cokus et al. 2008; Lister et al. 2008; Lister et al. 2009)(Brinkman et al. 2010; Illingworth et al. 2008; Li et al. 2010). These methods have been extensively compared in terms of their genomic coverage, resolution, cost, consistency and context-specific bias (Bock et al. 2010; Harris, Wang, Coarfa, Nagarajan, Hong, Downey, Johnson, Fouse, Delaney, Zhao, Olshen, Ballinger, Zhou, Forsberg, Gu, Echipare, O'Geen, Lister, Pelizzola, Xi, Epstein, Bernstein, Hawkins, Ren, Chung, Gu, Bock, Gnirke, Zhang, and Haussler 2010).

Among them, bisulfite conversion method is a widely accepted, which includes whole-whole bisulfite sequencing (WGBS), Reduced-representative Bisulfite sequencing (RRBS) and Methyl450k array. RRBS is an economic base-resolution method using MspI to digest DNA and isolate 100-150bp fragments, which can include about 85% CpG island and 60% Refseq promoter. Methyl450k array samples 96% CpG islands, total 485000 CpGs, including 99% RefSeq promoter, UTR and first exon, gene body and shores. WGBS can get genome-wide methylation profiling for all Cytosine types (CpG, CHH and CHG) with base resolution and be not affected by the low CpG area (Harris, Wang, Coarfa, Nagarajan, Hong, Downey, Johnson, Fouse, Delaney, Zhao, Olshen, Ballinger, Zhou, Forsberg, Gu, Echipare, O'Geen, Lister, Pelizzola, Xi, Epstein, Bernstein, Hawkins, Ren, Chung, Gu, Bock, Gnirke, Zhang, Haussler, et al. 2010).

By integrating gene expression data and global DNA methylation profiles from these high-throughput methods, a general genome-wide negative correlation between promoter methylation and gene expression was observed in multiple species (Bell et al. 2011; Pai et al. 2011). However, substantial overlap exists in the distributions of promoter methylation level between genes with low versus high expression (Weber et al. 2007; Bell et al. 2011; Pai et al. 2011). It was also suggested that for CpG island promoters, DNA methylation is sufficient but not necessary for their inactivation, while for promoters with low CpG content, hypermethylation does not preclude gene expression (Weber et al. 2007). The quantitative relationship between promoter methylation and gene expression is thus more complicated than once assumed (Jones 2012) and the details have not been fully worked out.

The high-throughput methods have also provided evidence that there is extensive DNA methylation at transcribable regions (Hellman and Chess 2007). Gene body methylation was observed to be positively correlated with gene expression in some cell types (Ball et al. 2009; Rauch et al. 2009), but not in others (Lister et al. 2009). It was suggested that the positive correlation could either be due to de novo methylation of internal CpG islands facilitated by transcription, in which case methylation was the consequence; or due to the repression of anti-sense transcripts that would down-regulate expression of the sense transcript, in which case methylation was the cause (Rauch et al. 2009). In contrast, it was also previously proposed that intragenic DNA methylation could reduce the efficiency of transcription elongation (Rountree and Selker 1997; Lorincz et al. 2004), which would result in a negative correlation between gene body methylation and expression. Furthermore, gene body methylation was reported to be related to the regulation of alternative promoters (Maunakea et al. 2010), and may play a role in RNA splicing (Choi et al. 2009). Whether these mechanisms co-exist and their relative importance in gene regulation remain not fully explored.

Most of the findings about promoter and gene body methylation described above were based on global trends. For instance, while promoter methylation has a general negative correlation with gene expression, huge variance exists between both the promoter activities and resulting expression levels of genes with similar methylation levels (Weber et al. 2007; Bell et al. 2011; Pai et al. 2011). Until now it has been unclear whether it is possible to construct a quantitative model that tells the expression level of an individual gene from its DNA methylation pattern alone or with additional information about histone modifications around its genomic region. Such quantitative modeling would be useful for understanding the combined effect of DNA methylation at different gene sub-elements, such as promoters, exons and introns, on gene expression. It could further help elucidate the relative roles of DNA methylation and other gene regulatory mechanisms in controlling gene expression, and estimate the degree of cooperation and redundancy between them. It could also provide a principal way to identify subsets of genes most affected by DNA methylation in particular cell types.

In recent studies, genomic regions hypo- or hyper-methylated in disease samples have been identified by applying high-throughput methods (Akalin et al. 2012; Irizarry et al. 2009; Ng et al. 2013; Toperoff et al. 2012). Having the ability to estimate the effect of DNA methylation on the expression of a gene, quantitative modeling could help identify the most biologically relevant events in disease states, from potentially long lists of differentially methylated regions, for downstream validation and functional studies.

We have presented our work in quantitatively modeling the relationships between DNA methylation and gene expression using high-throughput sequencing data that cover the methylome and transcriptome of three human samples and additional cell lines at single-base resolution (Lou et al. 2014). We show that DNA methylation is highly anti-correlated with gene expression only when the methylation or expression level of a gene is extremely high. We demonstrate that both promoter and gene body methylation are indicative of gene expression level, but gene body methylation has a stronger effect overall. Recently, there are also some other papers tried to elucidate the effect of methylation in gene body region. We will review the latest update on the evidences for the regulation of gene body methylation.

2. Methylation Data and global patterns

The whole-methylome bisulfite sequencing data at single-base resolution and whole transcriptome sequencing data are from peripheral blood mononuclear cells (PBMCs) of three individuals in a trio family: Father (F), Mother (M) and Daughter (D) (Lou et al. 2014). After data preprocessing, about 95% of the reads were uniquely mapped to the human reference genome.

The total number of raw reads of father, mother and daughter range from 2.6 billion to 3.0 billion. After reads mapping, there are 70~80% reads mapped to genome with about 35X read coverage. The methylation levels of different types vary and the average methylation level in CpG is the highest. Take the daughter sample for example:

TABLE 1 Average methylation level of C, CpG, CHG and CHH (H = A, C, or T)

Pattern	C	CpG	CHG	CHH
Methylation level*	3.88	63.7	0.56	0.61

methylation level is defined as: 100* read count that supports methylation/ total reads count

Distribution of genome-wide methylated-C (mC) (mCpG, mCHG and mCHH) level in each type is species-specific, and form a specific methylation profiling under certain and biological condition. In the trio family data, mCG make up near 97% of all methylated cytosine. The distribution of methylation level for mCG, mCHG and mCHH are also quite different. The majority of all CpG sites are hyper-methylated (60%~100%), in contrast to non-CG (CHG and CHH), the majority of which are hypo-methylated (Varley et al. 2013).

The global patterns of DNA methylation for the three PBMC samples are quite similar as described in (Varley et al. 2013). We also added the data of hESC and IMR90 for comparison. Overall, both the absolute number of methylated cytosines within CpG dinucleotides (mCG) in 10 kb sliding windows and the density of methylated cytosines with respect to the total number of CpG dinucleotides within the window (mCG/CG) are highly correlated among the five samples (Fig. 1 for chromosome 1 as an example). The methylation measure mCG/CG has been commonly used in various methylome studies to quantify DNA methylation level (Lister et al. 2009; Li et al. 2010; Ball et al. 2009). All the five samples, except some regions of IMR90, have very similar methylation patterns with most of CpG sites highly methylated. However, when we look through the raw data, those regions of IMR90 may result from low read coverage because our analysis pipeline will automatically treat the low read coverage region as non-methylated.

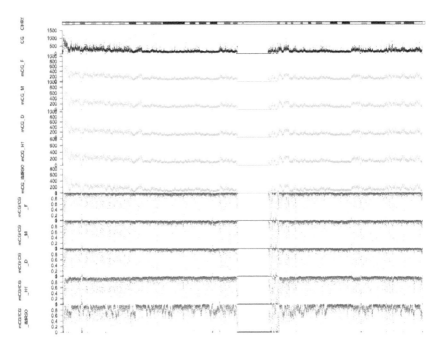

FIGURE 1 DNA methylation profiles of the five samples based on 10 kb sliding windows on chromosome 1. Abbreviations: CG: number of CpG dinucleotides in each window; mCG: number of methylated cytosines within CpG dinucleotides in each window; (F): Father; (M): Mother;(D): Daughter; (H1): hESC.

3. Global correlation and patterns between gene methylation and expression

Many studies have tried to explore the relationship between gene methylation and expression. In our study, we used the single-base resolution WGBS data to further investigate and compare the pattern for promoter and gene body methylation and expression. We computed the average methylation level along each gene, considering the gene body, upstream 2K regions and both respectively , and plotted these methylation levels against the corresponding expression levels. The resulting scatterplot displays different patterns for different methylation measurement. For raw methylated CpG(mCG), DNA methylation displays a very clear "L" shape, in which genes with very high expression levels all display very low methylation levels, and genes with very high methylation levels all show very low expression levels. This pattern suggests that for these extreme cases, there is a negative correlation between DNA methylation and gene expression. However, the global correlation (Pearson correlation $=-0.0486$, Spearman correlation $= 0.0709$) is very weak because the majority of genes have both low methylation and expression level, despite significant p-value of the Pearson correlation due to the large number of genes involved. In contrast, the plot based on the normalized measure, mCG/CG, shows a more global negative correlation with gene expression (Pearson correlation $=-0.1293$, Spearman correlation $=-0.3705$).

These observed differences led us to check whether we could find positive correlations between gene body methylation and expression levels as reported in some previous studies (Ball et al. 2009; Rauch et al. 2009). To do that, instead of considering both upstream regions and gene bodies at the same time, we made separate scatterplots for upstream regions and gene bodies. For mCG, L-shaped patterns were observed for both upstream regions and gene bodies. We also checked exons and introns separately, and found no significant differences between the global patterns of these plots and those in which they were taken together as gene bodies.

These results indicate that the relationship between DNA methylation and gene expression is complex and non-linear, and also include noise and redundant information. Yang etc. make use of a 5-aza-2′deoxycytidine (5-Aza-CdR, a cytosine analog), which can induce the gene body demethylation and alter gene expression in cancer, to investigate the causal relationship between gene body methylation and gene expression (Yang et al. 2014). After 4-5 days, the DNAs are maximum demethylated and then rebound gradually, until fully recovery after 42 days. They defined four different groups with distinct patterns based on the methylation pattern after the treatment of 5-Aza-CdR. The group with the fastest and slowest rebound of methylaition level after the treatment of 5-Aza-CdR was analyzed in details. The promoter methylation for both groups indicate a negative correlation between methylation and gene expression. However, it is more complex for gene body. The fastest rebound group has a large proportion of sites with significantly positive correlation with gene expression. However, for the lowest rebound group, both significantly positive and negative correlation between gene body methylation are found.

However, we need to consider the sampling bias of Methyl450K array. Though, it can detect the methylation of exact CG site, it has some limitations for the probe design. The probe length for Methyl450K array is 50bp, which means the distance from left to right closest CpG sites should equal or be greater than 50bp. We want to ask whether the relationship disclosed by the CpG sites on Methyl450K array can represent the site in its flanking regions. Zhang recently proposed a data mining method to predict DNA methylation (Zhang et al. 2015). They found the methylation can be predicted with very high accuracy, which is up to 92%. The upstream and downstream methylation status and distance are the most important factors. If we only consider the nearest up- and down-stream CpG sites, and use the methylation status of the two nearest CpG site to predict methylation level, we find the prediction accuracy decays very soon as the distance increases between these two neighboring sites. We also extract the left and right CpG site on Methyl450K array and draw the distribution of the distances (Fig. 2B). We found the peak of distribution is around 70bp, where the corresponding accuracy prediction by neighboring CpG site is about 0.65 (Fig. 2A). That means by 35% chance, the up- or down-stream neighboring CpG sites have different methylation patterns from the CpG site on Methyl450K array. Hence, we should carefully draw a conclusion when using methyl450K data since the pattern discovered by methyl450K CpG site might not able to represent the global CpG regulation. Moreover, there might be a chance that CpG sites from the same gene display opposite methylation patterns.

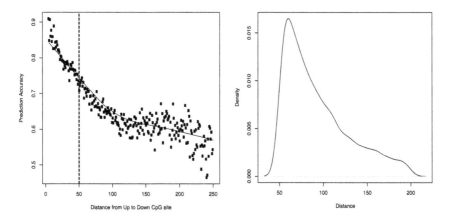

FIGURE 2   DNA methylation association with up- and downstream neighboring CpG site. (A) Prediction of methylation level based on up and down-neighboring CpG site using chromosome 1 WGBS data from IMR90 (Lister et al. 2009). (B). Distribution of distance from up- to down-stream neighboring CpG site of a Methyl450K CpG site.

The global correlation between gene body methylation and gene expression can also be discovered by exploring the relationship between promoter and gene body methylation since the promoter methylation are thought to be negatively correlated with gene expression. When plotting the DNA methylation levels at these two regions for all genes, the distributions based on mCG/CG displays a two-cluster pattern (Fig. 3). All genes were divided into two large clusters and both clusters display very high level of gene body methylation, but one with very high and the other with very low promoter methylation. The two clusters also suggest that the co-exist of positively and negatively correlation between gene body methylation and gene expression.

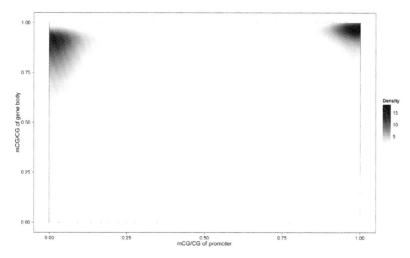

FIGURE 3 Relationship between DNA methylation at the upstream and transcribed region of transcripts.

4. Statistical modeling and data mining

4.1 Quantitative modeling

To systematically study the quantitative relationships between DNA methylation and gene expression, especially for different subregions, we performed statistical modeling by means of machine learning. We want to ask the quantitative relationship between methylation for different subregions and gene expression. We defined 16 subregions that surrounding gene body (Fig. 4). The subregions are in fixed or variable length: the upstream and downstream was classified into 5 consecutive subregions(Up5-Up1, Dw1-Dw5) with a fixed length while the gene body was grouped into 6 bins according to a composite gene structure: first exon , first intron, internal exons, internal introns, last exon, and last intron, as shown in a previous study (Li et al. 2010). Meanwhile, all genes were divided into four equal size classes according to the quantile of expression level. Using the methylation level in these 16 bins, we try to predict the gene expression classes.

We constructed models with all DNA methylation features from the 16 subregions of each gene, using the mCG/CG as methylation measure. We used the Random Forest method (Breiman 2001) as a proxy of how indicative of gene expression the methylation features are. Based on the AUC measure (area under the receiver operator characteristic curve), the accuracy of the one-class-against-all models for the four expression classes ranged from 0.63 to 0.82, where a random assignment of genes to expression classes would result in an AUC value of 0.5, indicating that the methylation features were able to partially separate genes from different expression classes. Among the four expression classes, the Lowest expression class had the highest accuracy, followed by the Highest, Medium-high and Medium-low classes. These results are consistent with what we observed from the scatterplots, that many genes with the lowest expression levels have very high methylation patterns, which can separate them from genes with higher expression levels. The genes with the highest expression levels are slightly more difficult to identify since their signature of low methylation is also shared by many genes from other expression classes.

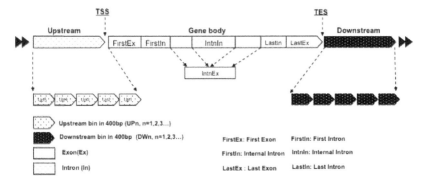

FIGURE 4 Sub-regions defined for each gene. The transcribed region (body) of a gene is divided into 6 variable-length sub-regions according to its exons and introns, namely first exon (FirstEx), first intron (FirstIn), last exon (LastEx), last intron (LastIn), internal exons (IntnEx) and internal introns (IntnIn). The 2 kb upstream region is divided into 5 fixed-length sub-regions Up1-Up5, each of 400 bp. Downstream sub-regions Dw1-Dw5 are defined analogously. In some analyses these sub-regions are further grouped into meta sub-regions, such as Upstream (Up1-Up5), Body (all the exonic and intronic sub-regions) and Downstream (Dw1-Dw5).

4.2   Gene body methylation is a stronger indicator of expression class than promoter methylation

We then compared the models constructed using features from either the upstream regions, gene bodies or downstream regions alone. Methylation levels at gene bodies were more capable of telling the expression class of a gene than upstream and downstream regions, for all four expression classes. Combining features from all sub-regions gave the best modeling accuracy, which shows that the features from the different sub-regions are not totally redundant, and may play different roles in gene regulation.

A potential confounding factor of the above analyses is that the upstream and downstream regions of a transcript could overlap with the body of another transcript (Maunakea et al. 2010). For instance, for a multi-transcript gene, DNA methylation at the promoter of some transcripts would be counted as gene body methylation of the gene, which may confuse the statistical models. To study how much this would affect the results, we repeated the statistical modeling using the subset of genes with only one annotated transcript isoform. Comparing the resulting models based on different feature sets, gene bodies still showed stronger modeling power than upstream and downstream regions, and the best accuracy is still obtained by combining features from all three sub-regions.

It was previously shown that DNA methylation of the first exon is linked to transcriptional silencing (Brenet et al. 2011). We checked whether the higher modeling accuracy of gene body feature was merely due to some strong features extended from the promoter to the first exon. Specifically, we considered two more sub-regions, namely gene bodies excluding the first exons (Genebody–FirstEx) and upstream regions including the first exons (Upstream+FirstEx). We observed that including the first exon in the upstream regions (Upstream+FirstEx) or gene bodies (Genebody) indeed increased the modeling accuracy as compared to having it excluded (Upstream and Genebody–FirstEx, respectively), thus confirming the important role of this sub-region in signifying expression levels (Fig. 5). On the other hand, when we compared upstream and gene body regions, we found that the modeling accuracy of Genebody–FirstEx was higher than Upstream, and that of Genebody was higher than Upstream+FirstEx when all annotated genes were considered. The same trends were also observed when only genes with one annotated transcript isoform were considered, except for a slightly higher accuracy of Upstream than Genebody–FirstEx when the mCG/len methylation measure was used. Altogether, our results show that in general, DNA methylation at gene bodies is a stronger indicator of the expression class than DNA methylation at promoters, and it is neither due to overlapping definitions of promoters and gene bodies for multi-transcript genes, nor signals coming from the first exon only.

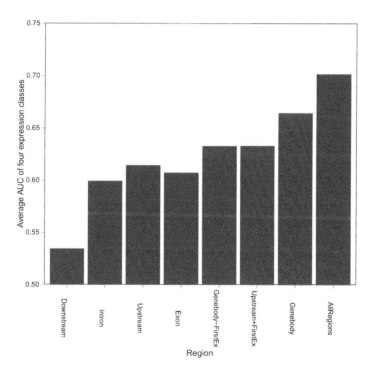

FIGURE 5 Accuracy of Random Forest expression models for all annotated genes when each expression class has the same number of genes. The different bars compare the accuracy of the models constructed from different feature sets. Body-FirstEx corresponds to the set of features from transcribed sub-regions excluding the first exon. Upstream+FirstEx corresponds to the set of features from both the 2kb upstream region and the first exon.

4.3   The quantitative relationship between gene body and promoter DNA methylation and gene expression are widely exist

All the results above were based on quantitative models both constructed and tested on the same individuals (albeit on different subsets of genes), using data from one single cell type (PBMC). To test if these models are generally useful for signifying expression classes, we collected single-base resolution bisulfite sequencing and RNA-seq data for two cell lines, H1 human embryonic stem cells (hESC) and the human lung fibroblast line IMR90, from the Roadmap Epigenomics Project (Bernstein et al.). We constructed models using DNA methylation and expression data from one individual/cell line, and applied the models to predict the expression class of genes in another individual/cell line based on its DNA methylation profile alone. To ensure the generality of the models, the genes used for training in the first individual/cell line and the genes used for testing in the second individual/cell line were mutually exclusive.

The results show that, for all combinations of training and testing individuals/cell lines, the prediction accuracy was much higher than the random predictions (which would have an AUC value of 0.5). Models constructed from any one of the three individuals were able to predict the expression classes of genes in another individual with an average AUC of about 0.9, which is expected as these samples all contained PBMC from individuals in the same family. More interestingly, the other data set combinations also have prediction accuracy of about 0.75 on average, which demonstrate the generality of the constructed models. These cross-sample results reconfirm our earlier findings that the more extreme expression classes are better indicated by methylation patterns.

---

5.   Gene body methylation and splicing

Our results offer several possible explanations for the apparent discrepancies among previous studies examining the relationships between gene body methylation and gene expression, that in some studies they were observed to be positively correlated (Lorincz et al. 2004; Maunakea et al. 2010) and in others, negatively correlated (Ball et al. 2009; Rauch et al. 2009; Cokus et al. 2008; Flanagan and Wild 2007).

Maayan and Gil found that the exon-intron GC content will affect exon selection in exon-skipping alternative splicing (Amit et al. 2012). They firstly defined two distinct GC content group between exons and flanking introns: differential GC and leveled GC group. The higher differential exon-intron GC content (differential GC group) will lead to more chance for exon exclusion. By flanking intron replacement, they proved the elevation of GC content in flanking intron, which decreased the exon-intron differential GC content, can increase the level of exon inclusion. The two groups have different nucleosome occupancy, because of the GC content are high correlated with occupancy. However, when comparing the methylation level of exon-intron differences, the two groups display specific trends: exons are higher methylated than flanking introns in both groups, the differential GC group has overall higher methylation level than leveled group, the increased level from flanking intron to exon is not as large as those in leveled group. The differential methylation pattern also happened on constitutive exons. It is not the absolute methylation level but relative changes can distinguish the consititutive exons from alternative ones. As to how alternative exons are defined for the leveled GC group, they proposed a mechanism that is an interlace interaction between Pol II stalling and nucleosome occupancy. However, in their study, they thought the GC content is the determinant of gene architecture, and DNA methylation affect splicing/alternative splicing in a dependent manner (Gelfman et al. 2013).

Many studies also discovered the co-transcriptional regulation by epigenetics factors, especially for DNA methylation. Two models were proposed: transcriptional blockage model and splicing factor co-recruitment model. In the former model, MeCP2 and CTCF, work as methylation-dependent factor to slow down the elongation of RNA Pol II (Maunakea et al. 2013). In the latter model, HP1 was found to be as a most supportive proof that directly regulates splicing by recruting the co-bounded splicing factor. They firstly found over 20% alternative exons are affected by the missing of DNA methylation, which act as either enhancer or repressor. Using targeted methylation of a single gene, they found the direct causal relationship that methylation can affect the inclusion level of alternative exon (Yearim et al. 2015).

Though it seems that the conclusions made by Gelfman and Yearim are contradictory, we notice that they defined their own gene group and can explain successfully for a certain group of genes. It is not surprising that both methylation dependent and independent regulatory will co-exist.

6. Discussion

Previous studies have examined high-level qualitative relationships between DNA methylation and gene expression. In this work, we have demonstrated that DNA methylation status alone can indicate the expression class of a gene with fairly high accuracy. The generality of our models has been confirmed by their cross-sample/cell line modeling capability. Our models provide a means to analyze the detailed quantitative relationships between DNA methylation and expression, with systematic assessments of the level of expression variations explainable by DNA methylation.

We showed that two groups of genes have particularly clear methylation profiles in our data, namely genes that lie on both ends of the spectrum – those with very high methylation and very low expression levels, and those with very high expression and very low methylation levels. If we apply a simple classification of genes into those with high or low expression and DNA methylation levels, among the four possible combinations, the one with both high expression and high DNA methylation is almost devoid of genes when three out of the four DNA methylation quantification measures were used. Our results indicate that on the one hand, strong DNA methylation is sufficient to indicate low expression of a gene, but on the other hand, while low DNA methylation is permissive of transcription, the actual expression level of a gene is largely determined by other factors.

A key finding of this study is that gene body methylation is a stronger indicator of expression class than promoter methylation for genes in all expression classes. Our results are consistent with the strong effects of gene body methylation on expression previously observed in plants (Hohn et al. 1996; Li et al. 2008). We provided evidence that the stronger modeling power of gene body methylation could not be explained by the effects of first exons alone or biases caused by the presence of multiple transcript isoforms in a single gene, nor was it affected by the quantification measure of DNA methylation levels. We also found that combining both promoter and gene body DNA methylation features resulted in a better modeling accuracy of gene expression classes. However, the correlation between gene body methylation and expression is complex, may either positive or negative correlate with gene expression. That's may also be the reason that there are no clear global trend can be observed using simple scatterplot, because of the mixer of two different directional effect. Studies also indicate the GC content define the gene structure, especially for exon and intron structures based on the gene architecture evolution. Though, studies don't come to an agreement that gene body methylation has dependent or independant effect on alternative splicing, All agree that methylation indeed plays an important role and effect a significant proportion of alternative exons. Our statistical modeling rely on a composite gene structure, and cannot be aware of splicing events. We need also recognize that the calculation of gene expression level is a challenging, especially when considering splicing and alternative splicing. The exclusion/inclusion level of alternative exons cannot be reflected from the expression level, this will limit the prediction power of our model.

Further studies will be needed to elucidate how promoter and gene body methylation of different transcripts of a gene are coordinated. Signals that cover a broad area, such as DNA methylation over whole transcript bodies, have a high chance of interfering with other transcripts. The coordination would be simple if promoter and gene body methylation both take on a repressive role, and different transcript isoforms of a gene co-express in a synchronized manner. In that case, DNA methylation would be mainly responsible for marking genes with all transcripts repressed. The co-expression of transcript isoforms was indeed observed in large-scale sequencing data from human cells (Djebali et al. 2012), although it is still unclear whether the different isoforms expressed simultaneously in the same cell, or actually different subsets of them were expressed in different subpopulations of the cells from which RNA was extracted and sequenced. Alternatively, intragenic DNA methylation that intersects promoters of some transcripts may be involved in regulating the use of alternative promoters (Maunakea et al. 2010). Whether other, more complex types of coordination exist is yet to be studied.

7. Methods

7.1 Sample collection

We collected DNA methylation and gene expression data from a family trio from our previous study (Lee HM et al., Discovery of type 2 diabetes genes using a multiomic analysis in a family trio, submitted). In the following section, we briefly describe sample collection, data generation and data processing. Blood samples were obtained from a Chinese family trio consisting of a father, a mother and a daughter, which we denote as *F, M* and *D*, respectively. Peripheral blood mononuclear cells (PBMCs) were isolated using Ficoll-Paque stepwise gradient centrifugation. The isolated PBMCs were divided for DNA and RNA extraction. Total DNA was prepared using proteinase K digestion and phenol extraction. Total RNA was extracted by Trizol (Life

Technologies, Carlsbad, CA, USA) following the manufacturer's protocol. The quality of the RNA samples was checked by Bioanalyzer (Agilent Technologies, Palo Alto, CA, USA) before they were subjected to sequencing.

7.2 Methylome sequencing and data processing

Bisulfite sequencing and data processing were carried out as described previously (Li et al. 2010). DNA was fragmented by sonication to 100 to 500 bp in size, followed by end-blunting, dA addition at the 3'end and ligation of adapters. The adapter sequence contained multiple methyl-cytosines(mC) to allow monitoring of the efficiency of the bisulfite conversion. Unmethylated cytosines were converted to uracils by bisulfite treatment using a modified protocol from Hayatsu (Hayatsu, Shiraishi, and Negishi 2008). DNA fragments in the size range of 320 to 380 bp were gel-purified for sequencing. All procedures were performed according to the manufacturer's instructions. Converted DNA was subjected to 50 bp paired-end sequencing using an Illumina Solexa GA sequencer (Illumina, San Diego, CA, USA). All the raw data were processed by the Illumina Pipeline v1.3.1 (Illumina, San Diego, CA, USA).

The cleaned reads generated were aligned to the reference human genome hg18 as follows. Since DNA methylation is strand-specific, the two strands of the reference human genome were modified separately in silico to convert all C's to T's, to generate a combined 6 Gbp target genome for aligning reads after bisulfite conversion. Correspondingly, the sequencing reads were also transformed using the following criteria: (1) observed C's in the forward reads were replaced by T's; and (2) observed G's in the reverse reads were converted to A's. The transformed reads were then aligned to the modified target genome using the SOAP2 aligner (Li et al. 2009). All the reads mapped to unique locations with minimum mismatches and clear strand information were defined as uniquely matched reads, and were used to determine the methylated Cytosines. According to the alignment results, the unconverted C's and G's from the original read sequences before the transformation were used to identify the methylated Cytosines. Bases with low quality scores were filtered to ensure accuracy of the results. The methylated Cytosines were defined as those having a significant number of reads supporting its methylated status, with less than 1% FDR according to a binomial distribution, as suggested previously (Li et al. 2010). All the Cytosine positions were then lifted over to the reference human genome hg19 by the LiftOver utility provided by the UCSC Genome Browser (Kent et al. 2002) for downstream analyses.

7.3 Transcriptome sequencing and data processing

Total RNA extracted from each sample was enriched by oligo-dT to get the polyA+ fraction for sequencing. The polyA+ mRNAs were then fragmented and converted to cDNA by reverse transcription. After ligation of the 5' and 3' sequencing adaptors to the cDNA, DNA fragments were size-selected for 75 bp paired-end sequencing by Illumina Genome Analyzer II using standard procedures. All the raw data were processed by the Illumina Pipeline v1.3.1. All sequencing reads were trimmed dynamically according to the algorithm provided by the -q option of the BWA tool (Li and Durbin 2009). After trimming, read pairs with both sides having at least 35 bp were retained and mapped to the human reference genome hg19 using TopHat (Trapnell, Pachter, and Salzberg 2009) (v.1.1.4) with the following parameters: micro-exon-search, butterfly-search and -r 20. The expression value of a gene was computed by the RPKM (reads per kilobase of exons per million mapped reads) measure (Mortazavi et al. 2008), defined as the number of reads that cover it (in million reads) normalized by its length (in kilobase) and the total number of reads in the data set.

Definition of the four DNA methylation quantification measures

We used two methylation measures based on methylated CpG sites. The first measure is the absolute number of methylated CpG sites in a region, denoted as mCG. The second measure is the density of methylated CpG sites relative to the total number of CpG sites in a region, denoted as mCG/CG.

Visualizing global DNA methylation patterns and computing local correlations between two individuals

We constructed global DNA methylation profiles of the three individuals as follows. We first divided up the human genome into 10 kb windows. In each window, we computed the DNA methylation level based on one of the four quantification measures. We then visualized the resulting global patterns using IGV (Robinson et al. 2011) and Circos (Krzywinski et al. 2009). To compute local correlations of DNA methylation profiles between two individuals, we divided up the genome into fixed-length windows (of size 10 kb, 50 kb, 100 kb or 250 kb), and computed the DNA methylation level in each window. For every 15 consecutive windows, we then computed the Pearson correlation between two individuals (F vs. M, F vs. D or M vs. D). The resulting distributions of correlation values were visualized using Box and Whisker plots.

7.4   Definition of gene sub-regions

For analyses involving genes, we considered the level 1 and level 2 protein-coding genes annotated in Gencode v7 (Harrow et al. 2012), based on composite gene models. We defined the body of a gene as the first transcription start site of its annotated transcripts to the last transcription termination site of its annotated transcripts. Within the gene body, we defined any region annotated as an exon in any of the associated transcripts as an exon of the gene. We then defined subregions of a gene as shown in Fig. 3 and explained in the Results and discussion section. We discarded genes with less than 4 exonic regions after merging overlapping exons from different transcripts, resulting in a set of 17,845 genes.

7.5   Definition of expression classes

By default we defined gene expression classes as follows. We first combined the genes from the three individuals into a set of 53,535 (17,845 ×3) genes. Each of them was then assigned to one of four expression classes, namely the "Highest", "Medium-high", "Medium-low" and "Lowest" classes, which contained genes with expression levels within the first, second, third and fourth quartiles on the list of expression values sorted in descending order. The Lowest expression class could contain genes with zero RPKM values Statistical modeling

We used 11 different methods to construct statistical models, including 5-Nearest Neighbors, 10-Nearest Neighbors, 20-Nearest Neighbors, Naive Bayes, Bayesian

Network, Decision Trees (C4.5), Random Forests, Logistic Regression, Support Vector Machine (SVM) with linear kernel, SVM with second-degree polynomial kernel, and SVM with Radial Basis Function (RBF) kernel. We used the implementation of all these methods in Weka (Hall et al. 2009). We constructed statistical models using these methods with features derived from DNA methylation and/or histone modification levels of the different genic sub-regions. We first constructed models for the three individuals using their combined data. We randomly sampled 1/3 of the genes as a left-out testing set. The remaining 2/3 of the genes were used to perform model training. The constructed model was then applied to the left-out set to compute the accuracy. For each setting, we repeated the process five times to compute an average accuracy of the five models.

We also tested the generality of our models by constructing models using the DNA methylation and gene expression data of a random set of 2/3 of the genes from one single individual/cell line for training, and applying the model to predict the expression levels of the remaining 1/3 of the genes in another individual/cell line based on the DNA methylation levels in this individual/cell line.

7.5 Collection and processing of cell line data

We downloaded data human embryonic stem cells and human lung fibroblast line IMR90 produced by Roadmap Epigenomics (Bernstein et al.) from the Gene Expression Omnibus (GEO) (Edgar, Domrchev, and Lash 2002) web site. The RPKM measure was used to compute the level of histone modification in each given region. For data sets with replicates, we used the mean values of the replicates for computing the histone modification signals.

7.6 Data availability

All raw sequencing reads have been deposited into NCBI Sequence Read Archive under entry SRP033491. All the processed data files used in this study can be found at http://yiplab.cse.cuhk.edu.hk/means/ website.

Acknowledgements

This work has been adapted from the original article "Whole-genome bisulfite sequencing of multiple individuals reveals complementary roles of promoter and gene body methylation in transcriptional regulation" by Shaoke Lou, Heung-Man Lee, Hao Qin, Jing-Woei Li, Zhibo Gao, Xin Liu, Landon L Chan, Vincent KL Lam, Wing-Yee So, Ying Wang, Si Lok, Jun Wang, Ronald CW Ma, Stephen Kwok-Wing Tsui, Juliana CN Chan*, Ting-Fung Chan* and Kevin Y Yip*. Genome Biology 2014, 15:408 (doi:10.1186/s13059-014-0408-0; http://www.genomebiolog. com/2014/15/7/408/). The original article is an open access article distributed under the terms of the Creative Commons Attribution License (http://creativecommons. org/licenses/by/2.0), which permits unrestricted use, distribution, and reproduction in any medium, provided the original work is properly cited.

Reference Cited

Akalin, A., Garrett-Bakelman F.E., Kormaksson M, et al. 2012. Base-pair resolution sequencing reveals profoundly divergent epigenetic landscapes in acute myeloid leukemia. *PLoS Genet* 8:e1002781.

Amit, M., Donyo M., Hollander D., et al. 2012. Differential GC content between exons and introns establishes distinct strategies of splice-site recognition. *Cell Rep.* 1 (5):543-56.

Ball, MP, JB Li, Y Gao, et al. 2009. Targeted and genome-scale strategies reveal gene-body methylation signatures in human cells. *Nat Biotechnol* 27:361 - 68.

Beck, S, and Rakyan VK. 2008. The methylome: approaches for global DNA methylation profiling. *Trends Genet.* 24:231 - 237.

Bell, JT, Pai AA, Pickrell JK, et al. 2011. DNA methylation patterns associate with genetic and gene expression variation in HapMap cell lines. *Genome Biol* 12:R10.

Bernstein, BE, Stamatoyannopoulos JA, Costello JF, et al. The NIH roadmap epigenomics mapping consortium. *Nat Biotechnol.* 28:1045 - 48.

Bird, A. 2002. DNA methylation patterns and epigenetic memory. *Genes Dev* 16 (1):6-21.

Bock, C, Tomazou EM, Brinkman AB, et al. 2010. Quantitative comparison of genome-wide DNA methylation mapping technologies. *Nat Biotechnol.* 28:1106 - 14.

Breiman, L. 2001. Random forests. *Mach Learn.* 45: 5 - 32.

Brenet, F, Moh M, Funk P, et al. 2011. Dna methylation of the first exon is tightly linked to transcriptional silencing. *PLOS ONE* 6:e14524.

Brinkman, AB, Simmer F, Ma K, Kaan A, Zhu J, and HG Stunnenberg. 2010. Whole-genome DNA methylation profiling using MethyCap-seq. *Methods.* 52:232 -36.

Busslinger, M., Hurst J., and R. A. Flavell. 1983. DNA methylation and the regulation of globin gene expression. *Cell* 34 (1):197-206.

Choi, JK, Bae J-B, Lyu J , Kim T-Y, and Y-J Kim. 2009. Nucleosome deposition and DNA methylation at coding region boundaries. *Genome Biol.* 10: R89.

Cokus, SJ, Feng S, Zhang X, et al. 2008. Shotgun bisulphite sequencing of the Arabidopsis genome reveals DNA methylation patterning. *Nature.* 452: 215 -19.

Djebali, S, Davis CA, Merkel A, et al. 2012. Landscape of transcription in human cells. *Nature.* 489:101 - 108.

Edgar, R, Domrchev M, and Lash AE. 2002. Gene expression omnibus: NCBI gene expres-sion and hybridization array data repository. *Nucleic Acids Res.* 30: 207 -10.

Ehrlich, M., Gama-Sosa M. A., Huang L. H., et al. 1982. Amount and distribution of 5-methylcytosine in human DNA from different types of tissues of cells. *Nucleic Acids Res.* 10 (8):2709-21.

Flanagan, JM, and L Wild. 2007. An epigenetic role for noncoding RNAs and intragenic DNA methylation. *Genome Biol* 8:307.

Gelfman, S., Cohen N., Yearim A., and G. Ast. 2013. DNA-methylation effect on cotran-scriptional splicing is dependent on GC architecture of the exon-intron structure. *Genome Res.* 23 (5):789-99.

Hall, M, Frank E, Holmes G, Pfahringer B, Reutemann P, and IH Witten. 2009. The WEKA data mining software: an update. *SIGKDD Explorations.* 11:10 - 18.

Harris, R. A., Wang T., Coarfa C., et al. 2010. Comparison of sequencing-based methods to profile DNA methylation and identification of monoallelic epigenetic modifications. *Nat Biotechnol.* 28 (10):1097-105.

Harris, RA, Wang T, Coarfa C, et al. 2010. Comparison of sequencing-based methods to profile DNA methylation and identification of monoallelic epigenetic modifications. *Nat Biotechnol.* 28:1097 - 1105.

Harrow, J, Frankish A, Gonzalez JM, et al. 2012. GENCODE: the reference human genome annotation for the ENCODE project. *Genome Res* 22:1760 - 74.

Hayatsu, H, Shiraishi M, and Negishi K. 2008. Bisulfite modification for analysis of DNA methylation. *Curr Protoc Nucleic Acid Chem.* 33:6.10.1 - 6.10.15.

Hellman, A and A Chess. 2007. Gene body-specific methylation on the active X chromosome. *Science* 315:1141 - 1143.

Hohn, T, Corsten S, Rieke S, Muller M, and H Rothnie. 1996. Methylation of coding region alone inhibits gene expression in plant protoplasts. *Proc Natl Acad Sci U S A.* 93:8334 - 339.

Illingworth, R, Kerr A, Desousa D, et al. 2008. A novel CpG island set identifies tissue-specific methylation at developmental gene loci. *PLoS Biol.* 6:e22.

Irizarry, RA, Ladd-Acosta C, Wen B, et al. 2009. The human colon cancer methylome shows similar hypo- and hypermethylation at conserved tissue-specific CpG island shores. *Nat Genet.* 41:178 - 186.

Jones, PA. 2012. Functions of DNA methylation: islands start sites gene bodies and beyond. *Nat Rev Genet* 13:484 - 92.

Kent, WJ, Sugnet CW, TS Furey, KM Roskin, TH Pringle, and AM Zahler. 2002. The human genome browser at UCSC. *Genome Res.* 12:996 - 1006.

Khulan, B, Thompson RF, K Ye, et al. 2006. Comparative isoschizomer profiling of cytosine methylation: the HELP assay. *Genome Res* 16:1046 - 1055.

Krzywinski, MI, Schein JE, Birol I, et al. 2009. Circos: an information aesthetic for comparative genomics. *Genome Res.* 19:1639 - 45.

Laird, PW. 2010. Principles and challenges of genomewide DNA methylation analysis. *Nat Rev Genet* 11:191 - 203.

Li, H and R Durbin. 2009. Fast and accurate short read alignment with Burrows-Wheeler transform. *Bioinformatics* 25:1754 - 60.

Li, R, Yu C, Li Y, et al. 2009. SOAP2: an improved ultrafast tool for short read alignment. *Bioinformatics* 25:1966 - 1967.

Li, X, Wang X, He K, et al. 2008. High-resolution mapping of epigenetic modifications of the rice genome uncovers interplay between DNA methylation histone methylation and gene expression. *Plant Cell* 25:259 - 276.

Li, Y, Zhu J, Tian G, et al. 2010. The DNA methylome of human peripheral blood mononuclear cells. *PLoS Biol* 8:e1.000533.

Lippman, Z, Gendrel A-V, Colot V, and R Martienssen. 2005. Profiling DNA methylation patterns using genomic tiling microarrays. *Nat Methods* 2:219 - 24.

Lister, R, O'Malley RC, Tonti-Filippini J, et al. 2008. Highly integrated single-base resolution maps of the epigenome in Arabidopsis. *Cell* 133:523 - 36.

Lister, R, Pelizzola M, Dowen RH, et al. 2009. Human DNA methylomes at base resolution show widespread epigenomic differences. *Nature* 462:315 - 22.

Lister, R., Pelizzola M., Dowen R. H., et al. 2009. Human DNA methylomes at base resolution show widespread epigenomic differences. *Nature* 462 (7271):315-22.

Lorincz, MC, Dickerson DR, Schmitt M, and M Groudine. 2004. Intragenic DNA methylation alters chromatin structure and elongation efficiency in mammalian cells. *Nat Struct Mol Biol* 11:1068 - 1075.

Lou, S , Lee H. M., Qin H,. et al. 2014. Whole-genome bisulfite sequencing of multiple individuals reveals complementary roles of promoter and gene body methylation in transcriptional regulation. *Genome Biol* 15 (7):408.

Maunakea, A. K., Chepelev I., Cui K., and K. Zhao. 2013. Intragenic DNA methylation modulates alternative splicing by recruiting MeCP2 to promote exon recognition. *Cell Res* 23 (11):1256-69.

Maunakea, AK, Nagarajan RP, Bilenky M, et al. 2010. Conserved role of intragenic DNA methylation in regulating alternative promoters. *Nature* 466:253 - 57.

Mortazavi, A, Williams BA, McCue K, Schaeffer L, and B Wold. 2008. Mapping and quantifying mammalian transcriptomes by RNA-Seq. *Nature Methods* 5:621 - 28.

Ng, CW, Yildirim F, Yap YS, et al. 2013. Extensive changes in DNA methylation are associated with expression of mutant huntingtin. *Proc Natl Acad Sci U S A.* 110:2354 - 59.

Pai, AA, Bell JT, Marioni JC, Pritchard JK, and Y Gilad. 2011. A genome-wide study of DNA methylation patterns and gene expression levels in multiple human and chimpanzee tissues. *PLoS Genet.* 7:e1001316.

Petkova, T. D., G. M. Seigel, and D. C. Otteson. 2011. A role for DNA methylation in regulation of EphA5 receptor expression in the mouse retina. *Vision Res.* 51 (2):260-68.

Rauch, TA, Wu X, Zhong X, Riggs AD, and GP Pfeifer. 2009. A human B cell methylome at 100-base pair resolution. *Proc Natl Acad Sci U S A* 106:671 - 678.

Robinson, JT, Thorvaldsdottir H, Winckler W, et al. 2011. Integrative genomics viewer. *Nat Biotechnol* 29:24 - 26.

Rountree, MR and EU Selker. 1997. DNA methylation inhibits elongation but not initiation of transcription in Neurospora crassa. *Genes Dev.* 11:2383 - 95.

Suzuki, MM and A Bird. 2008. DNA methylation landscapes: provocative insights from epigenomics. *Nat Rev Genet* 9:465 - 476.

Toperoff, G, Aran D, Kark JD, et al. 2012. Genome-wide survey reveals predisposing diabetes type 2-related DNA methylation variations in human peripheral blood. *Hum Mol Genet.* 21:371 - 383.

Trapnell, C, L Pachter, and SL Salzberg. 2009. TopHat: discovering splice junctions with RNA-Seq. *Bioinformatics* 25:1105 - 11.

Varley, K. E., Gertz J., Bowling K. M., et al. 2013. Dynamic DNA methylation across diverse human cell lines and tissues. *Genome Res* 23 (3):555-67.

Vrba, L., Jensen T. J., Garbe J. C., et al. 2010. Role for DNA methylation in the regulation of miR-200c and miR-141 expression in normal and cancer cells. *PLoS One* 5 (1):e8697.

Weber, M, Davies JJ, Wittig D, et al. 2005. Chromosome-wide and promoter-specific analyses identify sites of differential DNA methylation in normal and transformed human cells. *Nat Genet.* 37:853 - 862.

Weber, M, Hellmann I, MB Stadler, et al. 2007. Distribution silencing potential and evolutionary impact of promoter DNA methylation in the human genome. *Nat Genet.* 39:457 - 66.

Yang, X., Han H., De Carvalho D. D., Lay F. D., Jones P. A., and G. Liang. 2014. Gene body methylation can alter gene expression and is a therapeutic target in cancer. *Cancer Cell.* 26 (4):577-90.

Yearim, A., Gelfman S., Shayevitch R., et al. 2015. HP1 is involved in regulating the global impact of DNA methylation on alternative splicing. *Cell Rep.* 10 (7):1122-34.

Zhang, W., Spector T. D., Deloukas P., J. Bell T., and B. E. Engelhardt. 2015. Predicting genome-wide DNA methylation using methylation marks, genomic position, and DNA regulatory elements. *Genome Biol* 16:14.

Zhang, X, J Yazaki, A Sundaresan, et al. 2006. Genome-wide high-resolution mapping and functional analysis of DNA methylation in Arabidopsis. *Cell.* 126:1189 - 1201.

Section 5
Case Study

10

Computational Characterization of Non-small-cell Lung Cancer with EGFR Gene Mutations and its Applications to Drug Resistance Prediction

Debby D. Wang[1], Lichun Ma[2] and Hong Yan[2]*

Abstract

As a major type of lung cancer, non-small-cell lung cancer (NSCLC) is the leading cause of cancer deaths worldwide. Gene mutations affecting the catalytic activity of epidermal growth factor receptor (EGFR) normally promote NSCLC. In clinical treatments of NSCLC, tyrosine kinase inhibitors (TKIs) are broadly used to target the kinase domain of mutated EGFR. Although these drugs are effective initially, drug resistance rapidly emerges. Studies on the resistance mechanism can be guided by the dissection of EGFR signaling network, and computational approaches play a significant role. One general research focus is calculating the binding affinity of an EGFR mutant and a TKI, which can be characterized by the free energy of binding, interfacial hydrogen bonds, and shape complementation of interfaces. Another important research direction is investigating the modified molecular interactions in the EGFR downstream signaling pathways, and EGFR dimerization is a valuable segment. Specifically in our studies, techniques such as molecular dynamics (MD) simulations, three-dimensional molecular modeling and structural analysis, were applied. Binding affinities between each EGFR mutant and a partner were evaluated based on these techniques. Recently, dually-targeted TKIs (target dual receptors) and second-generation TKIs (covalently bound to the mutated EGFR kinase) are proposed, to combat drug resistance in clinical treatments of NSCLC. However, the risk of underlying drug resistance still exists. To decode these new mechanisms, computational modeling remains indispensable. Overall, these studies can encourage the development of new-generation and more sophisticated drugs, and further help the design of specialized therapies in NSCLC treatments.

[1] Caritas Institute of Higher Education, 18 Chui Ling Road, Tseung Kwan O, New Territories, Hong Kong.

[2] Department of Electronic Engineering, City University of Hong Kong, Kowloon, Hong Kong.

* Corresponding author : h.yan@cityu.edu.hk

1. Introduction

The epidermal growth factor receptor (EGFR) family, also termed ErbB family, is a group of receptor tyrosine kinases that fundamentally regulate epithelial tissue development and cancer progression(Marchetti et al. 2005; Haley 2008; Solca et al. 2012). Human EGFR family is composed of EGFR (Her1, ErbB-1), ErbB-2 (Her2), ErbB-3 (Her3) and ErbB-4 (Her4), all sharing a large structural homology (Haley 2008; Solca et al. 2012). Such a receptor comprises an extracellular domain that normally binds growth-factor ligands, a transmembrane domain, and an intracellular catalytic domain that activates downstream signaling through receptor dimerization (Yarden 2001; Haley 2008; Schlessinger 2002). Aside from the significant conservation among the EGFR-family receptors at the genetic or structural level, a functional diversity exists. EGFR and ErbB-4 both have intact intracellular tyrosine kinase (TK) domain, and are capable of binding ligands through their extracellular domains (Citri and Yarden 2006; Hynes and Lane 2005; Yarden and Sliwkowski 2001). ErbB-2 has an extracellular domain that is permanently in an active conformation, implying a lack of capability to bind ligands, while it possesses a powerful TK domain (Kallioniemi et al. 1992; Hynes and Lane 2005; Solca et al. 2012). Oppositely, ErbB-3 has an impaired TK domain, but can perfectly bind ligands (Haley 2008; Guy et al. 1994). In this regard, ErbB-2 and ErbB-3 mostly behave as a cooperator in receptor activation and signaling mediation.

A number of ligands, represented by epidermal growth factor (EGF) and neuroregulins (NRGs), can induce EGFR-family dimerization, upon which the catalytic capability of TK domains will be triggered (Haley 2008; Hynes and Lane 2005; Olayioye et al. 2000; Carraway et al. 1997). Specifically for EGFR, EGF, TGF-α and amphiregulin play a central role in the receptor-ligand interaction (Hynes and Lane 2005; Olayioye et al. 2000). Meanwhile, ErbB-2 has been considered as the preferred dimerization partner for other family members (GrausPorta et al. 1997; Solca et al. 2012). Receptor homo- or hetero-dimerization first results in the formation of an asymmetrical dimer, in a tail-to-head manner, of the partnered TK domains (Zhang et al. 2006; Haley 2008). This formation further activates the receptor catalytic capability and leads to a transphosphorylation of specific tyrosine residues within the participated TK domains (Zhang et al. 2006; Solca et al. 2012). Subsequently, adaptor proteins will be recruited by such phosphorylated domains, switching on various signaling pathways, such as the *Ras/Raf/Mek/Erk* and *PI3K/Akt/mTOR* pathways (Yarden 2001; Haley 2008; Solca et al. 2012). Overall, this signal transduction mechanism, profiled in Fig. 1, drives normal cell proliferation and differentiation, and thus aberrant signaling greatly influences the progression of multiple cancer types (Yarden 2001; Marchetti et al. 2005; Haley 2008; Kallioniemi et al. 1992).

In recent decades, activating mutations in the TK domain of the EGFR gene have been identified in several types of tumors (Santos, Shepherd, and Tsao 2011; Haley 2008; Solca et al. 2012). Non-small-cell lung cancer(NSCLC), a subtype of lung cancer, is the best-acknowledged one (Yasuda, Kobayashi, and Costa 2012; Marchetti et al. 2005; Wang, Zhou, et al. 2013). Most of the EGFR gene mutations that are connected with NSCLC are located in exons 18 to 21 (Yasuda, Kobayashi, and Costa 2012).

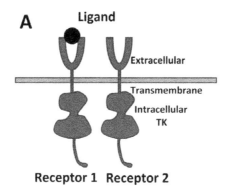

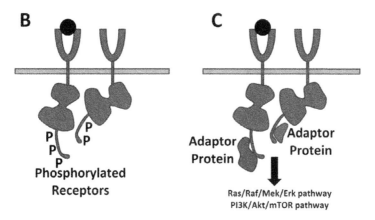

FIGURE 1 Working mechanism of EGFR signaling. (A) Binding between EGFR-family receptors and their ligands. (B) Formation of EGFR-family dimers (with an asymmetric mechanism in the TK domain) and the transphosphorylation. (C) Recruitment of adaptor proteins to switch on the downstream signaling pathways.

Among these gene mutations, the most frequently-occurred ones are termed *classic activation mutations*, including in-frame deletions of exon 19 (45%~50% of all) and point mutations in exon 21 (40%~45% of all) (Shigematsu et al. 2005; Sequist et al. 2007; Tokumo et al. 2005). Commonly, an EGFR gene mutation is translated to the protein sequence level, using the corresponding modifications in the wild-type (WT) EGFR sequence (Ma 2014; Ma et al. 2015; Wang, Zhou, et al. 2013). The notation principles of such mutations or mutants are as follows.

- Residue substitution can be denoted as *X*I*Y*, where *X* mutates into *Y* at position I. The well-known *mutation at exon 21, L858R*, is a representative.

- Deletion of residues *X* (at position I) to *Y* (at position II) is notated as del*X*I_ *Y*II. As an example, del*E*746_*A*750 describes one *deletion of exon 19* mutation that corresponds to the deletion of *Glu* at position 746 to *Ala* at position 750.

- Duplication of residues X (at position I) to Y (at position II) is represented by dul*XI_Y*II. For example, dul*S*768_*D*770 denotes the duplication of *Ser* at position 768 to *Asp* at position 770.

- Insertion of residue X or residue list **k** into position I can be notated as Iins*X* or Iins**k**. Specifically, 747ins*SK* represents inserting *Ser* and *Lys* into position 747.

- Modification of residues X (at position I) to Y (at position II) is defined by a combination of deletion and insertion as del*XI_Y*IIins**k**, where **k** is the above-mentioned residue list. Here delL747_A755insSKG is an instance.

- A double-point substitution of X with Y at residue site I and A with B at residue site II is named by combining the two single-point substitutions as XIY_AIIB, such as *T*854*A_L*858*R*.

In particular, these EGFR mutation-positive NSCLCs constitute a novel subgroup, which can cause aberrant EGFR signaling and thus be regarded as one of the best-described oncogenic mechanisms (Yasuda, Kobayashi, and Costa 2012; Solca et al. 2012). Increasing studies of EGFR family and its role in cancer progression have promoted the development of a number of ErbB-targeted therapeutic agents (Ma 2014; Ma et al. 2015; Paez et al. 2004; Haley 2008; Sordella et al. 2004; Wang, Zhou, et al. 2013; Solca et al. 2012). Among them, reversible tyrosine kinase inhibitors (TKIs), such as gefitinib (IRESSA™) and erlotinib (TARCEVA®), are one type of broadly-used agents that target the EGFR TK domain and block its kinase activity (Fig. 2A) (Ma 2014; Ma et al. 2015; Wang, Zhou, et al. 2013). Specifically, in gefitinib and erlotinib, their essential pharmacophore is a quinazoline ring, accounting for the major reaction with EGFR (Wang et al. 2015). Such TKIs were clinically verified to be especially efficient to the EGFR mutation-positive NSCLCs at early treatments (Paez et al. 2004; Sordella et al. 2004). However, an acquired resistance to them was frequently developed (Oxnard et al. 2011; Kobayashi et al. 2005; Yun et al. 2008). The resistance is normally correlated to the gatekeeper mutation *T*790*M*, resulting a lower binding affinity between the EGFR mutant and a TKI(Ma 2014; Ma et al. 2015; Wang, Zhou, et al. 2013; Yun et al. 2008). To overcome this drug-resistance problem, new strategies, such as dually-targeted strategies (Fig. 2B) (van der Veeken et al. 2009; Dienstmann et al. 2012; Wang, Yuan, et al. 2013; Tsang 2011) and irreversible TKIs (Fig. 2C) (Yang et al. 2012; de Antonellis 2014; Yap et al. 2010), have been proposed. Dually-targeted TKIs or antibodies aim to correct the EGFR mutation-induced aberrant signaling, through targeting the dimer formed by an EGFR-family member and it potential partner (van der Veeken et al. 2009; Haley 2008; Wang, Yuan, et al. 2013; Solca et al. 2012). This potential partner can be an EGFR-family protein, or other receptors importantly influencing cancer progression, such as IGF-1R and c-Met (Jo et al. 2000; van der Veeken et al. 2009; Morgillo et al. 2006; Wang, Yuan, et al. 2013). As another strategy, new-generation TKIs irreversibly bind to an EGFR TK domain, trying to permanently block the aberrant signaling in tumors (Yang et al. 2012; de Antonellis 2014; Yap et al. 2010). Overall, these therapeutic strategies are essentially correlated to the molecular binding affinities or interactions, in a receptor-inhibitor or receptor-receptor way (Ma 2014; Ma et al. 2015; Wang, Zhou, et al. 2013). Therefore, an efficient characterization of such molecular interactions can provide a deeper insight into the drug resistance mechanism, and further optimize the drug application in NSCLC treatment.

With the rapid development of computer techniques, modeling (Xiang, Soto, and Honig 2002; Xiang and Honig 2001; Leaver-Fay et al. 2011) and simulations (Case 2012) have become an indispensable alternative for investigating drug resistance (Cohen et al. 2010; Loo, Wu, and Altschuler 2007; Sneddon and Emonet 2012) and predicting the resistance level (Cao et al. 2005; Hou et al. 2009; Draghici and Potter 2003; Hao, Yang, and Zhan 2012; Zhou et al. 2013), in cancer research and drug discovery. These studies can be categorized into sequence-based or structure-based approaches, normally utilizing sequential or structural information of molecules (Sneddon and Emonet 2012; Hou et al. 2009; Draghici and Potter 2003). Particularly, structure-based methods have gained more attention and respect in recent years, for decoding proteins, nucleic acids,and complex assemblies from a molecular perspective. The protein data bank (**PDB**) (Berman, Westbrook, et al. 2000) is a main resource for such structure-based studies. It collects and stores three-dimensional (3D) structural information, such as atom/bond types and atomic positions, of biological macromolecules, which facilitates studies in molecular biology, structural biology, computational biology, and beyond (Berman, Westbrook, et al. 2000; Berman, Bhat, et al. 2000; Berman 2008). The derivation of such structural information benefits a lot from the techniques of X-ray crystallography (Smyth and Martin 2000) and nuclear magnetic resonance (NMR) spectroscopy (Harris 1986).

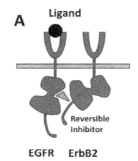

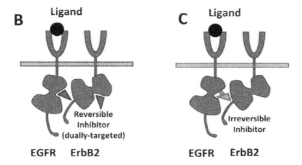

FIGURE 2 TKIs of EGFR-family receptors (A) A reversible TKI that targets the EGFR TK domain. (B) A dually-targeted TKI that blocks both EGFR and ErbB2 TK domains. (C) An irreversible TKI that covalently binds to the TK domain of EGFR.

Depending on data from **PDB**, molecular modeling, structural analysis, and molecular dynamics (MD) simulations are widely applied in above-mentioned research areas (Case 2012; Xiang, Soto, and Honig 2002; Xiang and Honig 2001; Leaver-Fay et al. 2011). Specifically for MD simulations, *AMBER* is a widely used and efficient tool (Case 2012). It includes a set of molecular mechanical force fields, which refer to the functional form and parameter sets used to calculate the potential energy of a system, for biomolecular simulations, and a suite of molecular simulation programs (Case 2012; Ma et al. 2015; Wang, Zhou, et al. 2013). In our studies, MD simulations were primarily applied to decode the receptor-inhibitor interactions for a variety of EGFR TK mutants and their inhibitors (Ma et al. 2015; Wang, Zhou, et al. 2013), which was later extended to the investigation of receptor-receptor interactions for those mutants and their potential partners. On top of that, local surface geometric properties of the mutants were extracted based on their 3D shapes and molecular topologies, to investigate their interactions with inhibitors from another point of view (Ma 2014). For the structural modeling of each mutant, the 3D alpha shape modeling (Ohbuchi and Takei 2003; Edelsbrunner 1994, 1992) is a favorable choice, since it has been considered as a strong shape modeling and surface reconstruction tool in molecular studies (Ma 2014; Zhou and Yan 2010, 2010). Extracted by this modeling technique, the local surface geometric properties of the examined mutants can successfully reveal important patterns of the mutant-inhibitor interactions (Ma 2014). At last, new-generation irreversible TKIs, represented by afatinib (Yang et al. 2012; de Antonellis 2014; Yap et al. 2010; Solca et al. 2012), for the EGFR-family were discussed. Overall, above structural studies and analyses can provide a better understanding of the molecular mechanism of drug resistance in the treatment of NSCLC or other tumors, which can greatly promote the improvement of personalized therapy design and innovative drug discovery.

2. Computational Modeling of Interaction between an EGFR Tyrosine Kinase and an Inhibitor

2.1 Mutant-inhibitor Binding Affinity Revealed by MD Simulations

Binding affinity between a receptor and its inhibitor, which is an important indicator of drug resistance or efficacy level, can be estimated by their binding free energy in a solvent environment (Ma et al. 2015; Wang, Zhou, et al. 2013). The *AMBER* software suite is a powerful tool for binding free energy calculation of such a receptor-inhibitor system (Case 2012). Specifically, the *Molecular Mechanics/Poisson Boltzmann (Generalized Born) Surface Area* (*MM/PB(GB)SA*) is the main module in *AMBER* that performs this energy calculation. Inspired by the thermodynamic cycle theory, the binding free energy calculation can be decomposed into the calculations of several free energy differences, which corresponds to either the difference between different states of a molecule in an environment or that between different environments for the same molecule. Coupled with the free energy in vacuum for the bound and unbound states of a receptor-inhibitor complex, the solvation free energies of the complex and separate molecules are primary components of the binding free energy. *MM/PB(GB) SA* solves the PB or GB model to capture these free energy differences based on the dynamics of involved molecules, leading to the derivation of several binding free energy components of Van der Waals Force (VDW), electrostatic interaction (EEL), polar (EPB or EGB) and nonpolar (ENPOLAR or ESURF) contributions of the solvation free energy.

In order to perform *MM/PB(GB)SA* on a receptor-inhibitor system, the dynamics of this system should be simulated in advance. Meanwhile, for a system comprising an EGFR TK mutant and an inhibitor, the 3D structure of the mutant must be modeled first, after which the mutant-inhibitor complex can be formed and its MD simulations can be implemented. In our studies, the structure of a mutant was obtained mainly based on homology modeling, utilizing the sequential information of the mutant and a prepared 3D structural template. Intuitively, the WT EGFR TK protein or mutant **L**858**R**, in **PDB:**2ITY or **PDB:**2ITZ downloaded from **PDB**, was regarded as this structural template. A number of structural modeling tools, namely *Scap* (Xiang and Honig 2001), *Loopy* (Xiang, Soto, and Honig 2002), and *Rosetta* (Leaver-Fay et al. 2011), were adopted for structural determination of various EGFR mutants. Simply, these tools first align the mutant sequence to the template sequence, and then computationally predict the mutation site based on spatial or energy constraints. *Scap* or *Rosetta* deals with residue substitution, besides, *Loopy* or *Rosetta* handles residue deletion or insertion. *Scap* provides multiple side-chain rotamer libraries for side-chain packing, and complies with principles such as steric feasibility and energy minimization in the rotamer selection (Xiang and Honig 2001). Similarly, *Loopy* or *Rosetta* predicts loops at the mutated neighborhood, based on a series of constraints for structure selection (Xiang, Soto, and Honig 2002; Leaver-Fay et al. 2011). Once a mutant structure is modeled, an *AMBER* minimization step can be implemented to optimize its topology and atomic positions. Specifically, a 5000-step minimization was carried out, with the first half of steepest descent steps, for the modeled mutant (Wang, Zhou, et al. 2013). To concentrate on the mutation site, the quantum mechanics/molecular mechanics (QM/MM) mechanism was used in the minimization, with the mutation site labeled as a QM region. A refined mutant structure was then aligned to an EGFR TK-inhibitor complex, **PDB:**2ITY for gefitinib and **PDB:**1M17 for erlotinib, to form the complex with an inhibitor. This process is roughly profiled in Fig. 3.

Next, MD simulations of the formed mutant-inhibitor system can be performed. At an early stage, a solvent environment, where the dynamics of the molecules are simulated, should be generated. A simple truncated octahedron water box (TIP3P model) was adopted, with a 10-angstrom (Å) buffer around the solute in each direction. For *AMBER* settings, the broadly-used *ff99SB* and *gaff* force fields were selected to deal with proteins and small ligands, respectively. The *tleap* program in *AMBER* preprocesses the solvated mutant-inhibitor complex prior to MD simulations, and based on the *tleap* products, the *sander* program can accomplish those simulations. A series of equilibration steps were first carried out before the key MD simulation, to guarantee its reliability. These steps are listed as follows:

- A short 1000-step minimization to remove bad contacts, with half of steepest descent steps and half of conjugate gradient steps.

- A heating phase of 50 picoseconds (ps) from 0 to 300 K, with a weak constraint on the complex. Here the SHAKE setting and Langevin temperature control are adopted, and the time step is 2 femtosecond (fs).

- A 50-ps density equilibration (time step of 2 fs), with the SHAKE setting and a weak constraint on the complex.
- A constant-pressure equilibration at 300 K, with the time step of 2 fs.

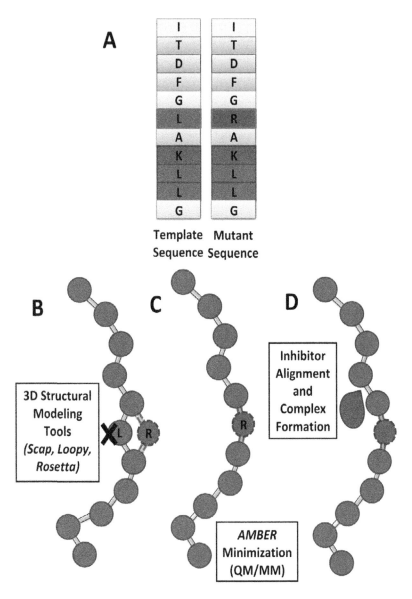

FIGURE 3 Formation of a complex comprising an EGFR TK mutant and an inhibitor. (A) Sequence alignment. (B) 3D structural modeling (*Scap*, *Loopy* or *Rosetta*) of a mutant based on a template structure. (C) *AMBER* minimization (QM/MM) on the modeled mutant structure for a structural optimization. (D) Alignment of the optimized structure to a receptor-inhibitor template to form its complex with a specific inhibitor.

The equilibration of a system can be verified by the stable backbone root-mean-square derivation (RMSD) curve of the solute. Relying on the equilibrated system, the key MD simulation can be implemented. We generated a simulation of 2 nanoseconds (ns) for the mutant-inhibitor system, and similarly verified the stable RMSD curve. Snapshots of the dynamics, also termed *the trajectory*, of the solvated system can be collected at a pre-defined time interval, and 200 frames were sampled in our simulations.

The resulted molecular trajectories are treated as the inputs of *MM/PB(GB)SA* in *AMBER*, leading to the calculation of the binding free energy and its components for the solvated protein-ligand system. Such binding free energies or energy components can be efficiently used to characterize the interaction between the participated mutant and inhibitor, and thus applied to the prediction or studies of inhibitor efficacy or resistance level. This characterization has been tested on a clinical patient dataset (Wang, Zhou, et al. 2013). This dataset was collected from the Queen Mary Hospital in Hong Kong, containing 168 NSCLC patients at stage IIIB or IV. They corresponded to 37 EGFR TK mutation types, and adotped a TKI-treatment of either gefitinib (137 patients) or erlotinib (31 patients). Screening of EGFR mutations were implemented based on formalin-fixed paraffin-embedded (PPFE) tumor biopsy samples of those patients. Such tumor samples assisted in the direct sequencing of EGFR TK domains, and the comparison between these mutated proteins and the normal EGFR led to the derivation of EGFR TK mutation types for the patients. Further, for each patient, the response levels (RL) to the TKIs were accordingly recorded as a study endpoint, and were categorized into four groups (RL = 1, 2, 3, 4). Combining the extracted mutant-inhibitor interaction features and specific patient personal features, classification strategies were applied to map these features to the resistance levels, resulting in high prediction accuracy. Commonly-used classification algorithms include support vector machines (SVMs), extreme learning machines (ELMs), decision trees, and complex neural networks. As a summary, the characterization of a mutant-inhibitor interaction pattern and its application to the drug resistance prediction can be briefly outlined in Fig. 4.

As a supplementary study, we gathered the patient data from a number of literatures (Gu et al. 2007; Wang, Zhou, et al. 2013; Yun et al. 2007), resulting in a thorough collection of 942 NSCLC patients with 112 EGFR TK mutation types. All the corresponding mutants were computationally determined using *Rosetta*, and their binding free energies (or energy components) with two inhibitors (gefitinib and erlotinib) were derived based on *AMBER*(Ma et al. 2015). Such mutant structural candidates and the binding free energies constitute an *EGFR TK Mutant Structural Database* (Ma et al. 2015), which can be a valuable resource for NSCLC studies and new drug/therapy design.

2.2 Hydrogen Bonding Analysis of Mutant-inhibitor Complexes

Besides the preceding binding free energy studies of EGFR TK mutant-inhibitor complexes, a simpler mechanism for investigating mutant-inhibitor interaction is the hydrogen bonding analysis of each complex. It is well acknowledged that hydrogen bonding plays an essential role in protein-ligand interaction (Zhou et al. 2013). Specifically for antibody-antigen and protease-inhibitor systems, hydrogen bonds control their binding specificities and stabilizations in a solvent environment (Babine and Bender 1997; Meyer, Wilson, and Schomburg 1996). In this regard, hydrogen bonding analysis is a potential and competent tool for studying mutant-inhibitor interaction and mutation-induced drug resistance in NSCLC treatments.

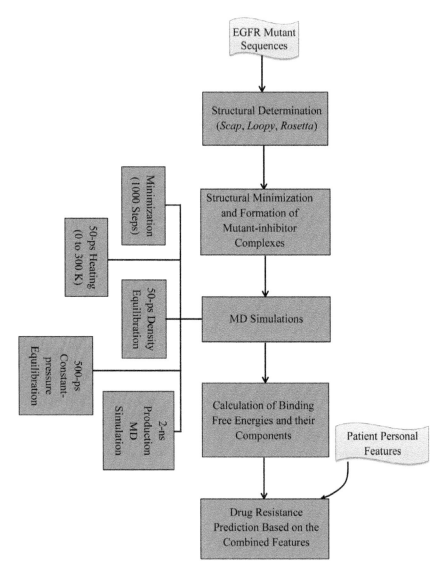

FIGURE 4 Computational characterization of interaction patterns between EGFR TK mutants and their inhibitors, coupled with its application to the drug resistance prediction.

As described in the precedingsection, an EGFR TK mutant can be computationally modeled by a number of structural modeling tools (Xiang, Soto, and Honig 2002; Xiang and Honig 2001; Leaver-Fay et al. 2011), based on its sequence and a structural template. Once the 3D structure of a mutant is derived, a refinement relying on *AMBER* simulations (Case 2012) can be implemented, followed by the alignment of an inhibitor (gefitinib, erlotinib, etc.) to the binding cave of the refined structure. Afterwards, searching inter-molecular hydrogen bonds in the mutant-inhibitor complex leads to a comprehensive understanding of the mutant-inhibitor interaction (Zhou et al. 2013).

Simply speaking, a hydrogen-bond system includes three primary components, namely donor atom (**D**), hydrogen atom (**H**), and acceptor atom (**A**). Atom types and spatial constraints can define hydrogen bonds. In other words, merely a number of specific atoms can serve as the donor atoms in the formation of a hydrogen-bond system, and only fixed atoms can accept hydrogen atoms in this bond system. Commonly-used rules for donor/acceptor definition (Meyer, Wilson, and Schomburg 1996) were adopted in our studies of mutant-inhibitor complexes. In detail, donors and acceptors for proteins are listed in Table 1.

TABLE 1 Donor and acceptor atomtypes for proteins to form hydrogen bonds.

	Atom Type
Donor	Main-chain N-H, His NE2, His ND1, Lys NZ, Asn ND2,Gln NE2, Arg NE, Arg NH1, Arg NH2, Ser OG, Thr OG1,Tyr OH, Trp NE1
Acceptor	Main-chain C=O, Asp OD1, Asp OD2, Glu OE1, Glu OE2, Asn OD1, Gln OE1, Ser OG, Thr OG1

For an inhibitor, we intuitively define nitrogen atoms bonded with hydrogen and sp3 hybridized oxygen atoms as donors, and all other oxygen/nitrogen atoms as acceptors. Aside from specific donors and acceptors, spatial constraints are required by the formation of a hydrogen-bond system. Let **D-H...A** be a hydrogen-bond system, then the following spatial constraints should be fulfilled to maintain the system (Baker and Hubbard 1984),

- The distance between **H** and **A** should be less than 2.5 Å;
- The distance between **D** and **A** should be less than 3.9 Å;
- The angle <**D-H...A** should be larger than 90°.

In our implementations, all missing hydrogen atoms were first added and adjusted by the *Reduce* program in *AMBER*. In the searching of inter-molecular hydrogen bonds for a mutant-inhibitor complex, the *hydrogen bond analysis tool* (*HBAT*) (Tiwari and Panigrahi 2007) program played an important role.

Consequently, for most mutants that develop drug resistance, there are no hydrogen bonds connecting them and an inhibitor. Oppositely, two hydrogen bonds were found between non-drug-resistant mutants and the inhibitor. Moreover, compared to a classic non-drug-resistant mutant *L858R*, most mutants without drug resistance have the same spatially-matched hydrogen bonds as *L858R*. Such hydrogen bonds can be a dispensable factor to maintain the stability of a mutant-inhibitor complex, thus partly explaining the EGFR mutation-induced drug resistance mechanism (Zhou et al. 2013).

2.3   Local Surface Geometric Properties of EGFR TK Mutants and Their Applications to Mutant-inhibitor Interaction Characterization

In this section, further investigation of mutation-induced drug resistance mechanism is elucidated, from the perspective of local surface geometric properties of EGFR TK mutants (Ma 2014). The simplified procedure is shown in Fig. 5. Similar as the preceding section, based on an EGFR mutant sequence and a template structure of WT EGFR TK domain (**PDB:2ITY**) (Berman, Westbrook, et al. 2000; Yun et al. 2007), the corresponding mutant structure can be computationally modeled using *Rosetta* (Leaver-Fay et al. 2011). *AMBER* (Case 2012) was subsequently employed to optimize the predicted structure. Later, the alpha shape model (Edelsbrunner 1994, 1992) of the optimized structure was computed, and solid angles (Zhou, Yan, and Hao 2012) of the surface atoms were calculated to express their curvatures. By comparing the local surface curvature, concerning the TKI-binding site, of a mutant alpha shape with that of the WT alpha shape, the changes of local geometric properties can be derived. Accordingly, a correlation analysis was conducted to correlate these changes with the progression-free survivals (PFSs) extracted from clinical patient data.

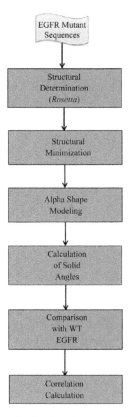

FIGURE 5   The procedure of extracting local surface geometric properties of EGFR TK mutants and applying them to the drug resistance investigation.

The 137 NSCLC patients treated with gefitinib, in the Queen Mary Hospital in Hong Kong, constituted our data set. All the patients harbored EGFR TK mutations and shared a total of 30 mutation types, such as del*E*709_*T*710ins*D*, del*E*746_*A*750, *L*858*R* and *S*768*I*_*V*774M. As previously illustrated, these mutation types can be categorized into several groups, according to the comparison between their corresponding sequences with the WT EGFR sequence. Such groups include residue substitution (single-point and double-point), deletion, duplication insertion, and modification(Wang, Zhou, et al. 2013). For each patient, the PFS of gefitinib was recorded to show its activity in the TKI-treatment (Lee et al. 2013). Specifically, for patients sharing the same mutation type, the median value of the corresponding PFSs was regarded as the PFS of gefitinib to the involved mutant.

Now we describe this study in detail, as follows. *Rosetta* (Leaver-Fay et al. 2011) was employed to generate the EGFR mutants, based on a template structure (WT EGFR) and the mutant sequences. The *ddg_monomer* protocol (Kellogg, Leaver-Fay, and Baker 2011) was adopted to deal with residue substitution, while the comparative modeling (*CM*) protocol (Marti-Renom et al. 2000) was used to handle the remaining types of mutations. After the 3D structures of the mutants were obtained, we applied *AMBER* for structural minimizations. Each refined structure was then aligned to the template (**PDB:**2ITY) to form the mutant-inhibitor complex. Additionally, a minimization procedure was conducted on each solvated complex, outputting an optimized structure for later analysis. The local surface geometric properties of a mutant structure, described by the surface curvature around its TKI-binding site, require to be extracted. In order to represent the geometric structure of a mutant (Fig. 6A), the *Computational Geometry Algorithms Library* (*CGAL*) (Fabri et al. 2000) assisted in the computation ofthe weighted alpha shape model for this mutant (Fig. 6B) (Edelsbrunner 1992).

Moreover, the solid angle (Fig. 6C) (Zhou, Yan, and Hao 2012) concerning each surface atom at the TKI-binding site (Fig. 7) was calculated, to represent the curvature of the corresponding atom. This solid angle was transformed into range [-1, 1] based on a cosine function for a normalization. The shape around each atom is defined to be concave if the transformed solid angle falls in [-1, 0], and to be convex if the solid angle belongs to (0, 1].

14 amino acids residues, including 102 atoms, are located at the TKI-binding site of the WT EGFR TK domain. We obtained local surface geometric changes by comparing the atom solid angles at mutant surfaces and those at the WT EGFR surface. These solid-angle changes, or local surface geometric changes, were grouped into four types, namely *Reverse, Degree Variation, Emergence*, and *Disappearance* (Fig. 8). If the tetrahedron topped at a binding-site atom of the WT EGFR has a concave shape, then *Reverse* represents a convex shapein the mutant. *Degree Variation* means the change of concavity or convexity, that is, the solid angle of the surface at the atom in the mutant. *Emergence* describes the appearance of new atoms at the mutant surface. *Disappearance* indicates that some surface atoms of the WT EGFR no longer belong to the mutant surface. Specifically for the 30 mutation types studied in this work, each mutant contains 3 to 9 binding-site atoms with the *Reverse* changes compared to the WT EGFR, 42 to 53 atoms with the *Degree Variation* changes, 6 to 13 atoms with the *Emergence* changes, and 1 to 9 atoms with the *Disappearance* changes.

Next, we analyzed the relationship between the four types of local surface geometric changes and the PFSs of gefitinib to the corresponding mutants. Results show that *Degree Variation* and *Disappearance* are closely related to PFS. The corresponding Spearman's rank correlation coefficients respectively are -0.63 and 0.63, indicating a valuable negative and positive correlation. Intuitively, the PFS may decrease if more atoms with concave or convex curvatures exist, and a longer PFS will be derived if more WT-surface atoms disappear at the mutant surface.

At last, we explored the relationship between the convex degree of a mutant TKI-binding site and the PFS of gefitinib to this mutant. Atoms with solid angles in the range of $[t, 1]$, where t is a pre-defined threshold, were summed and recorded. Results show that the Spearman's rank correlation coefficient between the PFS and the quantity of those atoms is -0.61, when $t = 0.71$ ($[0°, 180°]$), $t = 0.61$ ($[0°, 210°]$) or $t = 0.5$ ($[0°, 240°]$). The P-values in these cases are much smaller than 0.05, implying the significances of these cases. In this regard, if more mutant atoms have solid angles in above ranges, the patients harboring the corresponding mutation may experience a shorter PFS of gefitinb.

---

3. Characterization of EGFR or ErbB-3 Heterodimerization Using Computer Simulations

EGFR or ErbB-3 heterodimerization plays a significant role in regulating its downstream signaling pathways, and thus greatly influences cancer development and malignancy (Bae and Schlessinger 2010; Haley 2008; Frolov et al. 2007; Morgillo et al. 2006). In this regard, EGFR or ErbB-3 heterodimers and the interaction between each two dimer partners can be a valuable research focus, in the studies of drug resistance mechanism developed by NSCLC or other cancers(Wang et al. 2015).

EGFR brothers in its family, namely ErbB-2, ErbB-3 and ErbB-4, are favorable candidates for EGFR heterodimerization. While regarding NSCLC, merely ErbB-2 and ErbB-3 are considered, since both of them are primarily expressed in NSCLC (van der Veeken et al. 2009; Normanno et al. 2006; Abd El-Rehim et al. 2004). Recently, potential EGFR or ErbB-3 dimer partners, such as insulin-like growth factor 1 receptor (IGF-1R) (Morgillo et al. 2006) and c-Met (MET) (Jo et al. 2000), have been proposed, and their crosstalks with EGFR contribute a lot to the development of drug resistance in cancers. Especially, c-Met has been demonstrated to cause drug resistance by associating with ErbB-3, which can activate and strengthen the downstream PI3K/Akt signaling pathway (Tanizaki et al. 2011; Engelman et al. 2007). Accordingly, the interactions between EGFR mutants and their potential partners (ErbB-2, IGF-1R and c-Met) coupled with those between ErbB-3 and its partners (EGFR mutants, ErbB-2, IGF-1R and c-Met) were computationally explored in our study. These receptor-receptor interactions, together with the receptor-inhibitor interactions decoded in preceding section, can be an enriched characterization of an EGFR TK mutant and its performance in NSCLC progression.

The patient data discussed before, namely the 168 NSCLC patients from Queen Mary Hospital in Hong Kong, are used in this section. Likewise, the binding free energy between the two partners in each heterodimer was applied to estimate their binding affinity. As mentioned in **Introduction**, receptor dimerization leads to the formation of an asymmetrical dimer in the TK domain, according to a tail-to-head manner (Fig. 9A) (Zhang et al. 2006; Haley 2008). Besides, the EGFR mutations that the patients harbored were all located in the TK domain. Therefore, we merely considered the TK dimers in the computations. A dimer structure (Fig. 9B) from the **PDB** (**PDB:**4RIW) (Littlefield et al. 2014), composed of the TK domains of EGFR and ErbB-3, can be regarded as the template for EGFR or ErbB-3 heterodimers. Although ErbB-3 has an impaired TK activity, it behaves as a strong cooperator for other receptors. As a reference, the molecular surface of an EGFR/ErbB-3 TK domain heterodimer was generated by *Chimera* (Pettersen et al. 2004), and is now shown in Fig. 9C. Based on this structure we simulated the dynamics of each receptor-receptor dimer candidate using *AMBER* (Case 2012), and calculated their interaction (binding free energy). The MD simulation procedures are similar to those implemented in the previous section.

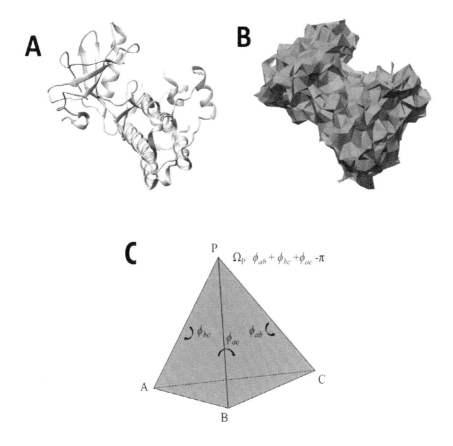

FIGURE 6 Alpha shape modeling and solid angle calculation. (A) The crystal structure of a WT EGFR TK domain (**PDB:**2ITY). (B) The alpha shape model of the TK domain in (A). (C) The solid angle at position P, where ϕ_{ab} is the dihedral angle between PAC and PBC, ϕ_{bc} is the dihedral angle between PAB and PAC, and ϕ_{ac} is that between PAB and PBC.

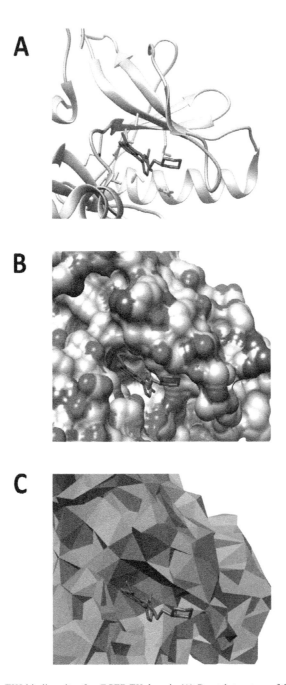

FIGURE 7   The TKI-binding site of an EGFR TK domain.(A) Crystal structure of the TKI-binding pocket of an EGFR TK domain. (B) Solvent-excluded molecular surface structure of a TKI-binding pocket. (C) Alpha shape model of the TKI-binding pocket of an EGFR TK domain. Inhibitors, in the binding pockets, are exhibited in (A) to (C).

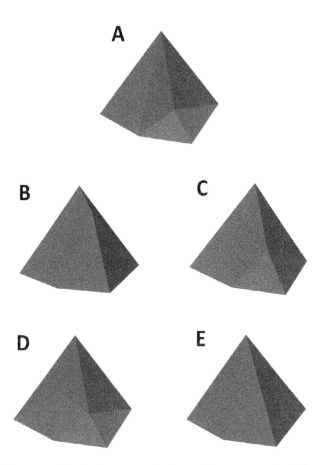

FIGURE 8 Four types of local surface geometric changes of a mutant, namely solid-angle changes of the tetrahedrons topped at binding-site atoms of the mutant, compared with the WT EGFR. (A) Concave shape in the WT EGFR. (B) *Reverse*. (C) *Degree Variation*. (D) *Emergence*. (E) *Disappearance*.

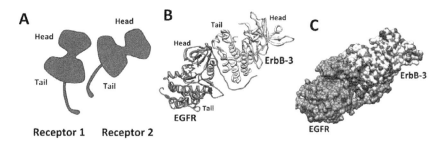

FIGURE 9 EGFR-family TK domain heterodimers. (A) Heterodimer formed by the TK domains of two EGFR-family receptors. (B) Crystal structure of an EGFR/ErbB-3 TK domain heterodimer (**PDB**:4RIW). (C) Molecular surface of the EGFR/ErbB-3 heterodimer in (B).

Since the receptor-receptor systems are more complicated than receptor-inhibitor systems, the settings of MD simulations were accordingly adjusted to achieve reliable results. In the structural minimization of each receptor-receptor system, two steps were adopted. One is a 1000-step minimization (half of steepest descent steps) with weak position restraints on the solute to remove bad contacts, and the other is an extra 1000-step minimization (half of steepest descent steps) without any restraints to further optimize the system. Subsequently, a 25-ps heating phase, a 25-ps density equilibration, and a 250-ps constant-pressure equilibration were sequentially implemented, with a short time step of 1 fs. The equilibration of each system was verified through its temperature, density, energy and the backbone RMSD curve of the solute. At last, a production MD simulation of 2 ns was conducted, with a stable backbone RMSD curve of the solute derived. Similarly as before, the binding free energy and its components for each receptor-receptor complex were resolved by the *MM/PB(GB) SA* program in *AMBER*, and they can be treated as important characteristics of the interaction pattern between two receptors.

These interaction patterns, together with the receptor-inhibitor interactions, were statistically analyzed, and a further regression analysis that connected the computed energy components with the PFS of an inhibitor (gefitinib or erlotinib) to a mutant was conducted. The statistical and regression analyses verified the important contribution of IGF-1R to the PFS of an inhibitor and to the drug resistance level. Additionally, a strong interaction between c-Met and ErbB-3, compared with those between EGFR mutants and ErbB-3, suggested a strengthened ErbB-3 signaling in EGFR mutation-positive NSCLCs.

4.   New-generation Irreversible EGFR TKIs

To improve the drug resistance situations in NSCLC treatments, a group of new-generation TKIs, which covalently bind to the TK domain of an EGFR-family receptor, were proposed and clinically tested. Unlike reversible TKIs (Fig. 10A) that attach themselves to proteins with non-covalent interactions (hydrogen bonds, hydrophobic interactions, ionic bonds, etc.), irreversible TKIs are attached to their targets through covalent binding and the inhibition can therefore not be reversed (Fig. 10B). Among irreversible TKIs, afatinib (BIBW 2992) (Yang et al. 2012; de Antonellis 2014; Yap et al. 2010; Solca et al. 2012) is a representative that targets both EGFR and ErbB-2 TK domains, and it showed potency, albeit lower, to the gatekeeper mutation *T790M* in clinical experiments (Solca et al. 2012). The primary reactive group, acrylamide, of afatinib is capable of a Michael reaction with the conserved cysteine residue (CYS797) within an EGFR TKI domain. This covalent reaction can subsequently result in an irreversible blocking of EGFR downstream signaling.

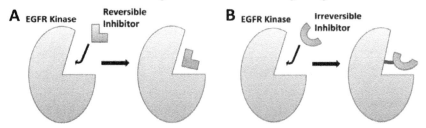

FIGURE 10   Inhibitors of EGFR TK domain. (A) A reversible TKI that binds to its target without covalent interactions. (B) An irreversible TKI that covalently binds to its target.

Despite the covalent binding between afatinib and a receptor, the potency of afatinib varies among patients harboring different EGFR TK mutants. Similar to the well-acknowledged reversible TKIs, afatinib produces better responses from patients with the *classic activation mutations* (Yang et al. 2012; de Antonellis 2014; Yap et al. 2010). Accordingly, to achieve an optimal application of afatinib and the analogs in NSCLC treatments, the selectivity of EGFR mutants should be explored and profiled for such inhibitors, based on the mutant-inhibitor binding modes.

Similarly, the mutant-inhibitor binding modes for EGFR TK mutants and irreversible inhibitors can be computationally estimated. In our study, calculating the free energy difference between a mutant-inhibitor complex and the two separate molecules, based on *AMBER* MD simulations and *MM/PB(GB)SA* free energy computations, is the initial idea for revealing the binding mode. By correlating such a binding mode with the PFS of an inhibitor to a specific mutant, a mutation-selectivity profile can be further constructed for the inhibitor, which is helpful to optimize the application of this inhibitor to cancer treatments.

6. Summary and Future Works

Binding free energy of a solvated protein-ligand or protein-protein system is an important indicator of the molecular binding affinity (Ma et al. 2015; Wang, Zhou, et al. 2013). Beneficial from the fast development of research areas such as bioinformatics and computational biology, the binding free energies can be computationally estimated based on MD simulations and structural modeling (Case 2012; Xiang, Soto, and Honig 2002; Xiang and Honig 2001; Leaver-Fay et al. 2011). Relying on these techniques, the receptor-inhibitor and receptor-receptor interactions were decoded for EGFR and its family members in our work. The derived interaction patterns were efficiently applied to characterize the EGFR mutation-positive NSCLCs, and were used for prediction of drug resistance levels or PFSs of specific inhibitors. Taking patient personal features into consideration, a favorable personalized prediction model of the drug resistance level was constructed. Importantly, these studies can support personalized therapy design and new drug discovery.

Based on the correlation between the surface geometric properties of a mutant TKI-binding site and the PFS of a TKI to this mutant, the mutation-induce drug resistance can be predicted for NSCLC patients (Ma 2014). For a mutant TKI-binding site, if the quantity of disappeared surface atoms increases compared with the WT TKI-binding site, then a higher binding affinity between this mutant and an inhibitor may be achieved. Specifically for gefitinib, a mutant most probably develops a drug resistance if more atoms, at the binding-site surface, experience *Degree Variation* or have solid angles in the range of [0.71, 1], [0.61, 1] or [0.5, 1]. Such a study, concerning local surface geometric properties of an EGFR TK mutant, indicates that these geometric properties play an important role in predicting mutation-induced drug resistance in NSCLC treatment. In addition, it can be extended to the studies of drug resistance mechanisms developed in other diseases.

The above-described computational modeling and structural analysis can also be applied to the investigation of new-generation TKIs such as afatinib (Yang et al. 2012; de Antonellis 2014; Yap et al. 2010; Solca et al. 2012). Based on multiple computational techniques, the binding modes between various EGFR TK mutants and afatinib can be derived and used in the prediction of resistance level or PFS of afatinib. These results will assist the profiling of EGFR mutation-selectivity for afatinib, encouraging a better application of such inhibitors in cancer treatments.

As one of our future works, more efficient strategies should be explored to further reduce the computational costs in structural modeling and energy calculation. Modern graphics processing units (GPUs) and field programmable gate arrays (FPGAs) will be our first candidates for fast parallel computing (Owens et al. 2008; Kruger and Westermann 2003; Bolz et al. 2003). Another work will be a thorough collection of patient data, which can refine our prediction model and verify its reliability. At last, EGFR signaling networks in cancer therapy (Yarden 2001; Haley 2008; Hynes and Lane 2005; Solca et al. 2012) should be carefully dissected, focusing on various pathways and the signaling nodes in the networks. Graph theories and probabilistic models can be strong tools in such analyses (Wagner and Wright 2004; Binder et al. 2006; Samaga et al. 2009). All these studies will greatly improve the understanding of cancer progression and drug resistance development.

Acknowledgements

This work is supported by the Hong Kong Health and Medical Research Fund (HMRF) (Project 01121986) and City University of Hong Kong (Project 9610326).

References

Abd El-Rehim, D. M., Pinder S. E., Paish C. E., *et al.* 2004. Expression and co-expression of the members of the epidermal growth factor receptor (EGFR) family in invasive breast carcinoma. *British Journal of Cancer.* 91 (8):1532-42.

Babine, R. E. and S. L. Bender. 1997. Molecular recognition of protein-ligand complexes: Applications to drug design. *Chemical Reviews.* 97 (5):1359-1472.

Bae, J. H. and J. Schlessinger. 2010. Asymmetric tyrosine kinase arrangements in activation or autophosphorylation of receptor tyrosine kinases. *Molecules and Cells.* 29 (5):443-48.

Baker, E. N. and R. E. Hubbard. 1984. Hydrogen-Bonding in Globular-Proteins. *Progress in Biophysics & Molecular Biology.* 44 (2):97-179.

Berman, H. M. 2008. The Protein Data Bank: a historical perspective. *Acta Crystallographica Section A.* 64:88-95.

Berman, H. M., Bhat T. N., Bourne P. E., *et al.* 2000. The Protein Data Bank and the challenge of structural genomics. *Nature Structural Biology.* 7:957-959.

Berman, H. M., Westbrook J., Feng Z., *et al.* 2000. The Protein Data Bank. *Nucleic Acids Research.* 28 (1):235-242.

Binder, B., Ebenhoh O., Hashimoto K., and R. Heinrich. 2006. Expansion of signal transduction networks. *Iee Proceedings Systems Biology.* 153 (5):364-368.

Bolz, J., I. Farmer, E. Grinspun, and P. Schroder. 2003. Sparse matrix solvers on the GPU: Conjugate gradients and multigrid. *Acm Transactions on Graphics* 22 (3):917-24.

Cao, Z. W., L. Y. Han, C. J. Zheng, et al. 2005. Computer prediction of drug resistance mutations in proteins. *Drug Discovery Today.* 10 (7):521-529.

Carraway, K. L., J. L. Weber, M. J. Unger, et al. 1997. Neuregulin-2, a new ligand of ErbB3/ErbB4-receptor tyrosine kinases. *Nature* 387 (6632):512-16.

Case, DA and Darden, TA and Cheatham III, Thomas E and Simmerling, CL and Wang, J and Duke, R.E. and Luo, R and Walker, RC and Zhang, W and Merz, KM and others. 2012. AMBER 12. *University of California, San Francisco.* 142.

Citri, A., and Y. Yarden. 2006. EGF-ERBB signalling: towards the systems level. *Nature Reviews Molecular Cell Biology.* 7 (7):505-516.

Cohen, A. R., Gomes F. L. A. F., Roysam B., and M. Cayouette. 2010. Computational prediction of neural progenitor cell fates. *Nature Methods.* 7 (3):213-U75.

de Antonellis, P. 2014. Afatinib, a lung cancer inhibitor of ErbB family. *Naunyn-Schmiedebergs Archives of Pharmacology.* 387 (6):503-504.

Dienstmann, R., De Dosso S., Felip E., and J. Tabernero. 2012. Drug development to overcome resistance to EGFR inhibitors in lung and colorectal cancer. *Molecular Oncology.* 6 (1):15-26.

Draghici, S. and R. B. Potter. 2003. Predicting HIV drug resistance with neural networks. *Bioinformatics* 19 (1):98-107.

Edelsbrunner, Herbert and P. Ernst 1992. *Weighted alpha shapes*: University of Illinois at Urbana-Champaign, Department of Computer Science.

Edelsbrunner, Herbert and P. Ernst 1994. Three-dimensional alpha shapes. *ACM Transactions on Graphics (TOG).* 13 (1):43--72.

Engelman, J. A., Zejnullahu K., Mitsudomi T., et al. 2007. MET amplification leads to gefitinib resistance in lung cancer by activating ERBB3 signaling. *Science.* 316 (5827):1039-1043.

Fabri, A., Giezeman G. J., Kettner L., Schirra S., and S. Schonherr. 2000. On the design of CGAL a computational geometry algorithms library. *Software-Practice & Experience.* 30 (11):1167-1202.

Frolov, A., Schuller K., C. W. D. Tzeng, et al. 2007. ErbB3 expression and dimerization with EGFR influence pancreatic cancer cell sensitivity to erlotinib. *Cancer Biology & Therapy.* 6 (4):548-54.

GrausPorta, D., Beerli R. R., Daly J. M., and N. E. Hynes. 1997. ErbB-2, the preferred heterodimerization partner of all ErbB receptors, is a mediator of lateral signaling. *Embo Journal.* 16 (7):1647-55.

Gu, D., W. A. Scaringe, K. Li, et al. 2007. Database of somatic mutations in EGFR with analyses revealing indel hotspots but no smoking-associated signature. *Human Mutation.* 28 (8):760-70.

Guy, P. M., Platko J. V., Cantley L. C., Cerione R. A., and K. L. Carraway. 1994. Insect Cell-Expressed P180(Erbb3) Possesses an Impaired Tyrosine Kinase-Activity. *Proceedings of the National Academy of Sciences of the United States of America.* 91 (17):8132-8136.

Haley, John D., William Gullick, John. 2008. *EGFR Signaling Networks in Cancer Therapy.* Edited by B. A. Teicher: Springer.

Hao, G. F., Yang G. F., and C. G. Zhan. 2012. Structure-based methods for predicting target mutation-induced drug resistance and rational drug design to overcome the problem. *Drug Discovery Today.* 17 (19-20):1121-26.

Harris, Robin Kingsley. 1986. *Nuclear magnetic resonance spectroscopy.* John Wiley and Sons Inc., New York, NY.

Hou, T. J., Zhang W., Wang J., and W. Wang. 2009. Predicting drug resistance of the HIV-1 protease using molecular interaction energy components. *Proteins-Structure Function and Bioinformatics.* 74 (4):837-46.

Hynes, N. E. and H. A. Lane. 2005. ERBB receptors and cancer: The complexity of targeted inhibitors (vol 5, pg 341, 2005). *Nature Reviews Cancer.* 5 (7).

Jo, M. J., Stolz D. B., Esplen J. E., Dorko K., Michalopoulos G. K., and S. C. Strom. 2000. Cross-talk between epidermal growth factor receptor and c-Met signal pathways in transformed cells. *Journal of Biological Chemistry.* 275 (12):8806-11.

Kallioniemi, O. P., Kallioniemi A., Kurisu W., et al. 1992. Erbb2 Amplification in Breast-Cancer Analyzed by Fluorescence Insitu Hybridization. *Proceedings of the National Academy of Sciences of the United States of America.* 89 (12):5321-25.

Kellogg, E. H., Leaver-Fay A., and D. Baker. 2011. Role of conformational sampling in computing mutation-induced changes in protein structure and stability. *Proteins-Structure Function and Bioinformatics.* 79 (3):830-38.

Kobayashi, S., Ji H. B., Yuza Y., *et al.* 2005. An alternative inhibitor overcomes resistance caused by a mutation of the epidermal growth factor receptor. *Cancer Research.* 65 (16):7096-101.

Kruger, J. and R. Westermann. 2003. Linear algebra operators for GPU implementation of numerical algorithms. *Acm Transactions on Graphics.* 22 (3):908-16.

Leaver-Fay, A., Tyka M., Lewis S. M., et al. 2011. Rosetta3: An Object-Oriented Software Suite for the Simulation and Design of Macromolecules. *Methods in Enzymology, Vol 487: Computer Methods, Pt C.* 545-74.

Lee, V. H. F., Tin V. P. C., Choy T. S., *et al.* 2013. Association of Exon 19 and 21 EGFR Mutation Patterns with Treatment Outcome after First-Line Tyrosine Kinase Inhibitor in Metastatic Non-Small-Cell Lung Cancer. *Journal of Thoracic Oncology.* 8 (9):1148-55.

Littlefield, P., Liu L. J., Mysore V., Shan Y. B., Shaw D. E., and N. Jura. 2014. Structural analysis of the EGFR/HER3 heterodimer reveals the molecular basis for activating HER3 mutations. *Science Signaling.* 7 (354).

Loo, L. H., Wu L. F., and S. J. Altschuler. 2007. Image-based multivariate profiling of drug responses from single cells. *Nature Methods.* 4 (5):445-53.

Ma, L. C., Wang D. D., Huang Y. Q., Yan H., Wong M. P., and V. H. F. Lee. 2015. EGFR Mutant Structural Database: computationally predicted 3D structures and the corresponding binding free energies with gefitinib and erlotinib. *BMC Bioinformatics.* 16:85

Ma, L. C., and Wang, D. D. and Yiqing Huang, Maria P Wong, Victor HF Lee, and Yan, Hong. 2015. Decoding the EGFR mutation-induced drug resistance in lung cancer treatment by local surface geometric properties. *Computers in Biology and Medicine.* 63: 293-300

Marchetti, A., Martella C., Felicioni L., *et al.* 2005. EGFR mutations in non-small-cell lung cancer: Analysis of a large series of cases and development of a rapid and sensitive method for diagnostic screening with potential implications on pharmacologic treatment. *Journal of Clinical Oncology.* 23 (4):857-65.

Marti-Renom, M. A., Stuart A. C., Fiser A., Sanchez R., Melo F., and A. Sali. 2000. Comparative protein structure modeling of genes and genomes. *Annual Review of Biophysics and Biomolecular Structure.* 29:291-325.

Meyer, M., Wilson P., and D. Schomburg. 1996. Hydrogen bonding and molecular surface shape complementarity as a basis for protein docking. *J Mol Biol.* 264 (1):199-210.

Morgillo, F., Woo J. K., Kim E. S., Hong W. K., and H. Y. Lee. 2006. Heterodimerization of insulin-like growth factor receptor/epidermal growth factor receptor and induction of survivin expression counteract the antitumor action of erlotinib. *Cancer Research.* 66 (20):10100-10111.

Normanno, N., De Luca A., Bianco C., et al. 2006. Epidermal growth factor receptor (EGFR) signaling in cancer. *Gene.* 366 (1):2-16.

Ohbuchi, R. and T. Takei. 2003. Shape-similarity comparison of 3D models using alpha shapes. *11th Pacific Conference on Computer Graphics and Applications, Proceedings*:293-302.

Olayioye, M. A., Neve R. M., Lane H. A., and N. E. Hynes. 2000. The ErbB signaling network: receptor heterodimerization in development and cancer. *Embo Journal.* 19 (13):3159-67.

Owens, J. D., Houston M., Luebke D., Green S., Stone J. E., and J. C. Phillips. 2008. GPU computing. *Proceedings of the Ieee.* 96 (5):879-99.

Oxnard, G. R., Arcila M. E., Sima C. S., *et al.* 2011. Acquired Resistance to EGFR Tyrosine Kinase Inhibitors in EGFR-Mutant Lung Cancer: Distinct Natural History of Patients with Tumors Harboring the T790M Mutation. *Clinical Cancer Research.* 17 (6):1616-22.

Paez, J. G., Janne P. A., Lee J. C., et al. 2004. EGFR mutations in lung cancer: Correlation with clinical response to gefitinib therapy. *Science.* 304 (5676):1497-1500.

Pettersen, E. F., Goddard T. D., Huang C. C., et al. 2004. UCSF chimera - A visualization system for exploratory research and analysis. *Journal of Computational Chemistry.* 25 (13):1605-12.

Samaga, R., Saez-Rodriguez J., Alexopoulos L. G., Sorger P. K., and S. Klamt. 2009. The Logic of EGFR/ErbB Signaling: Theoretical Properties and Analysis of High-Throughput Data. *Plos Computational Biology.* 5 (8).

Santos, G. D., Shepherd F. A., and M. S. Tsao. 2011. EGFR Mutations and Lung Cancer. *Annual Review of Pathology: Mechanisms of Disease, Vol. 6* 6:49-69.

Schlessinger, J. 2002. Ligand-induced, receptor-mediated dimerization and activation of EGF receptor. *Cell* 110 (6):669-72.

Sequist, L. V., Bell D. W., Lynch T. J., and D. A. Haber. 2007. Molecular predictors of response to epidermal growth factor receptor antagonists in non-small-cell lung cancer. *Journal of Clinical Oncology.* 25 (5):587-95.

Shigematsu, H., Lin L., Takahashi T., et al. 2005. Clinical and biological features associated with epidermal growth factor receptor gene mutations in lung cancers. *Journal of the National Cancer Institute.* 97 (5):339-46.

Smyth, M. S. and J. H. J. Martin. 2000. x Ray crystallography. *Journal of Clinical Pathology-Molecular Pathology.* 53 (1):8-14.

Sneddon, M. W. and T. Emonet. 2012. Modeling cellular signaling: taking space into the computation. *Nature Methods.* 9 (3):239-42.

Solca, F., Dahl, G., G. Dahl, A. Zoephel, *et al.* 2012. Target Binding Properties and Cellular Activity of Afatinib (BIBW 2992), an Irreversible ErbB Family Blocker. *Journal of Pharmacology and Experimental Therapeutics* 343 (2):342-50.

Sordella, R., Bell D. W., Haber D. A., and J. Settleman. 2004. Gefitinib-sensitizing EGFR mutations in lung cancer activate anti-apoptotic pathways. *Science* 305 (5687):1163-67.

Tanizaki, J., Okamoto I., Sakai K., and K. Nakagawa. 2011. Differential roles of trans-phosphorylated EGFR, HER2, HER3, and RET as heterodimerisation partners of MET in lung cancer with MET amplification. *British Journal of Cancer.* 105 (6):807-13.

Tiwari, A. and S. K. Panigrahi. 2007. HBAT: a complete package for analysing strong and weak hydrogen bonds in macromolecular crystal structures. *In Silico Biol.* 7 (6):651-61.

Tokumo, M., Toyooka S., Kiura K., *et al.* 2005. The relationship between epidermal growth factor receptor mutations and clinicopathologic features in non-small cell lung cancers. *Clinical Cancer Research* 11 (3):1167-73.

Tsang, Roger Y., Saeed Sadeghi, and Richard S. Finn, 2011. Lapatinib, a Dual-Targeted Small Molecule Inhibitor of Egfr and Her2, in Her2-Amplified Breast Cancer: From Bench to Bedside. *Clinical Medicine Insights: Therapeutics.*

van der Veeken, J., Oliveira S., Schiffelers R. M., Storm G., Henegouwen P. M. P. V. E., and R. C. Roovers. 2009. Crosstalk Between Epidermal Growth Factor Receptor- and Insulin-Like Growth Factor-1 Receptor Signaling: Implications for Cancer Therapy. *Current Cancer Drug Targets.* 9 (6):748-60.

Wagner, A. and J. Wright. 2004. Compactness and cycles in signal transduction and transcriptional regulation networks: A signature of natural selection? *Advances in Complex Systems.* 7 (3-4):419-32.

Wang, D. D., Ma L., Wong M. P., Lee V. H., and H. Yan. 2015. Contribution of EGFR and ErbB-3 Heterodimerization to the EGFR Mutation-Induced Gefitinib- and Erlotinib-Resistance in Non-Small-Cell Lung Carcinoma Treatments. *PLoS One.* 10 (5):e0128360.

Wang, D. D., Zhou W. Q., Yan H., Wong M., and V. Lee. 2013. Personalized prediction of EGFR mutation-induced drug resistance in lung cancer. *Scientific Reports.* 3.

Wang, Y., Yuan J. L., Zhang Y. T., et al. 2013. Inhibition of Both EGFR and IGF1R Sensitized Prostate Cancer Cells to Radiation by Synergistic Suppression of DNA Homologous Recombination Repair. *PloS One.* 8 (8).

Xiang, Z. X. and B. Honig. 2001. Extending the accuracy limits of prediction for side-chain conformations (vol 311, pg 421, 2001). *Journal of Molecular Biology.* 312 (2):419-19.

Xiang, Z. X., Soto C. S., and B. Honig. 2002. Evaluating conformational free energies: The colony energy and its application to the problem of loop prediction. *Proceedings of the National Academy of Sciences of the United States of America.* 99 (11):7432-437.

Yang, J. C. H., J. Y. Shih, W. C. Su, et al. 2012. Afatinib for patients with lung adenocarcinoma and epidermal growth factor receptor mutations (LUX-Lung 2): a phase 2 trial. *Lancet Oncology.* 13 (5):539-48.

Yap, T. A., Vidal L., Adam J., *et al.* 2010. Phase I Trial of the Irreversible EGFR and HER2 Kinase Inhibitor BIBW 2992 in Patients With Advanced Solid Tumors. *Journal of Clinical Oncology.* 28 (25):3965-72.

Yarden, Y. 2001. The EGFR family and its ligands in human cancer: signalling mechanisms and therapeutic opportunities. *European Journal of Cancer.* 37:S3-S8.

Yarden, Y., and Sliwkowski M. X.. 2001. Untangling the ErbB signalling network. *Nature Reviews Molecular Cell Biology.* 2 (2):127-37.

Yasuda, H., Kobayashi S., and Costa D. B. 2012. EGFR exon 20 insertion mutations in non-small-cell lung cancer: preclinical data and clinical implications. *Lancet Oncology* 13 (1):E23-E31.

Yun, C. H., Boggon T. J., Li Y. Q., *et al.* 2007. Structures of lung cancer-derived EGFR mutants and inhibitor complexes: Mechanism of activation and insights into differential inhibitor sensitivity. *Cancer Cell.* 11 (3):217-27.

Yun, C. H., Mengwasser K. E., Toms A. V., et al. 2008. The T790M mutation in EGFR kinase causes drug resistance by increasing the affinity for ATP. *Proceedings of the National Academy of Sciences of the United States of America.* 105 (6):2070-75.

Zhang, X. W., Gureasko J., Shen K., Cole P. A., and J. Kuriyan. 2006. An allosteric mechanism for activation of the kinase domain of epidermal growth factor receptor. *Cell.* 125 (6):1137-49.

Zhou, W. Q., Wang D. D., Yan H., Wong M., and V. Lee. 2013. Prediction of anti-EGFR drug resistance base on binding free energy and hydrogen bond analysis. *Proceedings of the 2013 Ieee Symposium on Computational Intelligence in Bioinformatics and Computational Biology (CIBCB)*:193-97.

Zhou, W. Q., and H. Yan. 2010. A discriminatory function for prediction of protein-DNA interactions based on alpha shape modeling. *Bioinformatics.* 26 (20):2541-48.

Zhou, W. Q., and H. Yan. 2010. Relationship between periodic dinucleotides and the nucleosome structure revealed by alpha shape modeling. *Chemical Physics Letters.* 489 (4-6):225-28.

Zhou, W. Q., Yan H., and Q. Hao. 2012. Analysis of surface structures of hydrogen bonding in protein-ligand interactions using the alpha shape model. *Chemical Physics Letters.* 545:125-31.

Section 6
Advanced Topics

11

Quality Assurance in Genome-Scale Bioinformatics Analyses

Eleni Giannoulatou[1], Amir Hossein Kamali[2], Andrian Yang[1],
Tsong Yueh Chen[3] and Joshua W. K. Ho[1]*

Abstract

The advent of Next Generation Sequencing (NGS) is transforming the landscape of biomedical research, ranging from disease gene discovery to clinical application of genomic medicine. NGS enables low-cost, high-throughput sequencing for a wide variety of genome-wide scale analysis of the genome, epigenome and transcriptome. However, with this vast quantity of data, we are faced with unprecedented technical challenges in terms of quality assurance of the computational analytical pipelines. In this chapter, we review current approaches used for bioinformatics validation and quality control in whole genome sequencing analysis for genomic medicine applications. We further discuss how state-of-the-art software testing techniques can be used to establish strong quality assurance measures in genome-scale bioinformatics.

1. Introduction

Bioinformatics is the application of computational, mathematical and statistical techniques to solve problems in biology and medicine. Arguably the main research focus has so far been on the computational and statistical basis of the algorithms. Surprisingly much less effort has been placed on the validation and quality assurance of the tools that implement these algorithms – even though correct design and implementation of the underlying algorithm is at least as important as the algorithm itself. Incorrectly computed results may lead to wrong biological conclusions, and subsequently misguide downstream experiments. The widespread problem of errors or mis-use of scientific computing in biology and medicine is highlighted by recent news and commentary articles in top-tier journals such as Nature and Science on this

---

[1] Victor Chang Cardiac Research Institute, Darlinghurst, NSW 2010, Australia
  The University of New South Wales, Sydney, NSW 2052, Australia
[2] Victor Chang Cardiac Research Institute, Darlinghurst, NSW 2010, Australia
  The University of Sydney, Sydney, NSW 2006, Australia
[3] Swinburne University of Technology, VIC, 3122, Australia
[*] Corresponding author : j.ho@victorchang.edu.au

issue (Joppa *et al.*, 2013; Merali, 2010), and it could be attributed to the lack of proper software verification and validation (Alden and Read, 2013; Hayden, 2013).

Lack of quality assurance is especially a critical problem if these bioinformatics tools are to be used in a translational clinical setting. Cost of DNA sequencing is no longer a limiting factor; the real bottleneck is the reliable and fast analysis of the massive amount of sequencing data produced by Next Generation Sequencing (NGS). Many bioinformatics analysis pipelines have been developed, however recent studies found the variants being called by different pipelines from the same sequencing data set can differ substantially (O'Rawe *et al.*, 2013). Given a whole genome sequencing (WGS) or whole exome sequencing (WES) analysis pipeline for identification of sequence variants, one must have high confidence that the resulting variant calls have high sensitivity and specificity. Although true positives can be distinguished from false positives easily through external validation, it is almost impossible to systematically distinguish false negatives from the vast amount of true negatives. There is therefore a critical need to ensure that only correct and validated algorithms are used, and that they are implemented correctly into computer programs.

There are several unique challenges in implementing rigorous validation and quality control in bioinformatics that are not addressed in the current national or international guidelines for genetic diagnostic laboratories:

1. *Validation before deployment.* How to generate a large number of diverse and realistic test cases to ensure a sufficient coverage of the input space?

2. *Quality Control during deployment.* How to determine the correctness of real data without a gold standard?

In the following sections, we will review standard bioinformatics pipelines used in genomic medicine, and methods that can be employed to establish quality assurance for these pipelines.

2.   Whole genome sequencing analysis for genomic medicine

A standard pipeline for whole genome sequencing analysis that has been used widely for the analysis of NGS data consists of the steps outlined below. These steps, or slight variations of them, are commonly followed to identify disease-causing variants from sequencing studies. A flowchart of the pipeline is shown in Fig. 1.

1. *Quality Control of raw sequencing data (FASTQ files):* The raw sequencing files are assessed using FastQC to identify any potential issues regarding the read sequence quality, large GC content biases, sequence length distribution, sequence duplication levels, overrepresented sequences or presence of adapter sequences (contamination).

2. *Read Mapping to the Human Reference Genome and Preprocessing:* The sequencing reads are aligned to the human reference genome (hg19) using the Burrows-Wheeler aligner (Li and Durbin, 2009) or fast gapped-read aligner Bowtie (Langmead and Salzberg, 2012). The mapping step can be performed in a parallel way to allow multiple samples to be processed at the same time. The resulting SAM/BAM files (Sequence Alignment/Map

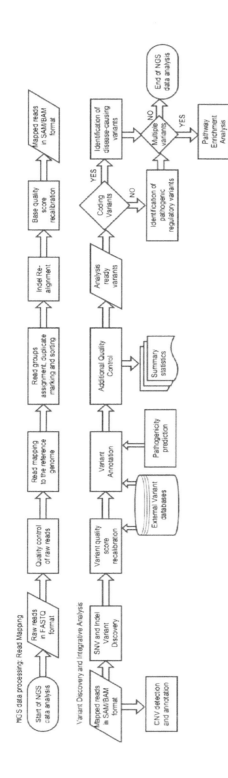

FIGURE 1 Flowchart of the analysis pipeline for discovery of disease-causing variants from whole genome next-generation sequencing data.

Format and its binary form) are subsequently processed so they contain accurate read groups, marked duplicates and sorted reads using Picard set of tools. Using the current version of Genome Analysis Toolkit (GATK) (McKenna *et al.*, 2010) base quality score recalibration, indel (insertions/deletions) realignment and removal of duplicate reads are performed. SAM/BAM files of samples ran in multiple lanes are merged.

3. *Variant Discovery*: Single Nucleotide Variants (SNV) and short insertions and deletions are discovered and genotyped across all samples simultaneously using standard filtering parameters or variant quality score recalibration according to GATK Best Practice recommendations (DePristo *et al.*, 2011). GATK is the most widely used variant calling algorithm, although other algorithms such as Platypus (Rimmer *et al.*, 2014) and VarScan (Koboldt *et al.*, 2009) have also been used in large sequencing studies.

4. *Variant Annotation*: Variants are annotated using the latest version of ANNOVAR (Wang *et al.*, 2010). Useful information is extracted such as the frequency of the variants in public databases including the 1000 Genomes database, the Exome Variant Server and dbSNP as well as predicting pathogenicity of variants located in coding regions, using MutationTaster (Schwarz *et al.*, 2014), PolyPhen2 (Adzhubei *et al.*, 2010) and CADD (Kircher *et al.*, 2014). Annotated VCF files are then further analyzed using custom scripts or reformatted and analyzed using VARSIFTER (Teer *et al.*, 2012) that enables easy filtering of the variants without the need for further script executing.

5. *Additional Quality Control*: Summary statistics are calculated for each sample including the total number of variants, number of heterozygous genotypes, number of singletons, average transition/transversion ratio and average quality for variants for which individual has a non-reference genotype. These statistics are examined together with coverage analysis statistics such as average read coverage for each gene. For family data, kinship coefficients are estimated, by applying KING (Manichaikul *et al.*, 2010) on the common SNPs (Minor Allele Frequency>1%) in order to verify the family relationships and flag potential pedigree errors or sample mix-ups. For case-control datasets, such methods can be used to identify cryptic relatedness between samples that would inflate the false positive rate. In addition, principal components analysis is performed to identify critical potential population stratification that would otherwise limit the genetic association analysis and lead to artificial findings (Price *et al.*, 2006) .

6. *Identification of candidate disease-causing DNA variants*: Analysis of WGS family data is undertaken by first categorizing the assumed nature of segregation of disease within each affected family member and applying various inheritance models, including dominant, recessive as well as *de novo* occurrence. For case-control samples, single-locus association tests or burden tests are performed. Single-locus tests interrogate each DNA variant for association with disease. All inheritance models can again be considered (allelic, dominant, recessive) but such analysis can be underpowered in the absence of large sample sizes. Burden tests assess the cumulative

effect of multiple variants in a genomic region (i.e. gene) by collapsing or summarizing rare variants within the region by a single value (i.e. number of rare variants in cases vs. controls). The count of case-unique rare alleles can be tested or other tests such as the variable threshold test (Price *et al.*, 2010) or the two-sided C-alpha test can be applied (Neale *et al.*, 2011).

6.1 *Coding Variants*: In the first stage, the analysis is restricted to protein coding genes after excluding variants with no impact on protein sequence or splicing efficacy/efficiency. Variants of interest are those not present in publicly available control databases such as the 1000 Genomes (The 1000 Genomes Project Consortium, 2012) and the Exome Variant Server. If novel variants are not found, a mean allele population frequency of less than 1% is used. Variants are assumed to be more significant if they occur within evolutionarily conserved sequences according to Genomic Evolutionary Rate Profiling (GERP score > 2.0). Candidate variants are then refined for those predicted not to be benign, identified according to MutationTaster and PolyPhen2. The candidate variants can also be interrogated within sequence data of large databases such as the Exome Aggregation Consortium (63,000 exomes available at ExAC Browser). In the case of *de novo* mutations detected in families, a statistical framework is applied to further evaluate excesses of the de novo mutations identified, compared to the levels of mutations expected by chance (Samocha *et al.*, 2014).

6.2 *Non-coding Variants*: Non-coding variants prioritization is performed using GWAVA (Ritchie *et al.*, 2014) which aims to predict the functional impact of noncoding genetic variants based on a wide range of annotations of noncoding elements (largely from the Encyclopedia of DNA Elements – ENCODE), along with genome-wide properties such as evolutionary conservation and GC-content. Regulatory variants can affect transcription factor binding sites, chromatin states, epigenetic modifications and regulatory non-coding RNAs. MutationTaster and CADD are also used here since they can provide predictions for non-coding variants especially for the ones that can be mapped to a transcript model, such as those in untranslated regions and introns.

7. *Detection of Copy Number Variation (CNV)*: CNVs can be detected using state-of-the-art algorithms for exome sequencing data, such as xHMM (Fromer and Purcell, 2014), cn.mops (Klambauer *et al.*, 2012) and exomeDepth (Plagnol *et al.*, 2012). Detecting CNVs from exome sequencing data is challenging, as high read depth and a large number of samples are required for normalization and calculation of read-depth baseline. For WGS data, the segmentation into CNV regions is more accurate, and specific algorithms are used for this, such as GenomeStrip (Handsaker *et al.*, 2011) CNVnator (Abyzov *et al.*, 2011) and Control-FREEC (Boeva *et al.*, 2011). If complex CNV patterns are observed, additional existing bioinformatics tools can be applied to detect complex chromosomal structural variations.

8. *Pathway Enrichment Analysis*: In the case of multiple candidate variants segregating with the disease, pathway enrichment analysis can also be performed using DAVID (Huang *et al.*, 2008) and Enrichr (Chen *et al.*, 2013)

that use a list of genes as input. For noncoding variation, GREAT (McLean *et al.*, 2010) is applied. This algorithm aims to assign biological meaning to a set of noncoding genomic regions by analyzing the annotations of the nearby genes.

3. The problem of quality assurance

The pipeline described above is used widely by bioinformaticians in order to analyse large amounts of DNA sequencing data produced every day. Each step consists of the application of specific set of bioinformatics tools that have been developed by academic researchers. Hence, although the algorithms, main concept and mathematical framework behind the software have been assessed as part of the peer review process of scientific articles, the software developed have not been systematically verified and validated (Hayden, 2015). Previous work on scientific software evaluation has shown that numerical disagreement between programs of scientific computation grows at around the rate of 1% in average absolute difference per 4000 lines of implemented code and that the nature of this disagreement is nonrandom (Hatton and Roberts, 1994). The software developers most often detect errors that could have significant effect in the results. However, errors that can occur in specific input cases are harder to detect and are expected to occur in any scientific program.

Lack of quality assurance of software pipelines used for analysis of WGS data is a critical problem that hinders the widespread adoption of genomic medicine in a clinical setting. A recent study found that the concordance of single nucleotide variants identified by five widely used variant calling pipelines (i.e., sequence alignment followed by variant calling) was less than 60%, while the concordance of the identified indels (insertions/deletions) was lower than 30% (O'Rawe *et al.*, 2013). Analysis of the discordant variant calls revealed a considerable false negative rate, suggesting that these pipelines might miss many genuine and possibly disease-relevant genetic variants. Besides variant calling, the use of different variant annotation software programs and transcript annotation files can also make a substantial difference in annotation results that are not commonly appreciated (McCarthy *et al.*, 2014). These reports highlight the need to ensure that any bioinformatics pipeline should be subjected to better validation and quality control, especially for clinical genomic medicine applications.

4. Standard validation and QC approaches used in diagnostic laboratories

As next-generation whole exome sequencing and whole genome sequencing are becoming poised by widespread applications in a clinical setting, several international regulatory and professional bodies have proposed guidelines for quality assurance for genetic testing in clinical laboratories. All existing guidelines have a clear definition of quality metrics, such as sensitivity, specificity, reproducibility, and reportable range as shown in Table 1. These guidelines also stress the importance of testing a variant calling pipeline against these metrics through proper and adequate testing using appropriate reference materials (Gargis *et al.*, 2012) before the deployment of the system (i.e., validation) and continuously throughout the operational lifespan of the pipeline (i.e., quality control and assurance).

TABLE 1 Definitions of quality metrics used for whole genome sequencing analysis for genomic medicine.

Quality metric	Definition
Accuracy	The degree of agreement between the nucleic acid sequences derived from the assay and a reference sequence.
Precision	The degree to which repeated sequence analyses give the same result- repeatability (within-run precision) and reproducibility (between-run precision).
Analytical sensitivity	The likelihood that the assay will detect the targeted sequence variations, if present.
Analytical specificity	The probability that the assay will not detect a sequence variation when none are present (the false positive rate is a useful measurement for sequencing assays).
Reportable range	The region of the genome in which sequence of an acceptable quality can be derived by the laboratory test.
Reference range	Reportable sequence variations the assay can detect that are expected to occur in an unaffected population.

These definitions exist in established guidelines for whole genome sequencing in clinical laboratory practise and originally developed by the Next-generation Sequencing: Standardization of Clinical Testing (Nex-StoCT) workgroup as reported in Gargis *et al.*, 2012.

In such guidelines, the computational tasks composing the bioinformatics pipeline are separated into primary, secondary and tertiary analysis. Primary analysis involves sequencing instrument-specific analysis such as base-calling. Secondary analysis involves sequence read quality control, read alignment, and variant calling. Tertiary analysis involves annotation and interpretation of the genetic variants. For simplicity, this chapter focuses on performing quality assurance on the secondary analysis pipeline. Nonetheless, the same principles can be applied to primary or tertiary analyses as well. These guidelines also specify requirements for documentation of various steps including version control to track software releases and updates to the analysis methods, quality metrics assessed during a test, results of pipeline validation and the process of data handling and storage.

In addition, these guidelines also specify the need for conducting a validation study, with the aim of providing objective evidence that the bioinformatics pipeline is fit for the intended purpose. It is suggested that the validation study must identify and rectify common sources of errors that may challenge the analytical validity of the bioinformatics pipeline. Analytical validity refers to the ability of a bioinformatics pipeline to correctly call and annotate a variant. In the proposed guidelines it is also strongly suggested that analytical validity must be achieved before clinical validity can be considered. Clinical validity refers to the ability of a test to detect or predict a phenotype of interest. Clinical validity must be established by external knowledge such as results from large-scale population studies or functional studies.

In the validation process, the bioinformatics pipeline must be benchmarked using reference materials that are chosen to be appropriate for assessing performance of the pipeline for its intended purpose. It is suggested that the validation study should compare the results from multiple pipelines, where possible, to allow identification of pipeline-specific artefacts. Following the quality control and quality assurance, it is strongly advised that the diagnostic laboratory should confirm any variant calls using experimental techniques. We should note that the term validation is used slightly differently in the software testing field, which involves only checking whether the right program is developed rather than checking whether the program is developed correctly (which is termed verification).

4.1   Genome in a Bottle and current frameworks for performance testing

The validation of a bioinformatics pipeline would require its application on specific input datasets where the correct status of the included variants is known. Such datasets are called reference materials (RM) and are used for benchmarking of the pipeline. Recently, a first accurate set of genotypes across a genome was developed by the Genome in a Bottle Consortium and the National Institute of Standards and Technology (Zook *et al.*, 2014). This dataset was created by integrating and arbitrating between 14 data sets from 5 sequencing technologies, 7 read mappers and 3 variant callers. Additionally, regions for which no confident genotype calls could be made were identified and classified based on reasons for uncertainty. It has been shown that high-confidence genotype calls from a well-characterized whole genome have been useful for assessing biases and rates of accurate and inaccurate genotype calls using different bioinformatics methods. This resource is publicly available and is continuously updated and integrated with other "gold standard" variant calls such as datasets from Real Time Genomics and Illumina Platinum Genomes that additionally utilize pedigree information. A genome comparison and analytic testing (GCAT) platform has also been developed to facilitate development of performance metrics and comparisons of analysis tools across these metrics. "An analytical framework for optimizing variant discovery from personal genomes" Gareth Highnam, Jason J. Wang, Dean Kusler, Justin Zook, Vinaya Vijayan, Nir Leibovich & David Mittelman Nature Communications 6, Article number: 6275, 2015. This platform gives access to multiple performance reports that are crowdsourced to encourage community involvement and input. It is constantly updated and developed to incorporate standard performance metrics and benchmarking tools being developed by the new Global Alliance for Genomic Health Benchmarking working group.

Although sources of error can occur at any stage of the bioinformatics pipeline, here we focus on the secondary analysis which consists of next-generation sequencing read mapping, variant calling and annotation. Despite the current efforts in methodological approaches in diagnostic laboratories and in the academic community, there are no clear guidelines on several important aspects that affect the comprehensiveness and effectiveness of the validation and quality assurance framework, including:

1. What are the characteristics of effective test cases (e.g., reference material)?
2. How do we measure the effectiveness of a set of test cases?

3. How many test cases are enough?
4. How can we check the correctness of test cases beyond the small set of gold standard reference material where the correctness is known?

5. Introduction to a software-testing framework

Software testing is the process of actively identifying potential faults in a computer program (Ammann and Offutt, 2008). This process can be used for two purposes: to ensure the program is correctly implemented against the specification (i.e., verification) and to ensure the correct specification is used against the desired user requirement (i.e., validation). In other words, verification asks, "Are we building the software right?" whereas validation seeks to answer, "Are we building the right software?". Software testing can be static or dynamic. Static testing involves code review or inspection (Hayden, 2013), whereas dynamic testing – the more common approach – involves execution of the program under test (PUT) given a set of test cases. In dynamic software testing, the PUT can be thought of an implementation of a (mathematical or computational) function $f(x) = y$ where x represents all valid input from the input domain and y represents all possible output.

The goal of verification is to show that for a given implementation PUT, namely $f_{PUT}(x) = f(x)$ for all possible x from the input domain. An input, $x_{failure}$ is a failure-causing input if $f_{PUT}(x_{failure}) \neq f(x_{failure})$, and the PUT is deemed to contain a fault. A PUT may fail to satisfy a user's need because of incorrect implementation of the algorithm (i.e., the verification problem), or a mismatch between the algorithm and the intended behaviour (i.e., the validation problem). It is often impossible to exhaustively transverse through the entire input space. An effective software testing strategy attempts to exploit information of the PUT to generate a set of test cases that can trigger as many distinct failures as possible. In order words, we want to maximise the chance of a test case being a failure-causing input.

The challenges faced when testing any scientific software have been found to occur mainly due the lack of an oracle. An oracle is a mechanism that decides if the software output is correct given an input. In bioinformatics, it is often impossible to obtain such an oracle for the entire input domain. Nonetheless, it is often possible to check the correctness of a subset inputs by the use of a "gold standard" data set where the expected outputs are known. The bioinformatics software can then be tested against the expected outputs. However, such gold standard data sets often do not exist. For example, we do not have a "gold standard" data set for aligned sequenced reads from next generation sequencing in a SAM/BAM format. Formally, an oracle problem is said to exist when: (1) "there does not exist an oracle" or (2) "it is theoretically possible, but practically too difficult to determine the correct output" (Chen *et al.*, 2003; Weyuker, 1982).

Challenges can also arise due to cultural differences between scientists and the software engineering community (Kanewala and Bieman, 2014). Scientists can often view the code and the model that it implements as inseparable entities. Therefore they test the code to assess the model and not necessarily to check for faults in the code. In addition, bioinformaticians are often not aware of any testing methodologies that can be used to ensure their software implementation is correct. In section 6 we will describe existing software testing approaches that can be utilised for bioinformatics software development.

Table 2 shows the definitions of software testing terms as defined by the Institute of Electrical and Electronics Engineers (IEEE) Standards Glossary and the International Software Testing Qualification Board Glossary. These terms are established in the software engineering field and are therefore used for the purposes of this Chapter.

TABLE 2 Description of terminology used in software testing as defined by the Institute of Electrical and Electronics Engineers (IEEE) Standards Glossary and the International Software Testing Qualification Board Glossary (ISTQB).

Term	Description
Validation	The process of evaluating a system or component during or at the end of the development process to determine whether it satisfies specified requirements.
Verification	The process of evaluating a system or component to determine whether the products of given development phase satisfy the conditions imposed at the start of that phase.
Quality Control	A set of activities designed to evaluate the quality of developed or manufactured products.
Quality Assurance	A planned and systematic pattern of all actions necessary to provide adequate confidence that an item or product conforms to established technical requirements.
Test Case	A set of test inputs developed for a particular objective, such as to exercise a particular program path or to verify compliance with a specific requirement.
Test Suite	A set of several test cases for a component or system under test, where the post condition of one test is often used as the precondition for the next one.
Regression Testing	Testing of a previously tested program following modification to ensure that defects have not been introduced or uncovered in unchanged areas of the software, as a result of the changes made. It is performed when the software or its environment is changed.
Test Coverage	The degree to which a given test or set of tests addresses all specified requirements for a given system or component.
Fault	An incorrect step, process, or data definition. For example, an incorrect instruction in a computer program.
Failure	Deviation of the component or system from its expected delivery, service or result.
Static Testing	Testing of a software development artifact, e.g., requirements, design or code, without execution of these artifacts, e.g., reviews or static analysis.
Dynamic Testing	Testing that involves the execution of the software of a component or system.

6. Software testing approaches, applications and evaluations

In this section we will describe traditional as well as more recently developed approaches for software testing and assess their advantages and disadvantages. These methods outline general principles that can be effectively utilized for bioinformatics software development.

6.1. Traditional software testing approaches

Current approaches for bioinformatics software testing can be broadly classified into three classes: special test cases, N-version programming, and valid range testing.

The use of special test cases is perhaps the most common approach of software testing in practice. The idea is that even though it is impossible to verify $f_{\text{PUT}}(x) = f(x)$ for all possible x, it is often possible to verify the correctness of a subset of input x. In the context of testing a variant calling pipeline, it is a standard practice to generate small input FASTQ files as test cases, where the output can be verified manually. Taking this idea further, this process can be automated by using a sequence read simulator to generate many test cases and to check the correctness of the output against the DNA sequence from which the input was simulated. The major shortcoming of this approach is that these test cases often do not represent the same diversity as real data. As a result, we are often left to infer the correctness of the untested input domain from the tested input domain.

N-version programming (Knight and Leveson, 1986) is also a very common approach in gaining confidence about the correctness of a bioinformatics program. The idea is that multiple implementations (or versions) of a program are compared to identify possible failures in one or more of the implementations. In fact, the method used to discover the high false negative rate in variant calling pipelines was none other than N-version programming (O'Rawe *et al.*, 2013). One advantage of this approach is that it can be used with any input as a test case and can be readily implemented if multiple versions of the same program already exist. In the absence of a test oracle, a main disadvantage is that it is difficult to judge what the correct output is, when different pipelines generate very different results. The observed substantial discordance between five commonly used variant calling pipelines highlight the problems of solely relying on this approach for quality assurance (O'Rawe *et al.*, 2013). Additionally, having multiple N-versions is computationally expensive.

Lastly, "valid range testing" is often employed to test whether $f_{\text{PUT}}(x)$ is within a reasonable range of $f(x)$. This is a very intuitive method that many scientists use to decide whether a computer program "looks" fine. This technique requires knowledge of the valid range of $f(x)$, and the usefulness of this approach depends on whether a tester can identify a tight range.

These approaches are implicitly used in current standard quality assurance criteria for genetic pathology testing, such as the use of "gold standard" reference materials for proper and adequate testing (special test case), validation by comparison of multiple computer programs (N-version programming), and defining the operational range of the output (valid range testing). Beyond pathology genetic testing, these basic testing methods are also widely used in other fields of bioinformatics. This problem is particularly relevant in the area of systems biology. The challenge of testing both deterministic and stochastic simulators has been realized by the Systems Biology Mark-up Language (SBML) community. To test the reliability of a new SBML capable simulator, the current practice involves executing it with multiple existing simulators on some well studied input models and compare the consistency of the simulation results (Bergmann and Sauro, 2008), or running a tool using a small test suite where the expected output is known (Evans *et al.*, 2008). The same idea is also applied to evaluate the performance of different gene network reconstruction algorithms – the so-called crowd approach (Marbach *et al.*, 2012). All these examples make use of special test cases, N-version programming, and/or valid range testing.

In addition to the aforementioned software testing techniques, there are some state-of-the-art techniques that are also suitable for testing bioinformatics programs.

In the following sections, we will review two such techniques: Metamorphic Testing, and Adaptive Random Testing.

6.2. Metamorphic Testing

Metamorphic Testing (MT) alleviates the oracle problem by using some algorithm or problem domain-specific properties, namely metamorphic relations (MRs), to verify the testing outputs. The central idea is that although it is impossible to directly test the correctness of any given test case, it may be possible to verify the relationships of the outputs generated by multiple executions of a program (Chen *et al.*, 1998; Zhou *et al.*, 2004). In other words, Metamorphic Testing tests for properties that users expect of a correct program. Some test cases (namely source test cases in the context of Metamorphic Testing) can be selected according to some traditional testing techniques. Further test cases (namely follow-up test cases in the context of Metamorphic Testing) can be generated based on the source test cases and according to the metamorphic relations. All test cases are executed, and then the outputs of the source and follow-up test cases are checked against the metamorphic relations. If any group of source and follow-up test cases violates (that is, does not satisfy) their corresponding metamorphic relation, the tester can say that a failure is detected and hence conclude that the program has bugs. In other words, a metamorphic relation serves two purposes: (1) generation of additional test cases by transforming the source input, and (2) checking the relationship between the outputs produced by the execution of the "source" and "follow-up" test cases. It should be noted that in general many follow-up test cases can be derived from a single source test case input based on one metamorphic relation. In addition, more than one source test case can be applied for a particular metamorphic relation.

As an example, let us consider how we can use Metamorphic Testing to test a program that computes the mathematical sine function. Some special test cases exist based on well known mathematical knowledge, such as $\sin(0) = 0$, $\sin(\pi/2)=1$, $\sin(\pi/6) = 0.5$, etc. Nonetheless, it is much harder to determine the value of any arbitrary x, such as $\sin(1.345)$. The key idea of Metamorphic Testing is that we can construct metamorphic relations based on well known properties of the mathematical sine function, such as $\sin(x + 2\pi) = \sin(x)$, $\sin(x + \pi) = -\sin(x)$, $\sin^2(x) + \sin^2(\pi/2 - x) = 1$, etc. After selecting the source test case, such as x = 1.345, we can use these metamorphic relations to generate additional follow-up test cases such as x = 1.345 $+ 2\pi$, 1.345 $+ \pi$, and $\pi/2 - 1.345$. The output of these follow-up test cases can then be compared to the output of the source test case to determine if the metamorphic relations hold. Using this method, the entire range of input domain can be tested, hence alleviating the oracle problem.

6.3. Adaptive Random Testing

Failure-causing inputs are not randomly distributed in the input space, but are usually clustered together to form distinct failure regions (Chan *et al.*, 1996). The implication is that non-failure regions are also contiguous; therefore after the execution of a non-failure-causing input x_i, one should select a random test case that is the furthest away from x_i in the input space. This is the basis of Adaptive Random Testing

(ART) (Chen *et al.*, 2004, 2010). Theoretical and empirical studies have shown that Adaptive Random Testing can be up to 50% more effective than traditional random testing in terms of failure detection ability (Chen and Merkel, 2008). The simplest implementation of Adaptive Random Testing involves first generating a random set of test cases in the input domain, from which the test case that is furthest away from all previous "successful" test cases is chosen. This process should continue until a pre-defined criterion is satisfied. Adaptive Random Testing provides a simple and rational approach to automatically select diverse test cases.

6.4. Case study: testing of NGS short-read alignment software

Metamorphic Testing has now been applied to many types of computer programs, including bioinformatics software (Chen *et al.*, 2009; Giannoulatou *et al.*, 2014; Sadi *et al.*, 2011; Xie *et al.*, 2011). In a recent study (Giannoulatou *et al.*, 2014) Metamorphic Testing was used to test three commonly used NGS short-read aligners: BWA (Li and Durbin, 2009), Bowtie (Langmead *et al.*, 2009) and Bowtie2 (Langmead and Salzberg, 2012). Nine metamorphic relations were developed that aim to capture the expected behaviour of a short-read alignment program. For example, the following three metamorphic relations (MRs) were applied for alignment of paired-end reads:

MR1: Random permutation of reads. The reads in the FASTQ files are reshuffled. The output mapping is expected to be the same as the original output.

MR5: Extension of reads. After initial mapping, each read is extended by 20 bp to the 3' or 5' end of the read, with high quality score, based on the reference genome sequence. The output mapping is expected to remain the same.

MR7: Mapped reads. After initial mapping, only the mapped reads are selected and remapped against the reference genome. It is expected that all of the reads will be mapped.

Using these metamorphic relations, many test cases were generated based on simulated data, and real WES data. None of the three aligners were found to satisfy all nine metamorphic relations, including the above three (Giannoulatou *et al.*, 2014). Failing to satisfy these metamorphic relations implies the presence of false positive and/or false negative alignment. In order to investigate the effect of these misalignments in downstream WGS or WES analysis, a traditional pipeline was applied that involves BWA alignment followed by using GATK for variant calling. The pipeline was applied to an exome sequencing run of the HapMap sample NA12872 and the analysis was repeated after considering only the uniquely mapped reads. It was found that prior to any filtering, the number of variants called was different when the original BAM file was used and after the application of metamorphic relations such as MR1, MR5 and MR7 (Fig. 2A). Non-uniquely mapped reads were subsequently filtered and the variant calling step was repeated. Surprisingly, there was still discordance between the numbers of variant calls made (Fig. 2B). In order to achieve concordance between the results higher quality threshold was needed while the specified tags were not sufficient to capture all of the non-uniquely mapped reads.

It is important to note that in the above example, the limitation of BWA was revealed by directly testing the individual pipeline instead of comparing it with other

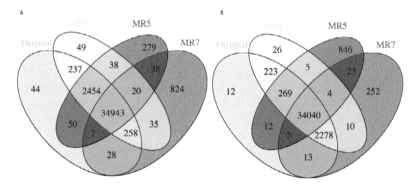

FIGURE 2   Number of variants called using original read mapping and mapping after the application of MR1, MR5 and MR7. (A) Using all the reads. (B) After removal of non-uniquely mapped reads. This figure is adapted with modification from (Giannoulatou *et al.*, 2014).

algorithms as in (O'Rawe *et al.*, 2013), and is based on a real WES data set from a HapMap individual (instead of smaller special test cases). This unique capability enables the correctness of a variant calling pipeline to be evaluated in a deployment environment on real WES or WGS data. Therefore it enables ongoing quality assurance monitoring of a pipeline in its operational environment on real data. Such an approach provides the first step towards automated testing of any NGS short read aligner and ultimately of any NGS bioinformatics pipeline.

6.5   Empirical evaluation of software testing strategies: Mutation Analysis

An important question that impacts the implementation of bioinformatics quality assurance is how can we demonstrate the effectiveness of a set of test cases? One approach is to artificially generate a number of fault-seeded versions of the program under test, and determine whether a given set of test cases can detect them. This process is referred to as mutation analysis (Andrews *et al.*, 2005; Woodward and Halewood, 1988). A program with a seeded fault is called a mutant (Fig. 3), and is said to be "killed" if a failure is detected by a set of test cases. Each mutant is generated by applying one of several very simple mutation operators to alter the source code of the original program. Previous studies have shown that, despite the simplicity of these mutation operators, the capability of detecting failures from the generated mutants is a good indicator of the effectiveness of a testing method (Andrews *et al.*, 2005). Mutants can either be generated by hand or by the a mutation analysis tool such as MuJava (Ma *et al.*, 2005). Therefore, it is possible to generate many distinct non-equivalent mutants of most programs.

FIGURE 3   A (mutant) program with a seeded fault.

Mutation analysis has been previously used to measure the fault-revealing ability of the test cases generated by Metamorphic Testing when applied to a short read sequencing aligner, a gene network simulator (Chen *et al.*, 2009), and a phylogenetic inference program (Sadi *et al.*, 2011). A general observation from these results is that test cases generated based on real data tend to have better fault-revealing ability, presumably because they better capture the complexity in the data that makes it prone to triggering a failure.

One unique feature of mutation analysis is that it provides a means to quantify the effectiveness and comprehensiveness of a test suite in terms of the proportion of mutants killed. In a study of the benefits and applicability of mutation analysis, it was concluded that the implementation of mutation operators in the studied mutation tools, provided a good approximation of the actual quality of a test suite and an advantage over conventional code coverage (Ramler and Kaspar, 2012). In addition, the information about which mutants cannot be killed will give insight into the limitation of a test suite, thus informing how additional test cases can be generated.

7. Cloud-based testing as a service (TaaS) for bioinformatics

Testing-as-a-Service (TaaS) is a model of software testing which provides automated software testing on the web (van der Aalst 2009). The development of TaaS is fueled by the recent growth in the cloud computing model — which provides scalable, on demand computing resources with relatively low cost — and the need for an automated software testing framework in order to reduce the barrier in implementing quality assurance in genomic medicine applications. Aside from helping software developers to write better software, TaaS application also extends to end users for testing software prior to deployment and as a quality certification service to provide objective assessment of software quality (Candea, Bucur, and Zamfir 2010).

There are a growing number of commercial TaaS providers targeting different software domains. Sauce Lab Selenium Testing is an example of TaaS which provides a cloud-based, cross browser testing of web application using Selenium. For testing, multiple virtual machines (VM) are set up with different operating systems and/or browsers to test the web application in various environments. User interaction within the web application is then simulated and the result of the interaction verified using Selenium. The result is then collected and the VMs are destroyed to ensure security of data. Zephyr, on the other hand, is a real-time test management platform for managing testing life cycles during software development. Using Zephyr's web interface, users are able to create, manage and plan tests to be performed. Tests are then executed in the Zephyr cloud and the results of testing are made available through the metrics and dashboard on the web interface.

There are also a growing number of TaaS frameworks developed by academic researchers. One example is EPFL (Ecole Polytechnique Federale de Lausanne) cloud9, which provides cloud-based automated testing using parallel symbolic execution (Ciortea et al. 2010). While symbolic execution is an effective automated testing technique, it is very poorly scalable due to the need to analyze all possible paths of a program's execution tree. Cloud9 implements a symbolic execution engine that can scale to a large cluster of VM depending on the testing task in order to enable software testing in a short amount of time. Another example of TaaS framework developed by academic researchers is University of York YETI (York Extensible Testing Infrastructure), which offers a cloud-based automated random testing tool (Oriol and Ullah 2010). This is achieved by executing the standalone version of YETI on a cloud-based deployment of Hadoop, Apache's implementation of Map Reduce framework. During the mapping stage, tests are executed depending on the arguments read as input. The result of the tests are then aggregated during the reduce step and written to disk to be made accessible for users. By default, most TaaS tools above are designed to test for major system failures, such as a computer crash, rather than the correctness of the software output. In order to incorporate correctness testing to the TaaS tools, domain-specific knowledge is required for creating test cases. In the field of genomic medicine, it is vital for bioinformatics software to be reliable in terms of producing correct outputs.

The key advantage of TaaS model is cost-effective software testing. Software testing commonly requires obtaining and maintaining testing infrastructure that can be quite costly, especially for large testing tasks. With TaaS, developers are able to deploy resources on the cloud only when required for testing and they only need to pay for the resources used, thus eliminating the need to purchase and set up testing infrastructure (Riungu-Kalliosaari, Taipale, and Smolander 2012). Another key advantage of the TaaS model is the flexibility of the tests execution and the allocation of resources. With TaaS, developers are able to perform testing at anytime without the need to reserve computing resources. Furthermore, the scalable nature of cloud computing allows the variable provisioning of resources depending on the complexity of tasks at different stages of testing (Ciortea et al. 2010). Finally, TaaS provides a more user-friendly means of automating software testing compared to traditional testing framework. This is designed to encourage developers, in particular beginners, to adopt software testing during software development.

Alhough there are significant advantages for TaaS model, there are a number of challenges. The main technical challenges are providing a continuous testing service and ensuring that test results are delivered on time. This requires the TaaS providers to provision enough computing resources for the service to run continuously and be able to cope with the varied testing workload, without over-provisioning to avoid loss. Aside from technical challenges due to resource allocation, there are also other challenges such as data security and privacy. To perform effective testing, some test data are required from the user of TaaS, which will need to be stored securely in order to avoid confidential breach of data. This is especially important when dealing with human genomic data (Riungu-Kalliosaari, Taipale, and Smolander 2012).

There is currently a need for bioinformatics-specific TaaS framework that would test the reliability and correctness of bioinformatics software for use in genomic medicine. Such a system would allow users – software developers, bioinformatics users, or technicians from a clinical diagnostic laboratory – to easily test their bioinformatics pipeline in an environment which realistically simulates the specific hardware and software requirement in which the pipeline is supposed to operate in. This would encourage validation and quality control to be taken much more seriously by bioinformatics software developers, hence lead to the improvement of the bioinformatics software quality. This in turns will facilitate the widespread adoption of genomic medicine in a clinical setting.

8. Future Directions

Building capability in genomics and bioinformatics is the key to translating the benefit of "omic" technologies into clinical practice. As genetic testing based on next-generation sequencing technology is already being used in research and poised for clinical application in the next 2-3 years, research into quality assurance of bioinformatics pipelines used in genomic medicine is needed. In this chapter, we argue that the adoption of state-of-the-art analytical framework and methodologies from the field of software testing can provide a foundation for developing evidence-based, effective, and readily deployable strategies for validation and quality control of bioinformatics pipelines in genomic medicine. The field of bioinformatics software testing is still young, so there is ample room for future research and development. Specifically, we have identified important research directions in the field of bioinformatics that are currently lacking. These include:

1. The characterization of the failure-causing inputs of common variant calling pipelines, and the development of automated test case generators for systematic pipeline validation.
2. The development of real-time quality control strategies that can verify the correctness of real input under the operational environment.
3. The implementation of cloud-based validation and quality control modules that will enable widespread use of quality assurance strategies.

Overall, we believe that further research will lead to development of new evidence-based quality assurance guidelines and computational tools that will benefit whole exome and whole genome sequencing projects being carried out by medical research institutions, hospitals, and clinical laboratories worldwide. Such techniques and tools can also be applicable to a wide range of bioinformatics or health informatics problems, such as biomedical text and image mining, or clinical decision support systems. Therefore, such research directions can also benefit the entire biomedical community.

Acknowledgements

This work was supported in part by funds from the New South Wales State Government Ministry of Health, a New South Wales Genomics Collaborative Grant, and a Microsoft Azure Research Award.

References

Abyzov, A. *et al.* 2011. CN Vnator: An approach to discover, genotype, and characterize typical and atypical CNVs from family and population genome sequencing. *Genome Res.*, **21**: 974–84.

Adzhubei, I.A. *et al.* 2010. A method and server for predicting damaging missense mutations. *Nat. Methods*, **7**, 248–249.

Alden, K. and M. Read, 2013. Computing: Scientific software needs quality control. *Nature*, **502**, 448–48.

Ammann, P. and J. Offutt, 2008. Introduction to Software Testing 1st ed. Cambridge University Press, New York, NY, USA.

Andrews, J.H. *et al.* 2005. Is mutation an appropriate tool for testing experiments? In, *27th International Conference on Software Engineering, 2005. ICSE 2005. Proceedings.*, pp. 402–11.

Bergmann, F.T. and H.M. Sauro, 2008. Comparing simulation results of SBML capable simulators. *Bioinformatics*. **24**: 1963–65.

Boeva, V. *et al.* 2011. Control-FREEC: a tool for assessing copy number and allelic content using next generation sequencing data. *Bioinformatics*. btr670.

Candea, G. *et al.* 2010. Automated software testing as a service. In, *Proceedings of the 1st ACM symposium on Cloud computing*. ACM, pp. 155–60.

Chan, F.T. *et al.* 1996. Proportional sampling strategy: guidelines for software testing practitioners. *Inf. Softw. Technol.*, **38**, 775–82.

Check Hayden", E. to "Hayden, EC 2015. Rule rewrite aims to clean up scientific software. *Nature*, **520**: 276–77.

Chen, E.Y. *et al.* 2013. Enrichr: interactive and collaborative HTML5 gene list enrichment analysis tool. *BMC Bioinformatics*, **14**: 128.

Chen, T.Y. *et al.* 2004. Adaptive Random Testing. In, Maher, M.J. (ed), *Advances in Computer Science - ASIAN 2004. Higher-Level Decision Making*, Lecture Notes in Computer Science. Springer Berlin Heidelberg, pp. 320–29.

Chen, T.Y. *et al.* 2010. Adaptive Random Testing: the ART of Test Case Diversity. *J. Syst. Softw.*, **83**: 60–66.

Chen, T.Y. *et al.* 2009. An innovative approach for testing bioinformatics programs using metamorphic testing. *BMC Bioinformatics*, **10**: 24.

Chen, T.Y. *et al.* 2003 Fault-based testing without the need of oracles. *Inf. Softw. Technol.* **45**: 1–9.

Chen, T.Y. *et al.* 1998 Metamorphic testing: a new approach for generating next test cases. *Tech. Rep. HKUST-CS98-01 Dep. Comput. Sci. Hong Kong Univ. Sci. Technol. Hong Kong.*

Chen, T.Y. and R. Merkel, 2008 An Upper Bound on Software Testing Effectiveness. *ACM Trans Softw Eng Methodol*, **17**, 16:1–16:27.

Ciortea, L. *et al.* 2010 Cloud9: A software testing service. *ACM SIGOPS Oper. Syst. Rev.*, **43**, 5–10.

DePristo, M.A. *et al.* 2011 A framework for variation discovery and genotyping using next-generation DNA sequencing data. *Nat. Genet.*, **43**: 491–98.

Evans, T.W. *et al.* 2008 The SBML discrete stochastic models test suite. *Bioinformatics*, **24**, 285–286.

Fromer, M. and S.M. Purcell, 2014 Using XHMM Software to Detect Copy Number Variation in Whole-Exome Sequencing Data. In, *Current Protocols in Human Genetics*. John Wiley & Sons, Inc.

Gargis, A.S. *et al.* 2012 Assuring the quality of next-generation sequencing in clinical laboratory practice. *Nat. Biotechnol.* **30**, 1033–36.

Giannoulatou, E. *et al.* 2014 Verification and validation of bioinformatics software without a gold standard: a case study of BWA and Bowtie. *BMC Bioinformatics*, **15**: S15.

Handsaker, R.E. *et al.* 2011. Discovery and genotyping of genome structural polymorphism by sequencing on a population scale. *Nat. Genet.*, **43**: 269–276.

Hatton, L. and Roberts, A. 1994. How accurate is scientific software? *IEEE Trans. Softw. Eng.* **20**: 785–797.

Hayden, E.C. 2013. Mozilla plan seeks to debug scientific code. *Nature.* **501**: 472.

Huang, D.W. *et al.* 2008. Systematic and integrative analysis of large gene lists using DAVID bioinformatics resources. *Nat. Protoc.* **4**, 44–57.

Joppa, L.N. *et al.* 2013 Troubling Trends in Scientific Software Use. *Science.* **340**, 814–15.

Kanewala, U. and J.M. Bieman, 2014. Testing scientific software: A systematic literature review. *Inf. Softw. Technol.* **56**: 1219–32.

Kircher, M. *et al.* 2014. A general framework for estimating the relative pathogenicity of human genetic variants. *Nat. Genet.* **46**, 310–15.

Klambauer, G. *et al.* 2012. cn.MOPS: mixture of Poissons for discovering copy number variations in next-generation sequencing data with a low false discovery rate. *Nucleic Acids Res.* gks003.

Knight, J.C. and N.G. Leveson, 1986. An Experimental Evaluation Of The Assumption Of Independence In Multi-Version Programming. *IEEE Trans. Softw. Eng.* **12**, 96–109.

Koboldt, D.C. *et al.* (2009) VarScan: variant detection in massively parallel sequencing of individual and pooled samples. *Bioinformatics*, **25**: 2283–85.

Langmead, B. *et al.* 2009 Ultrafast and memory-efficient alignment of short DNA sequences to the human genome. *Genome Biol.* **10**: R25.

Langmead, B. and S.L. Salzberg, 2012. Fast gapped-read alignment with Bowtie 2. *Nat. Methods.* **9**: 357–59.

Li, H. and Durbin, R. (2009) Fast and accurate short read alignment with Burrows–Wheeler transform. *Bioinformatics*, **25**, 1754–60.

Manichaikul, A. *et al.* 2010. Robust relationship inference in genome-wide association studies. *Bioinformatics*, **26**, 2867–873.

Marbach, D. *et al.* 2012. Wisdom of crowds for robust gene network inference. *Nat. Methods.* **9**: 796–804.

Ma,Y.-S. *et al.* 2005 MuJava: an automated class mutation system. *Softw. Test. Verification Reliab.*, **15**: 97–133.

McCarthy, D.J. *et al.* 2014. Choice of transcripts and software has a large effect on variant annotation. *Genome Med.*, **6**: 1–16.

McKenna, A. *et al.* 2010. The Genome Analysis Toolkit: A MapReduce framework for analyzing next-generation DNA sequencing data. *Genome Res.*, **20**: 1297–1303.

McLean, C.Y. *et al.* 2010. GREAT improves functional interpretation of cis-regulatory regions. *Nat. Biotechnol.*, **28**: 495–501.

Merali, Z. 2010. Computational science: ...Error. *Nat. News*, **467**: 775–777.

Neale, B.M. *et al.* 2011. Testing for an Unusual Distribution of Rare Variants. *PLoS Genet*, **7**, e1001322.

O'Rawe, J. *et al.* 2013. Low concordance of multiple variant-calling pipelines: practical implications for exome and genome sequencing. *Genome Med.*, **5**: 28.

Oriol, M. and Ullah, F. 2010. YETI on the Cloud. In, *2010 Third International Conference on Software Testing, Verification, and Validation Workshops (ICSTW).*, pp. 434–37.

Plagnol, V. *et al.* 2012. A robust model for read count data in exome sequencing experiments and implications for copy number variant calling. *Bioinformatics*, **28**: 2747–754.

Platinum Genomes.

Price, A.L. *et al.* (2010) Pooled Association Tests for Rare Variants in Exon-Resequencing Studies. *Am. J. Hum. Genet.*, **86**: 832–838.

Price, A.L. *et al.* (2006) Principal components analysis corrects for stratification in genome-wide association studies. *Nat. Genet.*, **38**: 904–909.

Ramler, R. and Kaspar, T. 2012. Applicability and benefits of mutation analysis as an aid for unit testing. In, *2012 7th International Conference on Computing and Convergence Technology (ICCCT).*, pp. 920–25.

Rimmer, A. *et al.* 2014. Integrating mapping-, assembly- and haplotype-based approaches for calling variants in clinical sequencing applications. *Nat. Genet.*, **46**: 912–18.

Ritchie, G.R.S. *et al.* 2014. Functional annotation of noncoding sequence variants. *Nat. Methods*, **11**: 294–96.

Riungu-Kalliosaari, L. *et al.* 2012. Testing in the Cloud: Exploring the Practice. *IEEE Softw.*, **29**: 46–51.

Sadi, M.S. *et al.* 2011. Verification of phylogenetic inference programs using metamorphic testing. *J. Bioinform. Comput. Biol.*, **9**: 729–47.

Samocha, K.E. *et al.* 2014. A framework for the interpretation of de novo mutation in human disease. *Nat. Genet.*, **46**: 944–50.

Schwarz, J.M. *et al.* 2014. MutationTaster2: mutation prediction for the deep-sequencing age. *Nat. Methods*, **11**: 361–62.

"The 1000 Genomes Consortium 1000 G.P. 2012 An integrated map of genetic variation from 1, 092 human genomes. *Nature*, **491**, 56–65.

Teer, J.K. *et al.* 2012. Var Sifter: Visualizing and analyzing exome-scale sequence variation data on a desktop computer. *Bioinformatics*, **28**: 599–600.

van der Aalst, L, 2009. Software testing as a service (staas)." *Sogeti Whitepaper. Available at* www.sogeti.com/staas. Last Accessed March 20120

Wang, K. *et al.* 2010. ANNOVAR: functional annotation of genetic variants from high-throughput sequencing data. *Nucleic Acids Res.*, **38**: e164–e164.

Weyuker, E.J. 1982. On Testing Non-Testable Programs. *Comput. J.*, **25**: 465–70.

Woodward, M.R. and K. Halewood, 1988. From weak to strong, dead or alive? an analysis of some mutation testing issues. In, , *Proceedings of the Second Workshop on Software Testing, Verification, and Analysis, 1988.*, pp. 152–58.

Xie, X. *et al.* 2011. Testing and Validating Machine Learning Classifiers by Metamorphic Testing. *J. Syst. Softw.*, **84**, 544–58.

Zhou, Z.Q. *et al.* 2004. Metamorphic Testing and its Applications. In, Proceedings of the 8th International Symposium on Future Software Technology (ISFST 2004). Software Engineers Association.

Zook, J.M. *et al.* 2014. Integrating human sequence data sets provides a resource of benchmark SNP and indel genotype calls. *Nat. Biotechnol.*, **32**: 246–51.

12

Recent Computational Trends in Biological Sequence Alignment

Mohamed Issa[1]*

1. Introduction

Sequence alignment between amino acids or nucleotides sequences represents the evolutionary history of the sequences. Also, it helps in understanding the function and information of biological sequences. Therefore, biological sequence alignment is one of the most important problems in bioinformatics due to it helps in finding the function of a newly discovering biological sequence, evolutionary relation between genes and predicting the structure and function of proteins. It is based on optimizing the number of matches between residues of sequences occurring in the same order in each sequence. There are two main classes of sequence alignment:

1. *Pairwise sequence alignments* which is used for aligning two biological sequences only.
2. *Multiple sequence alignments* involves the alignment of more than two biological sequences

This chapter will focus on pairwise sequence alignment algorithms. In addition, the development of it from the point of view of accelerating its execution using hardware accelerators like *Graphical Processing Unit and Field Programmable Gate Array.* Also, this chapter includes an application called Gene Tracer which is based on pairwise sequence alignment algorithms. Gene Tracer is used to trace the common subsequences of ancestors in the offspring. The chapter is organized as follows: The types of pairwise alignments algorithms are presented in section 2. In section 3, an overview on accelerating sequence alignment algorithms on hardware accelerators such as Field Programmable Array Gate, Graphical Processing Unit and Multi-Core. In section 4 Gene Tracer application will be explored and its acceleration on Graphical Processing Unit and Multi-Core architecture. Finally, in the last section, we present the conclusion of this chapter.

[1.] Computer and Systems Department, Faculty of Engineering, Zagazig University, Egypt
* Corresponding author: mohmed.issa@gmail.com

2.   Pairwise Sequence Alignment

Pairwise sequence alignment is divided into two approaches: (1) Dot Plots approach, (2) Computation scoring approaches. Computation scoring algorithms like Global pairwise   sequence alignment, Local pairwise sequence alignment, Semi-Global pairwise sequence alignment and other heuristic algorithms. In the following each kind will be discussed.

2.1   Dot Plots Approach

Dot plot is the simplest method to measure the similarity between genetic sequences. It represents the region of similarity using dot plots. The method is done by constructing a two dimension dot plot, the first sequence is assigned to the horizontal axis and the second sequence assigned to vertical axis. The nucleotides of horizontal axis at positions (1 to w), where w is the window size, are compared to the nucleotides of vertical axis at (1 to w). If the number of identical nucleotides more than certain cut off score a dot is plotted at position (1 , 1). Then the nucleotides at position (2 to w+1) in the horizontal axis is compared again with that at positions (1 to w) in the vertical axis. The process is repeated until each window size (w) in the horizontal compared to each window in the vertical. The similar region is indicated by a diagonal of dots as shown in Fig. 1. and its length represents the grade of similarity. The window size and the cutoff score can be both varied according the similarity of the two sequences being compared.

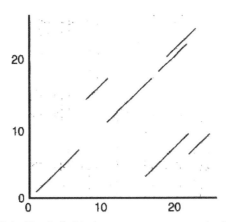

FIGURE 1   The similarities between two sequences using Dot plots.

2.2   Sequence Scoring Approach

A sequence alignment is the operation of pairwise matching between nucleotides in DNA or a mino acids in Proteins. The alignment is used as a measurement for the evolutionary relationships (sharing common ancestors) between biological sequences. Three movements control the alignment of correspondence residues in the two sequences: (1) a mutation that replace one residue with another. (2) an insertion of one or more positions, (3) deletion of one or more positions [Dayhoff *et al* .1987].

In nature, mutation has been oc.curred at higher frequency than insertion and deletion. Gaps are added in many positions in the two alignment sequences to adjust the correspondence position of similar residues. Besides, Gaps are used to increase the number of possible alignments between two or more sequences as shown in Fig. 2.

AATCTATA	AATCTATA	AATCTATA
AAG-AT-A	AA-G-ATA	AA--GATA

FIGURE 2   Three possible gapped alignment for two DNA sequences.

As seen in the dot plot approach, it gives only the visual inspection of the alignment's length or positions. So, a scored alignment was developed to reflect a value of the homology and evolutionary divergent. A simple scoring alignment including gap penalties can be computed as follows:

$$\sum_{i=1}^{n} \left\{ \begin{array}{l} gap \ penalty \ if \ seq1_i = \text{'} - \text{'} or \ seq2_i = \text{'} - \text{'} \\ match \ score; \ if \ no \ gaps \ and \ seq1_i = seq2_i \\ mismatch \ score; \ it \ no \ gaps \ and \ seq1_i \neq seq2_i \end{array} \right\}$$

For example, for the three different alignment in Fig. 2, if we assumed the match score is +1, mismatch score is 0 and gap is -1. So the alignment score from left to right would be 1, 3, 3. In nature there is a substitution score between nucleotides in DNA or amino acids in Proteins. These scores called substitution matrices and determined based on residue hydro-phobicity, charge, electro negativity and size. The common usage scoring matrices are PAM (*Percent Accepted Mutations*) and BLOSUM (*BLOcks SUbstitution Matrix*).

Pairwise alignment algorithm can be computed using two essential approaches, dynamic programming approach and heuristic methods. Dynamic programming consumes long time of execution but gives highly accurate alignment. In contrast, heuristic methods may have small execution time but don't guarantee accurate alignment.

Dynamic programming approach is a programming method like the divide-and-conquer approach, where the problem is solved by dividing it into sub problems and the optimal solution may be obtained [Cormen et al.]. There are three types of pairwise sequence alignment algorithms based on dynamic programming: Global pairwise sequence alignment, Local pairwise sequence alignment and Semi-Global pairwise sequence alignment.

2.2.1   *Global pairwise sequence alignment*

A Global pairwise sequence alignment involves the alignment of the entire of two sequences to finds the similar and different portions of the two sequences. The Needleman and Wunsch sequence alignment algorithm [Needleman and Wunsch

1970] is the essential dynamic programming sequence alignment method for computing the global alignment. This algorithm works as follow for aligning two sequences *S1 and S2* globally with lengths *n and m* respectively:

(a) **Alignment Computing**: a scoring matrix **M** of size (m+1)*(n+1) is constructed and initialized using a substitution matrix, such as PAM [Dayhoff et al 1987], BLOSUM [Henikoff 1992] for proteins or specified substitution matrix for DNA. Line by line scores are computed according to Eqn.1 starting from the left upper cell to the right lower cell [Elloumi and Zomaya 2011] .

$$M[i, f] = \max \left\{ \begin{array}{l} se(i, j) + M\,(i-1, j-1), \\ M\,[i-1, j] + p, \\ M\,[i, j-1] + p \end{array} \right\} \qquad (1)$$

Where *P* is a constant gap penalty, *se* is the divergent evolutionary score between the resduies or nucleotides at position *i* in S_1 and the one at position *j* in S_2 and is substituted from standard substitution matrices like PAM and BLOSUM.

(b) **Finding the Alignment:** this process aims to trace back the scoring matrix by building a path was called maximum score path such as in Fig. 3, which gives an optimal global pairwise sequence alignment. It starts from the lowest right cell to the upper left cell and three types of possible movements are allowed:

- *Diagonal movement*: This movement corresponds to the passage from a cell (*i,j*) to a cell (*i-1,j-1*).
- *Vertical movement*: This movement corresponds to the passage from a cell (*i,j*) to a cell (*i-1,j*).
- *Horizontal movement*: This movement corresponds to the passage from a cell (*i,j*) to a cell (*i,j-1*).

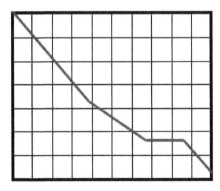

FIGURE 3   Back-Tracing Alignment scores' matrix to find the optimal alignment.

***Needleman_wunch** Alignment algorithm with constant gap*

Input :

Seq1	:	*sequence // 1st sequence*
Seq2	:	*sequence // 2nd sequence*
M	:	*length of Seq1*
N	:	*length of Seq2*
g	:	*constant gap cost*
S	:	*match score*
NS	:	*mismatch score*

Output :

Seq1_Align	:	*sequence //Seq1 Aligned*
Seq2_Align	:	*sequence //Seq2 Aligned*

Variable :

north :	*contain value of upper cell of current cell*
north_west :	*contain value of upper left cell of current cell*
west :	*contain value of left cell of current cell*
H :	*score matrix length of (M+1)(N+1)*

Begin
// Forward Trace :

```
Sub_mat ( Seq1 [i] , Seq2 [j] )
      {
      If ( Se[i] = = Seq2 [j] ) return  S
      If (Seq1 [i]  ! = Seq2 [j] ) return  NS
      }
For i  : = 0 to M do H[0,i] : = − i * g   end
For j := 1 to N do H[ j,0] : = − j * g   end
         For i  := 1 to M do
           For j  :=  1 to N do
              north = H[i−1,j] − g
              west = H[i,j−1] − g
              north_west = H [i−1,j−1] +  Sub_mat ( Seq1 [i] , Seq2 [i] )
                 H [i,j] :=  max ( north  , west , north_west )
           end
         end
```

// Backward Trace :

```
    i = ( M +1 ) , j = ( N +1 ) // Start from the lowest right cell.
    K = 0
    Align  ( ( M + 1 ) , ( N + 1 ) , k ) {
        If ( i = 0 , j = 0 )
        Return
        If ( H [ i−1 , j ] − g = H [ i , j ] )
        {
        Seq1_Align [k] = Seq1 [i] . Seq2_Align [k] = '_'
        K ++ ,  Align ( i−1 , j , k )
        }
```

If (H [i , j-1] – g = H [i , j])
{
Seq1_Align [k] = '_' , Seq2_Align [k] = Seq2 [i]
    K ++ , Align (i , j-1 , k)
}
If (H [i-1 , j-1] + score (Seq1[i] , Seq2[i]) = H [i,j])
{
    Seq1_Align [k] = Seq1 [i] , Seq2_Align [k] =Seq2 [i]
    K ++ , Align (i-1 , j-1 , k)
}
}
    Reverse Seq1_Align , Seq2_Align
    Return Seq1_Align , Seq2_Align

End

Time complexity of the algorithm of Needleman and Wunsch is $O(m*n)$ and space complexity is $O(m*n)$ where m, n are lengths of the two sequences. An example of global alignment algorithm as shown in the following:

Input : Seq1 = ATAT   ,   Seq2 = TATA

For S = 1 , NS = -1, g = 1

Scoring matrix (H) :

		T	A	T	A
	0	-1	-2	-3	-4
A	-1	-1	0	-1	2
T	-2	0	-1	1	0
A	-3	-1	1	0	2
T	-4	-2	0	2	1

The bold cells are the optimal sequence alignment path starting from the lower right cell and ending at the upper left cell. The first row and column are not included in matrix but to clarify the algorithm.

Output: Seq1_Align =   T   A   T   A   _
              Seq2_Align =   _   A   T   A   T

The common application of global alignment is comparing two genes with almost the same function, for example, comparison of human's gene versus mouse's gene.

2.2.2 Pairwise Local Sequence Alignment

A Pairwise Local Alignment is used to find the similarities between biological sequences. It involves the alignment of portions of two sequences as shown in Fig. 4, however global sequence alignment aligning the entire of two biological sequences.

Smith-Waterman [Smith and Waterman 1981] developed an algorithm for finding local sequence alignment between two biological sequences.

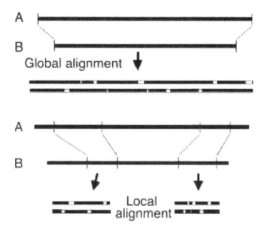

FIGURE 4   Global alignment versus Local alignment.

Smith's algorithm compute alignment the same as global alignment using two steps: computing the scores and backtracing the matrix to find the alignment portions. For computing alignment scores, the algorithm used Equation 2 [Smith and Waterman 1981] and used score zero to avoid negative scores since it aimed to find the alignment of portions not entire sequence.

$$M[i, f] = \max \left\{ \begin{array}{c} se(i, j) + M\,(i - 1, j - 1], \\ M\,[i - 1, j] + p, \\ M\,[i, j - 1] + p \\ 0 \end{array} \right\} \quad (2)$$

The alignment of portions is found by tracing back the matrix but starting from the cell contains maximum score to guarantee finding the optimal local sequence alignment. It resumes tracing back the matrix until find zero and stop.

Smith-Waterman Alignment algorithm with constant gap

Input :

Seq1	:	*sequence // 1st sequence*
Seq2	:	*sequence // 2nd sequence*
M	:	*length of Seq1*
N	:	*length of Seq2*
g	:	*constant gap cost*
S	:	*match score*
NS	:	*mismatch score*

Output :

Seq1_Align :	*sequence //Seq1 Aligned*
Seq2_Align :	*sequence //Seq2 Aligned*

Variable :
           north :   *contain value of upper cell of current cell*
           north_west :   *contain value of upper left cell of current cell*
           west :   *contain value of left cell of current cell*
           H :   *score matrix length of (M+1)(N+1)*
           Maxscore: *maximum score of H matrix*

Begin

// Forward Trace :

Sub_mat (Seq1 [i] , Seq2 [j])
           {
           If (Se[i] = = Seq2 [j]) return S
           If (Seq1 [i] ! = Seq2 [j]) return NS
           }
For i : = 0 to M do H [0 , i] : = 0   end
For j := 1 to N do H [j , 0] : = 0   end
Maxscore = 0;
           For i := 1 to M do
              For j := 1 to N do
              north = H[i−1 , j] − g
              west = H[i , j−1] − g
              north_west = H [i−1 , j-1] + Sub_mat (Seq1 [i] , Seq2 [i])
                 H [i , j] := max (north , west , north_west , 0)
                    If (H [i , j] > Maxscore)
                          {
                                      Maxscore = H[i , j]
                                      Pos_i = i , Pos_j = j
                          }
              end
           end

// Backward Trace :
K = 0 , i = Pos_i , j = Pos_i
Align ((M + 1) , (N + 1) , k) {
           If (i = 0 , j = 0)
                 Return
      If (H [i−1 , j] − g = H [i , j])
      {
                 Seq1_Align [k] = Seq1 [i] , Seq2_Align [k] = '_'
                 K ++ , Align (i-1 , j , k)
      }
      If (H [i , j-1] − g = H [i , j])
      {
                 Seq1_Align [k] = '_' , Seq2_Align [k] = Seq2 [i]
                 K ++ , Align (i , j-1 , k)
      }

If (H [i-1 , j-1] + score (Seq1[i] , Seq2[i]) = H [i,j])

   {

        Seq1_Align [k] = Seq1 [i] , Seq2_Align [k] =Seq2 [i]
        K ++ , Align (i-1 , j-1 , k)

   }

}

   Reverse Seq1_Align , Seq2_Align
   Return Seq1_Align , Seq2_Align

End

Time and space complexities of the algorithm of Smith and Waterman [Smith and Waterman 1981] is $O(m*n)$.

An example of pairwise local alignment is as shown in the following:

Input :   Seq1 = GACGG,   Seq2 = ACGA

For S = 1 , NS = -1 , g = 1

Scoring matrix H :

		G	A	C	G	G
	0	0	0	0	0	0
A	0	0	1	0	0	0
C	0	0	0	2	1	0
G	0	1	0	1	3	2
A	0	0	2	1	2	2

Starting from the maximum score and ending at the cell contain the first 0 in the optimal alignment path. The maximum score express the length of common sub sequence.

Output :

        Seq1_Align =   A   C   G

        Seq2_Align =   A   C   G

The most commonly usage of local alignment is searching for local similarities in large sequences (e.g., newly sequenced genomes).

2.2.3 *Pairwise Semi-Global Sequence Alignment*

A *pairwise semi-global alignment* is used for searching about the short sequence in a huge genomes. It don't penalize the gaps at the start or at the end of the two sequences. It is like global alignment but the difference is in back-trace. It starts from the maximum score value in the last row or column depends on which sequence is assumed to be prefix of the other. The main application of using semi-global sequence alignment is DNA fragment assembly [S.Henikoff and J.G. Henikoff 1992].

Semi-Global Sequence Alignment algorithm with constant gap

Input :

Seq1	:	sequence // 1st sequence
Seq2	:	sequence // 2nd sequence
M	:	length of Seq1
N	:	length of Seq2
g	:	constant gap cost
S	:	match score
NS	:	mismatch score

Output :

Seq1_Align	:	sequence //Seq1 Aligned
Seq2_Align	:	sequence //Seq2 Aligned

Variable :

north :	contain value of upper cell of current cell
north_west :	contain value of upper left cell of current cell
west :	contain value of left cell of current cell
H :	score matrix length of (M+1)(N+1)

Begin

// Forward Trace :

```
Sub_mat ( Seq1 [i] , Seq2 [j] )
        {
        If ( Se[i] = = Seq2 [j] ) return  S
        If (Seq1 [i]  ! = Seq2 [j]  ) return  NS
        }
For i  : = 0 to M do H[0,i] : = − i * g   end
For j := 1 to N do H[ j,0] : = − j * g   end
          For i  := 1 to M do
            For j  :=  1 to N do
              north = H[i−1,j] − g
              west = H[i,j−1] − g
              north_west =  H [i−1,j-1] +  Sub_mat ( Seq1 [i] , Seq2 [i] )
                H [i,j] :=  max ( north , west , north_west )
              end
          end
Maxscore = H[0,N]
For i := 1 to M
      If (H[i , N] > Maxscore)
        {
        Maxscore = H[i , j]
        Pos_i  = i , Pos_j = j
        }
 // Backward Trace :
K = 0 , i = Pos_i   , j = Pos_i
    Align ( ( M + 1 ) , ( N + 1 ) , k )  {
```

> *If (i = 0 , j = 0)*
> *Return*
>> *If (H [i−1 , j] − g = H [i , j])*
>> *{*
>> *Seq1_Align [k] = Seq1 [i] , Seq2_Align [k] = ' _ '*
>> *K ++ , Align (i-1 , j , k)*
>> *}*
>> *If (H [i , j-1] − g = H [i , j])*
>> *{*
>> *Seq1_Align [k] = ' _ ' , Seq2_Align [k] = Seq2 [i]*
>> *K ++ , Align (i , j-1 , k)*
>> *}*
>> *If (H [i-1 , j-1] + score (Seq1[i] , Seq2[i]) = H [i,j])*
>> *{*
>> *Seq1_Align [k] = Seq1 [i] , Seq2_Align [k] =Seq2 [i]*
>> *K ++ , Align (i-1 , j-1 , k)*
>> *}*
> *}*
>
> *Reverse Seq1_Align , Seq2_Align*
> *Return Seq1_Align , Seq2_Align*

End

Time complexity of Semi – Global alignment algorithm and also space complexity is $O(m*n)$. An example of Semi - Global alignment algorithm is as shown in the following:

Input :   Seq1 = CA   ,   Seq2 = GACAAG

For S = 1 , NS = -1 , g = 1,

Scoring matrix H is shown in the right.

		C	A
	0	0	0
G	0	-1	-1
A	0	-1	0
C	0	1	0
A	0	0	2
A	0	1	1
G	0	0	0

Output :

Seq1_Align =	G	A	C	A	A	G
Seq2_Align =		–	C	A	–	–

2.2.4   Heuristic methods

Heuristic approaches are used to speed up the alignment operation. It is very helpful in searching and scanning biological sequence database. For example, for finding the most similar protein from a database contains around 13 million proteins to another protein it will consume huge hours using dynamic programming sequence alignment approaches. Instead, heuristic methods will accelerate the operation. One of the most common tools for searching biological database based on heuristic approach is the BLAST (Basic Local Alignment Search Tool) [Pearson and Lipman 88]. BLAST becomes most common tool due to its efficient search result and it was developed to work with parallel architecture like Multi-Core and Graphical Processing Units. The algorithm is straightforward, it finding the similar sequences by finding sub-sequences from the database that are similar to sub-sequences in the query sequence. It starts by breaking down the query sequence into words (segments pair) of a fixed length (for example, 4 is the default value). All the possible words are computed by sliding a window has the same length of words on the query sequence. For example, the sequence AILVPTV would divided into different four words (AILV, ILVP, LVPT, VPTV). Then, the search process is occurred for each word by aligning it with each sequence and extend the alignment until the alignment' score pass certain threshold. The step of choosing the value of threshold is an important parameter due to it determine how likely the resulting sequences are to be biologically relevant homologs of the query sequence. There are a lot of sequence alignment and database search tools are developed for various specific types of sequence searches. For example, BLASTP searches protein databases, BLASTN allowing for searching nucleotides sequence databases and BLASTX allow translating from nucleotides sequence to proteins sequences prior to searching.

Another commonly used family of alignment and search tools is FASTA. The same as BLAST, it divides the sequence into words (4 - 6 for nucleotides and 1 - 2 for proteins). It constructs a table show positions of each word in the query sequence. For example, for a query protein FAMLGFIKYLPGCM, Table 1. (A) shows the position of each residue (assuming the word size is 1). For a targeted protein sequence TGFIKYLPGACT in the database, we construct another table as shown in Table 1. (B) it shows position of each residue in it with the distance or offset with the query sequence. Where the offset computed by subtracting the position of each residue in the targeted protein from the position of each residue in the query. If the residue in the target protein does not exist in the query protein it has no offset. The best position of the alignment is found by notice the offset that repeated many time, in this case is position 3. So, the alignment operation between these two proteins are as follows:

FAMLGFIKYLPGCM

TGFIKYLPGACT

TABLE 1(A): Positions of each residue in the query protein sequence

Word	A	C	D	E	F	G	H	I	K	L	M	N	P	Q	R	S	T	V	W	Y
Position	2	13			1	5		7	8	4	3		11							9
					6	12				10	14									

TABLE 1(B): Positions of each residue in the target protein sequence in the database.

1	2	3	4	5	6	7	8	9	10	11	12
T	G	F	I	K	Y	L	P	G	A	C	T
	3	-2	3	3	3	-3	3	-4	-8	2	
	10	3				3		3			

These heuristic methods constrain the alignment to a known region of similar sequence, so it is faster than performing a complete dynamic programming alignment between the sequence and all possible targets in the database.

3. Acceleration of Sequence Alignment Algorithms:

Heuristic alignment methods are faster than dynamic programming alignment approaches due to quadratic time complexity of dynamic programming methods. But in contrast dynamic programming methods more accurate than the heuristic methods. So, acceleration the sequence alignment dynamic programming approaches is the solution to overcome the slow speed with dynamic programming methods especially with the huge grow of biological sequences databases. In computing, hardware acceleration is the use of customized hardware for speeding up the massive scale of computation needed for sequence alignment computations. These hardware platforms like Field Programmable Gate Array and Graphical Processing Unit which are separates from Central Processing Units or Multi-Core which is done on Central Processing Units. This part will introduce the efforts of speeding up the pairwise sequence alignment using such hardware platforms. For dynamic programming alignment methods, the algorithm spends most of its execution time for computing the scoring matrix. These matrix' cells may be computed parallel by compute it diagonal by diagonal like Fig. 5.

Field Programmable Logic Array devices are re-configurable data processing devices where the algorithm is mapped directly to the processing logic nodes, like NAND logic gates. To get the advantage of using Field Programmable Logic Array, the algorithm must be implemented massively parallel on this re-configurable device. So, it is well suited for speeding up sequence alignment algorithms. [Shaw *et al.* 2006] explore the advantage of using Field Programmable Logic Array for speeding up the sequence alignment by implemented it software purely. Besides, they replaced the intensive computation section with an Field Programmable Logic Array custom instruction. They found the processing run time of using Field Programmable Logic Array implementation is 287 % speed up the purely software implementation. In [Maruyama *et al.* 2002] an approach was proposed to achieve the high speed for implementing pairwise sequence alignment algorithm using runtime reconfigurable. It demonstrates that using off-the-shelf Field Programmable Logic Array boards the high performance of sequence alignment can be realized. They compared their approach with ordinary implementation of Smith-Waterman local alignment and they achieved a 300 time speed up for aligning a sequence contains 2048 elements with a database contains around 6 million sequences. Another acceleration method for speeding up sequence alignment algorithms is Systolic Arrays. It is an arrangement of processors in the form of matrix where data flow to each processor from the north

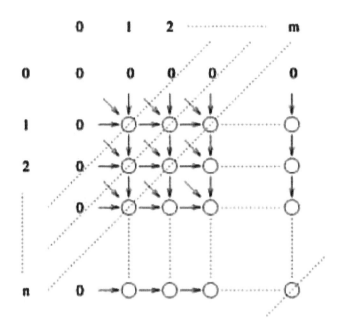

FIGURE 5   Parallelizing execution of alignment scoring matrix.

and west neighbors. After processing it the output is flowing to the south and east neighbors. In [Pedreira *et al.* 2004] the authors feed the systolic arrays with multiple data and writing several nucotides in a single bus write-cycle. In [Kreft *et al.* 2005], a concept for accelerating the sequence alignment algorithm using systolic arrays was achieved. The reasons of using this architecture is the simplicity and efficiency.

Another common parallel technique used for parallelizing the sequence alignment algorithm is Single Instruction Multiple Data (SIMD). It is a type of multiprocessor architecture where multiple sets of operands may be fetched to multiple processing units. Where, there are huge number of biological sequences but there are one operation needed on all sequences. So, the parallelization is done by devoting the cells in the same diagonal to a different processors. In [Borah et al 1994] an implementation of Smith–Waterman algorithm for local alignment is described using general purpose fine-grained architecture. They achieve a speed faster 5 times than the implementation of a sequence comparator on Field Programmable Logic Arrays [Daniel 1991]. The time complexity of the this method to align two biological sequences was O(MN). Where, M and N are the lengths of the two aligned sequences. So, to align K sequences, it would require O (MNK) steps. So, these massively parallel architecture may be used to solve computationally intensive problems in molecular biology efficiently and inexpensively. In [Farrar 2007] a parallel version of Smith-Waterman local alignment algorithm was developed on intel processor using SIMD technique. It achieved a speed up around 2-8 more than the earlier implementation.

Also, Kestrel parallel processor is another trial for speeding up the sequence alignment algorithm. It was a purpose processor was designed in University of California to be used on Human Genome Project and other biological applications using sequence analysis engine. It was able to efficiently analyze databases of billions of nucleotides for DNA and amino acids for Protein. In [Blas *et al.* 2005] the authors used this processor for implementing Smith Waterman algorithm on it for different query sizes [Blas *et al.* 2005]. They achieved speed up 287 times the implementation using Field Programmable Logic Arrays.

A Graphical Processing Unit is a device consisting of many multi-processors and a Dynamic Random Access Memory (DRAM). Each multiprocessor is coupled with a cache memory, large number of cores , Arithmatic Logic Unit (ALU) and control units. Mainly, Graphical Processing Units are used in the embedded systems, mobile phones, computers and game consoles. For example, in computer a Graphical Processing Unit can be found in the video adapter card or as an external unit in the motherboard such as in the notebook and new desktop computers. The reason behind evolution of Graphical Processing Unit is its powerful capabilities which is cleared in Fig. 5 and Fig. 6 which shows the Floating Point Operations Per Second (FLOPS) and memory bandwidth for Graphical Processing Unit versus CPU. Fig. 6 shows the powerful computation power of a Graphical Processing Unit over CPU in many versions such as Nvidia Graphical Processing Unit Single and Double Precision and Intel CPU Single and Double Precision. Besides, Fig. 7 shows the transfer bandwidth of the memory for many versions of Graphical Processing Unit which is faster versus CPU. So, Graphical Processing Units are well suited to solve many problems with data parallel processing. Single Instruction Multiple Data is the suitable parallel model to be implemented on Graphical Processing Unit. Where, the same program executed on different data in different cores. Each core is responsible for certain data, the portion that map this data into the core called thread. So, the program is executed in parallel

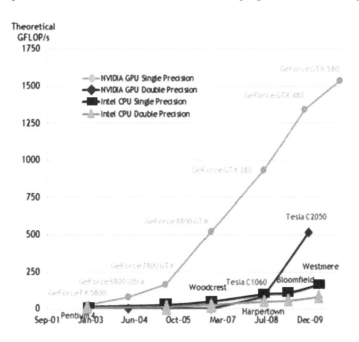

FIGURE 6   FLOPS for CPU and Graphical Processing Unit [Nvidia].

Theoretical GB/s

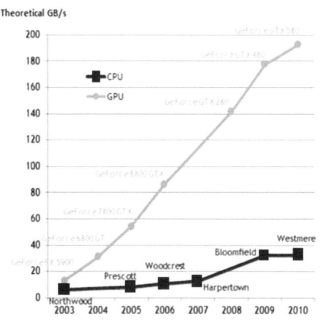

FIGURE 7   Memory bandwidth for CPU and Graphical Processing Unit [Nvidia].

on Graphical Processing Unit as many separate threads do the same operation on different data on separate cores at the same time. This way is efficient for accelerating many type of algorithms. Mainly Graphical Processing Units are designed for speeding up computer graphics applications. The high computational capabilities of Graphical Processing Units and their parallel structure allow speeding up of many algorithms in many fields such as scientific computing [Krüger and Westermann 03], computational geometry [Agarwal *et al.* 03] and bioinformatics [Charalambous *et al.* 05].

Thanks to the new sequencing technologies the number of biological sequences in databases such as GenBank [GenBank] and PubMed [PubMed] is increasing exponetionally. Besides, the length of one sequence pass thousands of bases (nucleotides or amino acids). So, comparing a query sequence to all sequences in the database is expensive operation in computing time and memory space. Therefore, acceleration of the pairwise sequence alignment algorithm is a vital role. Most of the acceleration development were devoted to local alignment algorithm. The idea for accelerating global alignment and semi-global alignment algorithms is the same. So, in the following we will focus on acceleration development of local alignment algorithm.

Two common programming languages on Graphical Processing Unit are OpenGL [Shreiner *et al.* 2005] and CUDA parallel programming languages that developed by Nvidia company [Nvidia]. Hence, there are two versions of local alignment algorithm on Graphical Processing Unit, either by using OpenGL or CUDA programming languages.

For accelerating Smith-Waterman local alignment algorithm by using OpenGL on Graphical Processing Unit the first trial was developed in [Liu *et al.* 2006a, Liu *et a* 2006b]. They used OpenGL [Shreiner *et al.*2005] for programming on Graphical

Processing Unit. Their work as follow: The query sequence and biological sequences are copied into the memory of Graphical Processing Unit as textures [Owens *et al.* 2008]. The score matrix was computed as anti-diagonal which mean computing diagonal by diagonal and the cells in the same diagonal are computed at the same time in parallel. Each cell is expressed as pixel where drawing each pixel execute small program called pixel shader that is responsible for computing matrix' scores. Now the implementation of [Liu *et al.2006*] searched merged databases of *Swiss-Prot* and Universal Protein Resourse (UniProt) [UniProt], and give an optimum speed of 650 Mega Cell Updates Per Second (MCUPS). This developed version Graphical Processing Unit program is faster than CPU version by 75 times.

The Cell Updates Per Second (CUPS) is computed as follows :

$$CUPS = \frac{(query\ sequence\ length\ *\ database\ size)}{run\ time}$$

Two implementation versions of [Liu *et al.* 2006b], the first one is with trace back and the second without trace back. The two version were executed on *Geforce* 7800 GTX Graphical Processing Unit on a database contains 983 sequences only. The version without trace back gave an optimum speed of 241 MCUPS and is faster than the version with trace back by 178 times and 120 times compared to the CPU version. The reason that the version without trace back is faster than one with trace back is tracing the matrix back which increased the complexity by O (M+N) where M is the length of query sequence and N is the length of each sequence in the database.

For speeding up the local alignment algorithm on Graphical Processing Unit using CUDA, the first implementation is SW-CUDA program that was developed by Manavski [Manavski and Valle 2008]. The developed algorithm works as follows: the whole query sequence was aligned with every sequence in the database. Where there are number of threads as number of sequences in the database and each thread responsible for aligning the query with each sequence. SW-CUDA reach speed of 3.5 GCUPS on a unit has two *Geforce 8800 GTX* Graphical Processing Units. Besides, a comparisons between SW-CUDA with BLAST [Pearson and Lipman 1988] and SSEARCH [Pearson 1991] had been done on a computer had 3 GHZ *Intel Pentium IV* Processor. SW-CUDA was also compared to Single Instruction Multiple Data (SIMD) implementation [Farrar 2007]. All these tests showed that SW-CUDA is faster than all previous implementation by speed from 2 to 30 times. MUMmerGPU [**Schatz** *et al.2007*] is another developed program for computing alignment on the Graphical Processing Unit by using CUDA. It aligns group of small DNA query sequences with a large number of sequences stored as a suffix tree [Ukkonen 1995], [Elloumi et al 2012]. MUMmerGPU reached a speed over 10 - fold more than serial CPU version of the sequence alignment kernel [Owens *et al.* 2008]. CUDAlign [Sandes and de Melo 2010] is another developed parallel alignment program on a Graphical Processing Unit using CUDA. CUDAlign was tested by making the alignment of the human chromosome 21 and the chimpanzee chromosome 22. It spent 21 hours on a Graphical Processing Unit of kind *GeForce* GTX 280 and it reached an optimum performance of 20.375 GCUPS. In [Striemer and Akoglu 2009] is another implementation for speeding up local sequence alignment algorithm for scanning database. It reach to speed faster than SSEARCH by 23 times. Smith-Waterman [Smith and Waterman 1981] computation matrix are computed purely on Graphical Processing Unit in Striemcr's Implementation. It works in three stages, 1) load databases contains biological sequences to Graphical Processing Unit's global memory. 2) Each thread is responsible for computing the alignment score between a query sequence and each biological sequence in the database. 3) The resulted alignment scores returned to the CPU to get the highest alignment score. This implementation finds only the highest alignment score or the most similar sequence to the query one but not find the alignment.

In contrast of SW-CUDA, Striemer's implementation do not include usage of CPU in partial computation of local alignment algorithm but it used Graphical Processing Unit only in the computation. In addition, Striemer's implementation depended on using Graphical Processing Units constant memory to save query sequence and substitution matrix. The reason is the access time of constant memory is the shortest access time for Graphical Processing Unit's memories.

Striemer's local alignment implementation works as following: for each sequence in the database stored in global memory of Graphical Processing Unit was compared to the query sequence. The alignment score of aligning query and each sequence was computed using BLOSUM substitution matrix. The scores are transferred to the CPU to get the maximum score of it and determine the most similar sequence of the database to the query.

4. Gene Tracer Application

Gene Tracer [Issa et al 2012.b] is an application was developed to trace genes alterations from ancestors sequences through offspring sequence. It finds the related parts between offsprings and its ancestors. Mainly, Gene Tracer based on local sequence alignment to do its function. Fig. 8 shows the function of Gene Tracer, it determines the similar parts between offspring sequence and its two ancestors. Besides it locates the location of similar parts in offspring and two ancestors.

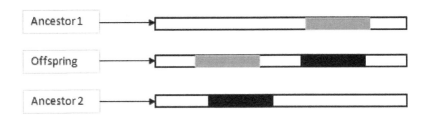

FIGURE 8   Determining location of similar parts using Gene Tracer.

The shaded parts with gray and black colors represent the similar subsequences. Where, the gray subsequence in Ancestor 1 is most similar to gray subsequence in offspring and the location is determined. Also , the contribution percentage of each Ancestor in offspring as division of subsequence length to the entire length of the offspring. The significance of the Gene Tracer application appeared for huge length sequences when the biologist want to determine common subsequences between Offspring and its known Ancestors. The developed Gene Tracer application based on local alignment algorithm as shown in the following.

Gene Tracer Application

Inputs

A_Seq1 : sequence // 1st Ancestor sequence
A_Seq2 : sequence // 2nd Ancestor sequence
Off_Seq : sequence // Offspring sequence

Outputs

Ancestor1 : sequence // *A_Seq1but common parts with Off_Seq in red* colored
Ancestor2 : sequence // *A_Seq2but common parts with Off_Seq in blue* colored
Off : sequence // *Off_Seq but common parts with A_Seq1 & A_Seq2 are //colored in red and blue*

Percent1, Percent2 : real // Percentages of common parts between ancestors   // and offspring length to an ancestor length (*A_Seq1* or *A_Seq2*)

Variables

L : integer // length of common part between Off_Seq & (A_Seq1 or A_Seq2)
i : integer // end position of common part in *A_Seq1* or *A_Seq2*
j : integer // end position of common part in *Off_Seq*
Match : integer // score of aligning two identical residues (characters)
NonMatch : integer // score of aligning two different residues
ConstGap : integer // score of aligning residue with gap.

Functions:

// *Local alignment between two sequences A_seq & Off_seq and determines //length and positions of common parts.*

Smith_Waterman (A_seq, Off_Seq, Match, NonMatch, ConstGap, L, i, j)

```
{
temp_score = 0
for ( k=0 ; k < Length  (Off_Seq) ) {
                    for ( z=0 ; z < Length (A_Seq) ) {
            north = H [z−1 , z]  −  ConstGap
                        west = H [k , k−1]  − ConstGap
if ( A_Seq [k] == Off_Seq [z] )
        {
                north_west =  H [ k−1 , z-1 ] +  Match
else
                north_west =  H [ k−1 , z-1 ] +  NonMatch
        }
        H[k,z] := max ( north , west, nort_west,0 )
SW_matrix[i][j] = H[k,z]; // assign SW computation matrix
                    }
If ( H > temp_score )
```

```
{
        temp_score = H ;
        temp_i = k ;
        temp_j = z ;
        }
}
L = temp_score
i= temp_i
j= temp_j
        }
```

Begin

// Step 1 : construct local alignment between *A_Seq1* and *Off_Seq*

Match = 1

NonMatch = 1

ConstGap = 1

   Smith_Waterman (A_Seq1, Off_Seq , Match , NonMatch , ConstGap, L, i, j)

Ancestor1:= Color_seq (A_Seq1, i-L , L)

Off := Color_Seq (Off_Seq , j-L , L)

Percent1 := L / length (A_Seq1)

// Step 2 : Construct local alignment between *A_Seq2* and *Off_Seq*

Smith_Waterman (A_Seq2, Off_Seq, Match, NonMatch, ConstGap, L, i, j)

Ancestor2 := Color_seq (A_Seq2 , i-L , L)

Off := Color_Seq (Off_Seq , j-L , L)

Percent2 := L / length (A_Seq2)

Return Ancestor1, Ancestor2, Percent1, Percent2

End

Gene Tracer application has time and space complexities of O (max (M,N)*P) where M,N and P are respectively length of Ancestor 1, Ancestor 2 and Offspring sequences. Gene Tracer application was implemented using PHP programming language on a computer has a 2.27 GHZ core i3, 4 GB Main memory. The application was tested on DNA and Protein short sequences and the result as follows:

   For DNA sequences the result as shown in Fig. 9, the following are short sequences used in the test.

Ancestor1: CGCCGGTCGCGGCTGCCCATGCAGG

Ancestor2: AGGCAGCGTGTCACGC

Offspring: CGCGGCAGGCA

For Protein sequences, the result is as shown in Fig. 10.

Ancestor 1 :

AKIKAYNLTVEGVEGFVRYSRVTKQHVAAFLKELRHSKQYEN VNLIHYIL

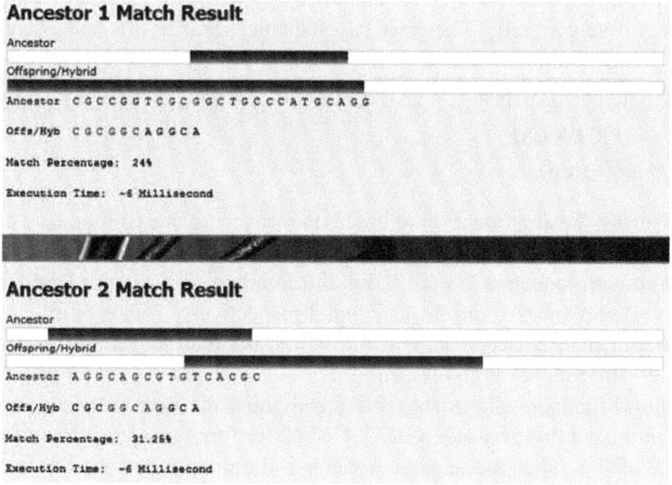

Ancestor 1 Match Result

Ancestor

Offspring/Hybrid

Ancestor C G C C G G T C G C G G C T G C C C A T G C A G G

Offs/Hyb C G C G G C A G G C A

Match Percentage: 24%

Execution Time: ~6 Millisecond

Ancestor 2 Match Result

Ancestor

Offspring/Hybrid

Ancestor A G G C A G C G T G T C A C G C

Offs/Hyb C G C G G C A G G C A

Match Percentage: 31.25%

Execution Time: ~6 Millisecond

FIGURE 9   Output of Gene Tracer for DNA sequences.

Ancestor 2: AERYCMRGVKNTAGELVSRVSSDADYNAMICPROG
RAMMINGAGGWCRKWYSAHRGPDQDAALGSFCIKNPGD

Offspring :
AGGWCRKWKQYENVNLIHYI

Ancestor 1 Match Result

Ancestor

Offspring/Hybrid

Ancestor A K I K A Y N L T V E G V E G F V R Y S R V T K Q N V A A F L K E L R H S K Q Y E N V N L I H Y I L

Offs/Hyb A G G W C R K W K Q Y E N V N L I H Y I

Match Percentage: 24%

Execution Time: ~10 Millisecond

Ancestor 2 Match Result

Ancestor

Offspring/Hybrid

Ancestor A E R Y C M R G V K N T A G E L V S R V S S D A D Y A G G W C R K W Y S A H R G P D Q D A A L G S F C
Ancestor I K N P C A A D

Offs/Hyb A G G W C R K W K Q Y E N V N L I H Y I

Match Percentage: 13.56%

Execution Time: ~10 Millisecond

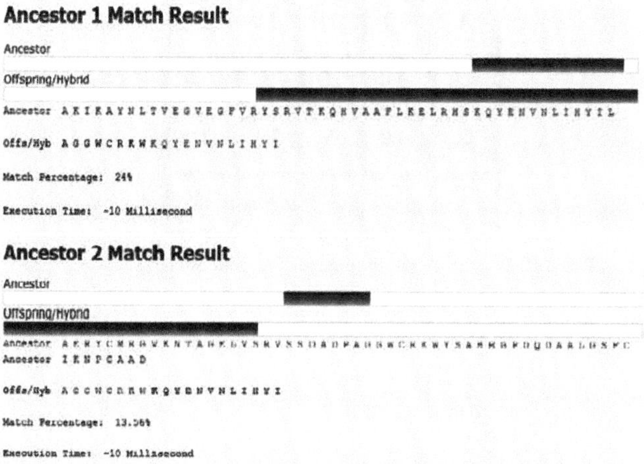

FIGURE 10   Output of Gene Tracer for Protein sequences.

In [Issa et al 2014.a], a development for Gene Tracer application was done to able to search biological database to find the most similar ancestors for unknown biological offspring sequence. The idea is to find the closest sequences biologically from a huge database based on measuring evolutionary divergent using substitution matrices. Then from those sequences find the two ones that have a longest common sub-sequences. In the following an example of why we cannot depend on measuring evolutionary divergent only. There are two sequences and an offspring sequence as follow:

Shuffling sequence (Offspring) = LMNCCH ,

Sequence 1 = **CCPKLM** ,

Sequence 2 = **LMNPA**

By measuring the alignment score based on measuring the evolutionary divergent using BLOSUM50 we found the score of alignment between offspring and sequence 1 was 26 and with sequence 2 was 19. But, the common subsequence length between offspring and sequences 1 and 2 was 2 and 3 respectively. This give indication that although offspring is closer to sequence 1 biologically than sequence 2, but sequence 2 has longer subsequence with offspring.

It computes the local alignment score for aligning the query sequence with each sequence in the database based on BLOSUM 62 for Protein sequences or the matrix in Fig. 11 for DNA. Then the sequences that has alignment score pass certain cut-off score will be realigned. again with the query sequence. But instead of using standard substitution matrices, it uses a scoring values +1 for match and -1 for mismatch. The reason of using such this values is to find the length of similar portions between the two sequences and location of each subsequence in each sequence (Offspring and Ancestors). The output of the modified application in log file as in Fig. 12, which called Tracer Format.

	A	**C**	**G**	**T**
A	10	-10	-10	5
C	-10	10	5	-10
G	-10	5	10	-10
T	5	-10	-10	10

FIGURE 11   DNA Evolutionary Divergent Scoring Matrix.

SEQ1
Q1,S1,L1
>sequence 1 name
SEQ2
Q2,S2,L2
>sequence 2 name
Sequence nucleotides or resudies
QUERY
Query nucleotides or resudies

FIGURE 12   Tracer Format.

Where:

Q1 :   position in query of the start of common substring between query and sequence 1.

S1 :   position in sequence 1 of start of common substring between query and sequence 1.

L1 :   the length of common substring between query and sequence 1.

Q2 :   position in query of the start of common substring between query and sequence 2.

S2 :   position in sequence 2 of start of common substring between query and sequence2.

L2 :   the length of common substring between query and sequence 2.

The modified method has time complexity $O(m*n*z)$ and space complexity is ($O(m*z)+ 3*O(z))$ where n is the length of query, m is the length of maximum sequence length in database and z is the number of sequences in the database. This application has disadvantage of slow execution time due to big time complexity. On the other hand the main advantages are it give the user the flexibility to use various substitution divergent matrices and is used of DNA or protein sequences. In addition it save the results in a log file.

The modification of Gene Tracer was implemented on Graphical Processing Unit and the test was done a *Swiss-Prot* biological sequences database contains 300000 Protein sequence and the speed up reach 140 times the implementation of the CPU.

Another development was done on Gene Tracer to speed up its execution on Graphical Processing Unit by optimization of occupancy. In [Issa et al 2014.a] , they improve the performance by accelerating it using maximization of Graphical Processing Unit's occupancy. The occupancy is a metric that is used to measure utilization of hardware called occupancy and is a key measure for Graphical Processing Unit efficiency [Nvidia]. Also, Occupancy mean keeping the processor busy as possible. Graphical Processing Units execute the programs as threads, where each thread is responsible for certain data. During execution, this threads are grouped in a warp like Fig. 13. As the number of warps is increased per multiprocessor, the multiprocessor being more busy since pausing a warp will allow the multiprocessor run another wrap. So, occupancy is the ration of active wraps per multiprocessor to the maximum allowable number of wraps per multiprocessor [Nvidia]. There are some equations to calculate the occupancy [Nvidia]. However, mainly there are three factors that allow controlling and maximizing the occupancy and efficiency of the Graphical Processing Unit: 1. The number of active threads per block, 2. The number of registers in the kernel and 3. The amount of shared memory allocated for each block in the kernel. Nvidia company develop an Excel sheet that control the value of occupancy based on changing the parameter values of number of threads, shared memory and number of registers.

In [Issa et al 2014.a], the decrease the execution time around 17 Sec for working on a database contains around 300000 biological sequence and Graphical Processing Unit contains 400 core. Therefore, a lot of development were worked on a sequence alignment on Graphical Processing Unit to speed up its execution and benefit from

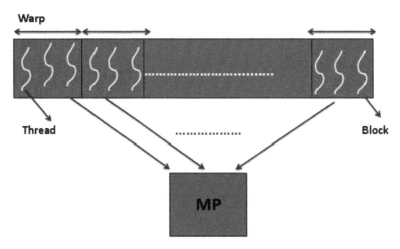

FIGURE 13   Executions of threads on the multiprocessors as wraps.

the massive computational capabilities of Graphical Processing Unit s. In [Issa et al 2014.c] another acceleration development for Gene Tracer on Multi-Core architecture using OpenMP [OpenMP 1997]. They depend on the idea in Fig. 5 for accelerarting the execution of the sequence alignment by execute it diagonal by diagonal. They reach to an improvement 150 % but the advantage of using multi-core execution that its ordinary hardware in most of modern laptops and computers, so it saves cost. Therefore, acceleration using multi-core suitable for small database around 20–40 thousands of biological sequence.

In conclusion, we assist of significance of sequence alignment algorithms and its vital role in many biological functions such as DNA assembly, Un-known Protein discoveries and evolutionary divergent measurements. In this chapter we briefly review the main pairwise sequence alignment algorithms such as Global pairwise sequence alignment, Local Sequence alignment and Semi-Global sequence alignment. In addition, we explored the recent development of acceleration the sequence alignment using modern hardware accelerators such as Graphical Processing Units and Field Programmable Array Gate, besides multi-processors. Also, a main important application of sequence alignment which called Gene Tracer was explored. This application is used to trace the modification of the ancestors in the offspring sequences.

_____

Acknowlededement

The author would like to thank his wife and daughter for encouraging him for releasing this work. Also, many thanks to author' supervisor Prof. Ahmed Alzohairy, Professor at Genetics Depratment, Faculty of Agriculture, Zagazig University and Head of Bioinformatics research node at Zagazig University, for give him support in working in this chapter.

References

A. Di Blas et. al. 2005. The UCSC Kestrel Parallel Processor. IEEE Transactions on Parallel and Distributed Systems, 16(1):80–92.

B Needleman, C. D. Wunsch, "A general method applicable to the search for similarities in the amino acid sequence of two proteins". Journal of molecular biology, vol. 48, no. 1, pp. 443-453. 1970.

C. Pedreira S.Bojanic, G.Caffarena and O. Nieto-Taladriz. 2004. High speed circuits for genetics applications. PROC. 24th International Conference on Microelectronics (MIEL 2004).

Daniel P. Lopresti. Rapid 1991 implementation of a genetic sequence comparator using eld programmable logic arrays. conference on Advanced research in VLSI pp 138-152.

D. Shreiner, M. Woo, J. Neider, and T. Davis, 2005. *OpenGL Programming Guide*, 5th edition. Reading, MA: Addison-Wesley (Publish.)

Elloumi, Mourad, Mohamed Al Sayed Issa, and Ahmed Mokaddem. Accelerating Pairwise Alignment Algorithms by Using Graphics Processor Units. *Biological Knowledge Discovery Handbook: Preprocessing, Mining, and Postprocessing of Biological Data*: pp. 969-80.

GenBank. http://www.ncbi.nlm.nih.gov/genbank/

H. Kreft G. Pfeiffer and M. Schimmler. Hardware enhanced biosequence alignment. International Conference on METMBS 2005.

J. Krüger and R. Westermann, *Linear algebra operators for GPU implementation of numerical algorithms*, ACM Transactions on Graphics (TOG), in Proc ACM SIGGRAPH'03, Vol. 22, Issue 3 : (July 2003), 908–16.

J. D. Owens, M. Houston, D. Luebke, S. Green, J. E. Stone, and J. C. Phillips, *GPU Computing*, Proceedings of the IEEE, (5) : (May 2008).

J. Shaw S. Seto J. Chiang, M. Studniberg and K. Truong. Hardware accelerator for genomic sequence alignment. Proceedings of the 28th IEEE EMBS. Annual International Conference New York City, USA, Aug 30-Sept 3, 2006.

Mohamed Issa, Hitham Abo Bakr, Ibrahim Ziedan and Ahmed Alzohairy. 2014. Sequence Analysis Algorithms for Bioinformatics Application, Master Thesis, Zagazig University.

M. Borah R. S. Bajwa S. Hannenhalli and M. J. Irwin. 1994. A SIMD Solution to the Sequence Comparison Problem on the MGAP . Proc. Int. Conf. on Application Specic Array Processors.

M. Charalambous, P. Trancoso, and A. Stamatakis. 2005 *Initial experiences porting a bioinformatics application to a graphics processor*, in *Proc. 10th Panhellenic Conference on Informatics* (PCI'05), Volos, Greece.

M. O. Dayhoff, R. M. Schwartz, and B. C. Orcutt, 1978 A model of evolutionary change in proteins, in Atlas of Protein Sequence and Structure", chapter 22, National Biomedical Research Foundation, Washington, DC: 345–58.

M. Elloumi , Y. Zomaya , 2011. *Algorithms in Computational Molecular Biology: Techniques, Approaches and Applications*, John Wiley & Sons.

M. Farrar, 2007. *Striped Smith-Waterman speeds database searches six times over other SIMD implementations*, Bioinformatics, 23 (2): pp.156-61.

Mohamed Issa, Hitham Abo Bakr, Ibrahim Ziedan and Ahmed Alzohairy «Tracking Genes Modifications in the Pedigree through GeneTracer Algorithm.» Database and Expert Systems Applications (DEXA), 2012 23rd International Workshop on. IEEE, 2012.

M. Issa, A. M. Alzohairy, TRACING ORIGINS OF UNKNOWN DNA/PROTEIN OFFSPRING SEQUENCES ON MULTI-PROCESSORS.

MP. Open «A proposed industry standard api for shared memory programming.» OpenMP Architecture Review Board (1997).

M. Striemer Gregory and Ali Akoglu,2009. Sequence Alignment with GPU: Performance and Design Challenges, IEEE Xplore, 2009.

M. C. Schatz, C. Trapnell, A. L Delcher, and A. Varshney, High-throughput sequence alignment using graphics processing units, BMC Bioinformatics, Vol. 8, 474 : (2007).

Nvidia: GPU programming guide. http://developer.Nvidia.com/Nvidia-gpu-programming-guide.

P. Agarwal, S. Krishnan, N. Mustafa, and S. Venkatasubramanian. 2003., *Streaming geometric optimization using graphics hardware*, in Proc. 11th Annual European Symposium on Algorithms (ESA'03), Budapest, Hungary.

PubMed. http://www.ncbi.nlm.nih.gov/pubmed/

S.Henikoff, J. G. Henikoff, 1992. "Amino acid substitution matrices from protein blocks", Proc. Natl. Acad. Sci. USA, (22): 10915-919.

T. Cormen , E.Rivest and C. Stelin,"introduction to algorithms ", MIT Press Cambridge, Massachusetts, London.

T . Maruyama Y . Y amaguchi, Y . Miyajima and A. Konagaya. High speed homology search using run-time reconguration. FPL 2002.

T. F. Smith, M. S. Waterman, Identification of common molecular subsequences, J. Molecular Biology, no. 147, pp. 195-97.

UniProt. http://www.uniprot.org/

W. R. Pearson, *Searching protein sequence libraries: comparison of the sensitivity and selectivity of the Smith-Waterman and FASTA algorithms*, Genomics, Volume 11, Issue 3 : (November 1991), p635-650.

W. R. Pearson, D. J. Lipman*, Improved tools for biological sequence comparison, Proc Natl Acad Sci U S A*, Vol. 85: (April 1988), p2444-2448

W. Liu, B. Schmidt, G. Voss, A. Schroder, and W. Muller-Wittig,2006. *Bio-Sequence Database Scanning on a GPU*, in Proc. 20th IEEE International Parallel & Distributed Processing Symposium (IPDYNAMIC PROGRAMMINGS'06), 5th IEEE International Workshop on High Performance Computational Biology) Workshop (HICOMB'06), Rhode Island, Greece.

Y. Liu, W. Huang, J. Johnson, S. Vaidya, *GPU accelerated Smith-Waterman*, in Proc. Computational Science (ICCS'06), Lecture Notes in Computer Science, Vol. 3994, Springer-Verlag, Berlin, Germany : (2006), p188-195.

13

State Estimation and Process Monitoring of Nonlinear Biological Phenomena Modeled by S-systems

Majdi Mansouri[1]*, Hazem Nounou*[1] *and Mohamed Nounou*[2]

1. Introduction

One of the major research activities in modern molecular biology is estimating the states associated with biological system models and monitoring the biological processes. Recent progress in measurement technologies such as the phosphorylation of protein kinase or mass spectrometry, nuclear magnetic resonance provided a wealth of comprehensive time profiles of metabolites that can be used for biochemical pathway modeling and proteomics. These metabolic profiles are simultaneous measurements of biochemicals that can be obtained as a sequence of snapshots or as simple snapshots. In order to mathematically represent this information, it is first required to estimate the model parameters using state estimation techniques and then to specify a mathematical modeling framework and to develop computational methods to fit the measured information to the selected modeling framework. In addition, due to consistent product quality demand and higher requirements in safety, the process monitoring performance has become a key factor in improving productivity and safety. Process systems are using large amounts of data from many variables that are monitored and recorded continuously every day. For these reasons, the problem of fault detection that responses effectively to faults that mislead the process and harm the system reliability represents a key process in such operation of these systems. Several multivariate statistical techniques for fault detection, analysis of process and diagnosis have been developed and used in practice. These techniques are useful since operation safety and the better quality products are some of the main goals in the industry applications. Faults detection has been performed manually using data visualization tools [1], however these tools takes a lot of time for real-time detection with continuous data. In the most recent years, researchers have proposed machine learning and automated statistical methods like: nearest neighbor [2, 3], clustering [4], minimum volume ellipsoid [5], convex pealing [6], neural network classifier [7], decision tree [8] and support vectormachine classifier [9]. These proposed techniques

[1]. Electrical and Computer Engineering Program, Texas A&M University at Qatar, Doha, QATAR,
[2]. Chemical Engineering Program, Texas A&M University at Qatar, Doha, QATAR
* Corresponding author: majdi.mansouri@qatar.tamu.edu

are quicker than other manual techniques, however there are disadvantages which make them inadequate for continuous fault detection for the cases of streaming data. More recently, principal component analysis (PCA) and multivariate statistical process control (MSPC) approach are proposed to overcome these problems. The authors in [10] have proposed PCA as a tool of MSPC.

Also, PCA was defined as a method which projects a high dimensional measurement space into a lower dimensional space [11]. PCA provides linear combinations of parameters which demonstrate most common trends in a data set. In mathematical terms, PCA relies on the orthogonal decomposition of the covariance matrix over the process variables along with the directions which give the maximum data variation. It is also mentioned that PCA is researched for two problems: the MSPC [12], and fault detection and isolation (FDI) problem [13]. The authors in [13] have listed diagnosis and fault detection techniques in three categories: (i) quantitative model-based schemes, (ii) qualitative model schemes and corresponding search strategies and (iii) process data based techniques. PCA falls into the third category since it utilizes databases in an attempt to obtain the statistical (PCA model). The main indices used with PCA methods are Hotelling statistic, T^2; sum of squared residuals, SPE; and/or Q statistics. The T^2 statistic is a way to measure the variation captured in the PCA model whereas the Q statistic is a way to measure the amount of variation which was not captured by the PCA model. PCA is known to be one of the most popular MSPC monitoring methods. Nevertheless, there are some disadvantages of it. One disadvantage is that the PCA is not suitable for monitoring processes that show non-stationary behavior. The other shortcoming of the PCA model is that most of the processes run under different circumstances. The use of standard PCA solution in this kind of processes might produce too many missed faults, since the grade transitions from one operation mode to another operation mode might damage the correlation existing between various parameters. In addition, the disturbances that are measured may be treated as faults.

In the area of state estimation in biological systems, several techniques have been developed and include the extended Kalman filter (EKF), the unscented Kalman filter (UKF), and more recently the particle filter (PF). The classical Kalman filter (KF) was developed in the 1960s, and has been widely applied in various engineering and science areas, including communications, control, machine learning, neuroscience, and many others. In the case where the model describing the system is assumed to be linear and Gaussian, the KF provides an optimal solution [21]. The KF has also been formulated in the context of Takagi-Sugeno fuzzy systems, which can be described by a convex set of multiple linear models [22]. It is known that KF is computationally efficient. However, it is limited by the non-universal linear and Gaussian modeling assumptions. To relax such assumptions, the extended Kalman filter [23, 24, 25] and the unscented Kalman filter [26, 27, 28] have been developed. In extended Kalman filtering, the model describing the system is linearized at every time sample (which means that the model is assumed to be differentiable). Therefore, for highly nonlinear models EKF does not usually provide a satisfactory performance. The UKF, on the other hand, instead of linearizing the model to approximate the mean and covariance matrix of the state vector, uses the unscented transformation to approximate these moments. In the unscented transformation, a set of samples (called sigma points) are selected and propagated through the nonlinear model to improve the approximation of these moments and thus the accuracy of state estimation. The above mentioned techniques usually require large amounts of computational resources and time. One aspect of concern is that these methods do not sufficiently take the special structures of biological system models into consideration. However, the propositions in [29, 30] have shown that consideration of the model structure may simplify the parameter estimation problem. For example, a non-linear model can be linearized to a linear one and the EKF can be applied to estimate the unknown sensitive parameters of the model. More details about other estimation techniques can be found in [31, 32]. In addition, due to the fact that the number of unknown parameters is much more than the number of states (metabolites), the conventional EKF and UKF algorithms are not capable of estimating the unknown parameters of S-systems. To overcome these drawbacks, a non-parametric Monte Carlo sampling based method called particle filtering has recently gained popularity. PF approximates the posterior probability distribution by a set of weighted samples, called particles. Since real-world problems usually involve high-dimensional random variables with complex uncertainty, the non-parametric and sample-based estimation of uncertainty (provided by the PF) has thus become quite popular to capture and represent the complex distribution $p(z|y)$ for nonlinear and non-Gaussian process models.

In this chapter, we extend the fault detection problem to the state estimation and fault detection approaches. This is achieved by introducing a supplementary observer which together with the observer produces full state estimation and fault detection. Thus this hybrid scheme led to the development of a more general approach, which has immediate application to the two areas of: (i) state estimation; and (ii) fault detection. To illustrate the advantages of these approaches we apply the state estimation and fault detection approaches to a biology model representing a Cad System in *E. coli* processes.

The body of the chapter is divided into a background section, an introduction section, and four distinct but closely related sections, describing our studies related to designing Bayesian for estimating the states variables and detecting the faults for biology system representing a Cad System in *E. coli* processes. We now summarize the main sections contents as follows.

Section 2 provides a brief introduction to state evolution model. This model is more appropriate to practical non-linear and non-Gaussian situations where no *a priori* information on the state variable value is available.

Section 3 presents a Bayesian theory and its relevance to solve states estimation. We begin with an overview of several state-of-the-art estimation method descriptions. Then, we review the filtering framework, and some classic filtering algorithms that are widely employed in states estimation. We briefly introduce some probability calculus tools, providing the required basics for understanding the algorithms and the analysis presented in subsequent sections, as well as references for the interested reader.

Section 4 presents a generalized likelihood ratio (GLR)-based PCA for faults detection in biology model representing a Cad System in *E. coli* processes. PCA is used to create the model and find linear combinations of parameters which describe the major trends in a data set and GLR test. Both are utilized to improve faults detection. GLR test has been proposed in order to establish an adaptive system, which reaches three important problems; estimation, fault detection and magnitude compensation of jumps. Therefore, in this chapter, we propose to benefit from the advantages of the GLR test in order to improve the fault detection task.

Section 5 compares the state estimation and fault detection techniques through their utilization to estimate the states variables and detect the faults of the biology process model representing Cad System in *E. coli*. Firstly, a description of a biology process model representing Cad System in *E. coli* (CSEC) is presented. Then, the state estimation techniques are used to estimate the four state variables (the enzyme CadA, the transport protein CadB, the regulatory protein CadC and lysine Lys for a Model of the Cad System in *E. coli*) for the biological model. Finally, the performance of the PCA-based GLR fault detection method is evaluated and compared to the convential PCA fault detection method.

2. State Estimation in Non-linear Biological Systems

2.1 Problem formulation

In this section, the state estimation problem is formulated, and then a comparative performance analysis of states estimation using extended Kalman filter, unscented Kalman filter, and particle filter will be conducted state and parameter estimation for CSEC model.

Here, the estimation problem of interest is formulated for a general system model. Let a nonlinear state space model be described as follows:

$$\begin{cases} \dot{x} = g\left(x, u, \theta, w\right) \\ y = l\left(x, u, \theta, v\right) \end{cases} \tag{1}$$

where, $x \in R^n$ is a vector of the state variables, $u \in R^p$ is a vector of the input variables (which can be changed as desired), $\theta \in R^q$ is an unknown parameter vector, $y \in R^m$ is a vector of the measured variables, g and l are nonlinear differentiable functions, and $w \in R^n$ and $v \in R^m$ are process and measurement noise, which quantify randomness in the process and errors in the measurements, respectively.

Discretizing the state space model (1), the discrete model can be written as follows:

$$\begin{cases} x_k = f\left(x_k, u_{k-1}, \theta_{k-1}, w_{k-1}\right) \\ y_k = h\left(x_k, u_k, \theta_k, v_k\right) \end{cases} \tag{2}$$

which describes the state variables at some time step (k) in terms of their values at a previous time step ($k - 1$). Since we are interested to estimate the state vector x_k, as well as the parameter vector θ_k, let's assume that the parameter vector is described by the following model:

$$\theta_k = \theta_{k-1} + \gamma_{k-1} \tag{3}$$

where γ_{k-1} is white noise. In other words, the parameter vector model (3) corresponds to a stationary process, with an identity transition matrix, driven by white noise. We can define a new state vector that augments the two vectors together as follows:

$$z_k = \begin{bmatrix} x_k \\ \theta_k \end{bmatrix} = \begin{bmatrix} f\left(x_{k-1}, u_{k-1}, \theta_{k-1}, w_{k-1}\right) \\ \theta_{k-1} + \gamma_{k-1} \end{bmatrix} \tag{4}$$

where z_k is assumed to follow a Gaussian model as $z_k \sim N(\mu_k, \lambda_k)$, and where at any time k the expectation μ_k and the covariance matrix λ_k are both constants. Also, defining the augmented vector,

$$\varepsilon_{k-1} = \begin{bmatrix} w_{k-1} \\ \gamma_{k-1} \end{bmatrix} \tag{5}$$

the model (2) can be written as:

$$z_k = \Im\left(z_{k-1}, u_{k-1}, \varepsilon_{k-1}\right) \\ y_k = \Re\left(z_k, u_k, v_k\right) \tag{6}$$

The objective is to estimate the augmented state vector z_k, given the measurements vector y_k. Next, the estimation techniques will be described.

3. Description of State Estimation Techniques

Here, the three estimation techniques of interest (EKF, UKF and PF) are described.

3.1 Extended Kalman Filter (EKF)

The objective in state estimation is to find an estimate $\hat{z}_k$ of the state vector z_k that minimizes the covariance matrix of the estimation error, $P = E\left[(z_k - \hat{z}_k)(z_k - \hat{z}_k)^T\right]$. Such minimization can be achieved by minimizing the following objective function:

$$J = \frac{1}{2} Tr\left(E\left[\left(z_k - \hat{z}_k\right)\left(z_k - \hat{z}_k\right)^T\right] \right) \tag{7}$$

Minimizing the above objective function (7), the extended Kalman filter (EKF) estimates the state vector x_k using a two-step algorithm: prediction and estimation [23, 24].

3.2 Unscented Kalman Filter (UKF)

The reason for the limitations observed by the EKF is that EKF approximates the mean and covariance of the nonlinear state vector by linearizing the nonlinear model, which may not provide a satisfactory approximation of these moments. To provide better estimates of these moments, the Unscented Kalman Filter (UKF) relies on the unscented transformation. The unscented transformation is a method for calculating the statistics of a random variable which undergoes a nonlinear mapping. Assume that a random variable $z \in R^r$ with Mean $\bar{z}$ and covariance P_z is transformed by a nonlinear function, $y = f(z)$. In order to find the statistics of y, define $2r + 1$ sigma vectors as follows ([35]):

$$Z_o = \bar{z}$$

$$Z_i = \bar{z} + \left(\sqrt{(r + \lambda) P_z}\right)_i \qquad i = 1,..,r \qquad\qquad (8)$$

$$Z_i = \bar{z} - \left(\sqrt{(r + \lambda) P_z}\right)_i \qquad i = r+1,..,2r$$

Where, $\lambda = e^2 (r + \kappa) - r$ is a scaling parameter and $\left(\sqrt{(r + \lambda) P_z}\right)_i$ denotes the ith column of the matrix square root. The constant $10^{-4} < e < 1$ determines the spread of the sigma points around $\bar{z}$. The constant κ is a secondary scaling parameter which is usually set to zero or $3 - r$ ([35]). Then, these sigma points are propagated through the nonlinear function, i.e.,

$$Y_i = f(Z_i) \qquad i = 0,..,2r \qquad\qquad (9)$$

and the mean and covariance matrix of y can be approximated as weighted sample mean and covariance of the transformed sigma points of Y_1 as follows:

$$\bar{y} \approx \sum_{i=0}^{2r} W_i^{(m)} Y_i, \quad \text{and} \quad P_z \approx \sum_{i=0}^{2r} W_i^{(c)} \left(Y_i - \bar{y}\right)\left(Y_i - \bar{y}\right)^T \qquad (10)$$

where, the weights are given by:

$$W_0^{(m)} = \frac{\lambda}{\lambda + r}, \quad W_0^{(c)} = \frac{\lambda}{\lambda + r} + (1 - e^2 + \zeta),$$

and
$$W_i^{(m)} = W_i^{(c)} = \frac{1}{2(\lambda + r)} \qquad i = 1,..,2r \qquad\qquad (11)$$

The parameter ζ is used to incorporate prior knowledge about the distribution of z. It has been shown that for a Gaussian and non-Gaussian variables, the unscented transformation results in approximations that are accurate up to the third and second order, respectively ([35]).

3.3   Particle Filter (PF)

A particle filter is an implementation of a recursive Bayesian estimator [32, 36]. Bayesian estimation relies on computing the posterior $p(z_k \mid y_{1:k})$, which is the density function of the unobserved state vector, (z_k), given the sequence of the observed data $y_{1:k} \equiv \{y_1, y_2,.., y_k\}$. However, instead of describing the required posterior distribution in a functional form, in this particle filter scheme, it is represented approximately as a set of random samples of the posterior distribution. These random samples, which are called the particles of the filter, are propagated and updated according to the dynamics and measurement models (Doucet & Johansen, 2009). The advantage of the PF is that it is not restricted by the linear and Gaussian assumptions, which makes it applicable in a wide range of applications. The basic form of the PF is simple, but may be computationally expensive. Thus, the advent of cheap, powerful computers over the last ten years has been a key to the introduction and utilization of particle filters in various applications. For a given dynamical system describing the evolution of the states that we wish to estimate, the estimation problem can be viewed as an optimal filtering problem ([23]), in which the posterior distribution, $p(z_k \mid y_{1:k})$, is recursively updated Here, the dynamical system is characterized by a Markov state evolution model, $p(z_k \mid z_{1:k-1}) = p(z_k \mid z_{k-1})$, and an observation model, $p(y_k \mid z_k)$. In a Bayesian context, the task of state estimation can be formulated as recursively calculating the predictive distribution $p(z_k \mid y_{1:k-1})$ and the filtering distribution $p(z_k \mid y_{1:k})$ as follows,

$$p\left(z_k \mid y_{1:k-1}\right) = \int p\left(z_k \mid z_{k-1}\right) p\left(z_{k-1} \mid y_{1:k-1}\right) dz_{k-1} \tag{12}$$

and

$$p\left(z_k \mid y_{1:k}\right) = \frac{p\left(y_k \mid z_k\right) p\left(z_k \mid y_{1:k-1}\right)}{p\left(y_k \mid y_{1:k-1}\right)} \tag{13}$$

where

$$p\left(y_k \mid y_{1:k-1}\right) = \int p\left(y_k \mid z_k\right) p\left(z_k \mid y_{1:k-1}\right) dz_k \tag{14}$$

The state vector (z_k) at any instant, k, is assumed to follow a Gaussian model,

$$z_k \sim N\left(\mu_k, \lambda_k\right) \tag{15}$$

where, at any time instant, k, the expectation, μ_k, and the covariance matrix, λ_k are both constants. The marginal state distribution is obtained by integrating over the mean and precision matrix as follows,

$$p\left(z_k \mid z_{k-1}\right) = \int N\left(z_k \mid \mu_k, \lambda_k\right) p\left(\mu_k, \lambda_k \mid z_{k-1}\right) d\mu_k \, d\lambda_k \tag{16}$$

where the integration with respect to the covariance matrix, λ_k, leads to the known class of scale mixture distributions introduced by Barndorff-Nielsen [37] for the scalar case.

The nonlinear nature of the system model leads to intractable integrals when evaluating the marginal state distribution. Therefore, Monte Carlo approximation is utilized, where the joint posterior distribution $p\left(z_{0:k} \mid y_{1:k}\right)$ is approximated by the point-mass distribution of a set of weighted samples (called particles) $\left\{z_{0:k}^{(i)}, \ell_k^{(i)}\right\}_{i=1}^{N}$:

$$\hat{p}_N\left(z_{0:k} \mid y_{1:k}\right) = \sum_{i=1}^{N} \ell_k^{(i)} \delta_{z_{0:k}^{(i)}}\left(dz_{0:k}\right) / \sum_{i=1}^{N} \ell_k^{(i)} \tag{17}$$

where, $\delta_{z_{0:k}^{(i)}}\left(dz_{0:k}\right)$ denotes the dirac delta function. Based on the same set of particles, the marginal posterior (of interest), $p\left(z_k \mid y_{1:k}\right)$, can also be approximated as follows:

$$\hat{p}_N\left(z_k \mid y_{1:k}\right) = \sum_{i=1}^{N} \ell_k^{(i)} \delta_{z_k^{(i)}}\left(dz_k\right) / \sum_{i=1}^{N} \ell_k^{(i)} \tag{18}$$

Using Bayesian importance sampling (IS), the particles, $\left\{z_{0:k}^{(i)}, \ell_k^{(i)}\right\}_{i=1}^N$, are sampled according to a proposal distribution, $\pi\left(z_{0:k} \mid y_{1:k}\right)$. Then, the estimate of the state, $\hat{z}_k$, can be approximated using a Monte Carlo scheme as follows:

$$\hat{z}_k = \sum_{i=1}^N \ell_k^{(i)} z_k^{(i)} \tag{19}$$

where, $\ell_k^{(i)}$, and the corresponding importance weights:

$$\ell_k^{(i)} \propto \frac{p\left(y_{1:k} \mid z_{0:k}^{(i)}\right) p\left(z_{0:k}^{(i)}\right)}{\pi\left(z_{0:k}^{(i)} \mid y_{1:k}\right)} \tag{20}$$

Sequential Monte Carlo (SMC) consists of propagating the state vector, $\left\{z_{0:k}^{(i)}\right\}_{i=1}^N$, in time without modifying the past simulated particles. This is possible for the class of proposal distributions having the following form:

$$\pi\left(z_{0:k} \mid y_{1:k}\right) = \pi\left(z_{0:k-1} \mid y_{1:k-1}\right) \pi\left(z_k \mid z_{0:k-1}, y_{1:k}\right). \tag{21}$$

The importance weights are then recursively computed in time as follows:

$$\ell_k^{(i)} \propto \ell_{k-1}^{(i)} \frac{p\left(y_k \mid z_k^{(i)}\right) p\left(z_k^{(i)} \mid z_{0:k-1}^{(i)}\right)}{\pi\left(z_k^{(i)} \mid z_{0:k-1}^{(i)}, y_{1:k}\right)}. \tag{22}$$

The optimal choice of the importance function is $p\left(z_k \mid z_{k-1}, y_k\right)$, which minimizes the variance of the importance weights conditionally upon the simulated trajectory, $z_{0:k-1}^{(i)}$, and the observations, $y_{1:k}$. For the considered Markov nonlinear state-space model, one can adopt the transition prior, $p\left(z_k \mid z_{k-1}\right)$, as the proposal distribution:

$$\pi\left(z_k^{(i)} \mid z_{0:k-1}^{(i)}, y_{1:k}\right) = p\left(z_k \mid z_{k-1}\right) \tag{23}$$

in which the weights are updated according to the likelihood function:

$$\ell_k^{(i)} \propto \ell_{k-1}^{(i)} p\left(y_k \mid z_k^{(i)}\right) \tag{24}$$

The resulting PF algorithm is fully recursive and computationally efficient since the sampling-based approach avoids integration for obtaining the moments at each time step [38, 39]. The recursive nature implies that solving a nonlinear optimization problem in a moving window is not required. Furthermore, SMC does not rely on restrictive assumptions about the nature of the error or prior distributions and models, making it broadly applicable.

4. Faults Detection of Biological Systems Representing Continousily Stirred Tank Reactor Model

In this chapter, generalized likelihood ratio (GLR)-based PCA is proposed to detect the faults in biological systems representing Cad System in *E. coli* (CSEC). PCA is used to create the model and find linear combinations of parameter s which describe the major trends in a data set and GLR test. Both are utilized to improve faults detection. GLR test has been proposed in order to establish an adaptive system, which reaches three important problems; estimation, fault detection and magnitude compensation of jumps. GLR test is proposed for fault detection of different applications: geophysical signal segmentation [21], signals and dynamic systems [22], incident fault detection on freeways [23], missiles trajectory [24]. Therefore, in the current work it is proposed to benefit from the advantages of the GLR te st in order to improve the fault detection task in the cases where process model is not available.

The rest of this Section is organized as the following. In Section 4.1, an introduction to PCA is given, followed by descriptions of the two main detection indices, T^2 and Q, which are generally used with PCA for fault detection. Then, the GLR test which is utilized in composite hypothesis testing is discussed in Section 4.2. After that, the PCA- based GLR method used for detecting fault which integrates PCA modeling and GLR statistical testing, is shown in Section 4.3.

4.1 Principal component analysis (PCA)

Let $X_1 \in R^m$ denotes a sample vector of m number of sensors. Also, assume there are n samples dedicated to each sensor, a data matrix $X \in R^{nxm}$ is with each row, displaying a sample. Meanwhile,X matrix is scaled to zero mean for covariance-based PCA and at the same time, to unit variance for correlation-based PCA [43]. The X matrix can be divided into two matrices: a score matrix S and a loading matrix W through singular value decomposition (SVD):

$$X = SW^T \tag{27}$$

where $S = [s_1 s_2 ... s_m] \in R^{mxm}$ is a transformed variables matrix, $s_i \in R^n$, are the score vectors or principal components, and $W = [w_1 w_2 ... w_3] \in R^{mxm}$ is an orthogonal vectors matrix $w_i \in R^m$ which includes the eigenvectors associated with the covariance matrix of X, i.e., Σ, which is given by

$$\Sigma = \frac{1}{n-1} X^T X = W \Lambda W^T \text{ with } W \Lambda W^T = W^T \Lambda W = I_n \tag{28}$$

where, where, $\Lambda = \text{diag}(\lambda_1, \lambda_2, ..., \lambda_m)$ is a diagonal matrix containing the eigenvalues related to the m PCs, λ_m are simply the eigenvalues of the covariance matrix ($\lambda_1 \geq \lambda_2 \geq ... \geq \lambda_m$), and I_n is the identity matrix ([44]). It must be noted at this point that the PCA model yields same number of principal components as the number of original variables (m). Nevertheless, for collinear process variables, a smaller number of principal components (l) are required so that most of the variations in the data are captures. Most of the times, a small subset of the principal components (which correspond to the maximum eigenvalues) might carry the most of the crucial information in a data set, which simplifies the analysis.

The effectiveness of the PCA model depends on the number of principal components (PCs) are to be used for PCA. Selecting an appropriate number of PCs introduces a good performance of PCA in terms of processes mon- itoring. Several methods for determining the number of PCs have been proposed such as; the Scree plot ([45]), the cumulative percent variance (CPV), the cross validation ([46]), and the profile likelihood ([47]). In this study herein, the cumulative percent variance method is utilized to come up with the optimum number of retained principal components.

The cumulative percent variance is computed as follows:

$$CPV(l) = \frac{\sum_{i=1}^{l} \lambda_i}{trace(\Sigma)} x\,100 \tag{29}$$

When the number of principal components l is determined, then, the data matrix X is shown as the following:

$$X = SW = [\hat{S}\ \tilde{S}][\hat{W}\ \tilde{W}]^T \tag{30}$$

where $\hat{S} \in R^{nxl}$ and $\tilde{S} \in R^{nx(m-l)}$ are matrices of l retained principal components and the $(m - l)$ ignored principal components, respectively, and the matrices $\hat{W} \in R^{mxl}$ and $\tilde{W} \in R^{mx(m-l)}$ are matrices of l retained eigenvectors and the $(m - l)$ ignored eigenvectors, respectively. Using Eq. (30), the following can be written:

$$X = \hat{S}\hat{W}^T + \tilde{S}\tilde{W}^T \tag{31}$$

The matrix $\hat{X}$ represents the modeled variation of X based on first l components.

4.2   Fault detection indices

When using PCA in detecting faults, a PCA model is built utilizing fault-free data. The model is used for fault detection through one of the detection indices (the Hotelling's T^2 and Q statistics), which are presented next.

4.2.1   Hotelling's T^2 statistic

The T^2 statistic is a way of measuring the variation captured in the principal components at various time samples, and it is known as ([48]):

$$T^2 = X^T \hat{W} \hat{\Lambda}^{-1} \hat{W}^T X \tag{32}$$

Where $\hat{\Lambda}^{-1} = diag(\lambda_1, \lambda_2, ..., \lambda_l)$, is a diagonal matrix containing the eigenvalues related to the l retained PCs. For new real-time data, when the value of T^2 statistic exceeds the threshold, $T^2_{l,n,\alpha}$ calculated as in ([48]), a fault is detected.

The threshold number used for the T^2 statistic is computed as ([48]):

$$T^2_{l,n,\alpha} = \frac{l(n-1)}{n-1} F_{l,n-l,\alpha} \tag{33}$$

where α is the level of significance (α usually between 10% and 5%), n is the number of samples in data set, l is the number of retained PCs, and $F_{l,n-l,\alpha}$ is the Fisher F distribution with l and $n-1$ degrees of freedom. These thresholds are computed using faultless data. When the number of observations, n, is high, the T^2 statistic threshold is approximated with a χ^2 distribution with l degrees of freedom, i.e., $T_\alpha^2 = \chi_{l,\alpha}^2$.

4.2.2 Q statistic or squared prediction error (SPE)

It is possible to detect new events by computing the squared prediction error *SPE* or Q of the residuals for a new observation. Q statistic ([49]), is computed as the sum of squares of the residuals. Also, the Q statistic is a measure of the amount of variation not captured by the *PCA* model, it is defined as ([49]):

$$Q = \left\| \tilde{X} \right\|^2 = \left\| X - \hat{X} \right\|^2 = \left\| (I - \hat{W}\hat{W}^T)X \right\|^2 \tag{34}$$

The monitored system, meanwhile, is accepted to be in normal operation if:

$$Q \le Q_\alpha \tag{35}$$

The threshold Q_α used for the Q statistic can be computed as [10]

$$Q_\alpha = \phi_1 \left[\frac{h_0 c_\alpha \sqrt{2\phi_2}}{\phi_1} + 1 + \frac{\phi_2 h_0 (h_0 - 1)}{\phi_1^2} \right], \tag{36}$$

where $\phi_i = \sum_{j=l+1}^{m} \lambda_j^i, \{i=1,2,3\}$, $h_0 = 1 - \frac{2\phi_1\phi_3}{\phi_2^2}$ and c_α is the value of the normal distribution with α level of significance. at the instant of an unusual event; when there is a change in the covariance structure of the model, this change is going to be detected by a high value of Q. For new data, the Q statistic is computed and compared to the threshold Q_α ([44]). This means a fault is detected when the confidence lim it is violated. The threshold value is computed on the assumption that the measurements are independent of time and they are multivariate normally distributed. The Q fault detection index is highly sensitive to errors in modeling and the performance of it is dependent on the number of retained PCs, l, [50].

4.3 Generalized likelihood ratio test (GLRT)

The faults detection step is done using the residuals computed using PCA. Using the information about the noise distribution of the residuals, a GLR test statistic is formed. To make the decision if a fault is present or not, the test statistic is compared to a threshold from the chi-square distribution.

4.3.1 Test Statistic

The GLR test is famous to be a uniformly most powerful test among all invariant tests (shown in Equation (36)). It is basically a hypothesis testing technique which has been utilized successfully in model-based faults detection ([51, 52, 53, 54]). Focusing on the following fault detection problem, $Y \in R^n$ is an observation vector formed by one of the two Gaussian distributions: $N(0, \sigma^2 I_n)$ or, $N(\theta \neq 0, \sigma^2 I_n)$ where θ is the mean vector (which is the value of the fault) and $\sigma^2 \succ 0$ is the variance (assumed to be known in this problem). The hypothesis test can be shown as:

$$\begin{cases} H_0 = \{Y \sim N(0, \sigma^2 I_n)\} \ (\textbf{\textit{null hypothesis}}); \\ H_1 = \{Y \sim N(\theta, \sigma^2 I_n)\} (\textbf{\textit{alternative hypothesis}}). \end{cases} \tag{37}$$

Here, the GLR method replaces the unknown parameter, θ, by its maximum likelihood estimate. This estimate is computed by maximizing the generalized likelihood ratio T(Y) as shown below:

$$T(Y) = 2 \log \frac{\sup_{\theta \in R^n} f_\theta(Y)}{f_{\theta=0}(Y)}$$

$$= 2 \log \left\{ \frac{\sup_{\theta \in R^n} \exp\left(\frac{\|Y - \theta\|_2^2}{f_{\theta=0}(Y)}\right)}{\exp\left(\frac{\|Y\|_2^2}{2\sigma^2}\right)} \right\} \tag{38}$$

$$= \frac{1}{2\sigma^2} \left\{ \min_{\theta \in R^n} \|Y - \theta\|_2^2 + \|Y\|_2^2 \right\}$$

$$= \frac{1}{2\sigma^2} \left\{ \|Y\|_2^2 \right\}$$

where $\hat{\theta} = \arg\min \|Y - \theta\|_2^2 = Y$ is the maximum likelihood estimate of θ, the probability density function of Y is $\frac{1}{(2\pi)^{n/2} \sigma^n} \exp\left(-\frac{\|Y - \theta\|_2^2}{2\sigma^2}\right)$, $\|\cdot\|_2$ represents the Euclidean norm. Because the GLR test utilized the ratio of distributions of the faulty and faultless data; for the case of non-Gaussian variables, non-Gaussiandistributions are required to be utilized. It must be noted that, in the derivation mentioned above, maximizing the likelihood function is equivalent to maximizing its natural logarithm since the logarithmic function is a monotonic function. At this stage, the GLR test then decides between the hypotheses H_0 and H_1 as follows:

$$\begin{cases} H_0 & \textit{if } T(Y) \prec t_\alpha \\ H_1 & \textit{else.} \end{cases} \tag{39}$$

Since distribution of the decision function T(Y) under H_0 allows to design a statistical test with a desired false alarm rate, α, where the threshold $t\alpha$ is chosen to satisfy the following false alarm probability:

$$P_0(\Lambda(Y) \geq t_\alpha) = \alpha \tag{40}$$

where, $P_0(A)$ represent the probability of an event A when Y is distributed according to the null hypothesis H_0 and α is the desired probability of the false alarm. Since Y is normally distributed, the statistics T is distributed according to the χ^2 law with $(m-1)$ degrees of freedom.

4.3.2 Statistic

To select an appropriate thresholds for the test statistics shown above, it is crucial to find their distributions. For that purpose, with the Gaussian noise within, the test statistics will be chi-square distributed variables ([41]). The normalized residual $\overline{R}$ is distributed as

$$\overline{R} \sim N(\theta, \sigma^2 I_n), \tag{41}$$

where $\theta = 0$ under the null hypothesis (13). Then, the test statistic is distributed as the non-central chi-square distribution as shown below:

$$t_\alpha = \frac{1}{\sigma^2}\left\{\|Y\|_2^2\right\} \sim \chi_n^2, \tag{42}$$

and the test statistic is distributed through the central chi-square distribution χ_n^2 with degree of freedom n. The threshold is now chosen from the chi-square distribution therefore the fault-free hypothesis is erroneously rejected with only a small probability.

4.4 Fault detection using a GLR-based PCA test

In this section, a GLR test to detect faults is derived, and its explicit asymptotic statistics computed using PCA. The objective of the GLR-based PCA fault detection technique is to detect the additive fault, θ, with the maximum detection probability for a given false alarm. Here, the fault detection task can be considered as a hypothesis testing problem with consideration of two possible hypotheses: null hypothesis of no change H_0, where measurements vector X, is fault-free, and the change-point alternative hypothesis H_1, where X contains a fault, and thus X is no longer categorized by the fault-free PCA model (31). For new data, the method needs to pick between H_0 and H_1 for the most efficient detection performance.

In the absence of a fault, the residual can be calculated as follows,

$$R = X - \hat{X}, \tag{43}$$

while in the presence of an additive fault vector, θ, the residual is computed as

$$R = X - \hat{X}[+\theta] \tag{44}$$

It is assumed that the residual in Equation (41) is Gaussian. Hence, the fault detection problem consists of detecting the presence of an additive bias vector, θ, in the residual vector, R.

The residual vector can be considered as a hypothesis testing problem by focusing on two hypotheses: the null hypothesis H_0, where R is fault-free and the alternate hypothesis H_1, where R contains a fault. The formulation of the hypothesis testing problem can be written as

$$\begin{cases} H_0 = \{R \sim N(0, \sigma^2 I_n)\} \quad (\textbf{\textit{null hypothesis}}); \\ H_1 = \{R \sim N(\theta, \sigma^2 I_n)\} (\textbf{\textit{alternative hypothesis}}). \end{cases} \qquad (45)$$

The algorithm which studies the developed GLR-based PCA fault detection technique is presented in Algorithm 1. The GLR- based PCA is proposed to detect the faults in the residual vector obtained from the PCA model, through which the GLR test is used for each residual vector, R.

Algorithm 1: GLR-based PCA fault detection algorithm.

Input: Training fault-free data *Xtr*, Testing faulty data *Xtest*, Confidence interval α

Output: GLR statistic T, GLR Threshold t_α

- *Data preprocessing step:*
  Standardize: computes data's mean and standard deviation, and standardize it;

- *PCA running step:*
  Compute the covariance matrix, Σ ;

  Calculate the eigenvalues and eigenvectors of Σ and sort the eigenvalues in decreasing order;

  Compute the optimal number of principal components to be used using the *CPV* method;

  Compute the sum of approximate and residual matrices using Equation (31);

  Testing step:

  Standardize the new data;

  Generate a residual vector, R , using PCA;

  Compute the GLR statistic T using Equation (40) for the new data; Compute the GLR statistic threshold t_α;

- *Decision step:*

  if $T \geq t_\alpha$, then declare a fault.

---

5. Simulation Results Analysis

5.1 States estimation in biological systems representing Cad System in E. coli

In this section, the state estimation techniques described in Section 3 (i.e., EKF, UKF, and PF) are compared through their utilization to estimate the states variables of a continuously stirred tank reactor. First, a description of the CSEC process model is presented, and then two comparative studies are conducted to assess the performances of these state estimation techniques. In the first comparative study, the three state estimation techniques are used to estimate the four state variables (the enzyme CadA, the transport protein CadB, the regulatory protein CadC and lysine Lys for a Model of the Cad System in *E. coli*) from noisy measurers of these variables. In the second comparative study, the various state estimation techniques are compared when used to simultaneously estimate the state variables as well as the model parameters of the CSEC. The effect of the number estimated parameters on the performances of these state estimation techniques is also investigated. Next, the model of CSEC, that will be used in our analysis, will be described.

5.1.1 Model of the Cad System in E. coli

The Cad system is one of the conditional stress response modules in *E.coli*, that is induced only at low pH and a lysine-rich environment [40]. The major components of the Cad system are the enzyme CadA, the transport protein CadB, and the regulatory protein CadC. The decarboxylase CadA converts lysine Lys into cadaverine in a reaction which consumes H+. The transport protein CadB imports the substrate, lysine and exports the product, cadaverine. So, the intracellular H+ concentration is reduced and the cell returns back to pH homeostasis. The membrane protein CadC senses the external conditions and regulates the stress response by binding directly to the DNA and activating the transcription of cadBA. This ensures that CadA and CadB are produced only under the appropriate external conditions of low pH and lysine abundance. Furthermore, as presented in [41], CadC senses the external cadaverine and the accumulation of cadaverine in extracellular medium causes a delayed transcriptional down regulation of cadBA expression.

In the current work, we use the available time profile data set presented in [42] for parameters and states estimation of the S-system model. Based on the model and available pathway information, presented in [40], the S-system model can be written as:

$$\frac{dCadA}{dt} = \alpha_1 \, [Cadav]^{g_{13}} - \beta_1 [CadA]^{h_{11}}$$

$$\frac{dCadB}{dt} = \alpha_2 \, [CadA]^{g_{21}} - \beta_2 [CadBA]^{h_{22}} \tag{46}$$

$$\frac{dCadC}{dt} = \alpha_3 [CadBA]^{g_{32}} - \beta_3 [Cadav]^{h_{33}} [Lys]^{h_{34}}$$

$$\frac{dLys}{dt} = \alpha_4 [CadA]^{g_{41}} - \beta_4 [Lys]^{h_{44}}$$

where $\theta = [\alpha_1..., \alpha_4, \beta_1..., \beta_4, g_{13}, g_{15}, g_{32}, g_{41}, h_{11}, h_{22}, h_{33}, h_{34}, h_{44}]$ is a set of parameters. Discretizing the model (46) using a sampling interval of and incorporating random process noise (to account for any uncertainties in the CSEC process model), the model can be written as,

$$CadA_k = CadA_{k-1} + [\theta_1 [Cadav_k]^{\theta_9} - \theta_5 [CadA_k]^{\theta_{13}}] \Delta t + \omega_{k-1}$$

$$CadB_k = CadB_{k-1} + [\theta_2 [CadA_k]^{\theta_{10}} - \theta_6 [CadBA_k]^{\theta_{14}}] \Delta t + \omega_{k-1} \tag{47}$$

$$CadC_k = CadC_{k-1} + [\theta_3 [CadBA_k]^{\theta_{11}} - \theta_7 [Cadav_k]^{\theta_{15}} [Lys_k]^{\theta_{17}}] \Delta t + \omega_{k-1}$$

$$Lys_k = Lys_{k-1} + [\theta_4 [CadA_k]^{\theta_{12}} - \theta_8 [Lys_k]^{\theta_{17}}] \Delta t + \omega_{k-1}$$

where $\omega_k \sim N(0, \sigma_w^2)$ is the process noise with zero mean Gaussian noise.

5.1.2 Generation of Dynamic Data

At this point of the research, the model parameters are assumed to be constants, and at their nominal value presented in Table 1. Therefore, we consider the state vector that we wish to estimate as: $z_k = x_k = [CadA_k CadB_k CadC_k Lys_k]^T$.

To go further in the research, it appear now to own data on which running the model. Indeed, the results may depend on the details of the model, on the way/quality the data are generated/measured with and on the specific data that are used. To be independent of these considerations, we will generate dynamic data from the CSEC. The model is first used to simulate the responses of the enzyme CadA, the transport protein CadB, the regulatory protein CadC and lysine Lys for a Model of the Cad System in E. coli as functions of time as functions. These simulated states, which are assumed to be noise free, are then contaminated with zero mean Gaussian errors, i.e., a measurement noise $v_{k-1} \sim N(0, \sigma_v^2)$. Considering a value of $\sigma_v^2 = 0.01$ the following data set can be generated. Fig. 1 shows the changes in the four state variables. The model parameters of CSEC as well as other physical properties are shown in Table 1.

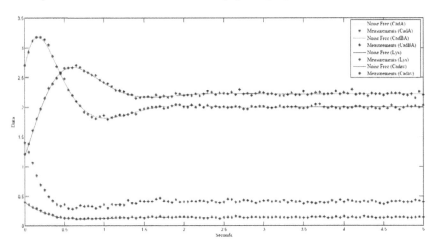

FIGURE 1   Simulated CSEC data used in estimation: state variables (CadA, CadB, CadC, Lys).

TABLE 1   True Values of Csec Parameters Model

Parameter	Value	Parameter	Value	Parameter	Value	Parameter	Value
α_1	12	β_1	10	g_{13}	-0.8	h_{11}	0.5
α_2	8	β_2	3	g_{13}	0.5	h_{22}	0.75
α_3	3	β_3	5	g_{13}	0.75	$h_{33} \& h_{34}$	0.5&0.2
α_4	2	β_4	6	g_{13}	0.5	h_{11}	0.8

5.1.3 Estimation of State Variables from Noisy Measurements using EKF, UKF and PF

In this comparative study, the objective is to compare the estimation accuracy of EKF, UKF and PF when they are used to estimate the four state variables of the CSEC process, i.e., the enzyme CadAk, the transport protein CadBk, the regulatory protein CadCk and lysine Lysk.

The simulation results of estimating the four states *CadA, CadB, CadC* and *Lys* using EKF, UKF and PF are shown in Fig. 2(a,b,c). Also, the estimation root mean square errors (RMSE) for the estimated states are shown in Table 2. It can be observed from Fig. 2 and Table VI that EKF resulted in the worst performance of all estimation techniques, which is expected due to the limited ability of EKF to accurately estimate the mean and covariance matrix of the estimated states through lineralization of the nonlinear process model. The results also show that the PF provides a significant improvement over the UKF, which is due to the fact that, by using UKF, linearizing the process model does not necessarily provide good estimates of the mean of the state vector and the covariance matrix of the estimation error which are used in state estimation.

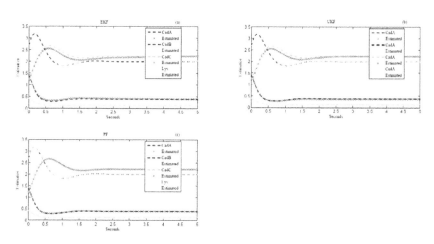

FIGURE 2   Estimation of state variables using various state estimation techniques.

TABLE 2   Root Mean Square Errors (Rmse) of Estimated States for Ekf, Ukf and Pf

Technique	$CadA_k$	$CadB_k$	$CadC_k$	Lys_k
EKF	0.0694	0.1160	0.1215	0.0311
UKF	0.0593	0.0937	0.1129	0.0195
PF	0.0009	0.0012	0.0009	0.0012

5.2   Faults detection of biological systems representing Cad System in E. coli

In the following section, the GLR-based PCA test algorithm performance will be assessed and compared to that of the conventional PCA set through two examples using simulated biological systems representing Cad System in *E. coli* data. The data set consists of 4 random variables, which are generated using the complex three degree of free domspring-mass-dashpot system. The generated data were arranged as a matrix *X* having 100 samples for discretization is 0.05 seconds and 4 Cad System in *E. coli* measurements. The responses of the 4 state variables *CadA, CadB, CadC* and *Lys* are shown in Fig. 3, where X_5 and X_6 are expressed as:

$$\begin{cases} X_5 = 0.5X_1 + 0.5X_2 \\ X_6 = 0.7X_3 + 0.3X_4 \end{cases} \tag{48}$$

5.2.1 Training of PCA model

As described in Algorithm 1, the PCA-based GLR fault detection method requires constructing a PCA model from fault-free data. Therefore, the fault-free Cad System in E. coli training data described earlier were used to construct a PCA reference model to be used in fault detection. The fault-free Cad System in E. coli system data were arranged as a matrix *Xtr* having 100 rows (samples) and 4 columns (Cad System in *E. coli* measurements). These data are first scaled (to have zero mean and unit variance), and then are used to construct the PCA model. The responses of the training fault-free data, are shown in Fig. 4. The training fault-free data matrix *Xtr* is used to construct a PCA model.

In PCA, most of the crucial variations in the data set are typically captured in the main principal components corresponding to the maximum eigenvalues as shown in Fig. 5. In this study herein, the cumulative percent variance (CPV) method is utilized to find out the optimum number of retained principal components. Utilizing a CPV threshold value of 90%, only the first two principal components of the total variations in the data as displayed in Fig. 5.) will be retained.

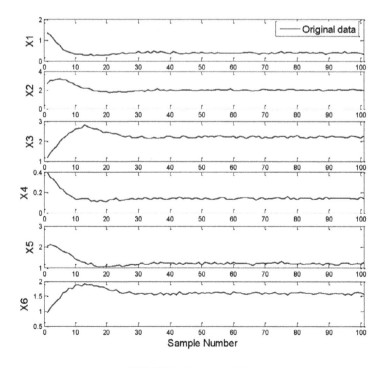

FIGURE 3 The original data.

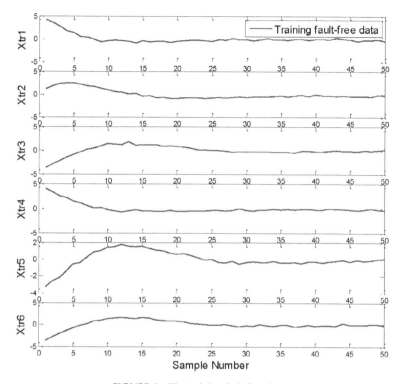

FIGURE 4 The training fault-free data.

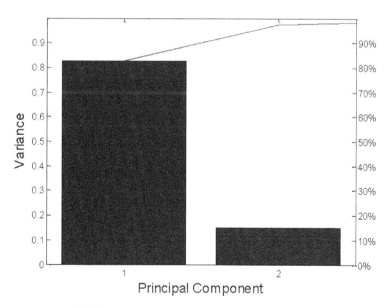

FIGURE 5 Variance captured by each principal component.

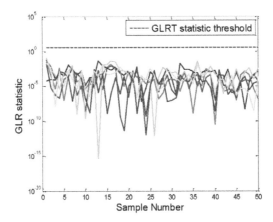

FIGURE 6 The time evolution of GLR decision function on a semi-logarithmic scale for the fault-free data.

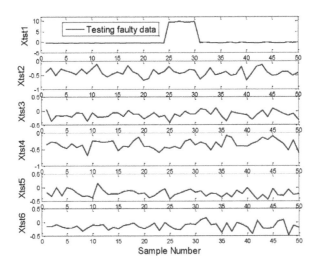

FIGURE 7 The single testing faulty data X_{tst}.

A plot of the decision function of the GLR test T (shown in Fig. 6) confirms that the process operates under normal conditions, where no faults are present.

5.2.2 Fault detection Cad System in E. coli processes

The PCA model formed utilizing the fault-free data is deployed in this section t o detect possible faults with unseen testing data. The data set from tests (which is simulated using Cad System in *E. coli* system) includes 50 data samples that are free of the training data. An additive fault was introduced in X_1. It consists of a bias of amplitude equal to 10% of the total variation in X_1, between sample numbers 25 and 30 (see Fig. 7).

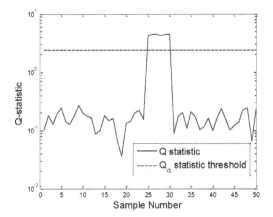

(a) Q statistic in the presence of simple fault.

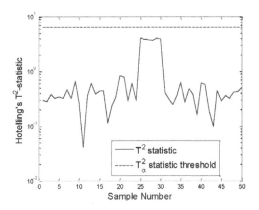

(b) Hotelling's T^2-statistic in the presence of simple fault.

The performance of the various faults detection methods will be compared. It is assumed that the sensor measuring one state variable is damaged by an additive fault.

The conventional PCA based monitoring technique is initially run using the training fault-free data. Based on the first four PCs, T^2 and Q statistics for the conventional PCA algorithm and the GLR-based PCA test algorithm are used for fault detection. Fig. 8(a) shows the testing faulty data (additive fault in $X1$) (see Fig. 7). The results of Q statistic are shown in Fig. 8(a), where the dotted line represents the detection threshold Q_α, which is found to be 2.426. Fig. 8(b) presents the results of the T^2 statistic, where the dotted line represents the detection threshold T^2_α which is found to be 6.514.

The process monitoring under fault using the PCA indices is presented in Fig. 8(a) and (b). Fig. 8(a) shows that the Q statistic at the time interval [25...30] is always above the threshold Q_α, which means that the data fit the PCA model well (since it could capture most of the variations in the data), and verifies that the data belongs to the normal operating region. However, using T^2 statistic the additive fault is not detected as shown in Fig. 8(b). In this case, the Q statistic detects this fault better

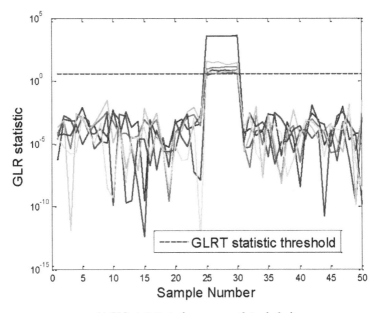

(c) GLR statistic in the presence of simple fault.

FIGURE 8   Fault detection in the presence of additive fault.

than T^2 statistic as these figures show. When the GLR test is applied using the same fault-free data, the GLR threshold value is found to be $t_\alpha = 3.841$ for a false alarm probability of $\alpha = 10\%$. A plot of the GLR and Q statistics (shown in Fig. 8) confirms that the process operates very well compared to the fault detection using T^2. We can show that the results which are shown in Fig. 8 (a), (b) and (c), show the ability of the GLR and Q statistics to detect this additive fault.

6.   Conclusions

In this chapter, first, state estimation techniques are used to simultaneously esti-mate the state variables for biological systems representing Cad System in *E. coli*. Various state estimation techniques, which include the extended Kalman filter (EKF), unscented Kalman filter (UKF), and particle filter (PF), are compared as they are used to achieve this objective. In the comparative study, EKF, UKF and PF are used to estimate the state variables for biological systems representing Cad System in *E. coli*. The simulation results of the comparative study show that the PF provides a higher accuracy than the EKF due to the limited ability of the EKF to deal with highly nonlinear process models. The results also show that the PF provides a significant improvement over the UKF. This is because the covariance is propagated through linearization of the underlying non-linear model, when the state transition and observation models are highly non-linear. Second, generalized likelihood ratio (GLR) based principal components analysis (PCA) is used for fault detection in biological systems representing Cad System in *E. coli*. The objective is to combine the GLR test with PCA model in order to improve fault detection performance. The PCA-based GLR is proposed to detect the faults of Cad System in *E. coli* process, in which, PCA is used to create the model and find a linear combinations of variables which describe major trends in data set, and the GLR test, is utilized to improve faults detection. It is demonstrated that the performance of faults detection can be improved by combining GLR test and PCA.

Acknowledgment

This work was made possible by NPRP grant NPRP7-1172-2-439 from the Qatar National Research Fund (a member of Qatar Foundation). The statements made herein are solely the responsibility of the authors.

References

Aidala, V., 1977 Parameter estimation via the kalman filter, IEEE Trans. on Automatic Control, 22(3): 471–72.

Barndorff-Nielsen, O. 1977. Exponentially decreasing distributions for the logarithm of particle size. Proceedings of the Royal Society of London. A. Mathematical and Physical Sciences, 353(1674), 401-19.

Benaicha, A., M. Guerfel, N. Boughila, and K. Benothman, New pca-based methodology for sensor fault detection and localization, in *MOSIM'10*, Hammamet - Tunisia, May 10-12.

Bolton, R. J., D. J. Hand *et al.*: Unsupervised profiling methods for fraud detection, *Credit Scoring and Credit Control VII*, pp. 235–255, 2001.

Bulut, A., A. K. Singh, P. Shin, T. Fountain, H. Jasso, L. Yan, and A. Elgamal, 2005 Real-time nondestructive structural health monitoring using support vector machines and wavelets," in *Nondestructive Evaulation for Health Monitoring and Diagnostics*. International Society for Optics and Photonics, 2005, pp. 180–89.

Chen, G., Q. Xie, and L. Shieh, "Fuzzy kalman filtering," Journal of Information Science, vol. 109, pp. 197–209, 1998.

Chiang, L. H., R. D. Braatz, and E. L. Russell, 2001 *Fault detection and diagnosis in industrial systems*. Springer.

Critchley, F., 1985 Influence in principal components analysis, *Biometrika*, 72(3) pp. 627–36.

David, Z., and B. Marta, 2008 From large chemical plant data to fault diagnosis integrated to decentralized fault-tolerant control: Pulp mill process application," *Industrial and Engineering Chemistry Research, vol47*, no. 4, pp. p1201–1220.

Dawdle, J. R., A. Willsky, and S. W. Gully, 1982 Nonlinear generalized likelihood ratio algorithms for maneuver detection and estimation, in *American Control Conference, 1982*. IEEE, pp. 985–87.

Diana, G. and C. Tommasi, 2002 Cross-validation methods in principal component analysis: A comparison," *Statistical Methods & Applications*, vol. 11, no. 1, pp. 71–82.

Gonzalez, F., D. Dasgupta, and R. Kozma, 2002 Combining negative selection and classification techniques for anomaly detection," in *Evolutionary Computation, 2002. CEC'02. Proceedings of the Congress on*, vol. 1. IEEE, 2002, pp. 705–710.

Grewal M., and A. Andrews, 2008 *Kalman Filtering: Theory and Practice using MATLAB*. John Wiley and Sons.

Gustafsson, F., 1996 The marginalized likelihood ratio test for detecting abrupt changes, *IEEE Transactions on Automatic Control*, 41(1):. 66–78.

Hotelling, H., Analysis of a complex of statistical variables into principal components," *Journal of Educational Psychology*, 24: 417–41.

Jackson J. E., and G. S. Mudholkar, 1979 Control procedures for residuals associated with principal component analysis," *Technometrics*, vol. 21, no. 3, pp. 341–49.

Jia, J. 2009. Parameter estimation for nonlinear biological system model based on global sensitivity analysis, in *3rd International Conference on Bioinformatics and Biomedical Engineering*, pp. 1–4.

John, G. H., 1995 Robust decision trees: Removing outliers from databases." in *KDD*, pp. 174–179.

Jolliffe, I., 2002 Principal component analysis," *second edition, Springer, Berlin*.

Julier, S. and Uhlmann, J. 1997). New extension of the kalman filter to nonlinear systems, *Proceedings of SPIE*, pp 182–93.

Kalman, R. E., 1960 "A new approach to linear filtering and prediction problem," Trans. ASME, Ser. D, J. *Basic Eng.*, 82: 34–45.

Kramer, M. A., 1991 Nonlinear principal component analysis using autoassociative neural networks, *AIChE journal*, 37(2) 233–243.

Ku, W., R. H. Storer, and C. Georgakis, 1995 Disturbance detection and isolation by dynamic principal component analysis, *Chemometrics and Intelligent laboratory Systems*, 26(9) 179–196, 1995.

Lane, S., E. Martin, A. Morris, and P. Gower, 2003 Application of exponentially weighted principal component analysis for the monitoring of a polymer film manufacturing process, *Transactions of the Institute of Measurement and Control*, 25(1): 17–35.

Lee, J.H., & Ricker, N.L. 1994. Extended kalman filter based nonlinear model predictive control. Industrial & engineering chemistry research, 33(6):1530–41.

Li, W., H. H. Yue, S. Valle-Cervantes, and S. J. Qin, 2001 Recursive pca for adaptive process monitoring, *Journal of process control*, 1–486.

Liu, J. and Chen, R., 1998. Sequential monte carlo methods for dynamic systems, *Journal of the American statistical association*, pp. 1032–1044.

MacGregor J., and T. Kourti, 1995 Statistical process control of multivariate processes, *Control Engineering Practice*, vol. 3, no. 3, pp. 403–414, 1995.

Mansouri, M. M., H. N. Nounou, M. N. Nounou, and A. A. Datta, 2014 State and parameter estimation for nonlinear biological phenomena modeled by s-systems, *Digital Signal Processing*, 281–17.

Mansouri, M., B. Dumont, V. Leemans, and M.-F. Destain, 2014 Bayesian methods for predicting lai and soil water content, *Precision agriculture*, v15(2): 184–201, 2014.

Mansouri, M., H. Snoussi, and C. Richard, 2009 A nonlinear estimation for target tracking in wireless sensor networks using quantized variational filtering," in Signals, Circuits and Systems (SCS), 2009 3rd International Conference on. *IEEE*, pp. 1–4.

Maronna, R., D. Martin, and V. Yohai, *Robust statistics*. John Wiley & Sons, Chichester. ISBN, 2006.

Matthies, L., T. Kanade, and R. Szeliski, 1989 Kalman filter-based algorithms for estimating depth from image sequences,"*International Journal of Computer Vision*, 3(3): 209–238.

Merwe, R. V. D. and E. Wan, 2001 The square-root unscented kalman filter for state and parameter-estimation, IEEE International Conference on Acoustics, Speech, and Signal Processing, 6, 3461–64.

Misra, M., H. H. Yue, S. J. Qin, and C. Ling, 2002 Multivariate process monitoring and fault diagnosis by multi-scale pca," *Computers & Chemical Engineering*, 26(9). 1281–1293.

Mourad, M. and J.-L. Bertrand-Krajewski, 2002 A method for automatic validation of long time series of data in urban hydrology, *Water Science & Technology*, 45(4–7): 263–270, 2002.

Nguyen, D. and B. Widrow, 1990 Improving the learning speed of 2-layer neural networks by choosing initial values of the adaptive weights. In *Neural Networks, IJCNN International Joint Conference on*. IEEE, 21–26.

Nounou, H., and M. Nounou, 2006 Multiscale fuzzy kalman filtering, *Engineering Applications of Artificial Intelligence*, 19 439–50.

Qin, S., 2003 Statistical process monitoring: Basics and beyond," *Journal of Chemometrics*, vol. 17, no. 8/9, pp. 480–502.

Ramaswamy, S., R. Rastogi, and K. Shim, 2000 Efficient algorithms for mining outliers from large data sets," in *ACM SIGMOD Record:* 29(2) ACM, pp. 427–438.

Rousseeuw, P. J. and I. Ruts, 1996 Algorithm as 307: Bivariate location depth, *Applied Statistics*, pp. 516–526.

Russell, E. L., L. H. Chiang, and R. D. Braatz, 2000 Fault detection in industrial processes using canonical variate analysis and dynamic principal component analysis," *Chemometrics and Intelligent Laboratory Systems*, vol. 51, no. 1, pp. 81–93.

Ruts, I. and P. J. Rousseeuw, 1996 Computing depth contours of bivariate point clouds, *Computational Statistics & Data Analysis*, 23(1): 153–168, 1996.

Ruymgaart, F., 1981 A robust principal analysis, vol. 1, p. 485–497.

Simon, D., 2003 Kalman filtering of fuzzy discrete time dynamic systems, Applied Soft Computing, 3 : 191–207.

Simon, D., 2006 *Optimal State Estimation: Kalman, H∞, and Nonlinear Approaches*. John Wiley and Sons.

Tamura, M. and S. Tsujita, 2007 A study on the number of principal components and sensitivity of fault detection using pca," *Computers & chemical engineering*, 31(9):. 1035–1046.

Tang, J., Z. Chen, A. W.-c. Fu, and D. Cheung, 2000 A robust outlier detection scheme for large data sets, in *In 6th Pacific-Asia Conf. on Knowledge Discovery and Data Mining*. Citeseer, 2001.

Wan, E., and R. V. D. Merwe, 2000 The unscented kalman filter for nonlinear estimation," *Adaptive Systems for Signal Processing, Communications, and Control Symposium*, pp. 153–58.

Wang, H. Qian, L. and Dougherty, E. 2007. Inference of gene regulatory networks using s-system: A unified approach, in *Computational Intelligence and Bioinformatics and Computational Biology*. pp.82–89.

Willsky, A. S., E. Chow, S. Gershwin, C. Greene, P. Houpt, and A. Kurkjian, 1980 Dynamic model-based techniques for the detection of incidents on freeways, *Automatic Control, IEEE Transactions on*, 25(3) 347–360.

Wu, F. 2007. Estimation of parameters in the linear-fractional models, in *Engineering in Medicine and Biology Society, 29th Annual International Conference of the IEEE*, 1086 89.

Xu, L. and A. L. Yuille 1995, Robust principal component analysis by self-organizing rules based on statistical physics approach, *Neural Networks, IEEE Transactions on*, vol. 6, no. 1, pp. 131–143.

Yang, N., Tian, W., Jin, Z., and Zhang, C., 2005. Particle filter for sensor fusion in a land vehicle navigation system, *Measurement science and technology*, p. 677.

Zhu, M. and A. Ghodsi, 2006 Automatic dimensionality selection from the scree plot via the use of profile likelihood, *Computational Statistics & Data Analysis*, 51: 918–930, 2006.

Zumoffen D., and M. Basualdo, 2008 From large chemical plant data to fault diagnosis integrated to decentralized fault-tolerant control: pulp mill process application," *Industrial & Engineering Chemistry Research*, $7(4). 4, 01–20.

14

Next-Generation Sequencing and Metagenomics

Nageswara Rao Reddy Neelapu[1]* and Challa Surekha[2]

Abstract

Isolating microbes such as bacteria, fungi and virues from different environments like soil, water, air, plants, animals, hot springs, cold climates and space in purest form of culture is of interest for any biologist. Despite of many new technologies, limitations have stalled to get the purest forms of genetic material of microbes without losing its nativity. Microbes' losing their nativity means losing their ability to express and grow in the virtue of its unique environment. Metagenomics is a powerful tool to recover sample/the genetic material of a microbe or entire communities of organisms directly from its environment without losing its nativity.

This review summarizes information and current opinions on metagenomics by providing more insights on how gene expression has shaped the microbe development and also how an environment can influence the genome expression. How different next generation technologies had offered a platform to sequence the complete metagenomes of the microbes from the environment; and subjected to bioinformatics analysis such as sequence prefiltering, assembly, gene prediction, species diversity, data integration, and comparative metagenomics to deliver holistic information on interactions among communities, ecosystems and populations on community metabolism and population genomics. In addition what way, application of metagenomics has solved many challenges and proved its practicality in the fields of agriculture, food, engineering, evolution, medicine and sustainability.

1. Introduction

Metagenomics is the broad field with combination of microbiology, environment science and genomics commonly known as environmental genomics, ecogenomics or

---

[1.] Department of Biochemistry and Bioinformatics, School of Life Sciences, GITAM Institute of Science, GITAM University, Rushikonda, Visakhapatnam 530045 (AP), India Phone: +91-891-2840464 Fax: +91-891-2790032 Email: nrneelapu@gmail.com

[2.] Department of Biochemistry and Bioinformatics, School of Life Sciences, GITAM Institute of Science, GITAM University, Rushikonda, Visakhapatnam- 530045 (AP), India Phone: +91-891-2840464 Fax: +91-891-2790032 Email: challa_surekha@yahoo.co.in

* Corresponding author: nrneelapu@gitam.edu

community genomics. In nature when a sample was isolated and profiled, it revealed microbial communities which are rich and diverse. Traditional way of understanding an organism is by isolating it from the sample collected from the environment. Then cultivating the organism for pure clonal culture and sequencing the genome of the microbe to characterize its virtue. During this process microbes and their communities may lose their nativity. Handelsman et al. (1998) for the first time used the term metagenomics to depict the idea of collection of genes sequenced directly from the environment. Chen and Patcher (2005) defined metagenomics as '...the application of modern genomics techinques to study the communities of microorganisms directly in their natural environments, by passing the need for isolation and lab cultivation of individual species'.

Metagenomics has its root in studies conducted on bacteria, eubacteria and archeabacteria based on rRNA sequences. These studies established the fact of existence of ribosome nucleotide sequence diversity along with microbial community diversity paving the way for present fields such as environmental genomics, ecogenomics or community genomics. Pace et al. (1991) proposed the idea of directly isolating and cloning bulk DNA from environmental sample. Healy et al. (1995) were successful in directly isolating and cloning bulk DNA from environmental sample consisting of microbial consortia by developing Zoo libraries. Stein et al. (1996) used marine samples to constructs marine libraries for sequencing of metagenome. Breitbart et al. (2002) used shot gun sequencing to establish the uncultured viral communities in 200 litres of sea water leading to the present day metagenomics. Subsequent studies cleared the twilight zone establishing new viruses in large numbers in human stool (Tyson et al. 2004), marine sediment (Tyson et al. 2004), and acid mine drainage system (Hugenhoz 2002). Developments in metagenomics have resulted in complete or nearly complete genome of bacteria/archeabacteria which were tough in obtaining pure cultures. Venter et al. (2004) as a part of Global Ocean Sampling Expedition (GOS) used shot gun metagenomics and identified nearly 2000 different species, which include 148 types of bacteria never known before. Many prestigious projects were undertaken by different groups around the world. Among them Human Microbiome Initiative is one of the most prestigious project. Successful metagenomics project implements the following practices: sampling and processing, sequencing, sequence prefiltering, assembly, binning, annotation, sharing and storage of metadata and data analysis (Fig. 1.)

1.1 Sampling and processing

Sample processing is the crucial step in a metagenomics project. The DNA isolated shall be representative of all cells in the sample and with sufficient amounts of high-quality nucleic acids. Sample is associated with a host (e.g. an animal or plant), then either fractionation or selective lysis is used (Burke et al. 2009; Thomas et al. 2010). Physical fractionation can be used if community sample is viruses in seawater. Direct lysis of the soil sample is quantifiable than indirect lysis. Biopsies or groundwater yield very small amounts of DNA. Thus, sampling and processing is an important step for getting required quantities of DNA for sequencing.

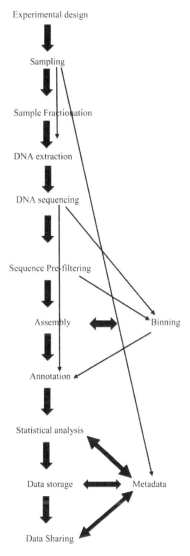

FIGURE 1   Flow diagram of metagenome project.

1.2   Metagenome sequencing technology

DNA double stranded structure was established in the year 1953 (Watson and Crick 1953), but it took decades to develop DNA sequencing methods. Wu (1970) developed the first method to determine the DNA sequence using location specific primer extension strategy. Synthetic location – specific primer was used to determine the sequence between the years 1970 and 1973 (Padmanabhan and Wu 1972; Wu et al. 1973; Jay et al. 1974; Padmanabhan et al., 1974) and Wu and Taylor (1971) subjected phage λ DNA to sequence 12 bp of cohesive ends. Chemical degradation was used by Walter Gilbert and Allan Maxam to sequence DNA (Maxam and Gilbert 1977). More rapid DNA sequencing method with chain-terminating inhibitors was developed by Frederick Sanger, which is now considered as the first generation sequencing method (Sanger et al. 1977). This method identifies linear randomly terminated nucleotide sequences, whereas automated Sanger's method uses fluorescent labeled terminators. The most accurate, most available and well defined technology is Sanger sequencing method. The throughput of this automated method can read 96 reactions in parallel. Using this method, complete human genome would take around 60 years to synthesize and cost approximately 5 to 30 million USD. First semi-automated DNA sequencing machine was announced in 1986 by Leroy E. Hood's laboratory followed by fully automated sequences by Applied Biosystem in 1987 and novel florescent labeling technique by Dupont's Genesis 2000 (Prober et al. 1987). Next generation high throughput DNA sequencing technologies capable of sequencing large number DNA sequences in a single reaction were developed. Next generation sequencing (NGS) technology provide high speed, throughput and monitors the sequential addition of nucleotides to DNA templates. The need and demand for low cost sequencing made the development of next generation sequencing (high-throughput sequencing) technologies. Overall high cost, reduction of sequencing errors, low reading accuracy are the limiting factors of the new technologies.

Several NGS methods like 454 pyrosequencing, llumina (Solexa) sequencing, SOLiD sequencing Massively parallel signature sequencing (MPSS), Polony sequencing, Ion Torrent semiconductor sequencing, DNA nanoball sequencing, Heliscope single molecule sequencing and Single molecule real time (SMRT) sequencing etc are the technologies used for sequencing metagenomes. Genomic DNA, cDNA, immunoprecipitated DNA can serve as DNA template for all NGS experiments (Fig. 2)

TABLE 1 Highlights of next generation sequencing technologies

Single-molecule real time sequencing (SMRT)	
Read Length	10,000bp to 15,000bp avg (14,000 bp N50); maximum read length >40,000 bases
Accuracy	99.99% consensus accuracy; 87% single read accuracy
Reads per run	50.000 per SMRT cell or 500-1000 megabases
Time per run	30 minutes to 4 hours
Cost per 1 million bases	$0.13-0.60
Advantages	Longest read length, Fast detects 4mC, 5mC, 6mA.
Disadvantages	Moderate throughput Equipment can be very expensive
Ion Semiconductor (Ion Torrent sequencing)	
Read Length	Upto 400bp
Accuracy	98%
Reads per run	Upto 80 million
Time per run	2 hours
Cost per 1 million bases	$1
Advantages	Less expensive equipment, Fast
Disadvantages	Hompolymer errors
Pyrosequencing (454)	
Read Length	750bp
Accuracy	99.9%
Reads per run	1million
Time per run	24hours

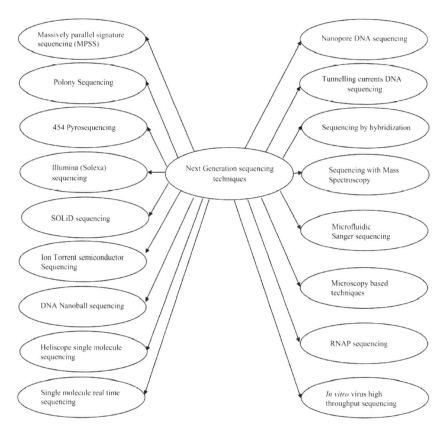

FIGURE 2 Next generation sequencing technologies.

Cost per 1 million bases	$10
Advantages	Long read size, Fast
Disadvantages	Runs are expensive, Homopolymer errors
Solexa Sequencing synthesis (Illumina)	
Read Length	500 to 300 bp
Accuracy	98%
Reads per run	Upto 3 million
Time per run	1 to 10days depending on sequence and specified read length
Cost per 1 million bases	$0.05 to 0.12
Advantages	Potential for high sequence yield, depending upon sequencer model and desired application
Disadvantages	Equipment can be very expensive and requires high concentrations of DNA.
Sequencing by Ligation (SOLiD sequencing)	
Read Length	50+35 or 50+ 50 bp
Accuracy	99%
Reads per run	1.2 to 1.4 billion
Time per run	1 to 2 weeks

Cost per 1 million bases	$0.13
Advantages	Low cost per base
Disadvantages	Slower than other methods, have issue sequencing palindromic sequence
Chain termination (Sanger sequencing)	
Read Length	400 to 900 bp
Accuracy	99.9%
Reads per run	N/A
Time per run	20 minutes to 3 hours
Cost per 1 million bases	$2400
Advantages	Long individual reads, Useful for many applications
Disadvantages	More expensive and impractical for larger sequencing projects.

1.2.1   454 pyrosequencing

454 pyrosequencing was first reported in 1988 using the principle of pyrophosphate detection and the technique was further developed (Ronaghi et al. 1996) in Stockholm to analyze 96 samples in parallel in a microtiter plate. The 454 GenomeSequencer FLX instrument was introduced in 2005 by 454 Life Sciences. Pyrosequencing method is based on principle 'sequencing by synthesis'. Template DNA is ligated with specific adapter to attach one fragment of DNA to a single primer coated on streptavidin bead to form a clonal colony. Emulsion PCR is carried out with water droplets in an oil solution. The droplets act as individual amplification reactors producing approximately 10^7 clonal copies per bead (Margulies et al. 2005). ATP sulfurylase and luciferase enzymes generate visible light upon incorporation of the complementary nucleotide. The amplified DNA is transferred into a picotiter plate and analyzed using a pyrosequencing reaction. The picotiter plate allows hundreds and thousands of parallel pyrosequencing reactions to be carried out increasing the sequencing throughput. As this approach is sequencing-by-synthesis, it measures the release of inorganic pyrophosphate detected by light enzyme luciferase (Tawfik and Griffiths 1998; Nyren et al. 1993). The light intensity is proportional to the pyrophosphate released. This technology provides intermediate read length and generates 80-120Mb of sequence in 4 hours.

1.2.2   Illumina (Solexa) sequencing

Illumina (Solexa) sequencing is based on reversible dye-terminators technology, and engineered polymerases (Bentley et al. 2008). This approach was invented by Balasubramanian and Klennerman from Cambridge University's chemistry department. This method is similar to Sanger sequencing, but uses modified dNTP's containing a terminator which blocks further polymerization. DNA molecules (single stranded) are attached to an immobilized surface by an adapter, subsequently bent and hybridized to complementary adapters for synthesis of their complementary strands. After the amplification, the templates are sequenced in a massive parallel form using sequencing-by-synthesis approach. To determine the sequence, four types of reversible terminator bases (RT-bases) are added in presence of special DNA polymerases and non-incorporated nucleotides are washed away (Ronaghi et al. 1996). A camera takes images of the fluorescently labelled terminator nucleotides and its position. Then the dye along with the terminal 3' blocker, are chemically removed from the DNA, allowing for the next cycle to begin. The Illumina sequencing is capable of generating 35bp reads and produce 1Gb sequence in 2-3 days. The method is highly accurate base by base sequencing by eliminating errors.

1.2.3   SOLiD sequencing

SOLiD sequencing employs the process by ligation and was introduced in autumn 2007. A library of DNA fragments are first ligated to magnetic bead with universal P1 adapter. Emulsion PCR takes place in microreactors with all reagents of PCR. The resulting beads, each containing single copies of the same DNA molecule, are deposited on a glass slide (Valouev et al. 2008). One DNA fragment per bead bound to an adapter is hybridized with a primer. These templates are characterized by fluorescent labels and detected by fluorescence. The sequencing process is continued in the same way with another primer and sequence reading length is about 35 bases. Sequences are determined in parallel for more than 50 million bead clusters, resulting in a very high throughput of the order of Gigabases per run. The new SOLiD instrument is capable of producing 1-3Gb of sequence in 8-day run and offers 99.94% accuracy.

1.2.4   Massively parallel signature sequencing (MPSS)

The first of the next-generation sequencing technologies, MPSS, was developed in 1990s at Lynx Therapeutics by Sydney Brenner and Sam Eletr. MPSS is a procedure used for identifying and quantifying mRNA transcripts similar to serial analysis of gene expression (SAGE). MPSS was a bead-based method that used a complex approach of adapter ligation followed by adapter decoding, reading the sequence in increments of four nucleotides. mRNA is reverse transcribed into cDNA, cDNA is amplified in microreactors as emulsion PCR. A sequence signature of ~16-20bp is determined from all the beads in parallel. Each signature sequence is cloned onto microbeads and then arrayed in a flow cell for sequencing and quantification. This method made it susceptible to sequence-specific bias or loss of specific sequences. The technology was so complex, MPSS was only performed 'in-house' by Lynx Therapeutics and no DNA sequencing machines were sold to independent laboratories. Lynx Therapeutics merged with Solexa (later acquired by Illumina) in 2004, leading to the development of sequencing-by-synthesis, a simpler approach acquired from Manteia Predictive Medicine, which rendered MPSS obsolete. Typically used for sequencing cDNA to measure gene expression levels (Brenner et al. 2000).

1.2.5   Polony sequencing

Polony sequencing was developed in the laboratory of George M Church at Harvard. This technique was among the first next-generation sequencing systems and was used to sequence a full genome in 2005. It combined an *in vitro* paired-tag library with emulsion PCR, an automated microscope, and ligation-based sequencing. The chemistry to sequence an *E. coli* genome at an accuracy of >99.9999% costed approximately 1/9 that of Sanger sequencing (Shendure et al. 2005). Emulsion PCR isolates individual DNA molecules along with primer-coated beads in aqueous droplets within an oil phase. PCR then coats each bead with clonal copies of the DNA molecule followed by immobilization for later sequencing.

1.2.6   Ion Torrent semiconductor sequencing

Ion Torrent Systems Inc. developed a system based on using standard sequencing chemistry, but with a novel, semiconductor based detection system. This method of sequencing is based on the detection of hydrogen ions that are released during the polymerisation of DNA, as opposed to the optical methods used in other sequencing systems. A microwell containing a template DNA strand to be sequenced is flooded with a single type of nucleotide. If the introduced nucleotide is complementary to the leading template nucleotide it is incorporated into the growing complementary strand. This causes the release of a hydrogen ion that triggers a hypersensitive ion sensor, which indicates that a reaction has occurred. If homopolymer repeats are present in the template sequence multiple nucleotides will be incorporated in a single cycle. This leads to a corresponding number of released hydrogens and a proportionally higher electronic signal (Rusk 2011).

1.2.7   DNA nanoball sequencing

DNA nanoball sequencing is a type of high throughput sequencing technology used to determine the entire genomic sequence of an organism. The method uses rolling circle replication to amplify small fragments of genomic DNA into DNA nanoballs. Fluorescent probes bound to complementary DNA are ligated to anchor sequences bound to known sequences on the DNA template. Unchained sequencing by ligation is then used to determine the nucleotide sequence (Drmanac et al. 2010). This method of DNA sequencing allows large numbers of DNA nanoballs to be sequenced per run and at low reagent costs compared to other next generation sequencing platforms (Porreca 2010). However, only short sequences of DNA are determined from each DNA nanoball which makes mapping the short reads to a reference genome difficult. This technology has been used for multiple genome sequencing projects and is scheduled to be used for more.

1.2.8   Heliscope single molecule sequencing

Heliscope sequencing is a method of single-molecule sequencing developed by Helicos Biosciences. This method helps in direct sequencing of cellular and extracellular nucleic acids in an unbiased manner. It uses DNA fragments with added poly-A tail adapters which are attached to the flow cell surface. The next steps involve extension-based sequencing with cyclic washes of the flow cell with fluorescently labeled nucleotides (one nucleotide type at a time, as with the Sanger method). The reads are performed by the Heliscope sequencer. The reads are short, up to 55 bases per run, but recent improvements allow for more accurate reads of stretches of one type of nucleotides (Heliscope gene sequencing; Thompson and Steinmann 2010). This sequencing method and equipment were used to sequence the genome of the M13 bacteriophage (Harris et al. 2008).

1.2.9   Single molecule real time (SMRT) sequencing

SMRT sequencing is based on sequencing by synthesis approach. The DNA is synthesized in zero-mode wave-guides (ZMWs) – small well-like containers with the capturing tools located at the bottom of the well. The sequencing is performed with use of unmodified polymerase (attached to the ZMW bottom) and fluorescently labelled nucleotides flowing freely in the solution. The wells are constructed in a way that only the fluorescence occurring by the bottom of the well is detected. The fluorescent label is detached from the nucleotide at its incorporation into the DNA strand, leaving an unmodified DNA strand. According to Pacific Biosciences, the SMRT technology developer, this methodology allows detection of nucleotide modifications (such as cytosine methylation). This happens through the observation of polymerase kinetics. This approach allows reads of 20,000 nucleotides or more, with average read lengths of 5 kilobases.

1.2.10   New Novel sequencing methods in development

There are a number of novel DNA sequencing methods which are still in development. The development is made in terms of reduction of reaction volumes, smaller amounts of reagents and low cost. Third generation technologies are aiming to increase the throughput, decrease time and cost, harnessing the processivity of DNA polymerase (Schadt et al. 2010). Methods in development are Nanopore DNA sequencing, Tunneling currents DNA sequencing, Sequencing by hybridization, Sequencing with mass spectrometry, Microfluids Sangers sequencing, Microscopy based technique, RNAP sequencing, *Invitro* virus high-throughput sequencing.

1.2.11   Applications of Next generation sequencing

Apart from metagenomics NGS can be applied to genome sequencing & resequencing, transcriptome profiling, DNA – protein interactions and epigenome characterization and resurrection of ancient genome (Fig. 3).

Transcriptome sequencing

Genome wide survey of gene expression levels were studied using qPCR, SAGE and microarray with limitations. Next generation sequencing techniques were implemented along with SAGE tags to sequence the RNA populations from the cells expressed. Noncoding RNA (ncRNA) are any RNA's that are transcribed and not translated to a protein. ncRNA in plants and animals are having an important role in regulation of gene expression. Next-generation sequencing technology has discovered many novel ncRNAs (Sanger and Coulson 1975; Venter et al. 2004; Nyren et al. 1993; Ewing 1998; Bainbridge et al. 2006; Hutchison 2007; Gowda et al. 2006; Mardis, 2006). These ncRNA are unique, diverse and regulate genes by a variety of mechanisms. The readouts from next generation technologies are quantitative, allowing to detect changes in expression levels due to changes in environment and onset of disease. Studying the roles of these new specific RNAs may help in uncover certain aspects of disease or cancer. Remarkable progress has been made in characterizing and understanding these molecules using next generation technologies. Discovering ncRNAs and sequencing of transcriptome would provide us new insights on the genome wide expression patterns of an organism.

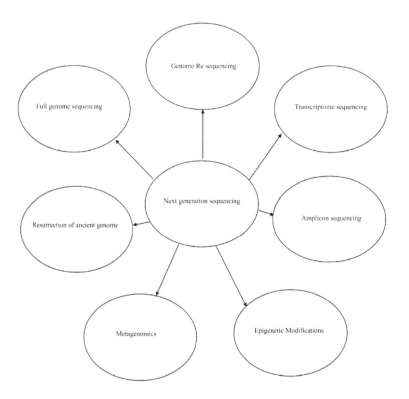

FIGURE 3   Applications of next generation technologies.

Resurrection of ancient genomes

Samples from fossils and ancient remnants are in a state of degradation. With the advent of molecular techniques such as PCR and DNA sequencing, deciphering of mitochondrial genomes from fossils and ancient remnants was possible. Several non-trivial technical complications arised from these cases, most notably DNA contamination. Next generation sequencing can be implemented to obtain sequence information from the degraded nature of the ancient genome. Sequence information from single fossil bone of Neanderthal genome was obtained using next generation sequencing technologies (Carrilho 2000). In addition to Neanderthal sequence, information was also obtained directly from the nuclear genomes of ancient remains of the cave bear (Robertson et al. 2007) and mammoth (Barski et al. 2007). In this way next generation sequencing technology can be used in resurrecting ancient genomes.

Analysis of epigenetic modifications of histones and DNA

Next generation sequencing technologies made it possible to study DNA methylation profile by bisulfite DNA sequencing, mapping histone modifications, mapping the locations of DNA-binding proteins, DNA accessibility and chromatin structure. The association between DNA and proteins is an interaction regulating gene expression and controlling the availability of DNA for transcription, replication, and other processes. Genome-wide chromatin immunoprecipitation (ChIP)-based studies for

DNA-protein interactions became possible in sequenced genomes (Korbel et al. 2007). ChIP-based approach and the Illumina platform provided insights into transcription factor binding sites in the human genome such as neuron-restrictive silencer factor (NRSF) (Tawfik and Griffiths 1998) and signal transducer and activator of transcription 1 (STAT1) (Bhinge et al. 2007). So, next generation sequencing technologies can be used for understanding of gene expression–based cellular responses due to DNA-protein interactions.

1.3   Sequence prefiltering

After obtaining metagenomic data, prefiltering of the sequence is the first step. Prefiltering includes removal of redundant, low quality sequences and sequences of eukaryotic origin (Adey et al. 2010; Bentley et al. 2008) using the methods EuDetect and DeConseq (Nakamura et al. 2011; Hess et al. 2011).

1.4   Assembly

Recovering the sequence from the environmental sample is the first important step followed by assembly of the recovered reads into metagenome. Two strategies are used for assembly of metagenomics samples: de novo assembly and reference-based assembly (co-assembly). De novo assembly tools are developed based on the de Bruijn graphs to compute large amounts of data. de Bruijn assemblers Velvet (Zerbino and Birney 2008) or SOAP (Li et al. 2008) are the tools used for De novo assembly. Reference based assembly can be used, if closely related reference genomes are available for assembly of the metagenomic dataset. The available reference based assembly software are Newbler (Roche), AMOS, or MIRA (Chevreux et al. 1999). Using the appropriate method for assembly of the metagenomic dataset is important.

1.5   Binning

Binning is….. 'the process of sorting DNA sequences into groups that might represent an individual genome or genomes from closely related organisms……'. Binning algorithms employ two types of information contained within a given DNA sequence. Firstly, compositional binning uses conserved nucleotide composition and secondly, the similarity with a reference database can be used to classify and bin the sequence. Phylopythia (McHardy et al. 2007), S-GSOM (Chan et al. 2008), PCAHIER (Diaz et al. 2009; Zheng and Wu 2010) and TACAO (Diaz et al. 2009) are the compositional-based binning algorithms, whereas IMG/M (Markowitz et al. 2008), MG-RAST (Glass et al. 2010), MEGAN (Huson et al. 2007), CARMA (Krause et al. 2008), SOrt-ITEMS (Haque et al. 2009) and MetaPhyler (Liu et al. 2009) are similarity-based binning softwares. PhymmBL (Brady and Salzberg 2009) and MetaCluster (Leung et al. 2011) are the binning algorithms that consider both composition and similarity.

1.6   Annotation

Annotation of metagenomes can be performed by two different initial pathways. RAST or IMG the existing pipelines for genome annotation are used if the genome is reconstructed or annotation on the entire community. Later metagenomic sequence data can be annotated in two steps: first, genes are identified and predicted; and second, putative gene functions are assigned. CDS of metagenome are predicted using tolls such as FragGeneScan (Rho et al. 2010), MetaGeneMark (McHardy et al. 2007), MetaGeneAnnotator (MGA)/ Metagene (Noguchi et al. 2008) and Orphelia (Hoff et al. 2009; Yok and Rosen 2011). BLAST-based searches are used for functional annotation. KEGG (Kanehisa et al. 2004), eggNOG (Muller et al. 2010), COG/KOG (Tatusov et al. 2003), PFAM (Finn et al. 2010), and TIGRFAM (Selengut et al. 2007) are the reference databases giving functional context to metagenomic datasets.

1.7   Sharing and storage of metadata

Genomic research has witnessed and is following the tradition of sharing genomic data as public databases. But, metagenomics field is yet to witness this tradition of sharing metagenomic data. At present NCBI, IMG/M, CAMERA and MG-RAST are the new level of organization and collaboration that provide metadata and centralized services.

1.8   Application of metagenomics

Metagenomics has been applied to solve many challenges in the fields of medicine, biofuel, enviromental remediation, biotechnology, agriculture and ecology (Fig. 4.)

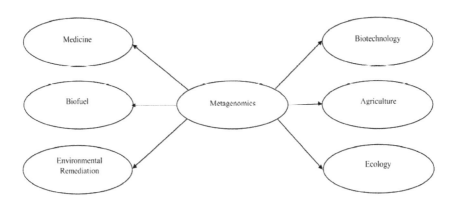

FIGURE 4   Applications of metagenomics.

1.8.1. Medicitne

Two metagenomic projects Human Microbiome initiative and Metagenomics of the Human Intestinal Tract (MetaHit) established the fact that microbial communities play an important role in human health. Human Microbiome initiative characterized microbial communities from 15-18 body sites of 250 individuals. The primary goal of this project is to correlate the human microbiome with human health. This project was somewhat successful in understanding the key role of microbial communities in preserving human health. Some more questions like composition of microbial

TABLE 2 Metagenome assemblers used for assembling reads into metagenomes

S. No	Assembler	Description	Reference
1	Orione	A Galaxy-based framework consisting of publicly available research software and specifically designed pipelines to build complex, reproducible workflows for next-generation sequencing microbiology data analysis.	Cuccuru et al., 2014
2	GARM	A new software pipeline to merge and reconcile assemblies from different algorithms or sequencing technologies.	Soto-Jimenez et al., 2014
3	GeneStitch	Novel way of using the de Bruijn graph assembly of metagenomes to improve the assembly of genes.	Wu et al., 2012
4	CLC Bio's denovo assembler (CLC Bio)	CLC bio's de novo assembly algorithm utilize de Bruijn graphs to represent overlapping reads which is a common approach for short read de novo assembly	http://www.clcbio.com/files/appnotes/CLC_bio_De_novo_Assembly.pdf
5	Ray Meta	A massively distributed metagenome assembler that is coupled with Ray Communities, which profiles microbiomes based on uniquely-colored k-mers	Boisvert et al., 2012
6	MAP	A de novo metagenomic assembly program for shotgun DNA reads	Lai et al., 2012
7	MetAMOS	A modular and open source metagenomic assembly and analysis pipeline	Treangen et al., 2013
8	Meta-IDBA	An iterative De Bruijn Graph De Novo short read assembler specially designed for de novo metagenomic assembly	Peng et al., 2007
9	MetaVelvet	Modified and extended a single-genome and de Bruijn-graph based assembler, Velvet, for de novo metagenome assembly	Namiki et al., 2012
10	Newbler	Newbler is for de novo DNA sequence assembly. It is designed specifically for assembling sequence data generated by the 454 GS-series of pyrosequencing platforms sold by 454 Life Sciences, a Roche Diagnostics company.	Nederbragt, 2014
11	MIRA	Assembler uses iterative multipass strategies centered on high-confidence regions within sequences and has a fallback strategy for using low-confidence regions when needed	Chevreux et al., 2004
12	SPADES	Assembler for both single-cell and standard (multicell) assembly	Bankevich et al., 2012

communities, core human microbiome etc are yet to be answered (Kristensen et al. 2009; Zimmer 2010).

The MetaHit project is based on 124 individuals related to healthy and diseased states like overweight and irritable bowel syndrome. The study established bacteriocides and firmicutes as the dominant distal gut bacteria and identified 1244 metagenomic clusters important for the health of the intestinal tract. These clusters were of two types : housekeeping and intestine specific. The housekeeping bacteria are required to influence the central metabolic pathways of the host like central carbon metabolism and aminoacid synthesis. The intestine/gut-specific bacteria functions include adhesion to host proteins or harvesting sugars of globoseries and glycolipids. Other outcomes of the project is that change in gut biome diversity was observed in patients with irritable bowel syndrome when compared with healthy individuals. Another new insight provided by this project was that only 7.6% of the known gut bacteria were captured and more research is necessary to identify novel bacteria from the gut biome.

1.8.2. Biofuel

Biomass such as corn and its stalk, grasses, sugarcane etc are used for the conversion of the cellulose into fuels which are known as biofuel. Different types of biofuels produced include ethanol, methane and hydrogen. Conversion of biomass into biofuel requires microbial consortia. This consortium converts/transforms various complex carbohydrates like cellulose into simple sugars. Later these sugars are fermented to produce ethanol, sometimes methane and hydrogen as main/by products.

Metagenomics is a poweful tool which helps us in understanding and providing new insights on how these microbial communities achieve the required function in the particular environment. This information can later be applied to control microbial communities for achieving the required function in controlled environment. Luen-Luen et al. (2009) applied metagenomics to screen enzymes like glycoside hydrolases from microbial consortia involved in biofuel production. Jaenicke et al. (2011) applied metagenomics to understand microbial consortia of biogas fermentors. Suen et al. (2010) applied metagenomics to understand the role of gut microbiome of insects-leaf cutter ants in converting biomass to simple substances in the gut of insect. So, metagenomics has helped in deciphering the role of microbial communities in converting complex biomass into simple molecules. Thus, this reinvented technology can be implemented in production of biofuels.

1.8.3. Enviromental remediation

Monitoring the impacts of pollutants on the environment and cleaning up the pollutants from the environment is the biggest task today. The amount of pollutants released in the environment is increasing leading to presence of huge amounts of xenobiotics. Some are easily degraded by the enzymes and the rest of amount is accumulated in the soil, water and living organisms. These accumulated pollutants are toxic and cause mutations finally leading to cancer in the organisms. Interestingly some of the organisms are resistant, rendering pollutants non-toxic and making the microbes adapt to polluted environment. The proteins, enzymes and genes of microbes had a capacity to break down the pollutants. Bioremediation can be *exsitu* by removing pollutants and *insitu*. Microbes remediating pollutants help to have an eco-friendly and cost effective way to have a health in polluted ecosystems. Metagenomics can be applied to isolate, assess and understand how microbial communities are coping up with pollutants and assess the improvement in contaminated sites (George et al. 2010).

1.8.4. Biotechnology

A vast array of biologically active compounds were identified and recovered from microbes. These compounds are actively being used as fine chemicals, agrochemicals and pharmaceuticals. Application of metagenomics helps in identification of desired trait or useful activity (Wong 2010). There are two types of bioprospecting metagenomic data: function-driven screening of expression data and sequence-driven screening of DNA sequence data (Patrick and Handelsman 2003). Function-driven screening analysis identifies clones with desirable traits or useful activity followed by biochemical characterization and sequence analysis. Whereas sequence-driven analysis uses PCR primers to screen clones for sequence of interest (Patrick and Handelsman 2003). Sometimes a combination of function and sequence-driven screening can be used for bioprospecting.

1.8.5. Agriculture

Plants are surrounded by a number of microbial communities both on it and in the soil. Much is not known about the microbial consortia inhabitating the soil despite of their economic importance. Microbial communities are known to fix atmospheric nitrogen, inducing plant growth, nutrient recycling, sequestering of iron and other metals and disease supression. Metagenomics can be used for exploring the microbial communties interacting with plants to improve crop health.

1.8.6. Ecology

Environmental communities is the well-known application for metagenomics. Community genomics can be used to understand the role of each community in an ecosystem. There are two types of microbial communties identified from metagenomic analysis : feast/famine and planktons. Metagenomics can provide more information and insights in functions of microbial communities (Raes et al. 2011). Meta genomics was applied to identify microbial consortia found in faeces of Australian sea lions. Australain sea lions faeces are rich with nutrients and microbial consortia. Microbial consortia present in the faeces break down the nutrients in the faeces and make it available to the food chain of the coastal ecosystem (Lavery et al. 2012). In addition, metagenomics also helps us in identifying microbial communities present in air, water and debris.

1.8.7. Metagenomics for species/strain Identification

Efforts to unambiguously classify metagenomic reads into species/strains/higher levels/clade-specific is challenging. Next generation technologies are used to capture DNA/RNA from the samples. These metagenome reads captured in the form of DNA/RNA are used to perform computationly extensive assembly to identify species/strains. Challenges in this process are sequencing errors, noise generated in reads during sequencing, ambiguity contributed by homology in the genome content of closely related species/strains, complex data preprocessing and assembly based on genome coding regions (Wang et al. 2012). Tu et al. (2014) considering both genome coding and non-coding data developed K- mer approach to identify species/strains from a metagenome data. *K*-mer-based approach, identifies genome-specific markers (GSMs) rapidly and comprehensively from all regions in the genome sequence and filter out non-specific sequences. The currently sequenced microbial metagenomes are searched against these GSMs, to determine the presence/absence and/or the relative abundance of each strain/species. Thus, species/strain identification is possible.

Acknowledgements

We would like to thank GITAM University, Visakhapatnam, India for providing the facility and support. The authors also thankful to Prof. I. Bhaskar Reddy and Dr Malla Rama Rao for constant support throughout the research work.

References

Adey, A., H.G. Morrison, X. Asan Xun, J.O. Kitzman, and E.H. Turner. 2010. Rapid, low-input, lowbias construction of shotgun fragment libraries by high-density *invitro* transposition. *Genome Biol.* 11(12): R119.

Bainbridge, M.N., R.L. Warren, M. Hirst, T. Romanuik, T. Zeng, A. Go, A. Delaney, M. Griffith, M. Hickenbotham, V. Magrini, E.R. Mardis, M.D. Sadar, A.S. Siddiqui, M.A. Marra, and S.J. Jones. 2006. Analysis of the prostate cancer cell line LNCaP transcriptome using a sequencing-by-synthesis approach. *BMC Genomics* 7 : 246.

Bankevich, A., S. Nurk, D. Antipov, A.A. Gurevich, M. Dvorkin, A.S. Kulikov, V.M. Lesin, S.I. Nikolenko, S. Pham, A.D. Prjibelski, A.V. Pyshkin, A.V. Sirotkin, N. Vyahhi, G. Tesler, M.A. Alekseyev, and P.A. Pevzner. 2012. SPAdes: a new genome assembly algorithm and its applications to single-cell sequencing. *J Comput Biol.* 19 (5) : 455-77.

Barski, A., S. Cuddapah, K. Cui. T.Y. Roh, D.E. Schones, Z. Wang, G. Wei, I. Chepelev, and K. Zhao. 2007. High-resolution profiling of histone methylations in the human genome. *Cell.* 129 : 823–37.

Bentley, D.R., S. Balasubramanian, H.P. Swerdlow, G.P. Smith, and J. Milton, et al. 2008. Accurate whole human genome sequencing using reversible terminator chemistry. *Nature*, 456 (7218): 53-59.

Bhinge, A.A., J. Ki, G.M. Euskirche, M. Snyder, and V.R. Iyer. 2007. Mapping the chromosomal targets of STAT1 by sequence tag analysis of genomic enrichment (STAGE). *Genome Res.* 17 : 910–916.

Boisvert, S., F. Raymond, E. Godzaridis, F. Laviolette, and J. Corbeil. 2012. Ray Meta: scalable de novo metagenome assembly and profiling. *Genome Biol.* 13 (12) : R122.

Brady, A., and S.L. Salzberg. 2009. Phymm and Phymm BL: metagenomic phylogenetic classification with interpolated Markov models. *Nat Methods.* 6 (9): 673-676.

Breitbart, M., P. Salamon, B. Andresen, J.M. Mahaffy, A.M. Segall, D. Mead, F. Azam, and F. Rohwer. 2002. Genomic analysis of uncultured marine viral communities. *Proc. Natl. Acad. Sci. U.S.A.* 99 (22): 14250–255.

Brenner, S., M. Johnson, J. Bridgham, G. Golda, D.H. Lloyd, D. Johnson, S. Luo, S. McCurdy, M. Foy, M. Ewan, R. Roth, D. George, S. Eletr, G. Albrecht, E. Vermaas, S.R. Williams, K. Moon, T. Burcham, M. Pallas, R.B. DuBridge, J. Kirchner, K. Fearon, J. Mao, and K. Corcoran. 2000. Gene expression analysis by massively parallel signature sequencing (MPSS) on microbead arrays. *Nat Biotechnol.* 18(6):630-34.

Burke, C., S. Kjelleberg, and T. Thomas. 2009. Selective extraction of bacterial DNA from the surfaces of macroalgae. *Appl Environ Microbiol.* 75(1):252-56.

Carrilho, E. 2000. DNA sequencing by capillary array electrophoresis and microfabricated array systems. *Electrophoresis* 21: 55–65.

Chan, C.K., A.L. Hsu, S.K. Halgamuge, S.L. Tang. 2008. Binning sequences using very sparse labels within a metagenome. *BMC Bioinformatics.* 9: 215.

Chen, K., and L. Pachter. 2005. Bioinformatics for whole-genome shotgun sequencing of microbial communities. *PLoS Comput. Biol.* 1 (2): e24.

Chevreux, B., T. Pfisterer, B. Drescher, A.J. Driesel, W.E. Müller, T. Wetter, and S. Suhai. 2004. Using the miraEST assembler for reliable and automated mRNA transcript assembly and SNP detection in sequenced ESTs. *Genome Res.* 14: 1147-1159.

Chevreux, B., T. Wetter, and S. Suhai. 1999. Genome sequence assembly using trace signals and additional sequence information computer science and biology. Proceedings of the German Conference on *Bioinformatics.* 99: 45-56.

CLC_bio_De_novo_Assembly. http://www.clcbio.com/files/appnotes/CLC_bio_De_novo_Assembly.pdf. Archived from the orginal on 30.06.2015

Cuccuru, G., M. Orsini, A. Pinna, A. Sbardellati, N. Soranzo, A. Travaglione, P. Uva, G. Zanetti, and G. Fotia. 2014. Orione, a web-based framework for NGS analysis in microbiology. *Bioinformatics.* 30 (13): 1928-29.

Diaz, D.N., L. Krause, A. Goesmann, K. Niehaus, and T.W. Nattkemper. 2009. TACOA: taxonomic classification of environmental genomic fragments using a kernelized nearest neighbor approach. BMC *Bioinformatics.* 10: 56.

Drmanac, R., A.B. Sparks, M.J. Callow, A.L. Halpern, N. L. Burns, and B.G. Kermani, et al. 2010. Human genome sequencing using unchained base reads in self-assembling DNA nanoarrays. *Science* 327 (5961): 78–81.

Ewing, B., and P. Green. 1998. Base-calling of automated sequencer traces using phred. II. Error probabilities. *Genome Res.* 8: 186–94.

Finn, R.D., J. Mistry, J. Tate, P. Coggill, A. Heger, J.E. Pollington, O.L. Gavin, A. Bateman. 2010. The Pfam protein families database. *Nucleic Acids Res.* 38: D211-D22.

George, I., B. Stenuit, and S. Agathos. 2010. Application of Metagenomics to Bioremediation. *Metagenomics: Theory, Methods and Applications.* Caister Academic Press.

Glass, E.M., J. Wilkening, A. Wilke, D. Antonopoulos, and F. Meyer. 2010. Using the metagenomics RAST server (MG-RAST) for analyzing shotgun metagenomes. Cold Spring Harb. Protoc., 1; pdb prot5368.

Gowda, M., H. Li, J. Alessi, F. Chen, R. Pratt, and G.L. Wang. 2006. Robust analysis of 5¢-transcript ends (5¢-RATE): a novel technique for transcriptome analysis and genome annotation. *Nucleic Acids Res.* 34 : e126.

Handelsman, J., M.R. Rondon, S.F. Brady, J. Clardy, and R.M. Goodman. 1998. Molecular biological access to the chemistry of unknown soil microbes: A new frontier for natural products. *Chem. Biol.* 5 (10): R245–R249.

Haque, M.M., T.S. Ghosh, D. Komanduri, and S.S. Mande. 2009. SOrt-ITEMS: Sequence orthology based approach for improved taxonomic estimation of metagenomic sequences. Bioinformatics. 25 (14): 1722-1730.

Harris. T.D., P.R. Buzby, H. Babcock, E. Beer, J. Bowers, and I. Braslavsky. et al. 2008. Single-molecule DNA sequencing of a viral genome. *Science* 320 (5872): 106–109.

Healy, F.G., R.M. Ray, H.C. Aldrich, A.C. Wilkie, L.O. Ingram, and K.T. Shanmugam. 1995. Direct isolation of functional genes encoding cellulases from the microbial consortia in a thermophilic, anaerobic digester maintained on lignocellulose. Appl. *Microbiol Biotechnol.* 43(4) : 667–74.

HeliScope Gene Sequencing / Genetic Analyzer System : Helicos BioSciences. United States1986. Patent 4,631,122

Hess, M., A. Sczyrba, R. Egan, T.W. Kim, H. Chokhawala, and G. Schroth. 2011. Metagenomic discovery of biomass degrading genes and genomes from cow rumen. *Science.* 331 (6016): 463-67.

Hoff, K.J., T. Lingner, P. Meinicke, and M. Tech. 2009. "Orphelia: predicting genes in metagenomic sequencing reads. *Nucleic Acids Res.* 37: W101-W105.

Hugenholz, P. 2002. Exploring prokaryotic diversity in the genomic era. *Genome Biology.* 3 (2): 1–8.

Huson, D.H., A.F. Auch, J. Qi, and S.C. Schuster. 2007. MEGAN analysis of metagenomic data. *Genome Res.* 17 (3): 377-86.

Hutchison, C.A. 2007. III, DNA sequencing: bench to bedside and beyond. *Nucleic Acids Res.* 35 : 6227–37.

Jaenicke. S., C. Ander, T. Bekel, R. Bisdorf, M. Dröge, K.H. Gartemann, S. Jünemann, O. Kaiser, L. Krause, F. Tille, M. Zakrzewski, A. Pühler, A. Schlüter, and A. Goesmann. 2011. Comparative and joint analysis of two metagenomic datasets from a biogas fermenter obtained by 454-pyrosequencing. *PLoS One.* 6(1): e14519.

Jay, E., R. Bambara, R. Padmanabhan, and R. Wu. 1974. DNA sequence analysis: a general, simple and rapid method for sequencing large oligodeoxyribonucleotide fragments by mapping. *Nucleic Acids Res.* 1 (3): 331–53.

Kanehisa, M., S. Goto, S. Kawashima, Y. Okuno, and M. Hattori. 2004. The KEGG resource for deciphering the genome. *Nucleic Acids Res.* 32 : D277-D80.

Kavita, K.S., C.L. Parsley, and M.R. Liles. 2010. Size does matter: application-driven approaches for soil metagenomics. *Soil. Biol. Biochem.* 42 (11): 1911–23.

Korbel, J.O., A.E. Urban, J.P. Affourtit, B. Godwin, F. Grubert, J.F. Simons, P.M. Kim, D. Palejev, N.J. Carriero, L. Du, B.E. Taillon, Z. Chen, A. Tanzer, A.C. Saunders, J. Chi, F. Yang, N.P. Carter, M.E. Hurles, S.M. Weissman, T.T. Harkins, M.B. Gerstein, M. Egholm, and M. Snyder. 2007. Paired-end mapping reveals extensive structural variation in the human genome. *Science* 318 : 420–26.

Krause, L., N.N. Diaz, A. Goesmann, S. Kelley, T.W. Nattkemper, F. Rohwer, R.A. Edwards, and J. Stoye. 2008. Phylogenetic classification of short environmental DNA fragments. *Nucleic Acids Res.* 36 (7): 2230-39.

Kristensen, D.M., A.R. Mushegian, V.V. Dolja, and E.V. Koonin. 2009. New dimensions of the virus world discovered through metagenomics. *Trends Microbiol.* 18 (1): 11–19.

Lai, B., R. Ding, Y. Li, L. Duan, H. Zhu. 2012. A de novo metagenomic assembly program for shotgun DNA reads. *Bioinformatics.* 28 (11) : 1455-62.

Lavery, T.J., B. Roudnew, J. Seymour, J.G. Mitchell, and T. Jeffries. 2012. High nutrient transport and cycling potential revealed in the microbial metagenome of australian sea lion (*Neophoca cinerea*) faeces. PLoS ONE. 7 (5): e36478.

Leung, H.C., S.M. Yiu, B. Yang, Y. Peng, Y. Wang, and Z. Liu. 2011. A robust and accurate binning algorithm for metagenomic sequences with arbitrary species abundance ratio. *Bioinformatics.* 27 (11): 1489-95.

Li., R., Y. Li, K. Kristiansen, and J. Wang. 2008. SOAP: short oligonucleotide alignment program. *Bioinformatics.* 24 (5): 713-14.

Liu, B., T. Gibbons, M. Ghodsi, and T. Treangen. 2011. Pop M: accurate and fast estimation of taxonomic profiles from metagenomic shotgun sequences. *BMC Genomics.* 12 (S 2): S4.

Luen-Luen, Li., S.R. McCorkle, S. Monchy, S. Taghavi, and D. van der Lelie. 2009. Bioprospecting metagenomes: glycosyl hydrolases for converting biomass. Biotechnol. *Biofuels* 2: 10.

Mardis, E.R. 2006. Anticipating the 1,000 dollar genome. *Genome Biol.* 7 : 112.

Margulies, M., M. Egholm, W.E. Altman, S. Attiya, and J.S. Bader, et al. 2005. Genome sequencing in microfabricated high-density picolitre reactors. *Nature*. 437: 376–80.

Markowitz, V.M., N.N. Ivanova, E. Szeto, K. Palaniappan, and K. Chu. 2008. IMG/M: a data management and analysis system for metagenomes. *Nucleic Acids Res*. 36 : D534-D38.

Maxam, A.M., and W. Gilbert. 1977. A new method for sequencing DNA. *Proc. Natl. Acad. Sci. U.S.A.* 74 (2): 560–64.

McHardy, A.C., H.G. Martin, A. Tsirigos, P. Hugenholtz, and I. Rigoutsos. 2007. Accurate phylogenetic classification of variable-length DNA fragments. *Nat. Methods*. 4 (1): 63-72.

Muller, J., D. Szklarczyk, P. Julien, I. Letunic, A. Roth, and M. Kuhn. 2010. eggNOG v2.0: extending the evolutionary genealogy of genes with enhanced non-supervised orthologous groups, species and functional annotations. *Nucleic Acids Res*. 38 : D190-D195.

Nakamura, K., T. Oshima, T. Morimoto, S. Ikeda, H. Yoshikawa, Y. Shiwa, S. Ishikawa, M.C. Linak, and A. Hirai, et al. 2011. Sequence-specific error profile of Illumina sequencers. *Nucleic Acids Res*. 39 (13): e90.

Namiki, T., T. Hachiya, H. Tanaka, and. Y. Sakakibara. 2012. MetaVelvet: an extension of Velvet assembler to de novo metagenome assembly from short sequence reads. *Nucleic Acids Res*. 40 (20) : e155.

Nederbragt , A.J. 2014. On the middle ground between open source and commercial software - the case of the Newbler program. *Genome Biol*. 15 (4): 113.

Noguchi, H., T. Taniguchi, and T. Itoh. 2008. MetaGeneAnnotator: detecting species specific patterns of ribosomal binding site for precise gene prediction in anonymous prokaryotic and phage genomes. *DNA Res*. 15 (6): 387-96.

Nyren, P., B. Pettersson, and M. Uhlen. 1993. Solid phase DNA minisequencing by an enzymatic luminometric inorganic pyrophosphate detection assay. *Anal. Biochem*. 208. 171–75.

Pace, N.R., E.F. Delong, and N.R. Pace. 1991. Analysis of a marine picoplankton community by 16S rRNA gene cloning and sequencing. *J. Bacteriol*. 173 (14): 4371–78.

Pace, N.R., D.A. Stahl, D.J. Lane, and G.J. Olsen. 1985. Analyzing natural microbial populations by rRNA sequences. *ASM News*. 51: 4–12.

Padmanabhan, R., and R. Wu. 1972. Nucleotide sequence analysis of DNA. IX. Use of oligonucleotides of defined sequence as primers in DNA sequence analysis. Biochem. Biophys. *Res. Commun*. 48 (5): 1295–302.

Padmanabhan, R., R. Wu, and E. Jay. 1974. Chemical synthesis of a primer and its use in the sequence analysis of the lysozyme gene of bacteriophage T4. *Proc. Nat. Aca. Sci*. 71 (6): 2510–14.

Patrick, D.S., and J. Handelsman. 2003. Biotechnological prospects from metagenomics. *Curr. Opin. Biotechnol*. 14 (3): 303–10.

Peng, Y., H.C. Leung., S.M. Yiu, F.Y. Chin. 2013. Meta-IDBA: a de Novo assembler for metagenomic data. *Bioinformatics*. 27 (13) : 194-101.

Porreca, G.J. 2010. Genome Sequencing on Nanoballs. *Nature Biotechnol*. 28 (1): 43–44.

Prober, J.M., G.L. Trainor, G.L. Dam, F.W. Hobbs, C.W. Robertson, and R.J. Zagursky, et al. 1987. A system for rapid DNA sequencing with fluorescent chain-terminating dideoxynucleotides. *Science*. 238 (4825): 336–341.

Raes, J., I. Letunic, T. Yamada, L. J. Jensen, and P. Bork. 2011. Toward molecular trait-based ecology through integration of biogeochemical, geographical and metagenomic data. *Mol. Sys. Bio*. 7: 473.

Rho, M., H. Tang, and Y. Ye. 2010. FragGeneScan: predicting genes in short and error-prone reads. *Nucleic Acids Res.* 38 (20) : e191.

Robertson, G., M. Hirst, M. Bainbridge, M. Bilenky, Y. Zhao, T. Zeng, G. Euskirchen, B. Bernier, R. Varhol, A. Delaney, N. Thiessen, O.L. Griffith, A. He, M. Marra, M. Snyder, and S. Jones. 2007. Genome-wide profiles of STAT1 DNA association using chromatin immunoprecipitation and massively parallel sequencing. *Nat. Methods.* 4 : 651–657.

Ronaghi, M., S. Karamohamed, B. Pettersson, M. Uhlen, and P. Nyren. 1996. Real-time DNA sequencing using detection of pyrophosphate release. *Anal Biochem.* 242: 84–89.

Rusk, N. 2011. Torrents of sequence. *Nat. Meth.* 8 (1): 44–44.

Sanger, F., and A.R. Coulson. 1975. A rapid method for determining sequences in DNA by primed synthesis with DNA polymerase. *J. Mol. Biol.* 94 (1975): 441–448.

Sanger, F., S. Nicklen, and A.R. Coulson. 1977. DNA sequencing with chain-terminating inhibitors. *Proc. Natl. Acad. Sci. U.S.A.* 74 (12): 5463–5467.

Schadt, E.E., S. Turner, and A. Kasarskis. 2010. A window into third-generation sequencing. *Hum. Mol. Gen.* 19 (R2): R227–40.

Selengut, J.D., D.H. Haft, T. Davidsen, A. Ganapathy, M. Gwinn-Giglio, and W.C. Nelson. 2007. TIGRFAMs and genome properties: tools for the assignment of molecular function and biological process in prokaryotic genomes. *Nucleic Acids Res.* 35 : D260-D264.

Shendure, J., G.J. Porreca, N.B. Reppas, X. Lin, J.P. McCutcheon, A.M. Rosenbaum, M.D. Wang, K. Zhang, R.D. Mitra, and G.M. Church. 2005. Accurate multiplex polony sequencing of an evolved bacterial genome. *Science* 309 (5741): 1728–1732.

Soto-Jimenez, L.M., K. Estrada, and A. Sanchez-Flores. 2014. GARM: genome assembly, reconciliation and merging pipeline. *Curr Top Med Chem.* 14 (3) :418-424.

Stein, J.L., T.L. Marsh, K.Y. Wu, H. Shizuya, and E.F. DeLong. 1996. Characterization of uncultivated prokaryotes: isolation and analysis of a 40-kilobase-pair genome fragment from a planktonic marine archaeon. *J. Bacteriol.* 178 (3): 591–599.

Suen, G., J.J. Scott, F.O. Aylward, S.M. Adams, S.G. Tringe, A.A. Pinto-Tomás, and C.E. Foster, et al. 2010. An insect herbivore microbiome with high plant biomass-degrading capacity. *PLoS Genet.* 6(9) : e1001129.

Tatusov, R.L., N.D. Fedorova, J.D. Jackson, A.R. Jacobs, B. Kiryutin, E.V. Koonin, D.M. Krylov, R. Mazumder, S.L. Mekhedov, A.N. Nikolskaya, B.S. Rao, S. Smirnov, A.V. Sverdlov, S. Vasudevan, Y.I. Wolf, J.J. Yin, and D.A. Natale. 2003. The COG database: an updated version includes eukaryotes. *BMC Bioinformatics.* 4 : 41.

Tawfik, D.S. and A.D. Griffiths. 1998. Man-made cell-like compartments for molecular evolution. *Nat. Biotechnol.* 16 : 652–656.

Thomas, T., D.M.Z. Rusch, P.Y. DeMaere Yung, M. Lewis, A. Halpern, K.B. Heidelberg, S.P.D. Egan Steinberg, and S. Kjelleberg. 2010. Functional genomic signatures of sponge bacteria reveal unique and shared features of symbiosis. *ISME J.* 4(12): 1557-1567.

Thompson, J.F., and K.E. Steinmann. 2010. Single molecule sequencing with a HeliScope genetic analysis system. *Current Protocols in Molecular Biology.* Chapter 7: Unit7.10.

Treangen, T.J., S. Koren, D.D. Sommer, B. Liu, I. Astrovskaya, B. Ondov, A.E. Darling, A.M. Phillippy, and M. Pop. 2013. MetAMOS: a modular and open source metagenomic assembly and analysis pipeline. *Genome Biol.* 14 (1): R2.

Tu, Q., Z. He, and J. Zhou. 2014. Strain/species identification in metagenomes using genome-specific markers. *Nucleic Acids Res.* 42(8): e67.

Tyson, G.W., J. Chapman, P. Hugenholtz, E.E. Allen, R.J. Ram, P.M. Richardson, V.V. Solovyev, E.M. Rubin, D.S. Rokhsar, and J.F. Banfield. 2004. Insights into community structure and metabolism by reconstruction of microbial genomes from the environment. Nature 428 (6978): 37–43.

Valouev, A., J. Ichikawa, T. Tonthat, J. Stuart, S. Ranade, and H. Peckham, et al. 2008. A high-resolution, nucleosome position map of *C. elegans* reveals a lack of universal sequence-dictated positioning. *Genome Res.* 18 (7): 1051–1063.

Venter, J.C., K. Remington, J.F. Heidelberg, A.L. Halpern, D. Rusch, and J.A. Eisen, et al. 2004. Environmental genome shotgun sequencing of the Sargasso Sea. *Science.* 304 (5667): 66-74.

Wang, X.V., N. Blades, J. Ding, R. Sultana, and G. Parmigiani. 2012. Estimation of sequencing error rates in short reads. *BMC Bioinformatics.* 13: 185.

Watson, J.D., and F.H. Crick. 1953. The structure of DNA. Cold. Spring. Harb. *Symp. Quant. Biol.* 18: 123–31.

Wong, D. 2010. Applications of Metagenomics for Industrial Bioproducts. *Metagenomics: Theory, Methods and Applications.* Caister Academic Press.

Wu R. *Faculty Profile.* Cornell University. Archived from the original on 2009-03-04.

Wu, R., and E. Taylor. 1971. Nucleotide sequence analysis of DNA. II. Complete nucleotide sequence of the cohesive ends of bacteriophage lambda DNA. *J. Mol. Biol.* 57: 491–11.

Wu, R., C.D. Tu, and R. Padmanabhan. 1973. Nucleotide sequence analysis of DNA. XII. The chemical synthesis and sequence analysis of a dodecadeoxynucleotide which binds to the endolysin gene of bacteriophage lambda. Biochem. *Biophys. Res. Commun.* 55 (4): 1092–99.

Wu, Y.W., M. Rho, T.G. Doak, and Y. Ye. 2012. Stitching gene fragments with a network matching algorithm improves gene assembly for metagenomics. *Bioinformatics.* 28 (18) : 1363-1369.

Yok, N.G., and G.L. *Rosen.* 2011. Combining gene prediction methods to improve metagenomic gene annotation. *BMC Bioinformatics.* 12: 20.

Zerbino, D.R., and E. Birney. 2008. Velvet: algorithms for de novo short read assembly using de Bruijn graphs. *Genome Res.* 18 (5): 821-29.

Zheng, H., and H. Wu. 2010. Short prokaryotic DNA fragment binning using a hierarchical classifier based on linear discriminant analysis and principal component analysis. *J. Bioinform Comput. Biol.* 8 (6): 995-1011.

Zimmer, C. (13 July 2010). How Microbes Defend and Define Us. *New York Times.* Retrieved 29 December 2011.

15

Metabolic Engineering: Dimensions and Applications

Nitin Thukral[1] and Yasha Hasija[1]

Abstract

Metabolic engineering, initially conceptualized as the manipulation of natural processes to improve or enhance the productivity of valuable products using genetic engineering, has now transformed into a field where microbial host can be engineered to produce almost any organic compound. Indeed, the concept is old one but development of molecular and systems biology has introduced a new dimension to this science of pathway regulation. Until recently metabolic engineering has found its application in microorganisms and plants: improving production of materials already synthesized by them; addition of new activities for bioremediation and production of industrially/commercially important chemicals, hormones, and proteins. Studying metabolic engineering could provide an insight into tissue and organ network and functioning; strongly influencing the areas of medicine and therapeutics. Aiding metabolic engineering with new computational and mathematical tools will help us recognize its true potential. Still, our understanding of fundamental biology has been the limiting factor. Improving production metrics and the range of attainable products will eliminate the need for presumptions currently being used. We believe that metabolic engineering will soon evolve into a robust analytical field, owing to the rapidly emerging technologies.

1. Introduction

The term "metabolic engineering" is a combination of two terms: *metabolism* (the sum total of all enzymatic transformations of organic molecules that are occurring in the cell of an organism) and *engineering* (to manipulate things to enhance their value). Metabolic engineering is the directed improvement or/and modification of cellular activities or products formation by manipulating specific biochemical pathways using recombinant DNA technology.

---

[1] Department of Biotechnology, Delhi Technological University, Shahbad Daulatpur, Main Bawana Road, Delhi 110042, India.
E-mail address: yashahasija@gmail.com
* Corresponding author: yashahasija@gmail.com

Metabolic engineering aims at mathematically modelling these biosynthetic networks, calculating the yield of useful products, and identifying the constrain in the production of these products (Torres and Voit, 2002; Yang and Bennett, 1998).

To increase the productivity of a certain desired metabolite, chemical mutagens were used to genetically modify microorganism and finally select the mutant strain over expressing the desired metabolite. However in doing so, one of the main aspects that remained unnoticed was identifying the most important metabolic pathway for that metabolite's production. As a consequence of which, constraints to production and modification of important pathway enzymes remained unexamined (Torres and Voit, 2002). In early 1990s, emergence of metabolic engineering that analyze the metabolic pathway of a microorganisms, and determines the constraints and their effects on the production of desired compounds was seen as the sorted solution to the earlier problem. It then employed genetic engineering to relieve the pathways from these constraints.

The ultimate goal of metabolic engineering is to use these organisms to produce industrially valuable substances in a cost-effective manner. Currently production of pharmaceuticals, milk products, alcoholic drinks and other biotechnology products comes under this technique. Since cells use their metabolic networks for their survival. The cells ability to produce the desired substance and its natural needs for survival leads to trade-offs in metabolic engineering. Earlier,the focus was on the regulatory networks to efficiently engineer the metabolic pathway rather than directly knocking out and/or overexpressing the genes that encode for important metabolic enzymes (Vemuri and Aristidou, 2005).The focus has now shifted to expressing and fine tuning heterologous pathways as researchers are finding them to be more challenging than engineering metabolic pathways (Yuan *et al.*, 2013).

Metabolic engineering seems to overlap with many biological fields like systems biology, genetic engineering and molecular biology but its distinct focus differentiates it from them. It is concerned with investigating the properties of the metabolic pathways and genetic regulatory networks as opposed to genetic engineering which only deals with investigating the genes and enzymes (Stephanopoulos and Sinskey, 1993). Another distinguishing characteristic of metabolic engineering is that it aims at engineering microorganisms that can be used as biocatalysts for the economical production of biofuels, chemicals, and pharmaceuticals, a lot more than simply stitching genes together to build a basic functioning pathway. Just stitching the pathway genes together will only produce few milligrams of product but further optimization of the heterologous pathways is required to reach optimal yield, titer and productivity for cost-effective production. So, although building a functional pathway takes few months improving it from the angle of commercialization is a time taking and a tedious task. This outlines the basic elements of metabolic engineering i.e. pathway design, construction, and optimization. These elements include components from graph theory, chemical reaction engineering, biochemistry, and optimization (Woolston *et al.*, 2013).

In this chapter, we haveenlisteddifferent areas where metabolic engineering has been applied. The different aspects of metabolic engineering like pathway design, construction and optimization have been discussed without going into the mathematical details of each. All this has been supplemented with thelist of computational methods used in metabolic engineering, thus giving readers an edge over the currently available articles.

2. Metabolic flux balance analysis

The word *flux* comes from the Latin word *fluxus* meaning "flow". In transport phenomena, flux is defined as the rate of flow of a property per unity area and has dimensions of [amount] [time]$^{-1}$ [area]$^{-1}$. Metabolic flux analysis is based on fluxomics technique used to calculate metabolite production and consumption rates in a biological system. All the enzymes involved in a pathway regulate the flux (the movement of matter through networks connected by metabolites and cofactors). Flux, is analyzed via flux balance analysis (FBA), and is of great importance in metabolic network model building as its regulation is vital for regulating the pathway's activity under changing conditions within a cell. The flux of metabolites through a pathway can be described by summing up the individual reaction steps. It is represented as:

$$J = Vf - Vr$$

where J is the metabolic flux through each reaction, Vf is the rate of forward reaction and Vr is the rate of reverse reaction.

There is no flux at equilibrium. It is difficult to reach an ideal steady state so, calculations are done at quasi-steady state (almost equal to steady state). Flux is usually determined by the rate limiting step of the reaction. In the reconstructed metabolic networks, flux analysis can clearly illustrate the interconnection of different parts of the cellular metabolism, particularly due to common cofactors. The tight connections in metabolic networks point to the fact that changes in fluxes in one part of the metabolic pathway will be communicated to many different parts, resulting in a global response. Thus, we could retrieve valuable information related to functioning of the complete metabolic network by even measuring metabolic fluxes at few points.

Also, FBA has its drawbacks. One of the major drawbacks of FBA is that it does not include kinetic parameters thus cannot be used for modeling dynamic behavior (Raman and Chandra, 2009)

Generally, metabolic flux analysis uses ^{13}C isotope labelling to determine metabolic pathway flux. In this approach, metabolic pathway maps (stoichiometric model) are used for calculating intracellular fluxes for the major intracellular reactions by applying mass balances around intracellular metabolites. Input to the calculations is a set of measure extracellular fluxes, typically substrates uptake rates and metabolites' secretion rates. As a result of this, a metabolic flux map is constructed representing diagrammatically the biochemical reactions included in the calculations along with the steady state rate estimation (i.e., the flux) at which each reaction occurs(Stephanopoulos *et al.*, 1998).

Flux-balance analysis(FBA) can help address the following:

* The biomass yield and maximum growth rate
* The biochemical production capabilities
* The effectiveness of the metabolic pathway
* The thermodynamic information related to metabolic pathways
* The trade-off between overproduction of metabolites and biomass yield.

Though FBA is of foremost importance for pathway analysis but it too has some drawbacks. Sometimes disagreement with the experimental data may arise which is often accounted for when regulatory loops are considered. This cannot be used for modelling dynamic behavior and also does not uniquely specify the flux.

FBA is a mathematical approach for simulating metabolic pathways in metabolic networks at genome scale. In terms of input data required for constructing the metabolic model, it is computationally inexpensive. A very little information in terms of enzyme kinetics and concentration is required for FBA. FBA makes two assumptions, steady state (metabolite concentration is not changing) and optimality (organism has been optimized such as for optimal growth).

The material balance model applies to every cellular pathway which implies:

Input = Output + Accumulation

For the system of microbial cells at steady state accumulation term tends to zero and thematerial balance equations is reduced to:

Input = Output

In such a system, input to the system is substrate which is consumed and biomass produced becomes the output from the system. The material balance equation then is represented by

Input – Output = 0

- Mathematical representation of the algebraic equations is usually done as a dot product of a matrix of coefficients and a vector of the unknowns. Since the steady-state assumption puts the accumulation term to zero. The system can be written as:

A.x = 0

- Firstly all the metabolic reactions in a pathway are mathematically represented in the form of a numerical matrix, of each reaction's stoichiometric coefficients.

S.v = 0

- These stoichiometric coefficients impose constraints on the metabolic flux (flow of metabolites through the network). There are two ways of representing constraints, one as equations that balance reaction inputs and outputs as mentioned above and second as inequalities imposing bounds on the system. The matrix of stoichiometries imposes flux (that is, mass) balance constraints on the system under consideration. Lower and upper bounds can be designated to every reaction, which define the maximum and minimum allowable fluxes of the reactions. These balances and bounds define the rates at which every metabolite is utilized or produced by each reaction.
- The next step is defining a phenotype (a biological objective) that is relevant to the problem under consideration. For example, in the case of predicting growth, the objective would be the rate at which metabolic compounds are converted into biomass constituents (nucleic acids, proteins and lipids).

Biomass production is mathematically represented by adding an extra column of coefficients in the matrix of stoichiometries. Now, the objective is predicting the maximum growth rate which can be accomplished by evaluating conditions that result in the maximum flux through the reaction. In many cases, phenotype of interest is the result of more than one reaction. The contribution of each reaction to the phenotype is defined mathematically by an 'objective function'. Taken together, these equations aresolved using linear programming.

Suppose if we are interested in calculating the maximum aerobic growth of *E.coli* while assuming that glucose and not the oxygen uptake is the limiting factor for growth. We would set the maximum rate of glucose uptake to a physiologically realistic level (18.5 mmol glucose $gDW^{-1}h^{-1}$; DW, dry weight) and set the maximum rate of oxygen uptake to such an arbitrarily level that it does not limit growth. Then, using linear programming it is possible to determine the maximum possible flux through the biomass reaction.(Orth *et al.*, 2010)

3.   Metabolic engineering in microorganisms

A remarkable array of metabolic pathways exists in nature as exhibited by the diverse microorganisms. Though only some of the organisms are suitable for commercial application, most of them require genetic improvements guided by our current understanding of microbial metabolism and genetics (Stephanopoulos *et al.*, 1998). Metabolic engineering is such a field that has a potential of producing a large number of valuable organic compounds from simple and inexpensive starting materials using microorganisms (Keasling, 2010).As a result of this new and desirable functionalities in microbial cells, pharmaceutical, environmental, agricultural, food and chemical sectors are the added beneficiaries. Microbial cell factories (MCFs) is believed to be a new revolutionary platform for producing organic compounds and replacing traditional chemical factories (Fisher *et al.*, 2014).

Increasing population and growing industrialization is consuming fossil fuels at an unprecedented rate. A sustainable and economical alternative to the petroleum-based production is engineering microbes for the production of fuels and chemicals using renewable carbohydrate feedstocks. Global energy and environmental problems have generated a stir among scientists to look for synthesizing liquid biofuels from renewable resources such as cellulosic biomass that will offer advantages such as higher energy density, lower vapour pressure, less hygroscopic and compatibility with existing infrastructure. This calls for designing efficient and optimized metabolic pathways for carrying our economically competitive bioprocesses for the production of new generation fuels. Metabolic engineering offers an alternative approach of engineering synthetic pathways into user-friendly hosts for the production of the sustainable fuels and then manipulating these hosts for improved production efficiency. One organism that has captured the interests of metabolic engineers is the yeast (*Saccharomyces cerevisiae*) due to its convenience to be genetically engineered, extensive information on its physiology, fully clarified genetics and metabolism and its robustness to handle harsh industrial conditions. Optimizing its native cellular metabolic pathways and introduction of novel pathways have expanded its range of cell factory applications. Yeast is being applied on a large scale for biofuel production. Nielsen J *et al.*, 2013 has wonderfully reviewed the current advancements in metabolic engineering of *S. cerevisiae* for the production of bioethanol, new generation biofuels and other chemicals (Nielsen *et al.*, 2013). Several higher fuels such as 1-propanol, 2-methyl-1butanol, 1-butanol, isobutanol and 3-methyl-1-butanol, which possess fuel properties like fossil fuels can be efficiently produced using metabolic engineering approach. Choi *et al.*, 2014 has extensively reviewed production of higher alcohols from microbes (Choi *et al.*, 2014).The recent progress in the engineering of *Escherichia coli* has been summarized in detail by Atsumi and Liao (Atsumi and Liao, 2008). Recently, microorganisms have been engineered to convert simple sugars into several types of biofuels, such as alcohols, fatty acid alkyl esters, alkanes, and terpenes, with high titers and yields. Many researchers have engineered biosynthetic pathways for the production of several advanced biofuel in well-characterized microorganisms such as *Escherichia coli* and *Saccharomyces cerevisiae*. This has been reviewed extensively by Zhang *et al.* (Zhang *et al.*,2011).

Till date, there are around 2,00,000 plant products known and many out of them demonstrate pharmacological properties. However, most of them are hard to be isolated from plants due to their limited abundance, cellular compartmentalization of products and seasonal variations. The advancement in next-generation sequencing technology and recombinant DNA technology, offering opportunities for functional integration of plant biosynthetic pathways in different microorganisms. Currently many plant products like isoprenoids, phenylpropanoids, alkaloids and many secondary metabolites are being produced using microorganisms as the host organisms, thanks to metabolic engineering (Marienhagen and Bott, 2013).

Microbial production of plant natural products using microorganisms is a promising alternative. Microorganisms are already being used to produce broad-range of products ranging from nutritional items, industrial chemical compounds and pharmaceuticals (amino acids, vitamins, sugars, insulin, biofuels) (Choi et al., 2014; Shin and Lee, 2014). Microbial production allows many advantages like (Marienhagen and Bott, 2013):

- It is more environmental friendly as the process does not involve use the of organic solvents, heavy metals and acids or bases.
- Microbes can convert simple and inexpensive renewable products into desired products.
- The scalability of microbial fermentations is highly achievable.
- The genetic availability of industrially relevant microbes allows easy reconstruction of metabolic pathways.
- Recombinant microbes lack competing pathways as compared to heterologously expressed pathways in plants, so the desired chemical will be made as a chemically distinct molecule.
- Simplified downstream processing in comparison to product isolation and purification from plants.

Case study 1: An example of metabolic engineering in E.coli

In the past, the laboratory work has kept its focus on manipulating central metabolic pathways in different microorganisms to achieve enhanced production of desired products. Here, we are specially exemplifying the metabolic engineering to overcome the problem of acetate accumulation in *E.coli*. Achieving high expression level of the cloned gene is one of the major challenges in production of recombinant protein. As a result of these conditions, the acetate amount in the reactor continues to increase.

Various strategies have been proposed in the past but they only lead to an improvement in the performance level of the process but are difficult and expensive to implement. Our goal is to design a strategy using metabolic engineering to reduce the acetate accumulation (Aristos *et al.*, 1994).

The Fig. 1 shown below is the normal glycolytic pathway along with a major pathway leading to acetate formation. Anyone can figure out several potential points where we can control or reduce the acetate production/accumulation. One very simple approach is to block the pathways leading to acetate formation from acetyl-CoA. Other approaches include regulating the uptake of carbon source and thus reducing the pyruvate or acetyl-CoA accumulation. However, introducing a new pathway will perturb the existing pathway which is highly regulated.

Here we will investigate the only one strategy of redirecting carbon flux in a direction that is less inhibitory to byproducts production. In this case, heterologous expression of a bacterial gene acetolactate synthase (ALS) obtained from *Bacillus subtilis* is considered. The enzyme produced by this gene converts pyruvate to acetoin. Due to its central position, pyruvate branch point is the critical node where carbon flux partition occurs between acetyl-CoA, lactate and acetoin. The introduction and expression of the ALS gene in the bacteria changed the glycolytic fluxes drastically and minimized the acetate accumulation in the bioreactor. It was also observed that acetoin was 50 times less harmful than acetate. Furthermore, the effect of ALS on the normal physiology was evaluated and it was found to be exerting negligible effect on the normal reaction network.

Case Study 2:   Metabolic engineering in Methanococcus maripaludis generate genome scale metabolic model

Figure 3 illustrates metabolic engineering in *Methanococcus maripaludis*. Genomic annotations were obtained from sequencing studies (Hendrickson *et al.*2004). The physiological, biochemical and genetic information about *M. maripaludis* was mined from available literature and public databases like KEGG, BRENDA and METACYC. The information obtained from these databases was merged for unification. The reaction stoichiometries were checked for mass balances. Several ORFs based on known physiological information and comparative genomics were annotated. Then reactions were checked for the production of biomass precursors using FBA through MetaFluxNet. The score on the scale of 1-5 was assigned to each reaction based on our confidence in the available experimental data. An SBML file of the model was deposited in the BioModels database.

The data from Timothy *et al.* and Lupa*et al.* was used to validate our genomic model. MSH (MS medium with NaCl, MgCl$_2$, and KCl) media with and without organics was used by Timothy *et al.* to grow a *M. maripaludis* culture. The H$_2$–CO$_2$ (75 : 25 v/v) pressure of 300 kPa and a temperature of 25 °C was maintained in anaerobic pressure tubes. A hydrogen utilization rate of 28.8 mmol g$_{DCW}$$^{-1}$ h^{-1}(1.6 ng µg$_{DCP}$$^{-1}min^{-1)}$ and 45 mmolg$_{DCW}$$^{-1}$h^{-1} was measured during cell growth with and without the organics respectively (DCP = dry cell protein and is assumed to be 60% of DCW (dry cell weight)). The lower bound in the above model was the experimental value of

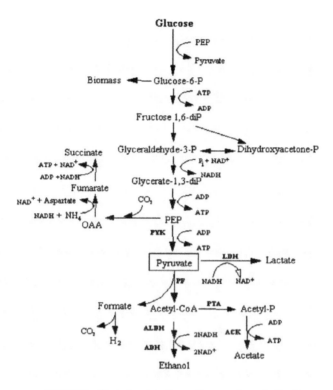

FIGURE 1 Central metabolism pathways scheme in *E.coli.*

hydrogen exchange and other reaction fluxes were calculated using FBA (Kral*et al.* 1998). On the other hand, Lupa *et al.* performed two experiments with wild-type *M. maripaludis* S2. The initial experiment, measured cell growth *via* absorbance at 600 nm and MERs *via* gas chromatography at 37 °C over a period of 25 h. The culturing conditions for this experiment were McNA (minimal medium supplemented with acetate) medium at 37 °C, H_2–CO_2 (80 : 20 v/v) under 276 kPa of the atmosphere. Sodium formate was used as a source for H_2 under the O_2-free N_2 atmosphere in the following experiment with culturing conditions to beN_2 : CO_2 (80 : 20 v/v) under 138 kPa of the atmosphere. The MER values so obtained were converted to mmol g_{DCW}^{-1} $h^{-1}$for simulating this experiment in our model, and set as upper bounds on methane exchange fluxes. The growth rates was predicted using FBA for the corresponding MERs(Lupa*et al.* 2008).For our model Growth Associated Maintenance (GAM) was set as 30 mmol ATP g_{DCW}^{-1} h^{-1} and O_2 uptake flux was set as zero for simulating anaerobic conditions. For each of these above predictions, a growth yield (gDCW mol_{CH4}^{-1}) was computed and then compared with average growth yields obtained by Lupa *et al.* for the two scenarios (Lupa*et al.* 2008). The predictions for the above metabolic model match closely with the experimentally observed growth yields.

After validation, phenotypical observations from Haydock *et al.*, Lin *et al.*, and Lie *et al.* were used to verify model predictions. Haydock *et al.* demonstrated the role of the *leuA* gene in the growth of *M. maripaludis*S2 by constructing a leucine-auxotrophic mutant (Haydock *et al.* 2004). Lin *et al.* on the other hand showed the

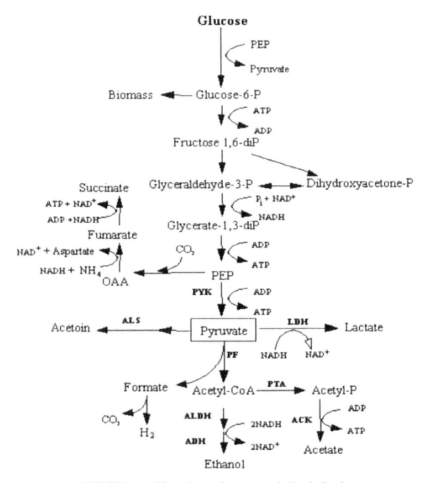

FIGURE 2   Modified scheme of central metabolism in *E.coli*

importance of *porE* genes on growth and oxidation of pyruvate (Lin *et al,* 2004). Lie *et al.* studied the effect of nitrogen on the regulation of *nif* (nitrogen fixation) and *glnA* (glutamine synthetase) operons (Lie and Leigh,2003). Under diazotrophic (nitrogen-rich) conditions, both *nif* and *glnA* exhibited high expressions and the expression was observable in the presence of alanine as a nitrogen source (Lie and Leigh, 2002).

4.   Metabolic engineering in plants

For years, humans have remained dependent on the products of plant metabolism for food, medicines, feed, fuel and fiber. As the world population is expected to approach 9 billion by 2050, unprecedented amounts of these products will be required, which calls for improved production strategies. Plants exhibit remarkable property of meta-bolic plasticity and diversity, but meeting the challenges of the growing population will also require engineering plants.Plants have the ability to produce most abundant

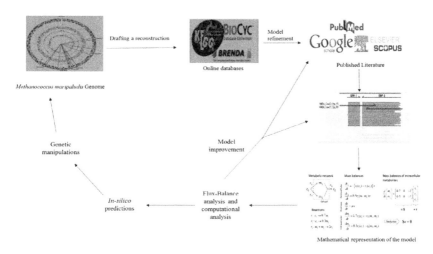

FIGURE 3   An adaptation from (Goyal *et al.*, 2014).

as well as other chemicals of high utility. As our knowledge about plants metabolism increases, the reconstruction of these pathways will become accessible.

As of the past few years, plant metabolic engineering was only confined to modifying regulatory or biosynthetic steps in metabolic pathways, often resulting in outcomes that are very different from what was expected (Carrari *et al.*, 2003). Though we are not dwelling deep into the available knowledge on plant metabolism but one can always go through some recent good reviews (Anarat-Cappillino and Sattely, 2014; Farré *et al.*, 2014; Shumskaya and Wurtzel, 2013; Vogt, 2010; Ziegler and Facchini, 2008). Having said that there are few key aspects of plant metabolism that should be considered for any successful metabolic engineering activity and therefore are worth mentioning:

- *"Promiscuous" nature of science*:   It appears to be true for most of the enzymes from central metabolism that they are "specialists" but it is becoming evident that a good majority of specialized metabolic enzymes are promiscuous (that they can act on a number of related or unrelated substrates in microbes (Nam *et al.*, 2012) and plants (Bar-Even and Salah Tawfik, 2013). This would mean that many pathways like phenylpropanoid biosynthesis (Bonawitz and Chapple, 2010) which were considered to be linear are actually complex pathways.

- *Specialized metabolic pathways are governed by inefficient enzymes*:   The process of gene duplication and neo-functionalization have given rise to many genes encoding enzymes involved in specialized metabolism from genes of central metabolism (Ober, 2005) However, as compared to their corresponding counterparts in central metabolism, specialized metabolism enzymes possess significantly lower k_{cat} values. This could be attributed to reduced selection pressures as compared to the enzymes of core metabolism, hence providing significant opportunities for enzyme optimization (Bar-Even and Salah Tawfik, 2013).

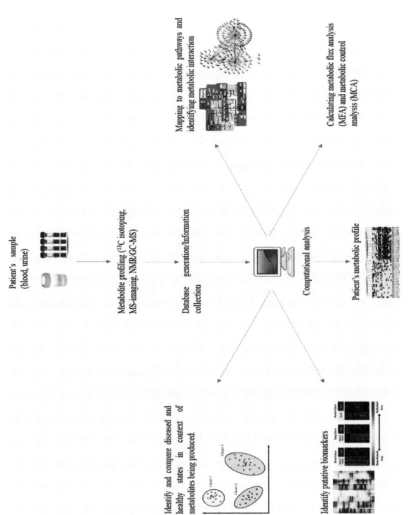

FIGURE 4 Metabolic engineering in humans

- *Gene clusters encoding metabolic enzymes:* Classical examples such as *lac* and *trp* operons have made us believe that co-regulated clusters play an important role in enabling bacterium to adapt quickly to the fluctuating environmental conditions (Lawrence, 2002). It is well known that operons, as found in bacteria are absent in plants (Osbourn and Field, 2009) but there is increasing evidence for the presence of clusters of genes up to several hundred kb in length (Chu *et al.*, 2011) and, can include more than 10 biosynthetic enzymes (Winzer *et al.*, 2012)arrests metaphase, and induces apoptosis in dividing human cells. Elucidation of the biosynthetic pathway will enable improvement in the commercial production of noscapine and related bioactive molecules. Transcriptomic analysis revealed the exclusive expression of 10 genes encoding five distinct enzyme classes in a high noscapine-producing poppy variety, HN1. Analysis of an F(2. Some specialized metabolites for which gene clusters were evident include oat avenacins(Qi *et al.*, 2004)which encodes the first committed enzyme in the avenacin biosynthetic pathway, is clearly distinct from other plant beta-amyrin synthases. Here we show that AsbAS1 has arisen by duplication and divergence of a cycloartenol synthase-like gene, and that its properties have been refined since the divergence of oats and wheat. Strikingly, we have also found that AsbAS1 is clustered with other genes required for distinct steps in avenacin biosynthesis in a region of the genome that is not conserved in other cereals. Because the components of this gene cluster are required for at least four clearly distinct enzymatic processes (2,3-oxidosqualene cyclization, beta-amyrin oxidation, glycosylation, and acylation, the *Papaver somniferum*noscapine alkaloid (Winzer *et al.*, 2012)arrests metaphase, and induces apoptosis in dividing human cells. Elucidation of the biosynthetic pathway will enable improvement in the commercial production of noscapine and related bioactive molecules. Transcriptomic analysis revealed the exclusive expression of 10 genes encoding five distinct enzyme classes in a high noscapine-producing poppy variety, HN1. Analysis of an F(2.

However, several bottlenecks exists for the manipulation of metabolic pathways that have slowed down this plant metabolic engineering in model systems such as Arabidopsis. The chief bottlenecks are challenges of introducing foreign genes into plants, optimization of localization of metabolites, etc.

5. Transcription factors vs. enzymes

There are numerous successful examples of utilizing pathway enzymes and their coding gene in engineering metabolic pathways. Alternatively, it has been studied that using transcription factors (TFs) for metabolic engineering is more effective in certain cases (Broun, 2004; Xie *et al.*, 2006). After analyzing the association of reported genes with domestication of crop plants, almost 70% of them were observed to be transcription regulatory genes while only 20% were coding enzymes. TFs commonly control downstream genes coding for multiple enzymes within a metabolic pathway and thus impacting profoundly phenotypic and metabolic outcomes. In metabolic engineering approaches, "push" and "pull" strategies are used in reference to TFs. Using TFs or early pathway enzymes genes to move the flux downstream is referred to as "push" strategy whereas using late pathway genes to draw the substrates towards the end products is referred to as "pull" strategy. For best results, combination of both strategies is used (Vanhercke et al., 2014).

6.   Metabolic trafficking and sequestration

The knowledge of the site of localization of metabolites in plants and trafficking plants responsible for their localization, is primitive. The accumulation of plant metabolic pathways in the wrong subcellular compartment tricks negative feedback mechanisms to prevent damage to the plants (Goodman et al., 2004). Thus, it is important to understand plant metabolite trafficking and sequestration, in order to successfully engineer plant metabolic pathways. This becomes more important when multiple cell types are part of a single metabolic pathway. For example, biosynthesis of monoterpene indole alkaloids involves multiple cell types like the chloroplast, the vacuole, the endoplasmic reticulum and the cytosol (Facchini, 2001; Verma et al., 2012).

The conventional methods to determine the metabolite accumulation involve cellular/subcellular fractionation followed by metabolite analysis. The new approaches like FRET-based nanosensors have been successfully applied to measure metabolites and other small molecules (De Michele et al., 2013; Fehr et al., 2004, 2002; Frommer et al., 2009; Ho and Frommer, 2014). Mass spectrometry imaging is another emerging powerful method to determine spatial distribution of metabolites between and within cells (Horn and Chapman, 2014; Lee et al., 2010; Matros and Mock, 2013).

7.   Genome editing

Genome editing uses site-specific nucleases to carry out targeted transgene integration experiments, and genome engineering in an efficient and precise manner. It is used for introducing targeted DNA double-strand breaks (DSBs) using an engineered nuclease further stimulating different cellular DNA repair mechanisms. Depending on the repair pathway and the availability of a repair template, modified genomes can be achieved in different ways. Two known and defined DSB repair pathways are non-homologous end joining (NHEJ) and homologous recombination (HR). NHEJ mostly causes random insertions or deletions (indels), causing frame shift mutations. They can effectively create a knockout organism if these indels occurs in the coding region of a gene. NHEJ can mediate the targeted introduction of a ds-DNA template with compatible overhangs, when the overhangs are created by DSB (Cristeaet al., 2013; Marescaet al., 2013). DNA damage repair by HR occurs when a template with regions of homology to the sequence surrounding the DSB is available. This mechanism is often used to achieve precise gene modifications/insertions. Even though the generation of breaks in both DNA strands induces recombination at specific genomic loci, NHEJ is by far the most common DSB repair mechanism in most organisms, including higher plants.

Zinc finger nucleases (ZNFs), synthetic proteins capable of cutting DNA at specific spots due to presence of DNA-binding domains, were developed by scientists in the early 2000s. Later, TALENs (transcription activator-like effector nucleases), another synthetic nuclease were developed as it provided an easier way to target a double-strand break to a specific locus. Until 2013, Zinc finger nucleases and TALENs, remained the dominant genome editing tools (Kim *et al.*, 1996; Christian *et al.*, 2010). They have been used successfully in many organisms including plants (Jankele and Svoboda, 2014). Both zinc-finger nucleases and TALENs depend on making custom proteins for each DNA target. Comparatively, CRISPRs are quite easier to design as canonical Watson-Crick base pairing of the guide RNA to the target site defines its specificity (Young, 2014).

CRISPRs (clustered regularly interspaced short palindromic repeats) are segments of prokaryotic DNA which contain short repetitive bases. It contain a short segment called spacer DNA interspersed by short repetitive sequences (Kramer, 2015). 40% of the known bacteria are found to carry CRISPRs while 90% of the sequenced archaea also carry it (Grissa*et al.,*2007). They are often associated with casgenes that code for proteins related to CRISPRs. The CRISPR/Cas system acts as the prokaryotic immune system conferring resistance to plasmidsand phages (Barrangou*et al.,*2007) and provides a form of acquired immunity. CRISPR spacers recognize and cut these exogenous genetic elements in a manner analogous to RNAi in eukaryotic organisms (Marraffini and Sontheimer, 2010). The CRISPR/Cas system has been extensively exploited since 2013 for gene editing (adding, disrupting or changing the sequence of specific genes) and gene regulation in species. The organism's genome can be cut at any desired spots by delivering the Cas9protein and appropriate guide RNAs into a cell.

Use of CRISPR has simplified many tasks such as creation of animals that mimic disease or creating knockout organisms to simulate the process of gene mutation/knockout. CRISPR may also be used at insert mutations in the germ-line cells (Rath*et al.,*2015), It has also found its use in functionally inactivating genes in cell lines and cells taken directly from humans, study *Candida albicans*, yeasts and make genetically modified crops (Ledford, 2015). In 2014, a patent was filed by a Chinese researcher Gao Caixia on the creation of powdery mildew resistant strain of wheat using TALENs and CRISPR gene editing tool (Gao *et al.*, 2014).

---

8.   Metabolic engineering in human disorders

Earlier metabolic techniques were applied to different physiological systems and organs in fasting, fed and diseased state to study metabolic patterns present. The nutritional statuses were assessed on some macroscopic properties like blood glucose, urea levels etc. Due to advent of improved diagnostic techniques and health care facilities, it has been possible to save people dying from stroke, trauma and other acute events. The field of metabolic engineering is thought to play a major role in designing future therapeutics.

Here, the metabolic engineering techniques for studying metabolic profiles of a human health condition is discussed,as in how metabolism will occur in normal physiological and pathophysiological state.

9. Principles and techniques of metabolic engineering in human diseases

Here we will go with the two important notions: Cellular metabolism is composed of an inter-related network and metabolic processes (systemic and cellular) are coupled. Two methodologies that are extremely helpful in human diseases for characterization and analysis of cellular metabolism are metabolic control analysis and metabolic flux balance analysis, as already mentioned. Mass isotope (extensively used to quantify fluxes in mammalian cells) and extracellular metabolite models are used for determination of flux. In isotope labelling, the labeling patterns are measured by NMR and analyzed using mathematical models to determine fluxes. Although it is a non-invasive method but instrumentation is quite expensive. Metabolic system states are characterized by metabolic flux data in terms of rate of intracellular reactions, as of course already mentioned.

Metabolic engineering also evaluates the rate controlling enzymes of the reactions. The most widely used analysis framework for this type of analysis is metabolic control analysis (Hogan, 1997). The changes in nutritional levels and other factors that effects the enzymatic activity are re-measured in terms of Flux control coefficient (FCC) by this analysis method. This type of analysis is complex and tedious and often relies on many assumptions. Despite of the drawbacks it can provide insight into mechanism governing metabolic adaptation to changing diseased state.

10. Heart Models

Metabolic engineers apply different approaches in studying normal and ischemic heart. Malloy *et al.* developed a heart model to investigate steady-state fluxes in the Kelvin cycle using isotope labelling method. The effect of ischemia-reperfusion injury on substrate was determined using this model. The study revealed that injured hearts exposed to the substrates (lactate/acetate/glucose) leads to decrease in contribution of lactate and an increase in contribution of acetate as a source of acetyl CoA, it also causes to an increase in reactions that leads to net synthesis of kelvin cycle intermediates (e.g. glutamate, aspartate). However, extracellularly added similar compounds are not metabolized. Thompson *et al.* developed another technique in which 1, 2-^{13}C-acetate and ^{13}C-lactate are used to perfuse hearts and the labelling of glutamate was analyzed (Sherry, *et al.,*1992). This method could be used to study rapidly changing metabolic conditions but it does not predict the anaplerotic flux. Lipoamide which is consider to enhance recovery after infarcts, was observed to prevent switching over from lactate to acetate utilization as induced by ischemia (Sumegi *et al.*, 1994). Lipoamide was also seen to enhance kelvin cycle rate by 64%, thereby positively effecting the heart's recovery during reperfusion.

The unsteady state procedure for determining substrate selection was extended to 4-substrate system containing acetoacetate, lactate, glucose and fatty acid physiological levels (Jeffrey *et al.*, 1995). Reperfusion-ischemia under starvation, produced an increase in acetoacetate utilization. The same technique was reciprocated to access the effect of perhexiline maleate on substrate utilization. It causes reduction in the fatty acid oxidation and leads to an increment in lactate utilization and no significant effect on oxygen concentration (Jeffrey *et al.*, 1995). These findings are in coherence with the hypothesis that shifting from fatty acid to carbohydrate oxidation exert a protective effect in the ischemic heart. This is true as we know that carbohydrate oxidation produces more ATPs per oxygen consumed than fatty acid oxidation pathway. Thus, metabolic engineering of heart models suggest potential avenues for therapeutics in treating myocardial infarction and other heart ailments.

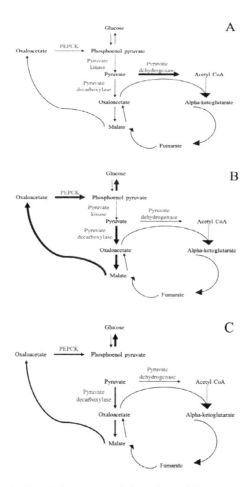

FIGURE 5   The mechanism of gluconeogenesis in fasting and diabetes. (A) In normal nutritional state, gluconeogenesis and glycolysis both can be seen to occur. Here a substantial amount of pyruvate enters kelvin cycle via acetyl-CoA. (B)In case of fasting, conversion of pyruvate to acetyl-CoA is reduced whereas a substantial increase in its conversion to oxaloacetate can be seen. Conversion to oxaloacetate leads to an increase in formation of phosphoenol pyruvate, some of which returns to the pyruvate in the glycolysis cycle. (C) In the state of diabetes, moderate increment in the conversion of pyruvate to oxaloacetate and eventually PEP can be seen. There is an inhibition of pyruvate kinase, and PEP is efficiently channelized into gluconeogenesis.

Parkinson's Disease (PD), which is the second most prevalent neurodegenerative disease in the world, is characterized by progressive loss of dopamine-producing neurons in the integral area of the brain called substantia nigra pars compacta. Though the exact cause of PD till now is not known but it is considered to be a complex disorder (Shastry, 2001). Currently no permanent cure exists for PD, but treatment using L-DOPA provides some relief. Also, L-DOPA is known to have many deleterious side effects. Currently, *in vivo* gene therapy techniques have been applied to deliver glial cell-line derived neurotrophic factor (GDNF), which slow down the destruction of nigro-striatal neurons, thus slowing down or reversing the symptoms of the disease (Gerin, 2002; Kordower *et al.*, 2000). Another approach used *ex vivo* gene therapy to engineer non-dopamine producing cells to produce dopamine for subsequent implantation. Dopamine production pathway is depicted in Fig. 6.

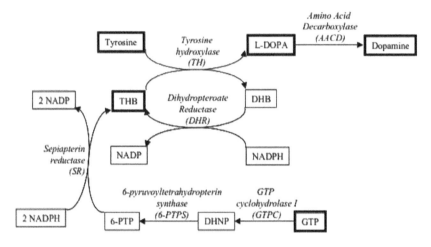

FIGURE 6 Metabolic pathway in humans depicting dopamine production by dopaminergic neurons.

It began by engineering peripheral cells using retrovirus-mediated expression of TH in immortalized rat fibroblasts to produce dopamine. However, this provided only limited improvement in rat models of PD(Horellou *et al.*, 1990).Long-term improvements were reported when herpes simplex I virus was used to express TH *in situ* in rat striatal neurons (During*et al.*, 1994). In order to gain further improvements, many different dopamine-producing metabolic pathways have been engineered into cells. It was observed that many of the non-dopamine producing cells used earlier to express TH also lack the expression of GTPC enzyme and as a result do not make THB. This lead to the introduction of both TH and GTPC into cells (striatal neurons), which yielded *in vivo* L-DOPA production(Mandel *et al.*, 1998) and lead to motor improvements in rats suffering from PD (Kirik *et al.*, 2002).

Investigations have reported the effect of AACD enzyme addition, to prevent the cells from ultimately releasing fully formed dopamine instead of the L-DOPA precursor. Three separate rAAVs were used for expressing TH, GTPC, and AACD together in the existing neurons of the rat striatum. This provided the evidence that AACD has beneficial effect on the behavioral recovery (Shen *et al.*, 2000). Similar enzymes have also recently been expressed together in a single multi-cistronic lentivirus, which led to sustained functional improvements, after intrastriatal injection in a rat model. Investigations implicated that oxidative stress is responsible for the loss of neurons in the substantia nigra pars compacta (Lotharius*et al.*, 2002). Nakamura *et al.* has shown that dopaminergic neurons could resist oxidative stresses in vitro through the THB cofactor's superoxide scavenging ability(Nakamura *et al.*, 2000). Therefore, it is believed that the expression of GTPC in cells would have dual function of providing THB for the TH reaction as well as protection from oxidative stress. This will open the door for engineering other pathways for oxidative stress intopotential graft cells. The fact that which and how many enzymes which should be expressed for optimal performance is still undetermined. Though it is believed that gaining insights from metabolic modelling and subsequent metabolic engineering will eventually help replace cell line to offer relief to PD patients.

Metabolomics and metabolic engineering plays a vital role in cancer diagnosis and treatment. It is believed that metabolomics or metabolic profile is the best closest thing to an individual's phenotype till now. Some putative metabolic biomarkers for cancer were detected, then assessment of efficacy of anti-cancer treatment is discovered in preclinical studies, followed by validation of these biomarkers. NMR is used to carry out metabolite detection and quantification and mass spectroscopy is used for metabolic profiling of the sample. Various metabolic biomarkers were reported and found to be associated with glycolysis, mitochondrial TCA cycle, fatty acid oxidation, to play an important role in treating cancer and drug responsiveness.

TABLE 1 Some of the medical applications to which metabolic engineering techniques have been applied.

Areas	Metabolic engineering applications
Brain	Novel pathways have been identified and engineered to produce the metabolite dopamine for patients affected by Parkinson's disease.
Heart	As mentioned in the above section of "heart models", heart's response to ischemia has been modelled, new treatments could be found for heart attack patients.
Liver	Many researchers have tried to model the liver's response to different disease states.
Pancreas	Cells (which are non-beta cells of Langerhans) have been produced using metabolic engineering that can secrete insulin on stimulation of glucose
Adipose	Cellular metabolism of adipose tissues has been evaluated using Flux balance analysis and metabolic control analysis
Skeletal muscle	Energy calculations has been done and modelled in skeletal muscles including modelling of mitochondrial disorders
Parasitic diseases	Metabolic control analysis has been applied to the cellular metabolism of *Trypanosomabrucei* which is known to cause sleeping sickness in humans

(Yarmush and Banta, 2003)

11. Computational Biology has come to aid metabolic engineering

Bioinformatics resources have brought a major shift in the bioscience research and mathematical tools are aiding in the development of new bioinformatics tools.

As metabolic engineering uses molecular biology techniques to modify and control microbial metabolism for producing desired compounds, computational biology can help identify new metabolic pathways and suggest required changes in the host metabolism to enhance production of desired chemicals. Traditional computational methods focused on identifying what all compounds can be made biologically. Development of new computational approaches have additionally been able to suggest different types of genetic modifications (gene deletion, gene regulation) as well as strategies to improve yield, productivity and substrate co-utilization(Long *et al.*, 2015). Molecular subsystems based mathematical models can be made for DNA replication, cell cycle control, transcription expression control, receptor trafficking processes which can simulate the subsystem responses. These models will play significant role in testing biological hypothesis thus reducing the cost and load on wet lab results.

One challenge to the mathematical models is the global gene expression. The molecular basis of these global gene expression is unknown. Those mathematical models that revolve around these phenomena will be precluded earlier from the list of available methods. Recently a model for metabolic network analysis was introduced called "cybernetic" that predict changes in enzyme expression based on efficient algorithm.

A number of methods for the phenotype simulation of microorganisms have been proposed based on different environmental and genetic conditions. However, these methods have restricted use to only one section of researcher. OptFlux provides a user-friendly computational tool for metabolic engineering applications. Features of OptFlux are:

- OptFlux is an open-source and modular software
- It allows the user to load a genome-scale model of a given organism which will serve as the basis to simulate the wild type and mutants. The simulation of these strains will be conducted using a number of approaches (e.g. Flux-Balance Analysis, Minimization of Metabolic Adjustment or Regulatory On/Off Minimization of metabolic fluxes) that allows determination of the set of fluxes in the organism's metabolism under the given set of environmental constraints. The software also includes optimization methods like Evolutionary Algorithms, Simulated Annealing etc. to reach the best set of gene deletions given an objective function, typically related with a given industrial goal.
- Identifies metabolic engineering targets.
- Can perform metabolic flux and pathway analysis.
- It also has inbuilt coding for reducing the search space.
- Supports importing/exporting to several flat file formats
- Also shows compatibility with the SBML standard.
- Has visualization module integrated from BioVisualizer application

Though conceptually it is possible to reduce metabolic pathways to linear sequences of reactions, in reality it is far complex, especially in plants and higher eukaryotes as they involve multiple layers of regulation and also the pathways are interconnected

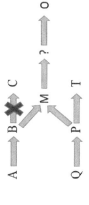

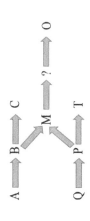

Using available knowledge to understand the pathways

Identify genes using NGS/Computational biology

Identify and characterize metabolites using Metabolomics

Proteomics/RNA-seq for identifying common proteins

Determining impact of different interventions using modelling or virtual designs.

Applying information/knowledge gained from analysis

Identify positive and negative interventions

Flux balance analysis (FBA)

Confirming desired metabolic yields using targeted metabolism to determine the effect of metabolic engineering on phenotype.

Looping for refinement

FIGURE 7 This example describes finding the best intervention strategy in the metabolic pathway leading to compound 'O'. The conventional approach applies trail-and-error approach by introducing or deleting single or multiple genes encoding the corresponding enzymes. The knowledge-driven approach as shown here involves the use of genomics, transcriptomics, proteomics and metabolomics to obtain relevant information regarding pathway reactions. Rather than using *ad hoc* interventions, strategies can be developed to generate metabolic libraries to gain detailed information about metabolites and pathways. These can then be used to fine-tune interventions. (Twyman *et al.*, 2015)

TABLE 2 Useful online resources for metabolic engineering, systems biology and metabolomics

Resource	Website link	Brief Description
Portals		
	http://dbkgroup.org/metabol.htm#links	Metabolomics links
	http://dbkgroup.org/sysbio.htm#links	Systems biology links
Metabolic Pathways		
BRENDA	http://www.brenda-enzymes.info/	Freely available enzyme functional database
Expasy	http://www.expasy.org/	ExPASy is the SIB Bioinformatics Resource Portal providing access to scientific databases and software tools in different areas of life sciences including proteomics, genomics, phylogeny, systems biology, population genetics, and transcriptomics.
Metacyc	http://www.metacyc.org/	MetaCyc is a curated database of experimentally elucidated metabolic pathways from all domains of life. MetaCyc contains 2310 pathways from 2668 different organisms.Predict metabolic pathways in sequenced genomes Support metabolic engineering via enzyme database Metabolite database aids metabolomics research
KEGG	http://www.genome.jp/kegg/	Many metabolic pathways included from different organisms
BioCyc	http://www.biocyc.org	Pathway databases for several organisms
Reactome	http://www.reactome.org/	Curated database of biological processes in humans
Biomodels.net	http://www.biomodels.net/	Kinetic models of pathways, many published models from literature
SABIO-RK Database	http://sabio.villa-bosch.de/	System for the analysis of biochemical pathways – reaction kinetics
Constraint-based reconstruction and analysis (COBRA) toolbox	http://www.bioeng.ucsd.edu/research/research_groups/gcrg/downloads/COBRAToolbox/	Interfaces with MATLAB for extensive analysis of networks using FBA; performs gene deletions – single and multiple (can interface with LINDO, GLPK, CPLEX)
MetaFluxNet	http://mbel.kaist.ac.kr/lab/mfn/	Metabolic flux analysis
CellNetAnalyzer	http://www.mpi-magdeburg.mpg.de/projects/cna/cna.html	Structural and functional analysis of cellular networks
SNA: Stoichiometric network analysis	http://www.bioinformatics.org/project/?group_id=546	Mathematica toolbox for stoichiometric network analysis

TABLE 2 (contd...)

Resource	Website link	Brief Description
Yana	http://yana.bioapps.biozentrum.uni-wuerzburg.de/	Network reconstruction, visualization and analysis
Pathway Analyser	http://sourceforge.net/projects/pathwayanalyser	FBA and MoMA of metabolic networks; gene deletion studies
Solvers for FBA/MoMA		
LINDO	http://www.lindo.com/	Commercial solver for optimization problems
CPLEX	http://www.ilog.com/products/cplex/	Commercial optimization software package
GNU linear programming toolkit (GLPK)	http://www.gnu.org/software/glpk/	Solver for LP problems
Object oriented quadratic programming (OOQP)	http://pages.cs.wisc.edu/~swright/ooqp/	Solver for QP problems
Metabolomics		
	http://www.metabolomics.ca/	Human metabolome database
	http://www.husermet.org/	Human serum metabolome project
Modelling		
SBML	http://www.sbml.org/	The Systems Biology Markup Language (SBML) is a representation format, based on XML, for communicating and storing computational models of biological processes.
SBGN	http://www.sbgn.org/	A model for SBML models visualization
Metabolic models		
	http://www.jjj.biochem.sun.ac.za/	Triple-J site
	http://www.ei.ac.uk/biomodels/	Bio models

(Caspi*et al.*, 2013). Thus, consequences are difficult to predict. Now, the traditional trial and error methods are being supplemented by a knowledge-driven approaches using information from genomics, transcriptomics, proteomics, metabolomics, metabolic libraries to define priorities before designing and introduction of streamlined gene cassettes (Figure 7)

As mentioned under the heading "Metabolic engineering in plants", that metabolic pathways in plants are complex with enzymes in different subcellular compartments and difficulty in product isolation, which calls for identifying key enzymes and intermediates in the pathways. This requires isolating the corresponding genes. The Plant Metabolic Network (PMN), is an attempt to catalog enzymes and enzyme-catalyzed reactions from a number of plants but limited availability of the biochemical data makes the resource of modest utility.

The computational tools that facilitate in evaluating and improving strains for metabolic engineering are developed and expanded at a fast pace. They can provide a wide range of experimental measurements to provide an enhanced understanding of metabolic states and current limitations, and they can be used to new intervention strategies for improving chemical production.

13. Computational Biology in identify metabolic pathways

Pathway assignment is the first step towards quantitative modeling of metabolism where genes are placed in metabolic pathways in their larger biological context. A recent study by Romero *et al.* (2004) carried out pathway assignment, computationally to assign enzymes encoded in the complete human genome. It successfully assigned 2,709 human enzymes to 896 biochemical reactions; out of which 622 enzymes were assigned roles in 135 predicted metabolic pathways, which closely matched the known nutritional pathways required of humans. This analysis also identified 203 pathway holes i.e. absence of some enzymes in the predicted pathways. Though they were only able to identify 25 putative genes to fill some of the gaps. HumanCYc, a pathway database, was used to describe the predicted human metabolic map. It provides a genome-based view of human nutrition which represents the association of the key dietary factors essential for humans with a set of validated metabolic pathways. It facilitates analysis of gene expression, proteomics, and metabolomics datasets by placing human genes in a pathways context (Romero *et al.*,2004).

A large amount of data on metabolic pathways have been generated in the last decade owing to extensive experimental studies, advancements in molecular biology and improved computational methods. As a result, many specialized databases such as KEGG and MetaCyc have been developed to store and organize this data. This data is usually represented in the form of many small sub-pathways. These sub-pathways are manually divided based on either function or by organism. However, it can be tedious to find connections between compounds by navigating through these sub-pathways especially for discovering novel or non-standard pathways spanning multiple organisms. Many applications such as metabolic engineering, metabolic network reconstruction exists for these type of pathways. Novel metabolic pathways for synthesizing important and useful compounds can be obtained by combining parts of pathways existing in different organisms. Here computational tools can provide an edge in finding novel biologically relevant pathways in metabolic data.

The primary problem where computational methods can help in identifying novel metabolic pathways is to try and find "biologicallyrealistic" pathways of enzymatic reactions that make the target compound from the start compound, given a substrate and the target compound. Owing to the availability of *atom mapping data*, defining which atoms in the substrate correspond to which atoms in the product has helped track atoms through metabolic networks to identify pathways where certain atoms are conserved from a given start compound to the desired product. The important feature of this kind of methodology is that it eliminates spurious connections and reactions. Additionally, atom tracking enables handling branched metabolic pathways. Health *et al.* developed two new algorithms based on *atom tracking* namely, LPAT (Linear Pathfinding with Atom Tracking) and BPAT-S (Branched Pathfinding with Atom tracking and Seed Pathways). LPAT takes a starting substrate, a target product, the minimum number of atoms to be kept conserved during the metabolic reactions, the number of pathways to return and a specially defined data structure called *atom mapping graph*. LPAT when tested on a metabolic network containing large number of reactions (from KEGG) was able to efficiently and accurately identify known linear pathways. On the other hand, BPAT-S, first uses LPAT to obtain a set of linear metabolic pathways. It then annotates the linear pathways and stores information regarding the compounds through which atoms are gained or lost. Seed pathways is the term used to denote to these annotated pathways, which are further indexed for processing and branching. Multiple combinations are tested for attachment with these seed pathways, in order to obtain set of branched pathways. These set of branched pathways so obtained, are ranked first on the basis of number of atoms that are kept conserved and then the total number of reactions contained in them (Health *et al.,*2010).

^{13}C flux analysis studies are an essential component of ongoing research in the field of metabolic engineering. The gradual expansion in the scope of ^{13}C flux analysis allowed inclusion of both isotopically steady-state and transient labeling experiments. The first publicly available software capable of analyzing both steady state as well as isotopically non-stationary metabolic flux, is INCA (Isotopomer network compartmental analysis). It is capable of generating mass balance equations and their solution in an automated fashion on networks with high complexity.Thus providing a comprehensive framework for flux analysis in metabolic pathways (Young, 2014).

14. Bioinformatics in identifying genome editing elements such as CRISPRs

CRISPRs as mentioned in the section "Metabolic engineering in plants"are widely distributed amongst the bacteria and archaea and are observed to show some sequence similarities (Kunin*et al.,*2007). However, the repeating spacers and direct repeats is their most notable characteristic which makes them easily identifiable in long and large sets of DNA sequences. CRT (Bland,*et al.* 2007), PILER-CR (Edgar, 2007) and CRISPR finder (Grissa *et al.,*2007) are currently the three efficient and widely used programs for CRISPR repeat identification by search for regularly interspaced repeats in long sequences.

However, analysis of CRISPRs in metagenomic data is more challenging, as it is challenging and confusing to assemble CRISPR loci due to their repetitive nature or strain variations. PCR can be used to amplify CRISPR arrays and analyze spacer content when reference genomes are available. However, this approach only yield information for specifically targeted CRISPRs and for organism whose genome are available in databases to be used for PCR primer designing.

The alternative approach is the extraction and reconstruction of CRISPR arrays from shotgun metagenomic data. But this is also challenging especially when metagenomic sequencing is done using second generation technologies (Roche 454, Illumina) as they produce short reads and that too without more than 2-3 repeats in a single read. However, CRISPR identification in raw reads could be achieved by using any of the following ways: using purely *denovo* identification (Skennerton*et al.,*2013), or direct repeat sequences from published genomes (Stern*et al.,*2012) and from contigs by using direct repeat sequences in partially assembled CRISPR arrays (Rho*et al.,* 2012).

15. Concluding remarks

Metabolic engineering has emerged as an important field since its inception in early 1980s. It has rainbow of applications in different metabolic models varying from microorganisms to plants to mammals, covering almost all forms of life. In microorganisms, it is used for improved production of industrially important chemicals or hormones; in plants it is used to produce primary and secondary metabolites both having important use in the industry as well as drugs for treating different disorders. In mammalian cells, metabolic engineering is used to study human disorders and alternate pathways that cell follows under diseased state. This has been facilitated by a combination of efficient theories and algorithms on metabolic fluxes and network control and novel molecular biology tools, bioinformatics methods, systems and synthetic biology methods which can be applied in a rational or combinatorial manner. The growing concern about sustainability and the associated increasing interest in the products obtained from renewable resources has been the driving force behind all these developments.

Although numerous studies describe a series of elegant metabolic engineering cases, they provide limited guiding instructions. Despite its plethora of applications, metabolic engineering has some limitations and challenges which need to be overcome using modern computational models and mathematical algorithms. Then, soon it would be possible to extensively exploit its applications, as has been in the case with proteomics and genomics approaches.

References

Anarat-Cappillino, G., Sattely, E.S., 2014. The chemical logic of plant natural product biosynthesis. Curr. Opin. Plant Biol. 19C, 51–58. doi:10.1016/j.pbi.2014.03.007

Aristos, A., San, K.Y., and G.N., Bennett, 1994. Modification of central metabolic pathway in Escherichia coli to reduce acetate accumulation by heterologous expression of the Bacillus subtilis acetolactate synthase gene. Biotechnol. Bioeng. 44, 944–951. doi:10.1002/bit.260440810

Bar-Even, A. and D., Salah Tawfik, 2013. Engineering specialized metabolic pathways-is there a room for enzyme improvements? Curr. Opin. Biotechnol. 24, 310–319. doi:10.1016/j.copbio.2012.10.006

Bonawitz, N.D., and C., Chapple, 2010. The genetics of lignin biosynthesis: connecting genotype to phenotype. Annu. Rev. Genet. 44, 337–63. doi:10.1146/annurev-genet-102209-163508

Broun, P., 2004. Transcription factors as tools for metabolic engineering in plants. Curr. Opin. Plant Biol. 7, 202–209. doi:10.1016/j.pbi.2004.01.013

Carrari, F., Urbanczyk-Wochniak, E., Willmitzer, L., Fernie, A.R., 2003. Engineering central metabolism in crop species: Learning the system. Metab. Eng. 5, 191–200. doi:10.1016/S1096-7176(03)00028-4

Choi, Y.J., Lee, J., Jang, Y.-S., Lee, S.Y., 2014. Metabolic engineering of microorganisms for the production of higher alcohols. MBio 5, e01524–14. doi:10.1128/mBio.01524-14

Chu, H.Y., Wegel, E. and A. Osbourn, 2011. From hormones to secondary metabolism: The emergence of metabolic gene clusters in plants. Plant J. 66, 66–79. doi:10.1111/j.1365-313X.2011.04503.x

De Michele, R., Ast, C., Loqué, D., Ho, C.H., Andrade, S.L. a, Lanquar, V., Grossmann, G., Gehne, S., Kumke, M.U., Frommer, W.B., 2013. Fluorescent sensors reporting the activity of ammonium transceptors in live cells. Elife, 1–22. doi:10.7554/eLife.00800

Facchini, P.J.P.J., 2001. Alkaloid biosynthesis in plants: biochemistry, cell biology, molecular regulation, and metabolic engineering applications. Annu. Rev. Plant Biol. 52, 29–66. doi:10.1146/annurev.arplant.52.1.29

Farré, G., Blancquaert, D., Capell, T., Van Der Straeten, D., Christou, P., Zhu, C., 2014. Engineering complex metabolic pathways in plants. Annu. Rev. Plant Biol. 65, 187–223. doi:10.1146/annurev-arplant-050213-035825

Fehr, M., Frommer, W.B. and Lalonde, S., 2002. Visualization of maltose uptake in living yeast cells by fluorescent nanosensors. Proc. Natl. Acad. Sci. U. S. A. 99, 9846–51. doi:10.1073/pnas.142089199

Fehr, M., Lalonde, S., Ehrhardt, D.W., Frommer, W.B., 2004. Live imaging of glucose homeostasis in nuclei of COS-7 cells. J. Fluoresc. 14, 603–609. doi:10.1023/B:JOFL.0000039347.94943.99

Fisher, A.K., Freedman, B.G., Bevan, D.R., Senger, R.S., 2014. A review of metabolic and enzymatic engineering strategies for designing and optimizing performance of microbial cell factories. Comput. Struct. Biotechnol. J. 11, 91–99. doi:10.1016/j.csbj.2014.08.010

Frommer, W.B., Davidson, M.W. and Campbell, R.E., 2009. Genetically encoded biosensors based on engineered fluorescent proteins. Chem. Soc. Rev. 38, 2833–841. doi:10.1039/b907749a

Gerin, C., 2002. Behavioral improvement and dopamine release in a Parkinsonian rat model. Neurosci. Lett. 330, 5–8. doi:10.1016/S0304-3940(02)00672-9

Goodman, C.D., Casati, P. and Walbot, V., 2004. A Multidrug Resistance – Associated Protein Involved in Anthocyanin Transport in *Zea mays*. Plant Cell 16, 1812–26. doi:10.1105/tpc.022574.weed

Goyal, N., Widiastuti, H., Karimi, I. a, Zhou, Z., 2014. A genome-scale metabolic model of Methanococcus maripaludis S2 for CO2 capture and conversion to methane. Mol. Biosyst. 10, 1043–54. doi:10.1039/c3mb70421a

Ho, C.H. and Frommer, W.B., 2014. Fluorescent sensors for activity and regulation of the nitrate transceptor CHL1/NRT1.1 and oligopeptide transporters. Elife, 1–21. doi:10.7554/eLife.01917

Hogan, Jr., 1997. Metabolic engineering and human disease. Nat. Biotechnol. 15, 328–330. doi:10.1038/nm0798-822

Horellou, P., Brundin, P., Kalén, P., Mallet, J., Björklund, a, 1990. *In vivo* release of dopa and dopamine from genetically engineered cells grafted to the denervated rat striatum. Neuron 5, 393–402. doi:10.1016/0896-6273(90)90078-T

Horn, P.J. and Chapman, K.D., 2014. Lipidomics in situ: Insights into plant lipid metabolism from high resolution spatial maps of metabolites. Prog. Lipid Res. 54, 32–52. doi:10.1016/j.plipres.2014.01.003

Jeffrey, F.M., Diczku, V., Sherry, A.D., Malloy, C.R., 1995. Substrate selection in the isolated working rat heart: effects of reperfusion, afterload, and concentration. Basic Res. Cardiol. 90, 388–96.

Keasling, J.D., 2010. Manufacturing molecules through metabolic engineering. Science 330, 1355–58. doi:10.1126/science.1193990

Kirik, D., Georgievska, B., Burger, C., Winkler, C., Muzyczka, N., Mandel, R.J., Bjorklund, A., 2002. Reversal of motor impairments in parkinsonian rats by continuous intrastriatal delivery of L-dopa using rAAV-mediated gene transfer. Proc. Natl. Acad. Sci. U. S. A. 99, 4708–4713. doi:10.1073/pnas.062047599

Kordower, J.H., Emborg, M.E., Bloch, J., Shuang, Y., Leventhal, L., Mcbride, J., Chen, E., Palfi, S., Roitberg, B.Z., Douglas, W., Taylor, M.D., Carvey, P., Ling, Z., Trono, D., Hantraye, P., Deglon, N., 2000. Neurodegeneration Prevented by Lentiviral Vector Delivery of GDNF in Primate Models of Parkinson`s Disease. Science 290, 767-773.

Lawrence, J.G., 2002. Shared strategies in gene organization among prokaryotes and eukaryotes. Cell 110, 407–13. doi:10.1016/S0092-8674(02)00900-5

Lee, D.Y., Bowen, B.P. and T.R. Northen, 2010. Mass spectrometry-based metabolomics, analysis of metabolite-protein interactions, and imaging. Biotechniques 49, 557–565. doi:10.2144/000113451

Long, M.R., Ong, W.K. and J.L. Reed, 2015. Computational methods in metabolic engineering for strain design. Curr. Opin. Biotechnol. 34, 135–141. doi:10.1016/j.copbio.2014.12.019

Mandel, R.J., Rendahl, K.G., Spratt, S.K., Snyder, R.O., Cohen, L.K., Leff, S.E., 1998. Characterization of intrastriatal recombinant adeno-associated virus-mediated gene transfer of human tyrosine hydroxylase and human GTP-cyclohydrolase I in a rat model of Parkinson's disease. J. Neurosci. 18, 4271–4284.

Marienhagen, J. and M. Bott, 2013. Metabolic engineering of microorganisms for the synthesis of plant natural products. J. Biotechnol. 163, 166–78. doi:10.1016/j.jbiotec.2012.06.001

Nam, H., Lewis, N.E., Lerman, J. A., Lee, D.-H., Chang, R.L., Kim, D., Palsson, B.O., 2012. Network Context and Selection in the Evolution to Enzyme Specificity. Science. 337, 1101–1104. doi:10.1126/science.1216861

Nielsen, J., Larsson, C., van Maris, A., Pronk, J., 2013. Metabolic engineering of yeast for production of fuels and chemicals. Curr. Opin. Biotechnol. 24, 398–404. doi:10.1016/j.copbio.2013.03.023

Ober, D., 2005. Seeing double: Gene duplication and diversification in plant secondary metabolism. Trends Plant Sci. 10, 444–449. doi:10.1016/j.tplants.2005.07.007

Orth, J.D., Thiele, I. and B.O. Palsson, 2010. What is flux balance analysis? Nat. Biotechnol. 28, 245–248. doi:10.1038/nbt.1614

Osbourn, A.E. and B. Field, 2009. Operons. Cell. Mol. Life Sci. 66, 3755–75. doi:10.1007/s00018-009-0114-3

Qi, X., Bakht, S., Leggett, M., Maxwell, C., Melton, R., Osbourn, a, 2004. A gene cluster for secondary metabolism in oat: implications for the evolution of metabolic diversity in plants. Proc. Natl. Acad. Sci. U. S. A. 101, 8233–38. doi:10.1073/pnas.0401301101

Shen, Y., Muramatsu, S.I., Ikeguchi, K., Fujimoto, K.I., Fan, D.S., Ogawa, M., Mizukami, H., Urabe, M., Kume, A., Nagatsu, I., Urano, F., Suzuki, T., Ichinose, H., Nagatsu, T., Monahan, J., Nakano, I., Ozawa, K., 2000. Triple transduction with adeno-associated virus vectors expressing tyrosine hydroxylase, aromatic-L-amino-acid decarboxylase, and GTP cyclohydrolase I for gene therapy of Parkinson's disease. Hum. Gene Ther. 11, 1509–19. doi:10.1089/10430340050083243

Shin, J.H. and S.Y. Lee, 2014. Metabolic engineering of microorganisms for the production of L-arginine and its derivatives. Microb. Cell Fact. 13, 1–11. doi:10.1186/s12934-014-0166-4

Shumskaya, M. and Wurtzel, E.T., 2013. The carotenoid biosynthetic pathway: Thinking in all dimensions. Plant Sci. 208, 58–63. doi:10.1016/j.plantsci.2013.03.012

Stephanopoulos, G. and a J. Sinskey, 1993. Metabolic engineering--methodologies and future prospects. Trends Biotechnol. 11, 392–396. doi:10.1016/0167-7799(93)90099-U

Stephanopoulos, G.N., Aristidou, A. and J. Nielsen, 1998. Examples of Pathway Manipulations: Metabolic Engineering in Practice. Metab. Eng. Princ. Methodol. 203–283. doi:http://dx.doi.org/10.1016/B978-012666260-3/50007-8

Sumegi, B., Butwell, N.B., Malloy, C.R., Sherry, a D., 1994. Lipoamide influences substrate selection in post-ischaemic perfused rat hearts. Biochem. J. 297, 109–113.

Torres, V. and E.O. Voit, 2002. Target: a useful model. Pathw. Anal. Optim. Metab. Eng. 1–17. doi:10.1017/CBO9780511546334

Twyman, R.M., Christou, P., Capell, T., Zhu, C., 2015. Knowledge-driven approaches for engineering complex metabolic pathways in plants. Curr. Opin. Biotechnol. 54–60. doi:10.1016/j.copbio.2014.11.004

Vanhercke, T., El Tahchy, A., Liu, Q., Zhou, X.R., Shrestha, P., Divi, U.K., Ral, J.P., Mansour, M.P., Nichols, P.D., James, C.N., Horn, P.J., Chapman, K.D., Beaudoin, F., Ruiz-López, N., Larkin, P.J., de Feyter, R.C., Singh, S.P., Petrie, J.R., 2014. Metabolic engineering of biomass for high energy density: Oilseed-like triacylglycerol yields from plant leaves. Plant Biotechnol. J. 12, 231–239. doi:10.1111/pbi.12131

Vemuri, G.N. and Aristidou, A. A, 2005. Metabolic Engineering in the -omics Era : Elucidating and Modulating Regulatory Networks. Microbiology and Molecular Biology Reviews 69, 197–216. doi:10.1128/MMBR.69.2.197

Verma, P., Mathur, A.K., Srivastava, A., Mathur, A., 2012. Emerging trends in research on spatial and temporal organization of terpenoid indole alkaloid pathway in *Catharanthus roseus*: A literature update. Protoplasma 249, 255–268. doi:10.1007/s00709-011-0291-4

Vogt, T., 2010. Phenylpropanoid biosynthesis. Mol. Plant 3, 2–20. doi:10.1093/mp/ssp106

Winzer, T., Gazda, V., He, Z., Kaminski, F., Kern, M., Larson, T.R., Li, Y., Meade, F., Teodor, R., Vaistij, F.E., Walker, C., Bowser, T. A., Graham, I. A., 2012. A *Papaver somniferum* 10-Gene Cluster for Synthesis of the Anticancer Alkaloid Noscapine. Science 336, 1704–1708. doi:10.1126/science.1220757

Woolston, B.M., Edgar, S., Stephanopoulos, G., 2013. Metabolic Engineering: Past and Future. Annu. Rev. Chem. Biomol. Eng. 4, 259–288. doi:10.1146/annurev-chembioeng-061312-103312

Xie, D.Y., Sharma, S.B., Wright, E., Wang, Z.Y., Dixon, R. A., 2006. Metabolic engineering of proanthocyanidins through co-expression of anthocyanidin reductase and the PAP1 MYB transcription factor. Plant J. 45, 895–907. doi:10.1111/j.1365-313X.2006.02655.x

Yang, Y., Bennett, G.N., 1998. Genetic and metabolic engineering. Electronic journal of Biotechnology 1, 134–141. doi:10.2225/vol1-issue3-fulltext-3

Yarmush, M.L. and Banta, S., 2003. Metabolic engineering: advances in modeling and intervention in health and disease. Annu. Rev. Biomed. Eng. 5, 349–381. doi:10.1146/annurev.bioeng.5.031003.163247

Yuan, Y., Du, J., Zhao, H., 2013. Customized optimization of metabolic pathways by combinatorial transcriptional engineering. Methods Mol. Biol. 985, 177–209. doi:10.1007/978-1-62703-299-5-10

Ziegler, J. and P.J. Facchini, 2008. Alkaloid biosynthesis: metabolism and trafficking. Annu. Rev. Plant Biol. 59, 735–769. doi:10.1146/annurev.arplant.59.032607.092730

16

Methods to Identify Evolutionary Conserved Regulatory Elements Using Molecular Phylogenetics in Microbes

Shishir K Gupta*1*, Mugdha Srivastava*1*, Suchi Smita*2*, Taruna Gupta*3* and Shailendra K Gupta*2**

Abstract

Phylogenetics describes the sequence of speciation events that lead to the forming of a set of current day species. In phylogenetic studies, the most convenient way of visually presenting evolutionary relationships among a group of organisms is through illustrations called phylogenetic trees. Until mid 1950's phylogenies were often constructed by the subjective criteria based on experts' opinion about evolution. With the exponential growth of high-throughput whole genome sequencing and data analysis using state of art computational methods and tools, the current era of phylogenetics attempts to determine the rates and patterns of change occurring in DNA and proteins. In this chapter, we will review state of art computational methods and tools available and present an integrative workflow to reconstruct the evolutionary history of regulatory elements for variety of biological applications.

1. Functional annotation of regulatory proteins

After the structural annotation of any genome the functional annotation of every gene is a fundamental goal of comparative genomics. Orthology based assignment of gene function is the widely used approach. Orthologs are defined as genes in different species that descend by speciation from the same gene in the last common ancestor

[1] Department of Bioinformatics, Biocenter, Am Hubland, University of Würzburg, 97074 Würzburg, Germany, Phone: +49 931 318 9747, Fax: +49 931 318 4552,
Email: shishir.gupta@uni-wuerzburg, mugdha.srivastava@uni-wuezburg.de.
[2] Department of Systems Biology and Bioinformatics, University of Rostock, 18051 Rostock, Germany, Phone: +49 381 498 7685, Fax: +49 381 498 7572,
Email: suchi.smita@uni-rostock.de
[3] Department of Bioinformatics, Systems Toxicology Group, CSIR-Indian Institute of Toxicology Research, 226001 Lucknow, India, Phone: +91 522 2294591,
Fax: +91 522 2628227, Email: taruna1606@gmail.com
[*] Corresponding author: shailendra.gupta@uni-rostock.de

(Fitch 1970). Therefore, they are likely to perform equivalent functions if they have diverged since the speciation event (Gabaldón and Koonin 2013). Orthologs are likely to be functional counterparts in different species (Price 2007) in contrast to paralogs, i.e., genes that diverged by gene duplication, and xenologs, i.e., homologous genes whose history of divergence includes one or more horizontal gene transfer (HGT) events. Because of high rates of HGT in bacteria, many genes are xenologs rather than orthologs (Price *et al.* 2007). Although the bacterial genomes are comparatively less complex than eukaryotic as the functional annotation of the genes becomes complex by HGT events. In contrast with vertical gene transfer i.e., the transfer of DNA from parent to offspring (deoxyribonucleic acid) the gene acquisition through HGTs involves movement of genetic material between different species by alternative mechanism such as transduction, conjugation, transformation and the uptake of free DNA from the environment into the bacterial cytosol. For the cell to benefit, the newly acquired genes have to be successfully integrated into the cellular regulatory system so that they are turned on and off at appropriate times (McAdams et al. 2004).

The occurrence of HGTs between species is challenging to molecular phylogeny as in the presence of frequent HGT events the notion of a unique organismal phylogeny could be misleading (Galtier 2007). However, several authors have also claimed that the core genes are much resistant to HGT and could serve to reconstruct the bacterial species tree (Brochier et al. 2002, Daubin et al. 2003, Kurland et al. 2003). Inparanoid (O'Brien *et al.* 2005) and OrthoMCL (Li *et al.* 2003) are the highly used algorithms for identification of orthologs. Despite the orthology often reflect the functional conservancy the study by Price *et al.* (2007) shows that even in more closely related bacteria, where the orthologous transcription factors (TFs) that have conserved functions, the annotation transfer of regulatory interactions are often incorrect. The modular organization of the regulatory circuitry enhances evolvability, because a simple change in the wiring of the regulatory circuitry can cause large changes in the organism's response to a signal (McAdams *et al.* 2004). For example, a mutation in the promoter of a master regulator gene that changes the regulatory protein controlling its activity could introduce radical changes in either the timing of expression of the master regulator or the conditions leading to its expression. This would ultimately change the pattern of expression of many downstream genes regulated by the master regulator. Considering all the facts mentioned above, the evolution of bacterial regulation should be analyzed with phylogenetic trees. In the chapter we will further describe about the fundamental of phylogenetic trees, state of art methods to identify functional equivalent proteins by using phylogenetic trees.

2. Structure of phylogenetic tree

Phylogenetic relationships are most commonly expressed in the form of networks (or trees) that summarize the divergence events that have occurred among sequences over time. Individual internal nodes correspond to distinct historical divergence events. Terminal nodes correspond to the actual sampled sequences. In addition, the phylogenetic distances between data can also be represented on these networks, typically by scaling the lengths of the branches that connect different nodes together.

In general, the phylogenetic trees can be of two broad types i.e., rooted and unrooted tree (Fig. 1). The rooted tree are the trees having a specific node from where the other nodes emerges while the unrooted tree represents the same phylogeny as the rooted tree but lacks the root node. The position of root can be estimated in a rooted tree by introducing outgroups which are the set of species that are definitely distant from all the species of interest.

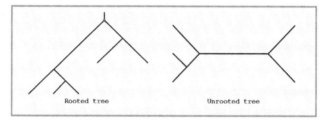

FIGURE 1   Sample framework of the rooted and unrooted phylogenetic tree. Brown dot is showing the root in the rooted tree.

3.   Methods of phylogenetics inference

Phylogenetic analysis presents a unique problem in biology, since the evolutionary history can never be known with certainty. Currently there appears to be no uniform way of performing a phylogenetic study or interpreting its results however, two broad methods have been developed by computational biologists that are frequently exploited for the phylogenetic inference. The first method is character-based method that uses the aligned characters, such as DNA or protein sequences, directly during tree inference while the second method is distance-based method that transforms the sequence data into pairwise distances, and uses the matrix during tree building.

3.1   Distance based methods

The reconstruction of a phylogenetics tree using distance based methods is a two-step process. In the initiative step all the sequences are aligned and the distances are calculated for all pairwise sequences using the appropriate distance. The distance is chosen depending on the characteristics of the sequence alignment for which a phylogeny is being reconstructed. Distances are arranged in the form of matrix. Further, the reconstruction of the phylogenetic tree from the distance matrix is performed in the final step. In the presence of limited phylogenetic signal in the data set being analyzed, it is valuable to add bootstrap analysis (Felsenstein 1985) for estimating the confidence that can be attached to a particular tree.

3.2   Character based methods

Whereas the distance based methods compress all sequence information into a single number, the character based methods attempt to infer the phylogeny based on all the individual characters. Characters can be derived from observable properties or can be derived from subsequences or even amino acids or nucleotides. Maximum parsimony, maximum likelihood and Bayesian inference are the methods, categorized within the character based methods.

While these methods are appealing because they have the promise of finding the optimal tree according to the applied distance and bootstrapping criterion, however, they can be computationally slow for even moderate numbers of taxa, to a point where the amount of time required for an exhaustive search is prohibitive. Therefore, this limitation has led to the development of the other improved computational methods that attempt to reliably get as close to the optimal tree as possible, in a reasonable amount of computational time.

4. Computational methods to identify conserved regulatory elements in microbes

Regulatory elements (REs) in microbes can be identified by bioinformatics tools.

4.1 Regulogger

Regulogger is a computational method for the identification of sets of genes having conserved sequence and regulatory signal among multiple organisms. These sets of genes were termed as regulogs. Regulogger is based on comparative genomics which results in the detection of quantitatively ranked conserved regulons (Alkema *et al.*, 2004). It follows the methodology in which starts from the identification of protein-coding regions in the genomic sequence. For each query protein, a set of orthologous genes in other genomes was obtained using the COGs (clusters of orthologous groups) database (Tatusov *et al.*, 2001). From the upstream regions of these sets of orthologous genes, conserved cis-REs were detected using Gibbs motif sampler (Thompson et al., 2003). The Gibbs sampler calculates a maximum a posteriori (MAP) value for each pattern. The most significant pattern from a sequence set was obtained by repeating the algorithm ten times and the highest average MAP-value was retained for each pattern. After that clustering of the patterns obtained by phylogenetic footprinting was done using Needleman-Wunsch (Needleman and Wunsch, 1970) and UPGMA algorithm. To validate the phylogenetic footprinting and regulogger methods two reference sets of transcription factors was constructed from *Bacillus subtilis* and *Escherichia coli* for which both the transcription-factor binding site and an experimentally verified regulon has been described. Finally, the regulogs (conserved regulons) could be constructed by calculating a relative conservation score (RCS) for each predicted regulon member. If geneA is a regulon member predicted to be under the control of the cis-RE, the RCS is given by

$$RCS_{geneA} = \text{orthologs}_{observed}/\text{orthologs}_{expected}$$

Where, orthologs $_{observed}$ = number of orthologs that are under the control of the same cis-RE, and orthologs$_{expected}$ = total number of orthologs present in the genomes considered in the study.

4.2 RegPredict

RegPredict a user-friendly and interactive web-server can be used for rapid and accurate analysis of known regulatory motif as well as prediction of novel regulons in varied groups of microorganisms following comparative genomics (Novichkov et al., 2010). The server uses the information of genomic data, predicted operons and orthologs from phylogenetic analysis available at MicrobesOnline database (Dehal et al., 2010). This approach not only emphasize on elaboration of regulon information from model organism to others but also in silico prediction of novel regulons and motifs in a group of micro-organisms. The genome of interest initially selected by the user remains constant throughout the entire procedure of regulon inference. After that the workflow proceeds in one of the two main methods of regulon inference either by availability of positional weight matrices (PWMs) for known regulatory motifs or by de novo regulon inference. The first method uses the information of known regulatory motifs and PWMs from three web resources namely RegPrecise database (Novichkov et al., 2010), RegTransBase (Kazakov et al., 2010) and RegulonDB database (Gama-Castro et al., 2008). It follows the 'Run Profile' procedure by scanning the genome with selected PWM profile and generates CRONs (Clusters of co-Regulated Orthologous operoNs). The second method follows the 'Discover Profile' procedure and starts with the set of co-regulated genes that are functionally important or homologous to a well-known regulon or having similar expression profile or derived from conserved or orthologous genes, resulting in the generation of candidate profiles which were further analyzed by following the 'Run Profile' procedure. All generated CRONS were further filtered on the basis of level of conservation and detail functional and genomic analysis was performed. The users can also export the information of resultant CRON analysis to a text file.

4.3 PhyloCon

PhyloCon (PhylogeneticConsensus) method can be used to identify regulatory motifs by integrating data of co-regulated genes in a single species and sequence conservation in multiple species (Wang and Stormo, 2003). It uses a novel ALLR (Average Log Likelihood Ratio) statistic for sequence profile comparison and a greedy method for searching common subprofiles. The three major steps of this algorithm are initial profile generation, profile comparison and profile merging. Initially multiple sequence alignment (MSA) was performed in the orthologous sequences using Wconsensus program (Hertz and Stormo, 1999) and obtained conserved regions were then transformed into sequence profiles. Then the profiles were compared using ALLR static and common regions between two profiles were identified and merged to create a new profile for further comparison. This step is repeated likewise so that all profiles were compared and new profiles were generated and ranked on the basis of ALLR score. Finally, the top ranked profiles were reported as regulatory motifs conserved in all orthologous sequences.

4.4   cis-Regulatory Non-coding RNA (ncRNA) identification pipeline

Yao and coworkers (2007) proposed an efficient and automated pipeline that utilizes structural information for identifying cis-regulatory ncRNAs in prokaryotes (Yao et al., 2007). The key feature of this pipeline is that it performs RNA motif inference with low sequence conservation and also improves the quality of RNA motifs by integrating the prediction with RNA homolog search and functional analysis. The pipeline proceeds in the following steps

1. Identification of homologous gene sets from NCBI's Conserved Domain Database (CDD) (Marchler-Bauer et al., 2005). CDD groups containing, less than four or more than seventy members, were removed to make the prediction more reliable and less expensive.
2. Collection of nc sequences upstream to each gene in the CDD group using MicroFootprinter (Neph and Tompa, 2006). tRNAs, rRNAs and highly similar sequences were removed.
3. A phylogenetic footprinting tool FootPrinter (Blanchette and Tompa, 2003) was used to select and ranked dataset that contain the conserved sequence motifs in unaligned homologous sequences set.
4. Inference of RNA motif in unaligned sequences was performed using CMfinder (Yao et al., 2006) and predicted motifs were ranked by a scoring function that considers motifs with stable secondary structure, local sequence conservation and present in diverged species.
5. The predicted motifs were post processed and scanned in the prokaryotic genome database to find additional homologs which facilitates the construction of more accurate motif models.
6. Finally, analysis of genomic context and literature search was done manually to rank the top motifs.

---

5.   Phylogenetic-based methods to identify regulatory proteins

Transcription in bacteria shaped by the interactions between TFs that bind cis regulatory elements in DNA in the vicinity of the structural portion of a gene that are required for gene expression, additional co-factors and influence the chromatin structure (Wasserman and Sandelin 2004). Trans-acting proteins control the rate of transcription or gene expression at the level of the individual gene that bind to crucial cis-regulatory sequences (Wasserman and Sandelin 2004). As we defined above the phylogenetic-based methods should be prioritize to identify regulatory proteins in contrast to reverse blast hit (RBH) based orthology predictions methods (Tatusov et al. 1997). Below, we introduce the algorithms and online resources that can be used to identify functional orthologs of regulatory proteins based on phylogeny.

5.1   PhyloFacts FAT-CAT web server

The PhyloFacts FAT-CAT (Fast Approximate Tree Classification) is a web server for protein functional annotation and identification of orthologs using hidden Markov models (HMMs) at every node of every tree which allows the very flexible prediction of function at all levels of a functional hierarchy (Afrasiabi et al. 2013). It uses the pre-calculated phylogenetic trees in the PhyloFacts database (Krishnamurthy et al. 2006). The server can distinguish between the paralogs and orthologs. User is allowed to submit the protein sequence (with maximum length of 2000 amino acids) as input. To handle different types of inputs four different parameters options i.e., high recall, high precision, remote homolog detection and partial sequence search are provided. The pipeline proceeds through four stages namely, family HMM scoring, subtree HMM scoring, ortholog selection, and functional annotation hierarchy (Afrasiabi et al. 2013). The output for each query is organized into a web page with separate tabs for family matches, predicted orthologs and functional annotations. The pipeline gives higher weight to manually curated and close ortholog than the annotations that are derived computationally or from more distant orthologs.

5.2 MetaPhOrs

MetaPhOrs is a comprehensive global repository of highly accurate, phylogeny-based orthology and paralogy predictions that were computed by combining phylogenetic information derived from PhylomeDB (Huerta-Cepas et al. 2014), EnsemblCompara (Vilella et al, 2009), EggNOG (Powell et al. 2014), OrthoMCL (Li et al. 2003), COG (Tatusov et al. 2007, Wolf and Koonin 2012), Fungal Orthogroups (Wapinski et al. 2007), and TreeFam (Li et al. 2006) databases. In the first step of MetaPhOrs pipeline (Pryszcz et al. 2011) all the phylogenetic trees for any given pair of sequences are retrieved then after filtering step it discards the phylogenetic trees made with suboptimal evolutionary models. Moreover, species-overlap algorithm, is implemented on every single tree to predict the type of homology relationship between this sequence pair and a consistency score for an orthology prediction is calculated, which lies between 1 (all trees predict an orthology relationship) and 0 (all trees predict a paralogy relationship between the sequences). Besides all, evidence level index is provided which indicates how many open sources databases have been used for the prediction. User can retrieve the orthology and paralogy predictions by searching for a particular protein, for a pair of species or for multiple proteins.

5.3 PHOG

PHOG (Phylogenetic Orthologous Groups) is completely automated method which builds clusters of orthologous groups at each node of the taxonomy tree (Merkeev et al., 2006). The resulting cluster is represented by an ancestral sequence obtained from the multiple sequence alignments of orthologous and paralogous genes. This database demonstrates the possibility to process any number of sequenced genomes to reconstruct orthologous and paralogous relationships among multiple genomes. PHOG includes many steps to complete the procedure, such as obtaining a supergene from a PHOG multiple alignments, running PHOG-BLAST, splitting procedure, multiple sequence alignments of core sequences in the orthologous group, and finding the paralogs.

5.4   PhylomeDB

PhylomeDB is a public database of complete collection of evolutionary histories of all genes in a genome (Huerta-Cepas et al. 2014). The trees can be easily accessed, queried and downloaded through web interface. The workflow proceeds by searching homologs for the input protein against the corresponding proteome dataset, after that the sets of homologous sequences are aligned using MUSCLE aligner. Subsequently the phylogenetic trees are derived from the resulting reliable alignments by using several tree generations methods such as NJ method, ML method or Bayesian phylogenetic reconstruction. The implemented species-overlap algorithm decides whether the nodes in the tree represent gene duplication or speciation event (Huerta-Cepas et al. 2007).

5.5   LOFT

LOFT (Levels of Orthology From Trees) is a JAVA based software which implements 'levels of orthology' concept to determine high resolution phylogeny based orthology. LOFT first decides the root of the tree using outgroup, then to discriminate speciation from duplication events for each node using the recommended species-overlap rule. It was shown that within COG clusters LOFT could assign the orthologous groups with 95% accuracy (van der Heijden et al. 2007). The executable file of software is available at http://www.cmbi.ru.nl/LOFT/.

5.6 QuartetS

QuartetS pipeline provides accurate and high-throughput ortholog predictions for large-scale applications. The program integrates the three steps procedure to automatically predict orthologous clusters for the input genomes. It uses Blast for sequence similarity search then QuartetS to analyze the evolutionary relationship of the two genes and infers possible evolutionary events to determine if they should be considered to be orthologs and finally Single Linkage Cluster (SLC) and Markov Cluster Algorithm (MCL) (Enright et al. 2002) to cluster genes in reliable ortholog groups. The program was used to determine orthology among 624 bacterial genomes. The executables for QuartetS and the pre-computed results for 624 bacteria are available at http://bhsai.org/downloads/quartets/

5.7   GepTop

GepTop tool (Wei et al. 2013) is not exactly created for functional annotation but the method implements phylogeny weighted orthology approach which could be exploited in standalone version for functional annotation after updating the database with the reference bacterial proteomes. The standalone program uses the reciprocal best hit (RBH) (Tatusov et al. 1997) method to estimate orthology of given protein sequences and the composition vector (CV) method (Xu and Hao 2009) to estimate the evolutionary distance between the query sequences and user defined bacterial proteome database. If the query protein is predicted to have cutoff score ≥ 0.15 the protein could be functionally annotated by annotation transfer form that reference database sequence which is the reciprocal best hit of the query sequence. The method is not suitable for large scale predictions and could not distinguish between paralogs and orthologs therefore, should be used as additional check of predictions.

6. Tools and web-servers for phylogenetic analysis

Several tools and servers are available to analyse the ingredient of phylogenetics analysis. These tools can be used to establish evolutionary relationships of any gene of proteins including regulatory transcription factors of microbes. We have further reviewed the broad methods of the construction of phylogenetic trees followed by the available tools with few demonstrations for analyzing phylogeny which can be easily applied by a biologist without any expertise in Bioinformatics. The protocol can be used to determine the degree of evolutionary similarity among proteins with any biological functions including the signalling, gene regulation and metabolism. Some of the frequently used useful tools for phylogenetic analysis are listed in Table 1.

TABLE 1 Different tools for the phylogenetic analysis, categorized on the basis of methods they implements.

Parsimony programs	Distance matrix methods	Maximum Likelihood method	Bayesian Inference
Phylip	Phylip	Phylip	Paml
Paup	Paup	Paup	Bambe
Mega	Mega	Phyml	Dambe
Malign	BioNJ	aLRT	Mr. Bayes
Past	Bioinformatics Toolbox	Treefinder	BEST
Emboss	Darwin	fastDNAml	PHYLLAB

6.1 Phylip

Phylip (Felsenstein 1989) includes programs to carry out parsimony, distance matrix methods, maximum likelihood, and other methods on a variety of types of data, including DNA and RNA sequences, protein sequences, restriction sites, 0/1 discrete characters data, gene frequencies, continuous characters and distance matrices. It may be the most widely-distributed phylogeny package, third after Paup and MrBayes in the competition to be the program responsible for the most published trees high impact research. PHYLIP is freely distributed at the PHYLIP web site at http://evolution.gs.washington.edu/phylip.html

6.2 PAUP

PAUP (Phylogenetic Analysis using Parsimony) package includes parsimony, distance matrix, invariants, and maximum likelihood methods and many indices and statistical tests (Swofford 2002). The principle of maximum parsimony is to find the evolutionary tree that requires the smallest number of character changes to generate the differences observed in operational taxonomic units. The package different operating systems is available at web page at http://paup.csit.fsu.edu/

6.3   MEGA

MEGA (Molecular Evolutionary Genetic Analysis) carries out parsimony, distance matrix and likelihood methods for molecular data (Tamura *et al.* 2013). Mega can perform bootstrapping, consensus trees, and a variety of distance measures, with NJ, Minimum Evolution (ME), UPGMA, and parsimony tree methods, as a well as a large variety of data editing tasks, sequence alignment using an implementation of ClustalW, tests of the molecular clock, and single-branch tests of significance of groups. It is freely available at its web site at http://www.megasoftware.net as windows executables, with a downloadable manual.

6.4   MALIGN

MALIGN is a parsimony-based alignment program for molecular data including both the nucleic acid sequences and amino-acid sequences (Wheeler and Gladstein 1994). A good alignment is the basic need for the phylogenetics analysis. Malign implements the original suggestion by Sankoff and co-workers (1973), that the alignment and phylogenies could be done at the same time by finding that tree that minimizes the total alignment score along the tree (Sankoff *et al.* 1973).

6.5   PAST

PAST (*PA*leontological *ST*atistics) is a package which carries out many kinds of paleontological data analyses, including stratigraphic and morphometric statistics (Hammer *et al.* 2001). It also does parsimony analysis, including exhaustive, branch-and-bound and heuristic algorithms for Wagner, Fitch and Dollo parsimony. It does bootstrap methods, strict and majority rule consensus trees, and consistency and retention indices. It calculates three stratigraphic congruency indices with permutation tests. It also does many other statistics and curve fitting. Past is available from its web site at http://folk.uio.no/ohammer/past/index.html as a windows executable. Manuals can be read online or downloaded from the web site.

6.6   EMBOSS

EMBOSS (European Molecular Biology Open Software Suite) a package of programs for general sequence analysis with some phylogeny and alignment programs (Rice et al. 2000). EMBOSS, developed by many developers, is a general suite of programs for sequence analysis. It is a full-featured sequence analysis program developed intended to provide the same functionality as GCG. In addition to its own programs, it also has a suite of other programs, EMBASSY, that are configured to work with EMBOSS. These include ClustalW and most PHYLIP programs. It can be downloaded from its web site at http://emboss.sourceforge.net

6.7 BioNJ

BioNJ, an improved version of NJ based on a simple model of sequence data which follows the same agglomerative scheme as NJ but uses a simple, first-order model of the variances and covariances of evolutionary distance estimates (Gascuel 1997). This model is appropriate when these estimates are obtained from aligned sequences. It retains the speed advantages of NJ while using a slightly different criterion to select pairs of taxa to join, one which will perform better when distances between taxa are large.

6.8 DARWIN

DARWIN (Data Analysis and Retrieval with Indexed Nucleotide/peptide sequences) is an environment which enables the user to carry out a variety of kinds of analysis with sequences, including phylogeny methods (Gonnet et al. 2000). These seem to include distance matrix, split decomposition, and a form of likelihood method. Darwin is available as executables for solaris, intel-compatible linux, irix, and HP/compaq/digital alpha machines. These are freely available if the user registers by filling out a form at the download page at the web page. The executables can then be transferred to the user by ftp or by e-mail of encoded files.

6.9 Bioinformatics Toolbox

The module of phylogenetics in MATLAB resides within the toolbox for bioinformatics. It has many functions for sequence analysis and microarray data, including multiple sequence alignment and consensus sequences. For this listing, the relevant ones are that it enables the user to create and edit phylogenetic trees. The user can calculate pairwise distances between aligned or unaligned nucleotide or amino acid sequences using a broad range of similarity metrics, such as Jukes-Cantor, p-distance, alignment-score, or a user-defined distance method. Phylogenetic trees are constructed using hierarchical linkage with a variety of techniques, including neighbor joining, single and complete linkage, and UPGMA. Bioinformatics Toolbox includes tools for weighting and rerooting trees, calculating subtrees, and calculating canonical forms of trees. Through the graphical user interface, user can prune, reorder, and rename branches; explore distances; and read or write Newick-formatted files. Moreover, user can also use the annotation tools in MATLAB to create presentation-quality trees. The toolbox is available as a MATLAB package.

6.10 PhyML

PhyML is a fast and accurate maximum likelihood program to estimate large phylogenies for nucleotide or protein sequence data (Guindon et al. 2010). It has six possible DNA substitution models, five amino acid substitution models, allowing estimation of many of the model parameters, and can allow for a gamma distribution of rates among sites and a proportion of invariable sites. It can also perform robust bootstrapping of the trees.

6.11 Treefinder

It is a maximum likelihood program for nucleotide sequence data. It makes available a variety of models of base change, including codon-position-specific models (Jobb et al. 2004). It carries out search for best trees by its own method of tree rearrangement, and can assess statistical support for groups by either bootstrap or a local paired-sites method. All parameters of the models can be optimized by searching for the values that maximize the likelihood. The program is fast, and has both a graphical user interface and a general language in which its operation can be programmed. Trees can be interactively manipulated and constrained in various ways.

6.12 PhyML-aLRT

PhyML-aLRT (Approximate Likelihood Ratio Test) is a program to carry out likelihood ratio tests of the presence of branches in a phylogeny. PhyML-aLRT is a modification of the original PhyML program, and is designed to compute test of the reality of branches in a known phylogeny. Five branch support tests are available: (1) the bootstrap, (2) aLRT statistics, (3) aLRT parametric (Chi^2-based) branch support, (4) aLRT non-parametric branch support based on a Shimodaira-Hasegawa-like procedure, and (5) a combination of these two latters supports, that is, the minimum value of both. The methods are described elsewhere (Anisimova et al. 2006). The program was also implemented to reconstruct genome wide phylogeny of seven mycobacteria (Mignard and Flandrois 2008).

6.13 PAML

PAML is a package of programs for the ML analysis of nucleotide or protein sequences, including codon-based methods that take into account both amino acids and nucleotides (Yang 1997). The programs can estimate branch lengths in a phylogenetic tree and parameters in the evolutionary model such as the transition/transversion rate ratio, the gamma parameter for variable substitution rates among sites, rate parameters for different genes, and synonymous and nonsynonymous substitution rates. They can also test evolutionary models, calculate substitution rates at particular sites, reconstruct ancestral nucleotide or amino acid sequences, simulate DNA and protein sequence evolution, compute distances based on the synonymous and nonsynonymous changes, and of course do phylogenetic tree reconstruction by ML and Bayesian Markov Chain Monte Carlo methods. PAML is a widely used platform for analyzing selection pressure on regulatory genes (Castillo-Davis et al. 2004, Takeuchi et al. 2012, Zhao et al. 2014, Flores et al. 2015).

6.14 BAMBE

BAMBE (Bayesian Analysis in Molecular Biology and Evolution) is a program for Bayesian analysis of phylogenies with DNA sequence data. It uses a prior distribution of trees and the arrangement mechanism has been introduced in the literature (Mau et al. 1997). The trees and parameter values are sampled by a Metropolis algorithm

Markov Chain Monte Carlo sampling. The resulting posterior distribution can be used to characterize the uncertainty about not only the tree, but the parameters of the substitution model as well.

6.15   DAMBE

DAMBE (Data Analysis in Molecular Biology and Evolution) is a general-purpose package for DNA and protein sequence phylogenies, and also gene frequencies (Xia 2013). It can read and convert a number of file formats, and has many features for descriptive statistics. It can compute a number of commonly-used distance matrix measures and infer phylogenies by parsimony, distance, or likelihood methods, including bootstrapping (by sites or by codons) and jackknifing. There are a number of kinds of statistical tests of trees available, and many other features. It can also display phylogenies. DAMBE includes a copy of ClustalW; there is also code from PHYLIP. An interesting feature is a simple web browser that allows sequences to be fetched over the web while running DAMBE.

6.16   MrBayes

MrBayes, a program for Bayesian inference of phylogenies from nucleic acid sequences, protein sequences, and morphological characters (Ronquist et al. 2012). It assumes a prior distribution of tree topologies and uses Markov Chain Monte Carlo (MCMC) methods to search tree space and infer the posterior distribution of topologies. It reads sequence data in the NEXUS file format, and outputs posterior distribution estimates of trees and parameters. It can also use a hierarchical Bayesian framework to infer sites that are under natural selection. It allows for rate variation among sites and a variety of models of sequence evolution.

6.17   BEST

BEST finds the joint posterior distribution of coalescent gene trees and the species tree for multi-locus data under a hierarchical Bayesian model (Liu and Pearl 2007, Edwards et al. 2007). Proposal gene trees are made using a gene tree MCMC procedure chosen by the user in MrBayes. This vector of gene trees is then paired with a species tree chosen under the constraint that the gene trees be consistent with the species tree. An MCMC importance sampling is then used to sample the species trees.

6.18   RAxML

RAxML (Randomized Axelerated Maximum Likelihood) uses rapid bootstrap algorithm to compute ML based phylogenetic trees (Stamatakis, 2014). It also provides the options to user to automatically select the best substitution model for their data. Genome scale phylogeny can be performed in RAxML using supermatrix approach of constructing species tree. This approach uses the concatenated orthologous clusters for inferring robust species trees.

Some important web-servers used for phylogenetic analysis are listed in Table 2. This next provides a brief demonstration of the constructing phylogenetic tree of protein sequence data using the Phylogeny.fr and PhyML and BioNJ.

TABLE 2 Robust web-servers for the phylogenetic analysis

Server	Application	URL
Phylogeny.fr	Robust Phylogenetic Analysis for the Non-Specialist	http://www.phylogeny.fr/
DendroUPGMA	The program calculates a similarity matrix (only for option a), transforms similarity coefficients into distances and makes a clustering using the Unweighted Pair Group Method with Arithmetic mean (UPGMA) algorithm.	http://genomes.urv.cat/UPGMA/
PHYLIP	Server for phylogenetic analysis using the PHYLIP package	http://bioweb2.pasteur.fr/ phylogeny/intro-en.html
PhyML	Server for Maximum Likelihood phylogenetic analysis	http://atgc.lirmm.fr/phyml/
The PhylOgenetic Web Repeater (POWER)	Performs phylogenetic analysis	http://power.nhri.org.tw/power/ home.htm
Evolutionary Trace Server (TraceSuite II)	Maps evolutionary traces to structures	http://mordred.bioc.cam. ac.uk/~jiye/evoltrace/evoltrace. html
BioNJ	Server for Neighbour-Joining phylogenetic analysis	http://mobyle.pasteur.fr/cgi-bin/ portal.py?#forms::bionj

7. Protocol for phylogenetics analysis with Phylogeny.fr

Phylogeny.fr is a free, simple to use web service dedicated to constructing and analysing phylogenetic relationships between molecular sequences. Phylogeny.fr runs and connects various bioinformatics programs to construct a robust phylogenetic tree from a set of sequences (Dereeper et al. 2008, Dereeper et al. 2010). Direct accesses to the individual tools are available on this server. Phylogenetic analysis in Phylogeny.fr can be performed in single click or in advanced mode.

7.1   One Click mode

1. Open the server homepage (http://www.phylogeny.fr/version2_cgi/index. cgi) and click on one click (Fig. 2a).

2. Prepare the sequence file in FASTA format. The sequence input can also be given in the EMBL or NEXUS format.

3. The sequence file can be uploaded or pasted in the textbox (Fig 2b).

4. To improve the alignment and remove the divergent sequences Gblocks (Castresana 2000) program can be used by checking the box.

FIGURE 2 Snapshots of phylogenetics analysis in Phylogeny.fr. (a) Snapshot of Phylogeny.fr one click mode window. (b) Snapshot of Phylogeny.fr with pasted sequences in fasta format. (c) A fraction of alignment results in Phylogeny.fr. The web version of alignment shows the colors according to the conservancy of the columns.

5. Click the submit button. The resulting phylogenetic tree can be downloaded in the PDF, PNG, TXT, NEWICK, SVG and TGF format.

6. Individual results of the alignment and phylogeny can also be obtained

7. To obtain the results of alignment click on the alignment button. The default tool used for the alignment is Muscle. Alignment results are predicted using the BLOSUM62 matrix.

8. The alignment results (Fig. 2c) can be viewed in Fasta, Phylip or Clustal format. The alignment can also be edited (trimmed), in Jalview for which the link is available in the same window.

7.2 Display options

The phylogenetic tree can also be viewed in ATV java enabled browser which contains different options for display. These options can be checked or unchecked to get the required results. The font size, x-axis or y-axis can be increased or decreased as per our requirement. The default tool for the prediction of phylogenetic tree is PhyML. The tree style can be changed to Cladogram, Radial or Circular. However, radial and circular tree are not suitable for very large dataset.

7.2.1 Cladogram

It is a diagram that depicts evolutionary relationships among groups. Each branch on a cladogram is referred to as a "clade" and can have two or more arms. Taxa sharing arms branching from the same clade are referred to as "sister groups" or "sister taxa." All taxa that can be traced directly to one node (that is they are "upstream of a node") are said to be members of a monophyletic group. Fig. 2d is depicting the cladogram tree.

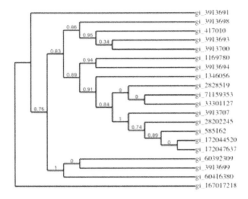

FIGURE 2(D)   A sample view of cladogram tree.

7.2.2   *Radial tree*

A radial tree is a method of displaying a tree structure in a way that expands out-wards, radially. It is one of many ways to visually display a tree. The radial tree graph also solves the problem of drawing a tree so that nodes are evenly distributed. A binary tree that is drawn linearly, so that the root is on one end, and nodes of the same level line up, will grow crowded very quickly but a radial tree will spread the larger number of nodes over a larger area as the levels increase. Fig. 2e is depicting the radial tree.

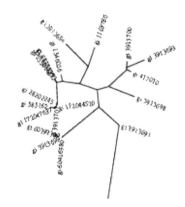

FIGURE 2(E)   A sample view of radial tree.

7.2.3   Circular tree

In the circular tree layout a single node is placed at the center of the display and all the other nodes are laid around it. The entire graph is like a tree rooted at the central node. The central node is referred to as the focus node and all the other nodes are arranged on concentric rings around it. Each node lies on the ring corresponding to its shortest network distance from the focus. Any two nodes joined by an edge in the graph is referred to as neighbors. Immediate neighbors of the focus lie on the smallest inner ring, their neighbors lie on the second smallest ring, and so on. Fig. 2f is depicting the cladogram tree.

7.3   Advanced Mode

The workflow settings can be done in the advanced mode. Each step can be performed all at once or one by one. Any box can be unchecked or checked as shown in the window according to the need.

1. Open the server homepage (http://www.phylogeny.fr/version2_cgi/index. cgi) and click on advanced (Fig. 2g).

2. The alignment of sequences with MUSCLE (Edgar, 2004) can be done in any of the following four mode (Fig. 2h). (a) *Full Mode*: The Full mode consists of draft progressive alignment, improved progressive alignment, alignment refinement. However, the full mode takes more time. (b) *Progressive Mode*: Progressive mode include draft progressive alignment and improved progressive alignment. This mode is faster than the full mode. (c) *Fastest Mode*: The fastest mode is draft alignment and is the fastest to perform and (d) Default Mode: Analysis based on all the default parameters. Minimal user input is required.

3. The number of iteration steps can be selected as required. Three iterations are done, and as we increase the iteration steps, iteration is done till the maximum convergence of sequences is reached. The default number of iteration is 16. Iteration improves the alignment but slows the process.

4. The find diagonals box is checked to find short regions of high similarity in two aligned sequences. A trick used in algorithms such as BLAST is to reduce the size of this matrix by using fast methods to find diagonals, i.e. short regions of high similarity between the two sequences. Creating a pair-wise alignment by dynamic programming requires computing an L1 x L2 matrix, where L1 and L2 are the sequence lengths. This speeds up the algorithm at the expense of some reduction in accuracy. MUSCLE uses a technique called k-mer extension to find diagonals.

5. Result of alignment is obtained in FASTA, PHYLIP or Clustal format. The results also include a Guide tree. However, Guide tree is given as an indication and is not a significant phylogenetic tree.

6. Next is the Curation step (Fig. 2i) which can be performed or skipped based on the requirement of the program outputs for integrating third party software.

7. The input data (i.e., the aligned sequences) for the phylogeny can be edited in integrated Jalview software (Fig. 2j).

The server also contains individual tools (Table 3) for sequence alignment, alignment editor and phylogeny.

TABLE 3 Different tools integrated in Phylogeny.fr server

Multiple Alignment	Phylogeny	Tree viewers	Utilities
Muscle	PhyML	TreeDyn	Gblocks
T-Coffee	TNT	Drawgram	Jalview
ClustalW	BioNJ	Drawtree	Readseq
ProbCons	MrBayes	ATV	Format converter

8. Protocol of phylogenetics analysis using PhyML

The phylogenetic analysis with Phyml can be done in the following steps:

1. Select the Datatype i.e., the provided sequences are of amino acids or nucleotides (DNA/RNA). The datatype can also be left to auto select.

2. Upload the sequences in FASTA, Phylip, Clustal, EMBL or NEXUS format. Alternatively sequence set can be pasted in the box.

3. The statistical tests for branches of evolutionary trees can be constructed on the basis of aLRT (Approximate likelihood Ratio test) or by using the bootstrap analysis. The aLRT is shown to be an accurate and powerful tool. It is implemented within the algorithm used by the recent fast maximum likelihood tree estimation program PhyML (Guindon and Gascuel, 2003). Bootstrapping is a way of testing the reliability of the dataset. It is the creation of pseudoreplicate datasets which are generated by randomly sampling the original character matrix to create new matrices of the same size as the original. Bootstrap analysis is used to examine how often a particular cluster in a tree appears when nucleotides or amino acids are resampled. However, to get quick results the server itself motivates for the aLRT analysis.

4. Any of the substitution models can be selected based on the datatype of the sequence i.e., Nucleotides or Proteins.

5. The gamma distribution is a two-parameter family of continuous probability distributions i.e., scale parameter and shape parameter. The gamma distribution represents the sum of n exponentially distributed random variables. Both the shape and scale parameters can have non-integer values. Typically, the gamma distribution is defined in terms of a scale factor and

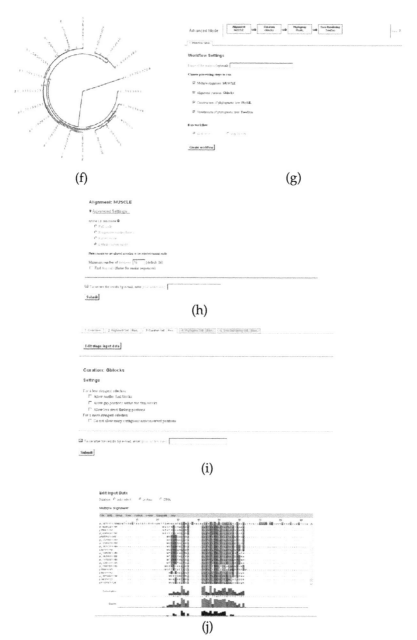

FIGURE 2 (F) A sample view of circular tree. (g) Snapshot of Phylogeny.fr advanced mode window. (h) Snapshot of Phylogeny.fr advanced mode window. (i) Snapshot of Phylogeny.fr curation step window. (j) Snapshot of aligned sequences in integrated Jalview in Phylogeny.fr.

a shape factor. When used to describe the sum of a series of exponentially distributed variables, the shape factor represents the number of variables and the scale factor is the mean of the exponential distribution. This is apparent when the profile of an exponential distribution with mean set to one is compared to a gamma distribution with a shape factor of one and a mean of one.

9. Protocol of phylogenetics analysis using BioNJ

BioNJ is a distance based phylogeny reconstruction algorithm, which is well suited for distances estimated from DNA or protein sequences. The steps are as follows:

1. The alignment file (Fig. 3a) can be uploaded in the FASTA, PHYLIP, Clustal, EMBL or NEXUS format) or distance matrix (Fig. 3b) in (PHYLIP or NEXUS format).
2. For protein analysis, bootstrapping is done by combining Seqboot (PHYLIP package) to perform bootstrap and Consense (PHYLIP package) to obtain the bootstrap tree from the BioNJ output. For nucleic acid analysis, bootstrapping is done with FastDist, and consensus generated by Consense (PHYLIP package). The limit for both protein and nucleic acid analysis is maximum one thousand steps.
3. Substitution matrix according to the requirement. A substitution matrix describes the rate at which one character in a sequence changes to other character states over time. For protein sequences two types of substitution matrix are available on the server:

(A)   (B)

FIGURE 3   Snapshots of phylogenetics analysis in BioNJ. (a) Snapshot of BioNJ data and setting tab. Amino acid sequences are submitted in the phylip format. (b) Snapshot of BioNJ data and setting tab with distance matrix of aligned proteins generated by phylogeny.fr server between organisms of interest.

9.1   Dayhoff PAM matrix

PAM (Point Accepted Mutation) matrix was developed by Margaret Dayhoff in the 1970s. PAM matrices are the description of the changes in amino acid composition that are expected after a given number of mutations that can be derived from the data used in creating the matrices. Thus the highest scoring alignment is the statistically most likely to have been generated by evolution rather than by chance.

9.2 Jones-Taylor-Thorntorn matrix

It is a matrix that computes a distance measure for protein sequences. This is similar to the Dayhoff PAM model, except that it is based on a recounting of the number of observed changes in amino acids by Jones, Taylor, and Thornton (1992). The distances can also be corrected for gamma-distributed and gamma-plus-invariant-sites-distributed rates of change in different sites.

The substitution matrices for nucleotides are as following:

9.3 Kimura 2-Parameter

Kimura's two parameter model (1980) corrects for multiple hits, taking into account transitional and transversional substitution rates, while assuming that the four nucleotide frequencies are the same and that rates of substitution do not vary among sites.

9.4 Jukes Cantor Model

In the Jukes and Cantor (1969) model, the rate of nucleotide substitution is the same for all pairs of the four nucleotides A, T, C, and G. It assumes an equality of substitution rates among sites equal nucleotide frequencies, and it does not correct for higher rate of transitional substitutions as compared to transversional substitutions.

9.5 Hamming Model

The Hamming distance is a simple position by position comparison. The Hamming distance is simple to calculate but it ignores a large amount of information about the evolutionary relationship among the sequences. The main reason is that the character differences are not the same as distances: the differences between two sequences are easy to measure, but the genetic distance involves mutations that cannot be observed directly.

Acknowledgements

SKG and TG would like to acknowledge support from Council of Scientific and Industrial Research (CSIR) Network Projects GENESIS (BSC0121) and INDEPTH (BSC0111).

References

Afrasiabi, C., Samad B., Dineen D., Meacham C. and K. Sjölander. 2013. The PhyloFacts FAT-CAT web server: ortholog identification and function prediction using fast approximate tree classification. *Nucleic Acids Res*. 41: W242-48.

Alkema, W.B.L., Lenhard B. and W.W. Wasserman. 2004. Regulog Analysis: Detection of Conserved Regulatory Networks Across Bacteria: Application to *Staphylococcus aureus*. *Genome Research*. 14:1362–73.

Anisimova, M., and O. Gascuel. 2006. Approximate likelihood ratio test for branches: A fast, accurate and powerful alternative. *Systematic Biol.* 55: 539-52.

Blanchette. M. and M. Tompa. 2003. FootPrinter: A program designed for phylogenetic footprinting. *Nucleic Acids Res.* 31:3840–42.

Brochier, C., Bapteste E., Moreira D. and H. Philippe. 2002. Eubacterial phylogeny based on translational apparatus proteins. *Trends Genet.* 18: 1-5.

Castillo-Davis, C.I., Hartl D.L. and G. Achaz. 2004. cis-Regulatory and protein evolution in orthologous and duplicate genes. *Genome Res.* 14: 1530-36.

Castresana, J. 2000. Selection of conserved blocks from multiple alignments for their use in phylogenetic analysis. *Mol. Biol. Evol.* 17: 540-52.

Daubin, V., Moran N.A. and H. Ochman. 2003. Phylogenetics and the cohesion of bacterial genomes. *Science* 301: 829-832.

Dehal, P.S., Joachimiak M.P., Price M.N., Bates J.T., Baumohl J.K., Chivian D., Friedland G.D., K.H. Huang, Keller K., Novichkov P.S., Dubchak I.L., Alm E.J. and A.P. Arkin. 2010. Microbes Online: an integrated portal for comparative and functional genomics. *Nucleic Acids Res.* 38:D396–D400.

Dereeper, A., Audic S., Claverie J.M. and G. Blanc. 2010. BLAST-EXPLORER helps you building datasets for phylogenetic analysis. *BMC Evol. Biol.* 10: 8.

Dereeper, A., Guignon V., Blanc G., Audic S., Buffet S., Chevenet F., Dufayard J.F., Guindon S., Lefort V., Lescot M., Claverie J.M. and O. Gascuel. 2008. Phylogeny. fr: robust phylogenetic analysis for the non-specialist. *Nucleic Acids Res.* 2008 ;36: W465-469.

Edgar, R.C. 2004. MUSCLE: multiple sequence alignment with high accuracy and high throughput. *Nucleic Acids Res.* 32: 1792-97.

Edwards, S.V., Liu L. and Pearl D.K. 2007. High resolution species trees without concatenation. *PNAS*, 104: 5936-41.

Enright, A.J., Dongen S. V., and C.A. Ouzounis. 2002. An efficient algorithm for large-scale detection of protein families, *Nucleic Acids Res.* 30: 1575-84.

Felsenstein, J. 1985. Confidence limits on phylogenies: an approach using the bootstrap. *Evolution* 39: 783-91.

Felsenstein, J. 1989. PHYLIP - Phylogeny Inference Package (Version 3.2). *Cladistics* 5: 164-166.

Fitch, W.M. 1970. Distinguishing homologous from analogous proteins. Syst. Zool. 19: 99-113.

Flores, H.A., DuMont V.L., Fatoo A., Hubbard D., Hijji M., Barbash D.A. and C.F. Aquadro. 2015. Adaptive evolution of genes involved in the regulation of germline stem cells in *Drosophila melanogaster* and *D. simulans*. G3 (Bethesda) 5: 583-592.

Gabaldón, T. and E.V. Koonin. 2013. Functional and evolutionary implications of gene orthology. *Nat. Rev. Genet.* 14: 360-66.

Galtier, N. 2007. A model of horizontal gene transfer and the bacterial phylogeny problem. Syst Biol. 56: 633-642.

Gama-Castro, S., Jimenez-Jacinto V., Peralta-Gil M., Santos-Zavaleta A., Penaloza-Spinola M.I., Contreras-Moreira B., J. Segura-Salazar, L. Muniz-Rascado, I. Martinez-Flores, Salgado H, C. Bonavides-Martínez, Abreu-Goodger C., Rodríguez-Penagos C., Miranda-Ríos J., Morett E., Merino E., Huerta A.M., Treviño-Quintanilla L. and J. Collado-Vides. 2008. RegulonDB (version 6.0): gene regulation model of Escherichia coli K-12 beyond transcription, active (experimental) annotated promoters and Textpresso navigation. *Nucleic Acids Res.* 36:D120–D124.

Gascuel, O. 1997. BIONJ: an improved version of the NJ algorithm based on a simple model of sequence data. *Mol. Biol. Evol.* 14: 685-95.

Gonnet, G. H., Hallett M. T., Korostensky C. and L. Bernardin. 2000. Darwin v. 2.0: an interpreted computer language for the biosciences *Bioinformatics* 16: 101-103.

Guindon S., Dufayard J.F., Lefort V., Anisimova M., Hordijk W. and O. Gascuel. 2010. New Algorithms and methods to estimate maximum-likelihood phylogenies: assessing the performance of PhyML 3.0. *Systematic* Biol. 59: 307-21.

Guindon, S., Lethiec F., Duroux P. and O. Gascuel. 2005. PHYML Online--a web server for fast maximum likelihood-based phylogenetic inference. *Nucleic Acids Res.* 33: W557-59.

Hammer, Ø., Harper D.A.T. and Ryan P.D. 2001. PAST: Paleontological statistics software package for education and data analysis. *Palaeontol. Electron.* 4: 9.

Hertz, G.Z. and G.D. Stormo. 1999. Identifying DNA and protein patterns with statistically significant alignments of multiple sequences. *Bioinformatics.* 15:563–577.

Huerta-Cepas, J., Dopazo H., Dopazo J. and T. Gabaldón. 2007. The human phylome. Genome Biol. 8: R109.

Huerta-Cepas, J., Capella-Gutiérrez S., Pryszcz L.P., Marcet-Houben M. and T. Gabaldón. 2014. PhylomeDB v4: zooming into the plurality of evolutionary histories of a genome. Nucleic Acids Res. 42: D897-902.

Jobb, G., von Haeseler A. and K. Strimmer. 2004. TREEFINDER: a powerful graphical analysis environment for molecular phylogenetics. BMC Evol. Biol. Jun 28: 4:18.

Jones, D.T., Taylor W.R. and J.M Thornton. 1992. The rapid generation of mutation data matrices from protein sequences. Comput Appl. Biosci. 8: 275-82.

Jukes, T.H. and C.R. Cantor. Evolution of protein molecules. In Munro (ed.), Mammalian Protein Metabolism, Academic Press, New York pp. 21-132.

Kazakov, A.E., Cipriano M.J., Novichkov P.S., Minovitsky S., Vinogradov D.V., Arkin A., Mironov A.A., Gelfand M.S. and I. Dubchak. 2007. RegTransBase–a database of regulatory sequences and interactions in a wide range of prokaryotic genomes. *Nucleic Acids Res.* 35:D407–D412.

Kimura, M. 1980. A simple method for estimating evolutionary rates of base substitutions through comparative studies of nucleotide sequences. *J. Mol Evol.* 16: 111-120.

Krishnamurthy, N., Brown D.P., Kirshner D. and K. Sjölander. 2006. PhyloFacts: an online structural phylogenomic encyclopedia for protein functional and structural classification. Genome Biol. 7: R83.

Kurland, C.G., Canback B. and O.G. Berg. 2003. Horizontal gene transfer: a critical view. Proc Natl Acad Sci U S A. 100: 9658-9662.

Li, H., A. Coghlan, Ruan J., Coin L.J., Hériché J.K., Osmotherly L., Li R., Liu T., Zhang Z., Bolund L., Wong G.K., Zheng W., Dehal P., Wang J. and R. Durbin. 2006. TreeFam: a curated database of phylogenetic trees of animal gene families. *Nucleic Acids Res.* 34: D572-80.

Li, L., C.J. Jr. Stoeckert and D.S. Roos. 2003. OrthoMCL: identification of ortholog groups for eukaryotic genomes. *Genome Res.* 13: 2178-2189.

Liu, L. and D.K. Pearl. 2007. Species Trees from Gene Trees: reconstructing Bayesian posterior distributions of a species phylogeny using estimated gene tree distributions. Systematic Biol. 56: 504-514.

Marchler-Bauer, A., Anderson J.B., Cherukuri P.F., DeWeese-Scott C., Geer L.Y., Gwadz M., He S., D.I. Hurwitz, Jackson J.D., Ke Z., Lanczycki C.J., Liebert C.A., Liu C., Lu F., Marchler G.H., Mullokandov M., B.A. Shoemaker, Simonyan V., Song

J.S., Thiessen P.A., Yamashita R.A., Yin J.J., Zhang D. and S.H. Bryant. 2005. CDD: A Conserved Domain Database for protein classification. Nucleic Acids Res. 33:D192–D196.

Mau, B., Newton M.A. and B. Larget. 1997. Bayesian Phylogenetic Inference via. Markov Chain Monte Carlo Methods. *Mol. Biol. Evol.* 14: 717-724.

McAdams, H.H., B. Srinivasan and A.P. Arkin. 2004. The evolution of genetic regulatory systems in bacteria. *Nat. Rev. Genet.* 5: 169-178.

Merkeev, I.V., Novichkov P.S. and A.A. Mironov. 2006. PHOG: a database of supergenomes built from proteome complements. *BMC Evol. Biol.* 2006. 6: 52.

Mignard, S. and J.P. Flandrois. 2008. A seven-gene, multilocus, genus-wide approach to the phylogeny of mycobacteria using supertrees. *Int. J. Syst. Evol. Microbiol.* 58: 1432-1441.

Needleman, S.B. and C.D. Wunsch. 1970. A general method applicable to the search for similarities in the amino acid sequence of two proteins. *Journal of Molecular Biology* 48 (3): 443–53.

Neph, S. and M. Tompa. 2006. MicroFootPrinter: A tool for phylogenetic footprinting in prokaryotic genomes. *Nucleic Acids Res.* 34:W366–W368.

Novichkov, P.S., Rodionov D.A., Stavrovskaya E.D., Novichkova E.S., Kazakov A.E., Gelfand M.S., Arkin A.P., Mironov A.A. and I. Dubchak. 2010. RegPredict: an integrated system for regulon inference in prokaryotes by comparative genomics approach. *Nucleic Acids Res.* 38(Web Server issue): W299–W307.

Novichkov, P.S., Laikova O.N., Novichkova E.S., Gelfand M.S., Arkin A.P., Dubchak I. and D.A. Rodionov. 2010. RegPrecise: a database of curated genomic inferences of transcriptional regulatory interactions in prokaryotes. *Nucleic Acids Res.* 38:D111–D118.

O'Brien, K.P., M. Remm and E.L. Sonnhammer. 2005. Inparanoid: a comprehensive database of eukaryotic orthologs. *Nucleic Acids Res.* 33: D476-480.

Powell, S., Forslund K., Szklarczyk D., Trachana K., Roth A., Huerta-Cepas J., Gabaldón T., Rattei T., C. Creevey, Kuhn M., Jensen L.J., von Mering C. and P. Bork. 2014. eggNOG v4.0: nested orthology inference across 3686 organisms. *Nucleic Acids Res.* 42: D231-239.

Price, M.N., Dehal P.S. and A.P. Arkin. 2007. Orthologous transcription factors in bacteria have different functions and regulate different genes. *PLoS Comput Biol.* 3: 1739-1750.

Pryszcz, L.P., Huerta-Cepas J. and T. Gabaldón. 2011. MetaPhOrs: orthology and paralogy predictions from multiple phylogenetic evidence using a consistency-based confidence score. Nucleic Acids Res. 39: e32.

Rice, P., Longden I. and A. Bleasby. 2000. EMBOSS: The European Molecular Biology Open Software Suite. *Trends in Genetics* 16: 276–277.

Ronquist, F., Teslenko M., van der Mark P., Ayres D.L., Darling A., Höhna S., Larget B., Liu L., M.A. Suchard and J.P. Huelsenbeck. 2012. MrBayes 3.2: efficient Bayesian phylogenetic inference and model choice across a large model space. *Systmatic Biol.* 61: 539-542.

Sankoff, D., Morel C., and R.J. Cedergren. 1973. Evolution of 5S RNA and the non-randomness of base replacement. *Nature New Biol.* 245: 232-234.

Stamatakis, A. 2014. RAxML version 8: a tool for phylogenetic analysis and post-analysis of large phylogenies. *Bioinformatics.* 30: 1312-1313.

Swofford, D.L. 2002. LPAUP. "Phylogenetic analysis using parsimony (* and other methods). Version 4." Sunderland, MA: Sinauer Associates.

Takeuchi, N., Wolf Y.I., Makarova K.S. and E.V. Koonin. 2012. Nature and Intensity of Selection Pressure on CRISPR-Associated Genes. *J Bacteriol.* 194: 1216-1225.

Tamura, K., Stecher G., Peterson D., Filipski A. and S. Kumar S. 2013. MEGA6: Molecular Evolutionary Genetics Analysis Version 6.0. *Molecular Biology and Evolution* 30: 2725-2729.

Tatusov , R.L., Koonin E.V. and D.J. Lipman. 1997. A genomic perspective on protein families. Science. 278: 631-637.

Tatusov, R.L., Natale D.A., Garkavtsev I.V., Tatusova T.A., Shankavaram U.T., Rao B.S., Kiryutin B., M.Y. Galperin, Fedorova N.D. and E.V. Koonin. 2001. The COG database: New developments in phylogenetic classification of proteins from complete genomes. Nucleic Acids Res. 29: 22–28.

Tatusov, R.L., Koonin E.V. and D.J. Lipman. 1997. A genomic perspective on protein families, Science 278: 631-637.

Thompson, W., Rouchka E.C. and C.E. Lawrence. 2003. Gibbs Recursive Sampler: Finding transcription factor binding sites. Nucleic Acids Res. 31: 3580–3585.

van der Heijden, R.T., Snel B., van Noort V. and M.A. Huynen. 2007. Orthology prediction at scalable resolution by phylogenetic tree analysis. BMC Bioinformatics 8: 83.

Vilella, A.J., Severin J., Ureta-Vidal A., Heng L., Durbin R. and E. Birney. 2009. EnsemblCompara GeneTrees: Complete, duplication-aware phylogenetic trees in vertebrates. Genome Res. 19: 327-35.

Wang, T. and G.D. Stormo. 2003. Combining phylogenetic data with co-regulated genes to identify regulatory motifs. Bioinformatics. 19:2369–2380.

Wapinski, I., Pfeffer A., Friedman N. and A. Regev. 2007. Natural history and evolutionary principles of gene duplication in fungi. Nature 449: 54-61.

Wasserman, W.W. and A. Sandelin. 2004. Applied bioinformatics for the identification of regulatory elements. *Nat Rev Genet.* 5: 276-287.

Wei, W., Ning L.-W., Ye Y.-N. and F.-B. Guo. 2013. Geptop: a gene essentiality prediction tool for sequenced bacterial genomes based on orthology and phylogeny. PloS One 8.

Wheeler, W. C. and D. S. Gladstein. 1994. MALIGN: A Multiple Sequence Alignment Program. J. Hered 85: 417-418.

Wolf, Y.I. and E.V. Koonin. 2012. A tight link between orthologs and bidirectional best hits in bacterial and archaeal genomes. *Genome Biol Evol.* 4: 1286-1294.

Xia, X. 2013. DAMBE5: a comprehensive software package for data analysis in molecular biology and evolution. *Mol Biol Evol.* 30: 1720-1728.

Xu, Z. and B.L. Hao. 2009. CVTree update: a newly designed phylogenetic study platform using composition vectors and whole genomes. *Nucleic Acids Res.* 37: W174-W178.

Yang, Z. 1997. PAML: a program package for phylogenetic analysis by maximum likelihood. *Comput Appl Biosci.* 13: 555-556.

Yao, Z., Barrick J., Weinberg Z., Neph S., Breaker R., Tompa M. and W.L. Ruzzo. 2007. A Computational Pipeline for High- Throughput Discovery of cis-Regulatory Noncoding RNA in Prokaryotes. *PLoS Comput Biol.* 3(7): e126.

Yao, Z., Weinberg Z. and W.L. Ruzzo. 2006. CMfinder—A covariance model based RNA motif finding algorithm. *Bioinformatics.* 22:445–452.

Zhao, X., Yu Q., L. Huang and Q.X. Liu. 2014. Patterns of Positive Selection of the Myogenic Regulatory Factor Gene Family in Vertebrates. *PLoS One* 9: e92873.

17

Improved Protein Model Ranking through Topological Assessment

Shikhin Garg[1], Smarth Kakkar[1] and Ashish Runthala[1]*

Abstract

Contrary to heavy genome as well as proteome sequencing rates, the pace of experimental solving of protein structures is quite low and the sequence-structure gap is constantly increasing. Detailed structural knowledge of a protein is essential to understand its native function in a cell and it consequently helps us to learn the expression profile of all the genes in that cell. Numerous computational algorithms for protein structure prediction have been developed to quickly construct the protein models and to bridge this ever-increasing sequence-structure gap. However, the current prediction methodologies often fail to select the true and acceptable conformation from the generated decoy structures. Also, the currently popular model ranking schemes are not efficacious in resolving very close structural models to identify the actual model even when the experimental structure of the considered sequence is available to aid in the decision making. This chapter extensively investigates the current best model assessment measures, including those which evaluate structure and geometry, such as GDT_TS, GDT_HA, Spheregrinder, TM-Score and RMSD scores, and illustrates their inaccuracies in effectively resolving and ranking the protein models. It further presents a new method of ranking the constructed protein models on the basis of topological differences in the dihedral angles and the distances between successive Cα atoms. The developed methodology is inspected by employing it for re-ranking the top five models for five TBM-Easy targets and five TBM-Hard targets from the CASP10 database, and it proves the inefficiency of the currently used model assessment measures in selecting the accurate structure for a target protein sequence. The ranking result is further verified by evaluating the atomic clashes in these models. The resultant model ranking is intelligently utilized to apprehend the coherent shortcomings of contemporary assessment measures and expound the improved model assessment-cum-ranking algorithm. Furtherance of model evaluation and ranking measures will certainly help us to consistently select the best predicted conformations for understanding a protein's functional role in the biological pathways of a cell system.

---

[1] Department of Biological Sciences, Faculty Division III, Birla Institute of Technology and Science, Pilani-333031, Rajasthan, India.
* Corresponding author: ashish.runthala@gmail.com, Phone: +91-759 797 1146.

1. Introduction

Owing to the pronounced efforts in genome sequencing and the subsequent translation of the discovered nucleotide sequences into amino acid sequences, there has been a spurt in the number of available protein sequences. A widespread focus on sequence analysis of proteins has led to several advanced techniques being invented in the last few years for making protein sequencing more efficient, resulting in the huge rise that has been observed in the number of solved protein sequences in various protein databases around the world. These sequences play a fundamental role in determining the functions of these proteins in various biological systems. The biochemical function of a protein can predict its involvement in biological pathways and the part played by it in the development or the cure of a disease (Gherardini and Helmer-Citterich 2008). It was propounded in the 1950s, that a protein's sequence encodes the information necessary to direct its conformational folding to the functionally active three-dimensional structure. A protein structure furnishes significant information about its function in cellular processes. Thus, important inferences about the protein's function could be drawn from its sequence using the structure as an intermediary (Sela et al. 1957).

Conserved protein sequence chunks play an important role in maintaining the structural as well as functional similarity of proteins along the evolutionary networks. However, even when protein sequences diverge during the evolution, their structures can still be retained and hence the structural information is crucial to unravel their native functions. Thus, the structural topology of a cellular protein is essential to infer its phylogenetic as well as functional relationship with another species, be it a close or a distant one (Gherardini and Helmer-Citterich 2008). The protein structures are also essential requisites to scrutinize and establish their interaction network in a cell for properly understanding the entire genetic regulation in a cell. The functional analysis of ligands binding to certain specific sites in proteins is also dependent on the structural topology of active sites encoded in these proteins (Baker and Sali 2001).

Sadly, for most of the protein sequences, we still do not have the experimentally solved structures and so their native functions are also not well understood. A significant gap between the available number of protein sequences and their experimentally solved structures has thus arisen and this sequence-structure gap is constantly increasing. In its release on 9 December 2015, the UniProt Knowledgebase and Translated European Molecular Biology Laboratory (UniProtKB/TrEMBL) database contained 55,270,679 sequences, which are substantially larger in number than the negligible 111,511 protein structures experimentally determined and stored in the Protein Data Bank (PDB).

Protein sequences are experimentally solved through several techniques like X-ray and Cryo-electron microscopy. However, these methodologies are too costly and time consuming. Thus, computational algorithms have been developed to construct the reliable as well as accurate protein conformations. The most accurate of these methodologies is the Template Based Modelling (TBM) algorithm that employs the dependable set of experimentally solved protein structures (templates) to construct a trustworthy conformation for the considered protein sequence (target). For further validation and testing of the accuracy of these prediction algorithms, a community-wide blind test titled 'Critical Assessment of protein Structure Prediction (CASP)' is organized every two years. CASP tests the modelling accuracy for a specific set of target protein sequences whose structures are experimentally solved and frozen for the test. It therefore assesses the exactness of all the models predicted by the different research groups for all these targets, against their actual native conformations and ranks them through different scoring parameters, as explained further.

During the assessment of these models, it is observed that while in some cases, the predicted protein models are not accurate and exhibit almost nil similarity, at other times, the models may be very close in their structure, to each other as well as to the native target conformation. In such cases, selecting the model which is structurally closest to the native conformation becomes quite challenging (Runthala 2012). As a solution to this problem, more competent model assessment methods are needed.

Assessment schemes superimpose the considered structures for calculating their similarity score. However, superimposition of the complete protein structures can mask the topological similarity at some local segments that exhibit a significantly higher conformational similarity. This consequently requires the consideration of an aspect of local similarity in these schemes. Several tools have been invented that mutually compare the constructed protein models to identify the best predicted model among the numerous constructed decoy structures for a target sequence. Although, we have developed considerably efficient structural comparison based model assessment methods, a complete understanding of the protein dynamics and a fool-proof method that can select the accurate model among the generated decoys, still eludes us. Various assessment methods employ the secondary structure, solvent exposure and pairwise residue interaction information of the predicted protein models to evaluate their stereochemistry and molecular energy in the bid to select the most accurate structure (Kryshtafovych and Fidelis 2009). Among the numerous model assessment methods that have been developed till date, the mostly employed and the current best measures for evaluating a predicted conformation against an experimentally solved structure, are selected to analyse their biological credibility. These assessment schemes are individually explained below along with their logical and algorithmic drawbacks.

1.1 RMSD

RMSD (Root Mean Square Deviation) is one of the simplest assessment methods to appraise a predicted protein model against its native conformation or the templates employed to construct it. It is computed through the optimal superimposition of the two structures by employing the rotational and translational degrees of freedom of one structure in relation to the other one. For two protein structures encoding an equal number of L amino acids, it is simply calculated with the following formula:

$$\text{RMSD} = \sqrt{\frac{1}{L} \sum_{i=1}^{L} d_i^2}$$

where d_i is the distance between the ith pair of $C\alpha$ atoms in the two structures.

RMSD can be calculated, either using all the atoms or only the $C\alpha$ atoms or only the centre-of-mass of the side-chain atoms present in a protein model, against its actual native conformation or the employed template(s). Lower RMSD scores indicate higher topological similarity among two structures. As two, functionally as well as structurally different proteins can be superimposed to yield a lower RMSD score, its biological reliability is always doubtful. It equally weighs all the atoms and so the topological misorientation at some local segments or the incorrect topology of a few residue chunks like loop regions in a model can result in an abruptly higher RMSD score. Also, a dependency on the number of atoms included in the comparison prevents the score from giving the true picture of the similarity which is certainly irrespective of molecular size (Zhang and Skolnick 2005, Xu and Zhang 2010, Hung and Samudrala 2012).

1.2   GDT_TS

GDT_TS stands for Global Distance Test Total Score and is one of the most popular methods currently available for assessing the quality of the generated protein models. Initial structural superimpositions are generated by taking all the possible continuous segments of three, five and seven residues from the Cα atoms along the backbone from both, the target and the template. These superimpositions create an initial set of equivalent residue pairs (corresponding Cα atom pairs from target and the template that are supposed to have the same structural, functional and the biological role). These segments of residue pairs are then repeatedly extended further by including all residue pairs which lie below a specific distance cut-off (distance between the residues of each pair) to arrive at the largest set of residues from the two structures that obey a given distance threshold. The threshold is iteratively set to 1, 2, 4 and 8Å, and the largest residue set is computed for each of these thresholds (Kryshtafovych et al. 2005) to calculate the GDT_TS score.

$$\text{GDT\_TS} = \frac{1}{4}[n1 + n2 + n4 + n8]$$

where nx denotes the number of residues superimposed for a threshold of xÅ.

GDT_HA is a higher accuracy version of GDT_TS which differs only in the values of the thresholds which are used. It employs the 0.5, 1, 2 and 4Å thresholds to compute an average score. Thus, it is better suited to look out for local similarities which are of a superior quality than their average global counterparts. However, it should be cautiously used for the structural evaluation of a sequence alignment and the corresponding constructed model against the template. Once a constructed protein model is refined through addition of loops or refinement of the backbone topology, the GDT_TS score ceases to reflect only the sequence alignment that has been employed for computing it (Dunbrack 2006).

The GDT family scores always carry the negative aspect of all the distance values between two consecutive thresholds amounting to an equal contribution to the final score, e.g., two different residue pairs with distance values of 5Å and 7Å contribute equally to the value of $n8$. The dependency of GDT_TS on the number of compared residues leads to no concrete conclusion being obtained from its results in most of the cases. Such trivial scores usually imply the fluky insignificant topological similarity or the conserved nature only for a few short segments between the considered protein structures (Zhang and Skolnick 2004). Moreover, the application of GDT score entails the task of adjusting the threshold according to the difficulty level of a considered target subject to the reliable coverage span provided by the best possible set of templates available for this target (Kopp et al. 2007).

1.3   LGA

LGA is the acronym for Local-Global Alignment. It is a model assessment method used for comparing the predicted target model or its substructures against the template or its native conformation. This method provides the advantage of being applicable in situations where there is no predefined correspondence between the residues of the two protein structures (Zemla 2003).

The RMSD score has shortcomings and can give a large value even if there is just a small region with a large deviation, leading to a high overall RMSD in an otherwise similar structure. Thus, considering local regions for superimposing and assessing similarity becomes even more imperative. This constraint is addressed by LGA, which takes into account both local and global superimpositions of the two structures. Regions of local similarity can be neglected if a single best global superimposition is to be computed and so the LGA algorithm makes several local superimpositions to find out local regions that are structurally similar.

The LGA scoring scheme is divided into two components: LCS (Longest Continuous Segments) and GDT (Global Distance Test). While, the LCS determines locally similar regions, the GDT score estimates the global structural similarity of the proteins. LCS determines longest continuous superimposable segments of residues that satisfy a given RMSD cut-off. GDT, on the other hand searches for the largest set of equivalent residues which obey a particular distance cut-off. The residues can also be discontinuous in the case of GDT. Limiting the LCS analysis to a specified RMSD cut-off would not bring out all the structural similarities and some might escape from being noticed. Thus, the algorithm is repeated for different cutoffs (1Å, 2Å and 5Å). Similarly, for GDT, scanning is done for a different cut-off at every 0.5Å from 0.5Å to 10Å. GDT assigns every residue from one of the molecule to the largest set of possible residues from the second molecule keeping a specified distance cut-off as necessary. The percentage of residues (in continuity) that satisfy an RMSD cutoff of viÅ are denoted as LCS_vi. Similarly, calculations are done for GDT_vi which is the percentage of residues satisfying a given distance cutoff of viÅ. The scoring function LGA_S is then calculated using the formula:

$$\text{LGA\_S} = (w \times S(\text{GDT\_}vi)) + ((1 - w) \times S(\text{LCS\_}vi))$$

where w $(0.0 \leq w \leq 1.0)$ is a suitable weighting factor and S(F_vi) is found using the following procedure:

$$X = 0.0$$

For each vi $(v1, v2, \ldots, vk)$

$$\{$$

$$Y = (k - i + 1)/k$$
$$X = X + (Y \times \text{F\_}vi)$$

$$\}$$

$$S(F\_vi) = X \div \left((1 + k) \times \left(\frac{k}{2} \right) \right)$$

In sequence dependent analysis, different superimpositions need not be done initially and the given equivalence is used to carry out a single superimposition for which the LGA_S is calculated. CASP6 revealed the fact that in situations involving substantial difference between the structures of the predicted models and the native target conformations, sequence-independent alignment procedures fail to achieve the correct alignment between the model and the target. Instead, the sequence-dependent alignments fare better by succeeding to identify atleast a substructure of the predicted model which is acceptably similar to its native structure (Dunbrack 2006).

1.4 TM-Score

TM-Score stands for Template Modelling score and is a scoring function that improves on the technique of GDT to create a refined means of model assessment. It removes the size-dependency and thus provides a realistic comparison of random structural models. TM-Score is defined as follows:

$$\text{TM-Score} = \text{Max} \left[\frac{1}{L_{Target}} \sum_{i=1}^{L_{ali}} \frac{1}{1 + \left[\frac{d_i}{d_0} \right]^2} \right]$$

where L_{Target} is the length of the target protein; L_{ali} is the number of residues chosen for alignment; d_i is the distance between the ith pair of residues in the aligned region; $d_0 = 1.24 \sqrt[3]{L_{Target} - 15} - 1.8$ is the parameter used for normalizing the distances between the residue pairs.

The consideration of a normalizing factor makes the TM-Score size-independent so that it depicts the average topological deviation between the aligned residues even in randomly compared set of protein structures. A score called the 'raw TM-Score' (rTMscore) is obtained if a fixed value of $d_0 = 5\text{Å}$ is taken, which like GDT, has the size-dependence (Zhang and Skolnick 2004). By definition of TM-Score, a value greater than 0.5 indicates a model with an approximately correct topology, and a value less than 0.17 always indicates a random prediction. However, TM-Score does not clearly bring about the regions of higher local similarity or dissimilarity compared to the overall similarity between the complete structures. A local region of high resemblance may be masked by a mediocre TM-Score as its algorithm is designed for global assessment of protein structure. On the contrary, a passable TM-Score may be obtained due to a large perturbation in a small part of fairly similar structures. Unlike RMSD score, TM-Score holds the advantage of protein segments of the compared structures contributing to the final score in accordance with the quality of their alignments which averts the misleading assessments to some extent.

1.5 Spheregrinder

Spheregrinder (SG) score, introduced in CASP9, evaluates the similarity between the two protein structures through an algorithm based on the localized conformational similarity between the corresponding substructures. SG algorithm employs the spheres of a specific radius by centring them on the corresponding $C\alpha$ atoms of both the considered structures to compute an individual RMSD score for the specific set of atoms structurally constrained in each of these spheres.

Based on the radius of the sphere and the employed RMSD cutoff, percentage of spheres complying with the structural similarity thresholds are used to calculate an average SG-score (Kryshtafovych et al. 2013) for a predicted protein model. Thus, this scheme deploys a mechanism which focuses on local structural deviations to give a dependable picture of the similarity between the target structure and its native conformation, and is devoid of the systematic topological shifts that may be induced through direct global superimposition of the two structures.

It has also been well observed that the function of a protein sequence is majorly determined by the biochemical nature, location and structural depth of a few functionally specific residues which usually encompass the active site(s) (Gherardini and Helmer-Citterich 2008). Thus, an effective combination of global as well as local structural comparison of the two proteins is the most probable way to estimate their trustworthy structural similarity. However, majority of the other such model assessment methods also lag on the same logical grounds that are discussed so far in this study.

A new assessment procedure which adroitly evaluates both the global and local similarity among a pair of protein structures is precisely the need of the hour. A better assessment measure is thus developed in this study to correctly rank the generated set of target model decoys and select the accurate model among the decoys constructed for a target sequence. There are some assessment measures that use an alternative of evaluating the models on the basis of the energy of their interatomic interactions. However, like most of the prevalent model assessment measures today, this new method is also fundamentally based on the structure of the proteins.

2. Methodology

The developed methodology employs the vital topological features of the constructed set of target protein models to assess them against the native target conformation and rank them in the best possible way. For evaluating and testing the applicability and reliability of our designed model ranking algorithm on both the TBM-Easy targets as well as the TBM-Hard targets, five targets from each of these sets are randomly considered from the set of TBM targets considered in CASP10. Thus, the CASP10 targets chosen for the study include five TBM-Easy targets (T0645-D1, T0650-D1, T0659-D1, T0664-D1 and T0689-D1) and five TBM-Hard targets (T0663, T0674, T0676-D1, T0726 and T0726-D1) (Source: *www.predictioncenter.org/casp10/results. cgi*). For each of these selected targets or target domains, the top five predicted protein models are chosen. The selected models are ranked through the devised technique and the resultant accuracy order of these models is then compared with the order followed in CASP10 (based on the popular assessment measures viz. GDT_TS, GDT_HA, TM-Score and Spheregrinder) to arrive at some inferences describing how the developed ranking technique fares against the popular methods.

The secondary structure architecture of a protein, encoding functionally active domains, heavily affects its function (Matsui et al. 2004). Moreover, the secondary structures, viz. α-helices and β-sheets, form the fundamental topology of a protein conformation that is important to decode its functional details. The considered target models have the same number of residues and our methodology computes the dihedral angles phi (ϕ) and psi (ψ) for every residue present in the protein backbone to correlate these values with the topological similarity of the considered protein structures and to further utilize them for ranking the protein models. The phi angle and psi angles are the angles of right-handed rotation around the N-CA and CA-C bonds respectively. The cartesian distance between every pair of consecutive Cα atoms is further calculated and employed in the bid to define the topological similarity of the two protein structures.

At each of the residue locus for a protein structure, the coordinates of the carbon, nitrogen and oxygen atoms are used in a particular order to calculate the dihedral angles. The procedure used for calculation of the angles involves calculating the angle between the two successive planes. For instance, if we consider a_1, a_2, a_3 and a_4 as the position vectors for four consecutive atoms along the protein backbone, the vector set of a_1, a_2 and a_3 and the vector set of a_2, a_3 and a_4 form the two successive planes. By assigning the four position vectors a_1, a_2, a_3 and a_4 to various atoms of each residue in a specific order, the dihedral angles can be arrived at. The angle phi (ϕ) is computed by considering a_1, a_2, a_3 and a_4 as the position vectors for the C atom of the residue which is previous to the one being evaluated and the N, Cα and C atoms of the residue under consideration, respectively. The angle psi (ψ) is computed by assigning a_1, a_2, a_3 and a_4 as the position vectors for the N, Cα and C atoms of the current residue and the N atom of the next residue, respectively. v_1, v_2 and v_3 are the successive bond vectors joining the four atoms. m_1 and m_2 are the normals to the two planes which include the dihedral angle, say θ, between them. For every plane, they are two possible normal of opposite signs. The chosen normals are of the same sign which ensures a zero dihedral angle when they become parallel. Finally, the required angle is calculated using three orthogonal unit vectors, l_1, l_2 and l_3. The various vectors enumerated above have been illustrated through a front view which has the second bond vector v_2 in the plane of the paper (Fig. 1) and an orthogonal view which has the bond vector v_2 projecting out of the paper (Fig. 2).

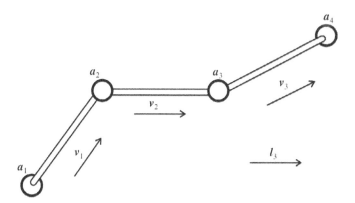

FIGURE 1   Front view of vectors used in calculating dihedral angles. Bond vector v_2 is in the plane of the paper.

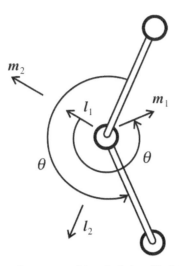

FIGURE 2 Orthogonal view of vectors used in calculating dihedral angles. Bond vector v_2 is projecting out of the paper.

$$v_1 = a_2 - a_1$$

$$v_2 = a_3 - a_2$$

$$v_3 = a_4 - a_3$$

$$m_1 = \frac{v_1 \times v_2}{|v_1 \times v_2|}$$

$$m_2 = \frac{v_2 \times v_3}{|v_2 \times v_3|}$$

$$l_1 = m_2$$

$$l_3 = \frac{v_2}{|v_2|}$$

$$l_2 = l_3 \times l_1$$

$$m_1 = (m_1 . l_1)l_1 + (m_1 . l_2) l_2$$

$$= (\cos\theta \times l_1) + (\sin\theta \times l_2)$$

$$\theta = -\tan^{-1}\frac{(m_1 . l_2)}{(m_1 . l_1)}$$

Thus, dihedral angles for each of the residue encoded in a protein structure are hereby computed for the native target conformation (solved target; parsed only for the residues assessed during the CASP10) and the selected CASP models for each of the considered targets.

The distances between consecutive $C\alpha$ atoms are computed using the coordinates provided in the PDB files and the formula for calculating the distance D, between two points in three-dimensional space:

$$D = \sqrt{(x_2 - x_1)^2 + (y_2 - y_1)^2 + (z_2 - z_1)^2}$$

where (x_1, y_1, z_1) and (x_2, y_2, z_2) are the coordinates of the two points.

Further, the angular difference is calculated between the dihedral angle values of a residue in the target and its corresponding residue in the model being assessed. This difference is obtained for all the residues present in the solved target. The summation of these angular differences across all the compared residues, viz. the total number of residues encoded in the target sequence, is calculated to give the total difference in the dihedral angle values between the model and the experimental structure.

$$Total\ difference = d_1 + d_2 + d_3 + ... + d_n$$

where d_1, d_2, d_3 ... and d_n denote the angular differences at each residue. Throughout the calculations, only the absolute values of the differences are used. This process is done twice, once for phi (ϕ) and psi (ψ), each. A similar methodology is implemented to calculate the difference between the distances between consecutive $C\alpha$ atoms, at every pair of corresponding residues in the target and the models and subsequently arrive at the total $C\alpha$ distance difference.

Using the total difference of a considered topological parameter between a model and the target, the mean of the difference is calculated using the formula:

$$Mean\ (\mu) = \frac{Total\ difference}{n}$$

where n is the number of residues for which the total difference is calculated. Further, the Standard Deviation (SD) of these differences is computed with the help of the following formula:

$$SD\ (\sigma) = \sqrt{\frac{\sum_{i=1}^{n} (d_i - \mu)^2}{n}}$$

The aforementioned procedures are repeated for all the models for a particular target. Based on the determined means and standard deviations, the models are ranked anew and the results are analysed and compared with the results of the popular schemes.

For each of the five selected TBM-Easy and TBM-Hard CASP10 targets, the selected CASP models are structurally evaluated against their native conformations by computing their TM-Score through the online tool of Zhang Lab (*zhanglab.ccmb.med. umich.edu/TM-Score*). The GDT_TS, GDT_HA, Spheregrinder and RMSD scores of all these models and the resulting model rankings are further considered through the official CASP10 portal (*www.predictioncenter.org/casp10/results.cgi*) and compared with our computed scores and rank order for each of the considered CASP10 targets. A supplementary assessment is further done by examining the selected CASP models for the atomic clashes. A severe clash is encountered in a model when the distance between two consecutive $C\alpha$ atoms is less than 1.9Å. A bump is said to occur, if this distance is less than 3.6Å and a model with four or more severe clashes or more than 50 bumps is defined as a clashed model (Tress et al. 2005). The best model identified through a ranking scheme hence gets further validated if it involves the least number of clashes or bumps and is the least clashed model.

The developed method is expected to have features which can possibly address the shortcomings that the presently popular scoring schemes possess and lead to a reliable ranking of protein models in the order of their biological accuracy.

3. Results

For each of the selected TBM-Easy and TBM-Hard CASP10 targets, the GDT_TS, GDT_HA, Spheregrinder, TM-Score and RMSD scores of all the selected models are enlisted in Table 1.

TABLE 1 Various scores of the best 5 models for the selected CASP10 targets

#	Group Name	GDT_TS	GDT_HA	Spheregrinder	TM-Score	RMSD
	TBM-Easy targets					
	T0645-D1 (40-537)					
1	Phyre2_A	80.12	62.45	64.46	0.9346	3.385
2	MULTICOM-NOVEL	80.07	62.55	71.29	0.9397	3.154
3	HHpredA	79.52	60.95	67.67	0.9345	3.145
4	HHpredAQ	79.52	60.95	67.67	0.9345	3.145
5	HHpred-thread	79.12	60.54	68.88	0.9328	3.29
	T0650-D1 (4-342)					
1	HHpredAQ	93.36	78.10	91.45	0.9737	1.981
2	HHpredA	93.36	78.10	91.45	0.9737	1.981
3	chunk-TASSER	93.36	78.32	93.81	0.9746	1.932
4	BAKER-ROSETTASERVER	93.22	78.03	94.40	0.9743	1.97
5	MATRIX	91.08	72.27	89.38	0.9676	2.136
	T0659-D1 (1-74)					
1	Phyre2_A	94.93	81.42	90.54	0.9233	2.029
2	MULTICOM-REFINE	94.93	81.76	97.30	0.9266	1.813
3	MULTICOM-NOVEL	94.93	82.09	94.59	0.9295	1.942
4	MATRIX	94.59	81.08	90.54	0.9236	1.915
5	FALCON-TOPO	94.59	81.76	85.14	0.9266	2.063
	T0664-D1 (43-540)					
1	PMS	84.54	68.07	76.91	0.9076	3.193
2	RaptorX-ZY	84.44	68.88	78.11	0.9053	3.376
3	Phyre2_A	83.83	67.27	75.90	0.9035	3.579
4	RaptorX	83.69	68.12	77.51	0.9030	3.614
5	HHpredA	83.64	66.82	76.91	0.9021	3.258

#	Group Name	GDT_TS	GDT_HA	Spheregrinder	TM-Score	RMSD
			T0689-D1 (23-130, 132-234)			
1	PconsM	88.74	73.34	86.26	0.9280	2.181
2	Jiang_Fold	88.39	74.05	88.15	0.9262	2.241
3	RaptorX	88.27	73.70	88.63	0.9255	2.234
4	YASARA	88.15	72.63	88.63	0.9230	2.365
5	MUFOLD-Server	87.92	72.63	91.94	0.9228	2.111
			TBM-Hard targets			
			T0663 (53-204)			
1	LEEcon	42.93	31.58	45.39	0.3277	11.79
2	BAKER	42.60	33.22	46.05	0.3314	10.738
3	MULTICOM-CLUSTER	41.94	31.25	40.13	0.3339	12.83
4	MULTICOM-CONSTRUCT	41.94	33.23	40.13	0.3320	12.811
5	Mufold	41.61	32.73	44.74	0.3324	13.605
			T0674 (1-340)			
1	MULTICOM-CONSTRUCT	39.91	29.92	43.39	0.4566	11.05
2	BAKER	39.91	29.41	56.61	0.4646	13.59
3	PconsQ	39.07	29.07	51.53	0.4384	16.797
4	CNIO	38.98	29.32	52.88	0.4356	16.192
5	Sternberg	38.81	30.51	49.15	0.4168	14.073
			T0676-D1 (32-204)			
1	zhang	43.21	23.27	15.03	0.5317	8.029
2	Jones-UCL	40.90	21.39	16.18	0.5133	8.225
3	QUARK	40.75	20.81	14.45	0.5178	8.179
4	Mufold	40.61	20.66	15.61	0.5158	8.12
5	Kim_Kihara	40.46	19.94	5.78	0.5195	8.275
			T0726 (1-597)			
1	LEE	37.86	21.76	31.69	0.7377	13.777
2	MUFOLD-Server	37.69	22.66	20.10	0.7245	7.252
3	LEEcon	36.97	21.72	33.90	0.7176	16.12
4	BAKER	36.88	21.59	31.86	0.7282	19.31
5	ProQ2clust2	36.84	21.46	17.04	0.7101	15.225
			T0726-D1 (1-447)			
1	MUFOLD-Server	48.94	29.59	27.52	0.7294	7.011
2	LEE	48.88	28.41	30.43	0.7400	6.472
3	LEEcon	48.32	28.52	36.02	0.7258	6.809
4	TASSER-VMT	48.27	28.52	29.08	0.7314	6.654
5	zhang	48.15	26.62	31.54	0.7480	6.093

The top five models selected for each of the considered targets are ranked using the means and standard deviations of the computed ϕ, ψ and $C\alpha$ distance difference scores. Ranking of the models is done on the basis of each of the difference scores, separately. The basis of these rankings is allotting a higher rank to the model with lesser mean and SD. If during a comparison, a model has a greater mean but a lower SD compared to another model, then it is allotted a higher rank only if the absolute value of the difference between the means is lesser than the absolute value of the difference between the SDs. Thus, three rankings are obtained for models of every target. One uses ϕ differences only, another uses ψ differences only and yet another uses $C\alpha$ distance differences only. For every model, the average of its rank in all the three rankings is computed and employed to arrive at the desired end ranking enlisted in Table 2.

No severe clashes are observed in the experimentally solved structures of the selected targets as well as the CASP models while screening for atomic clashes. Thus, only the bumps are reported in the following Table 2 for each of these selected CASP10 targets. The numbers in the brackets denote the range of residues that is assessed in each of these selected CASP10 targets as well as the corresponding models during CASP and by our ranking protocol.

As demonstrated in Table 2 and quite interestingly, some of the top five predicted models are found to have the same score for some of the employed assessment parameters and thus get the same rank. This might stem from the fact that such pairs of models might have been constructed through the same set of templates along with a similar model construction or sampling protocol.

4. Discussion

On the basis of Tables 1 and 2, it can be inferred that our devised method is able to re-rank the selected models in a more logical and biologically correct manner. The developed method is based on the dihedral angles and the distances between $C\alpha$ atoms, which play a huge role in determining the overall topology of the protein, including its secondary structure. A similar dihedral angle profile between the model and the template implies a similar structure throughout the length of the protein. Moreover, $C\alpha$ distances give an impression about the overall length of the protein. So if dihedral angle similarity is complemented by likeness in the separations between residues, an overall similarity at the local and global level is safely established. Further, considering the mathematically optimized weight for any of these parameters would most plausibly supersede the smaller, although significant score of some of the other considered parameters for a protein model. Therefore all these three considered parameters are given an equivalent weightage in our model-ranking methodology to define the overall topology for a protein model. Moreover, our methodology equally weighs all the residues of a protein model, irrespective of their biochemical nature, to compute its rank and is thus quite similar to the TM-Score, GDT and Spheregrinder algorithms. It is substantially perspicacious too as every residue of a protein structure is extremely significant to define its structure, function or activity and considering weightage for a certain specific set of residues would again mask the evocative significance of all the other residues.

Surprisingly, for most of the considered target domains, the selected models scoring the best with the other usually employed measures are shown as relatively inaccurate conformations by our model-ranking algorithm. The developed methodology efficiently empowers the model assessment method to robustly discriminate even the highly similar conformations with identical and indistinguishable scores on the popular schemes. It would thus be extremely handy for assessing the high accuracy

TABLE 2 Rank order of the models on the basis of ϕ, ψ and Cα distance differences

#	Group Name	Mean (Phi differences)	SD (Phi differences)	Mean (Psi differences)	SD (Psi differences)	Mean (Cα distance differences)	SD (Cα distance differences)	Rank	No. of Bumps
				TBM-Easy targets					
				T0645-D1 (40-537)					
				No. of Bumps: 3					
1	Phyre2_A	20.8992	33.9175	28.9123	43.9392	0.24052	4.02868	5	9
2	MULTICOM-NOVEL	19.0146	31.8910	24.7386	39.9536	0.23198	3.97319	1	5
3	HHpredA	19.5319	34.9805	25.5589	42.0575	0.22516	4.12280	3 or 4	10
4	HHpredAQ	19.5319	34.9805	25.5589	42.0575	0.22516	4.12280	3 or 4	10
5	HHpred-thread	18.9600	35.7137	22.8394	38.9712	0.24238	4.53343	2	9
				T0650-D1 (4-342)					
				No. of Bumps: 2					
1	HHpredAQ	11.8153	19.3288	14.2131	24.8368	0.11079	0.93156	1 or 2	2
2	HHpredA	11.8153	19.3288	14.2131	24.8368	0.11079	0.93156	1 or 2	2
3	chunk-TASSER	16.3321	22.0047	19.1148	26.7637	0.41207	6.32109	5	2
4	BAKER-ROSETTASERVER	11.7935	15.7917	14.3204	19.1078	0.38764	6.21587	3	2
5	MATRIX	15.7645	20.5135	18.4406	27.2998	0.17778	2.39204	4	1
				T0659-D1 (1-74)					
				No. of Bumps: 1					
1	Phyre2_A	23.9671	36.8182	25.1476	36.3831	0.44975	3.11407	5	2
2	MULTICOM-REFINE	21.8414	37.002	18.6581	25.2892	0.40886	2.79116	1	1
3	MULTICOM-NOVEL	21.8317	36.1603	17.6779	26.3883	0.41755	2.93279	2	1
4	MATRIX	24.0315	41.541	19.0006	30.0968	0.31721	2.43026	4	1
5	FALCON-TOPO	24.3489	40.7947	23.7929	31.9272	0.07544	0.03171	3	1
				T0664-D1 (43-540)					
				No. of Bumps: 19					
1	PMS	39.0135	46.6247	71.9037	64.9247	0.32890	4.31501	1	2
2	RaptorX-ZY	39.6552	46.7053	72.3853	64.5695	0.34762	4.31576	2	19
3	Phyre2_A	39.7247	46.7206	74.2571	64.8504	0.35205	4.31742	3	5
4	RaptorX	39.8503	48.4584	71.7763	65.6894	0.34808	4.31386	5	3
5	HHpredA	41.2712	46.437	73.2748	64.0882	0.36924	4.31429	4	44

#	Group Name	Mean (Phi differences)	SD (Phi differences)	Mean (Psi differences)	SD (Psi differences)	Mean (Cα distance differences)	SD (Cα distance differences)	Rank	No. of Bumps
				T0689-D1 (23-130, 132-234) No. of Bumps: 5					
1	PconsM	42.4085	41.9167	59.9916	61.2096	0.36688	3.33598	5	2
2	Jiang_Fold	40.9151	40.69	58.8727	60.7372	0.35888	3.33067	3	1
3	RaptorX	41.647	42.0259	58.1554	60.1264	0.35757	3.33045	2	1
4	YASARA	38.8172	39.9287	56.3319	58.681	0.3215	3.32658	1	1
5	MUFOLD-Server	40.6819	40.9578	59.0635	61.3916	0.36376	3.33198	4	1
				TBM-Hard targets					
				T0663 (53-204) No. of Bumps: 5					
1	LEEcon	48.3802	40.7034	94.1935	66.2767	0.61573	6.26614	3	0
2	BAKER	45.6974	40.8617	88.6319	67.9529	0.60416	6.27179	1	1
3	MULTICOM-CLUSTER	46.9862	40.9618	91.3043	67.4976	0.63724	6.27047	4	1
4	MULTICOM-CONSTRUCT	46.8872	40.5531	82.0514	67.3651	0.63256	6.27004	2	0
5	Mufold	56.5503	50.8732	97.1479	65.2258	0.62661	6.26949	5	1
				T0674 (1-340) No. of Bumps: 7					
1	MULTICOM-CONSTRUCT	45.0558	49.1453	72.2645	64.7042	0.36911	4.08519	5	2
2	BAKER	42.6356	49.5145	65.0671	63.1255	0.33266	4.08643	1	1
3	PconsQ	48.9776	50.483	65.2055	60.1234	0.32640	4.08564	2	3
4	CNIO	44.4623	50.7663	65.8895	63.8832	0.3504	4.085	3	2
5	Sternberg	43.1118	49.0961	69.9502	65.1353	0.37300	4.0849	4	3
				T0676-D1 (32-204) No. of Bumps: 2					
1	zhang	35.8014	40.6034	43.1298	50.978	0.43511	5.16622	2	1
2	Jones-UCL	28.1974	36.328	40.276	52.9035	0.24506	2.18989	1	3
3	QUARK	37.3586	44.3856	47.2693	53.6975	0.36701	4.13275	4	1
4	Mufold	35.6473	44.1072	45.5506	53.7829	0.43727	4.7559	3	3
5	Kim_Kihara	43.0491	45.593	55.4191	49.986	0.26453	2.66868	5	8

#	Group Name	Mean (Phi differences)	SD (Phi differences)	Mean (Psi differences)	SD (Psi differences)	Mean (Cα distance differences)	SD (Cα distance differences)	Rank	No. of Bumps
				T0726 (1-597) No. of Bumps: 8					
1	LEE	35.1226	39.2376	56.7576	59.5964	0.31627	5.36341	2	1
2	MUFOLD-Server	37.9647	41.2117	60.8632	62.1783	0.34017	5.36634	5	1
3	LEEcon	36.6432	40.5783	59.5864	60.5942	0.31661	5.36204	3	0
4	BAKER	36.9444	40.3322	56.928	58.7586	0.30778	5.36409	1	5
5	ProQ2clust2	37.8395	41.7683	57.0159	60.5119	0.33386	5.36967	4	1
				T0726-DI (1-447) No. of Bumps: 8					
1	MUFOLD-Server	37.9704	41.3005	58.5911	61.1522	0.38748	6.2677	5	1
2	LEE	34.7522	38.8416	55.1498	59.0432	0.36367	6.26829	1	1
3	LEEcon	36.4604	40.7174	57.0443	59.2705	0.36419	6.26898	2	0
4	TASSER-VMT	45.9748	46.5419	56.649	54.0926	0.37113	6.26695	3	0
5	zhang	39.4394	40.3038	56.5121	55.3987	0.37845	6.26358	4	0

CASP targets where the submitted protein models are highly close to each other and where the manual assessment becomes a cumbersome as well as pain-staking exercise.

The large deviations in dihedral angles and $C\alpha$-$C\alpha$ distances between the selected CASP10 models and their native target structures have been observed throughout this study and it implies that an incorrect set of templates or an incorrectly constructed target-template alignment might have been possibly employed by even the top-ranked models.

The assessment of atomic clashes further reveals that the model that is ranked the best with our assessment methodology, does manage to score one of the lower counts of atomic clashes. However, in some instances, another model is observed to score the lowest count of such atomic clashes. This can be resolved through refinement of models which could remove the small number of clashes that exist. The developed methodology is thus capable enough to skilfully assess the predicted protein models at the local and global levels for discerning the highly alike structures and supplementing it with a judicious ranking mechanism. Although here the protein models are evaluated against their experimentally solved native structures, the developed algorithm is equally applicable to rank the constructed models against the employed set of templates.

5. Conclusion

A method which assesses both the local as well as global structural similarity of models has been developed and it provides an unconventional option of ranking the predicted protein models through the evaluation of their overall conformational topology. The method is empowered to effectively order the predicted models in the quest for identifying the best constructed structure. Accurate model assessment is the key to unravelling a correctly predicted protein structure and to decipher its function. Proteins play a pivotal role in gene regulation serving as transcription factors and awareness of their true structural as well as functional information bodes well for the formulation of a better understanding of how different proteins are biologically folded in a cell and how their function is likely to evolve in varying microenvironments created due to the presence of constantly evolving genome regulatory enzymes and inducers.

6. Limitations and Further Research Possibilities

Our method employs a one-to-one residue correspondence between the selected CASP10 models as well as their native structures, and then puts the entire difference-based scores to rank them. However, in such a case, there may be a possibility where two local regions which are not related through this correspondence possess a mutual similarity which is much better than the similarity they share with their counterpart regions with which the differences in their dihedral angles are being determined. In other words, the selected structures may be looking dissimilar overall with large mean and SD of the differences, but there may be a structural shift between the two which may be the reason for the dissimilarity. Thus small local regions may be compared with all the local segments in the other structure without any one-to-one residue correlation to identify the all the topologically similar substructures.

Another drawback that may hinder the successful employability of this technique in every situation is the computational burden it entails. It scans through every residue of a protein model and calculates dihedral angles to subsequently compute the variations between the compared structures. Though the procedure works well for small proteins, computing the required scores for large proteins might charge larger infrastructure. As an auxiliary check to the developed assessment scheme, the models can undergo clustering based on the same parameters that have been deployed in this formulated model assessment scheme. The model which takes away the top position should prove to be the most similar to all the other structures, although it would fail when the correct structure is considerably different from the considered set of models.

Moreover, only the Cα backbone has been utilized here for calculating the scoring parameters employed for ranking the protein models. However, a protein structure has the side-chain also, which is involved in its functioning and these side-chains should also be compared using similar suitable parameters for a wholesome comparison of the protein structures. Another observed deficiency is that whenever two models generate exactly same scores, our method fails to reliably select the better structure among them. Hence, the developed methodology requires further fine-tuning of the employed parameters through additional assessments of side-chains and other structural aspects of a protein. Finally, a method which incorporates every aspect of the protein structure can be arrived at for a thriving future of proteomics.

Acknowledgements

The authors would like to thank all of their mentors, especially Prof. Shibasish Chowdhury, Prof. AK Das and Prof. Rajesh Mehrotra, who clarified the doubts, the anonymous reviewers and the editor for his prestigious guidance, help and support.

References

Baker, D. and A. Sali. 2001. Protein Structure Prediction and Structural Genomics. *Science.* 294(5540): 93-96.

Dunbrack, Jr., R.L. 2006. Sequence comparison and protein structure prediction. *Current Opinion in Structural Biology.* 16(3): 374–384.

Gherardini, P.F. and M. Helmer-Citterich. 2008. Structure-based function prediction: approaches and applications. *Briefings in Functional Genomics and Proteomics.* 7(4): 291-302.

Hung, L. and R. Samudrala. 2012. Accelerated protein structure comparison using TM-Score-GPU. *Bioinformatics.* 28(16): 2191-2192.

Kopp, J., L. Bordoli, J.N. Battey, F. Kiefer and T. Schwede. 2007. Assessment of CASP7 predictions for template-based modeling targets. *Proteins: Structure, Function, and Bioinformatics.* 69(8): 38–56.

Kryshtafovych, A. and K. Fidelis. 2009. Protein structure prediction and model quality assessment. *Drug Discovery Today* 14(7-8): 386-393.

Kryshtafovych, A., B. Monastyrskyy, and K. Fidelis. 2013. CASP prediction centre infrastructure and evaluation measures in CASP10 and CASP ROLL. *Proteins: Structure, Function, and Bioinformatics.* 82(2): 7–13.

Kryshtafovych, A., C. Venclovas, K. Fidelis and J. Moult. 2005. Progress over the First Decade of CASP Experiments. *Proteins: Structure, Function, and Bioinformatics* 61(7): 225–236.

Matsui, E., J. Abe, H. Yokoyama and I. Matsui. 2004. Aromatic residues located close to the active center are essential for the catalytic reaction of flap endonuclease-1 from hyperthermophilic archaeon Pyrococcus horikoshii. *The Journal of Biological Chemistry* 279(16): 16687-16696.

Runthala, A. 2012. Protein structure prediction: Challenging targets for CASP10. *Journal of Biomolecular Structure and Dynamics* 30(5): 607-615.

Sela, M., F.H. White, Jr. and C.B. Anfinsen. 1957. Reductive cleavage of disulfide bridges in ribonuclease. *Science* 125(3250):691-92.

Tress, M., I. Ezkurdia, O. Graña, G. López and A. Valencia. 2005. Assessment of Predictions Submitted for the CASP6 Comparative Modeling Category. *Proteins: Structure, Function, and Bioinformatics* 61(7): 27-45.

Xu, J. and Y. Zhang. 2010. How significant is a protein structure similarity with TM-score=0.5?. *Bioinformatics* 26(7): 889-895.

Zemla, A. 2003. LGA: a method for finding 3D similarities in protein structures. *Nucleic Acids Research* 31(13): 3370-3374.

Zhang, Y. and J. Skolnick. 2004. Scoring Function for Automated Assessment of Protein Structure Template Quality. *Proteins: Structure, Function, and Bioinformatics* 57(4): 702–710.

Zhang, Y. and J. Skolnick. 2005. TM-align: a protein structure alignment algorithm based on the TM-Score. *Nucleic Acids Research* 33(7): 2302-2309.

Index

Printed and bound by CPI Group (UK) Ltd, Croydon, CR0 4YY
01/11/2024
01782622-0013